Progress in Mathematics
Volume 121

First
European Congress of Mathematics

Paris, July 6–10, 1992

Volume III
Round Tables

A. Joseph
F. Mignot
F. Murat
B. Prum
R. Rentschler
Editors

Birkhäuser Verlag
Basel · Boston · Berlin

Editors:
Anthony Joseph
Laboratoire de Mathématiques
Fondamentales
Université Pierre et Marie Curie
4, place Jussieu
75252 Paris Cedex 05, France
and
The Weizmann Institute of Science
Rehovot 76100, Israel

Fulbert Mignot
Mathématiques
Bâtiment 425
Université de Paris-Sud
91405 Orsay Cedex, France

François Murat
Laboratoire d'Analyse Numérique
Université Pierre et Marie Curie
4, place Jussieu
75252 Paris Cedex 05, France

Bernard Prum
Laboratoire de Statistique Médicale
Université Paris V
45, rue des Saints Pères
75270 Paris Cedex 06, France

Rudolf Rentschler
Laboratoire de Mathématiques
Fondamentales
Université Pierre et Marie Curie
4, place Jussieu
75252 Paris Cedex 05, France

A CIP catalogue record for this book is available from the Library of Congress, Washington D.C., USA

Deutsche Bibliothek Cataloging-in-Publication Data

European Congress of Mathematics <1, 1992, Paris>:
First European Congress of Mathematics : Paris, July 6–10, 1992 / A. Joseph ... ed. – Basel ; Boston ; Berlin : Birkhäuser.
NE: Joseph, Anthony [Hrsg.]; HST
Vol. 3. Round tables. – 1994
(Progress in mathematics ; Vol. 121)
ISBN 3-7643-2800-2 (Basel ...)
ISBN 0-8176-2800-2 (Boston)
NE: GT

Printed on acid-free paper produced of chlorine-free pulp
Layout, design, typesetting by TEXniques, Inc.

ISBN 3-7643-2798-7 (Vol. I/PM 119)
ISBN 3-7643-2799-5 (Vol. II/PM 120)
ISBN 3-7643-2800-2 (Vol. III/PM 121)
ISBN 3-7643-2801-0 (Set)

ISBN 0-8176-2798-7 (Vol. I/PM 119)
ISBN 0-8176-2799-5 (Vol. II/PM 120)
ISBN 0-8176-2800-2 (Vol. III/PM 121)
ISBN 0-8176-2801-0 (Set)

9 8 7 6 5 4 3 2 1

Table of Contents of Volume I

Invited Lectures (Part 1)

Preface xiii
Lists of Committees xv
List of Sponsors xvi
Lists of Lectures and Round Tables xix
Opening Ceremony
Discours de Fulbert Mignot xxiii
Discours d'Henri Cartan xxiv
Address of Friedrich Hirzebruch xxv
Discours de Vincent Courtillot xxix
Discours d'Hubert Curien xxxi
Closing Ceremony
Address of Fulbert Mignot xxxiv
Address of Friedrich Hirzebruch xxxv
Prizes Ceremony
Discours de Jacques Chirac xxxvi
Discours de Max Karoubi xxxix
Prizes Awarded by the City of Paris xli

Plenary Lectures

The Vassiliev Theory of Discriminants and Knots
V.I. Arnold 3

Transparent Proofs and Limits to Approximation
László Babai 31

Poisson Algebraic Groups and Representations
of Quantum Groups at Roots of 1
Corrado De Concini 93

Gauge Theory and Four-Manifold Topology
S. K. Donaldson 121

Spectral Theory and Geometry
Werner Müller 153

Pattern Theory: A Unifying Perspective
David Mumford 187

Brownian Motion and Obstacles
Alain-Sol Sznitman 225

Geometric Quantization and Equivariant Cohomology
Michèle Vergne 249

Parallel Lectures

The Power of Exponentiation in Arithmetic
Zofia Adamowicz 299

Subspace Arrangements
Anders Björner 321

Optimal Recovery of Functions and Integrals
Borislav Bojanov 371

Existence globale et diffusion pour les modèles discrets
de la cinétique des gaz
Jean-Michel Bony 391

Sporadic Groups and String Theory
Richard E. Borcherds 411

A Harmonic Analysis Approach to Problems in
Nonlinear Partial Differential Equations
J. Bourgain 423

(Some) Old and New Results on Algebraic Surfaces
Fabrizio Catanese 445

Evidence for a Cohomological Approach
to Analytic Number Theory
Christopher Deninger 491

Cauchy–Riemann Operators, Self-Duality,
and the Spectral Flow
Stamatis Dostoglou and Dietmar A. Salamon 511

Index of Authors of Volumes I and II 547

Table of Contents of Volume II

Invited Lectures (Part 2)

Parallel Lectures (continued)

Martingales, Arbitrage, and Portfolio Choice
Darrell Duffie 3

Mathematical Aspects of the Quantum Hall Effect
Jürg Fröhlich 23

Analytic and Geometric Aspects of Variational Problems
for Vector Valued Mappings
M. Giaquinta 49

Harmonic Measures for Leafwise Elliptic Operators
Along Foliations
Ursula Hamenstädt 73

Feynman Diagrams and Low-Dimensional Topology
Maxim Kontsevich 97

KAM-Theory for Partial Differential Equations
S.B. Kuksin 123

Paradoxical Decompositions: A Survey of Recent Results
Miklós Laczkovich 159

A Path-Valued Markov Process and its Connections
with Partial Differential Equations
Jean-François Le Gall 185

The Cyclotomic Trace in Algebraic K-Theory
Ib Madsen 213

Algebraic K-Theory and Galois Cohomology
Alexander S. Merkurjev 243

Values of L-Functions and p-adic Cohomology
Jan Nekovář 257

Mantles, Trains and Representations
of Infinite Dimensional Groups
Yurii A. Neretin 293

The Evolutionary Dynamics of HIV Infections
Martin A. Nowak 311

On the Enumeration of Algebraic Curves
— from Circles to Instantons
Ragni Piene 327

Mathematical Aspects of Domain Decomposition Methods
Alfio Quarteroni 355

Paths in Graphs and Curves on Surfaces
Alexander Schrijver 381

Function Estimation and Functional Data Analysis
Bernard W. Silverman 407

Algebra and Complexity
V. Strassen 429

Generalizations of Fuchsian and Kleinian Groups
Pekka Tukia 447

Properties of Embedded Lagrange Manifolds
Claude Viterbo 463

Alternative Entropies in Operator Algebras
Dan Voiculescu 475

Algebraic K-Theory and Functional Analysis
Mariusz Wodzicki 485

Values of Zeta Functions and Their Applications
Don Zagier 497

Index of Authors of Volumes I and II 513

Table of Contents of Volume III

Round Tables

Preface xiii

Round Table A: Mathematics and the General Public

Jean-Pierre Kahane, Michèle Chouchan 1

Appendix: Replies to a Questionnaire

Jean-Michel Kantor 23

Round Table B: Women and Mathematics

Eva Bayer-Fluckiger 37

Round Table C: Mathematics and Educational Policy

Jacques Camus 75

Introduction 75

Mathematics in the School Curricula in Flemish Speaking Belgium

W. Dewilde 80

Mathematics in the French Educational System

Jacques Camus 89

The Role of Mathematics in Educational Policies in Germany

Paul Bungartz, Rudolf Strasser 101

Italy: Mathematics in Higher Education

Vinicio Villani 110

Mathematics and Educational Policy (United Kingdom)

Timothy Porter 115

Round Table D: Let's Cultivate Mathematics!

Yves Chevallard, André Rouchier 129

Introductory Remarks

Yves Chevallard 129

Institutional Dimension in the Definition
of Cultural Strategies in Mathematics

André Rouchier 131

Television Mathematics: Cultivation or Distortion?

Celia Hoyles 134

Mathematicians and the Need for Critical Self-Reflection

Christine Keitel 138

Mathematics in its Historical Context

Lucia Grugnetti 145

Ecology of Mathematical Knowledge:
An Alternative Vision of the Popularization of Mathematics
Juan D. Godino 150

Round Table E: Mathematical Europe, Myth or Historical Reality?
Catherine Goldstein, James Ritter 157
The Creation of (the History of) Algebra
Giovanna Cifoletti 161
The Perenniality of Programmes
Henk Bos 165
Mathematical Europe from Outside Europe
Karine Chemla 168
At the Beginning of our World: The Nineteenth Century
Hélène Gispert 172
Centers and Peripheries
Eduardo Ortiz 176
Conclusions 180

Round Table F: Philosophie des mathématiques : pourquoi ? comment ?
Hourya Sinaceur 183
Formes, opérations, objets en mathématiques et en philosophie
Gilles-Gaston Granger 185
Le platonisme transcendantal et le problème de l'objectivité
Jean Petitot 192
Reflexions on Formalism and Reductionism
in Logic and Computer Science
Giuseppe Longo 202
Pourquoi les mathématiques sont-elles difficiles ?
Jacques Dubucs 210
Le geste bien plutôt que la règle
Gilles Chatelet 217
Discussion 223

Round Table G: Mathématiques et sciences sociales
Pierre-André Chiappori, Roger Guesnerie 229
Introduction au débat
Roger Guesnerie 229
On the Use of Mathematics in Social Sciences: Some Remarks
Pierre-André Chiappori 232
La sociologie face à la mathématisation
Pierre-Michel Menger 238

Round Table H: Mathematics and Industry
Julian C.R. Hunt, Helmut Neunzert 257

Round Table I: Degree Harmonization and Student Exchange Programmes
Christian Berg, Hans J. Munkholm, Ivan Netuka, David Salinger, Vladimir Souček . . . 277

Round Table J: The Pythagoras Programme
A European Policy for Mathematics
Daniel Gabay . . . 321

Round Table K: Collaboration with Developing Countries
Pierre Bérard . . . 347
Appendix I: Contributions made by Mathematical Societies . . . 374
Appendix II: Individual Reports . . . 381
Appendix III: Centers in Collaboration . . . 394

Round Table L: Mathematical Libraries in Europe
Pierre Barrat, Geneviève Sureau, Liliane Zweig . . . 397
Introduction . . . 398
Financing of the Libraries . . . 400
The Situation of Libraries in Russia
S. Kuksin . . . 400
The Situation of Libraries in Poland
L. Goldstein . . . 401
The Financial Situation of Mathematical Libraries in Hungary
A. Szendrei . . . 403
Scientific Publishing and Problems of Documentation . . . 406
Documentary Inflation and the Rising Cost of Journals
C. Byrne . . . 407
New Technologies and Computerization . . . 411
Computerization of Spanish Libraries
P. Ibarrola . . . 411
The French Experience and Norms
P. Barrat . . . 412
Appendix 1: Results of the Questionnaire . . . 416
Appendix 2: Statistical Data . . . 419
Appendix 3: Manufacturing Costs
C. Byrne . . . 423
Appendix 4: Electronic Publishing: A Few Remarks
C. Byrne . . . 424
Questionnaire . . . 425

Round Table M: Mathematics and Economics
Bernard Cornet . . . 431

Round Table O: Mathématiques et chimie
Edgar Soulié 451
Mathématiques et modélisation en dynamique chimique
Patrick Hanusse 455
Mathématiques et science des matériaux
Claude Bernard 461
Knowledge Acquisition in Chemistry
Joachim Johann Gasteiger 466
Mathematical Methods for Determining Crystal Structure
Gérard Bricogne 473
La théorie des groupes en chimie
Michel Golfier 477
Mathématiques et chimie quantique
Marcel Allavena 489
Mathématiques et chimie en recherche pharmaceutique
Gilles Moreau 497
Two Global Methods for Molecular Geometry Optimization
Susana Gomez, David Romero 503
Géométrie fractale et équations différentielles
d'ordre non entier en chimie physique
Alain Le Méhauté, Frédéric Héliodore 510

Round Table P: Mathematics in Medicine and Biology
Robert Hiorns, Bernard Prum 515
Introduction 515
Medical Statistics: an Example from Diabetes Epidemiology
Niels Keiding 517
Point Process Approach of Life Histories: Causal Chains
Elja Arjas 520
Two Signal Cell Activation Explored with Simple Mathematics
Jonathan Swinton 528
A Theory of the Functional Organization of Biological Systems
Applied to Neural Networks
Gilbert Chauvet 536
Stochastic Vistas to the Analysis of Physical Mapping
of the Genome
Petre Tautu 550
Rhythms and Chaos in Biological Systems
Albert Goldbeter 559
Discussion and conclusions 564

Index of Authors of Volume III 573

Preface

The first European Congress of Mathematics was held in Paris from July 6 to July 10, 1992, at the Sorbonne and Panthéon-Sorbonne universities. It was hoped that the Congress would constitute a symbol of the development of the community of European nations. More than 1,300 persons attended the Congress.

The purpose of the Congress was twofold. On the one hand, there was a scientific facet which consisted of forty-nine invited mathematical lectures that were intended to establish the state of the art in the various branches of pure and applied mathematics. This scientific facet also included poster sessions where participants had the opportunity of presenting their work. Furthermore, twenty-four specialized meetings were held before and after the Congress.

The second facet of the Congress was more original. It consisted of sixteen round tables whose aim was to review the prospects for the interactions of mathematics, not only with other sciences, but also with society and in particular with education, European policy and industry.

In connection with this second goal, the Congress also succeeded in bringing mathematics to a broader public. In addition to the round tables specifically devoted to this question, there was a mini-festival of mathematical films and two mathematical exhibits. Moreover, a Junior Mathematical Congress was organized, in parallel with the Congress, which brought together two hundred high school students.

The present Proceedings of the first European Congress of Mathematics consist of three volumes. The first two volumes contain articles corresponding to the presentations by invited speakers. The third volume contains reports from the round tables. The first volume also includes the addresses delivered during the ceremonies and the lists of sponsors and committees of the Congress. All three volumes are illustrated with photographs taken during the Congress which hopefully capture something of its convivial atmosphere.

It is a great pleasure for me to conclude this preface by thanking all the persons who contributed to the organization of the Congress. In particular, I would like to thank the speakers, the organizers and participants of the round tables, the members of the committees and the technical staff. This list is by no means exhaustive and many other persons contributed to the success of the Congress, which could not have been organized without the dedication of all.

I would also like to thank the sponsors who provided generous financial support. The long list of these sponsors is given in Volume I. Our sincere thanks to each of them!

Finally, I would like to thank the management and staff of Birkhäuser Verlag for publishing these Proceedings in their widely distributed series, *Progress in Mathematics*, as well as for their excellent work.

Paris, July 10, 1993
F. Mignot
Chairman of the Organizing Committee

Acknowledgment

The English text of the Round Tables was translated/edited with the help of Catherine Rothman-Le Dret. The editors would like to acknowledge her contribution and thank her for her fine work.

J.-C. Risset, M. Emmer, T. F. Banchoff, A. Sossinski

J. Brette and M. Demazure

B. Roszkowska

I. Kersten

L. Tedeschini Lalli

C. Corrales

The audience

C. Bessenrodt

M.-F. Roy

P. Barrat

P. Bérard

H. Munkholm

J. Camus

Role of mathematics in educational policies

V. Villani, P. Bungartz, T. Porter, J. Camus, P. Legrand, W. Dewilde

Let's cultivate mathematics!

J. D. Godino, C. Hoyles, Y. Chevallard, C. Keitel, L. Grugnetti, A. Rouchier

Mathematical Europe, myth or historical fact?

C. Goldstein

E. Ortiz, H. Gispert, H. Mehrtens, K. Chemla, H. Bos, G. Cifoletti

Philosophy of mathematics: why and how?

J. Petitot, G.-G. Granger, H. Sinaceur, G. Longo, J. Dubucs, G. Chatelet

H. Sinaceur

Mathematics in social sciences

P.-M. Menger, F. Hahn, R. Guesnerie, J. Elster, P.-A. Chiappori

M. Andler

H. Neunzert

A question from the floor

Degrees harmonization and student exchange programmes

H. Munkholm, V. Souček, C. Berg, I. Netuka

What price the Eurodiploma?

H. Munkholm, P. Valentin, C. Lance, C. Cercignani, D. Gabay

C. Cercignani and D. Gabay

R. Labarca, J. Martins, A. O. Kuku, A. Verjovski, J.-M. Lemaire, P. Bérard, M.S. Narasimhan, J. Eells, R. Pelczar

J.-P. Raoult (left) in the audience

Mathematical libraries in Europe

C. Byrne, P. Barrat, A. Szendrei, S. Kuksin, S. Goldstein, G. Sureau, L. Zweig, P. Ibarrola

Mathematics and economics

B. Cornet, N. El Karoui, S. Sorin, J.-M. Lasry, R. Aumann, G. Debreu

Mathematics and computer science

H. Prodinger, K. Compton, B. Larrouturou, F. Morain

Mathematics and chemistry

E. Soulié

P. Hanusse

P. Tautu, N. Keiding, B. Prum, R. Hiorns,
A. Goldbeter, G. Chauvet, E. Arjas

F. Murat, Treasurer of the Congress

Round Table A: July 6, 1992

MATHEMATICS AND THE GENERAL PUBLIC

by Jean-Pierre Kahane[1] and Michèle Chouchan[2]

Université de Paris XI, Orsay (France)[1], Radio France Culture, Paris (France)[2]

1. The State of Affairs

In view of the diversity of the contributions sent before the two round tables, Mathematics and the General Public of the first European Congress of Mathematics, it appears that the relationships between mathematics, its developments, the working methods of researchers and the popularization of knowledge constitute a whole set of questions that has more to do with the world of mathematicians than is often thought, and on a more international basis as well.

There is a long tradition of remarkable experiments in Eastern countries, notably in the ex-USSR. As their political regimes crumble one after another, could it be that Western European countries are beginning to be concerned with consciousness-raising campaigns that have proved their worth, especially among youth?

We can acknowledge the Leeds meeting in England as one of the first important conferences on the subject [1]. This meeting was organized by ICMI (International Commission on Mathematical Instruction) from September 17 to September 22, 1989. During that week, academics, journalists, students and educators were able to meet and compare the situation in their respective countries, particularly regarding public appreciation of mathematics and the potential recruitment of future researchers. There were exhibits, competitions, lectures and quite a bit of committee work. Personal experience came as a welcome complement to reflections on communication on various levels. The meeting resulted in a synthesis that was meant to serve as a documentary basis for subsequent work. In the foreword of this document, Geoffrey Howson and Jean-Pierre Kahane wrote: "How do we know that we have a problem?" These words are still significant three years later in an internationally worrisome situation.

> There is no doubt about it. In most developed countries the public image of mathematics is bad. Jokes appear in the newspapers; stereotyped, incorrect views about mathematics abound. "All problems are already formulated"... "Mathematics is not creative"... "Mathematics is not a part of human culture"... "The only purpose of mathematics is to sort out students" ... "Mathematics may be important to other people, not to me"...

> Even when it seems positive, the image is usually wrong: Mathematics is always correct, providing absolute truth, solid and static.
>
> The image of mathematicians is still worse : arrogant, elitist, middle class, eccentric, male social misfits. They lack social antennae, common sense, and a sense of humour.
>
> This is not new. "Mark all mathematical heads which be wholly and only bent on these sciences, how solitary they be themselves, how unfit to live with others, how unapt to serve the world". This view of mathematicians, expressed by Roger Ascham, 16th century educator and tutor to Queen Elisabeth I of England, is one which is echoed in many later writings. Blaise Pascal, who was himself intimately concerned with mathematics, used to contrast "esprit de géométrie" (a mathematical mind) and "esprit de finesse" (an accurate mind). The latter was an attribute of "honnêtes gens" (nobility and high bourgeoisie), whereas the former was poorly regarded. This contrast has been a favourite theme for dissertations in French high schools, and has contributed to the view of mathematicians as strange characters, divorced from the real world.
>
> It happens that mathematicians often reinforce this view by their behaviour or their writings; there is no shortage of examples, many of them famous. However there are figures of a very different type (think only of Sonia Kovalevskaya) and, of course, most mathematicians do not stand out from the crowd.
>
> People may have very strong feelings about mathematics and their relation to it. Mathematics yields emotional experiences. It can be seen as fun and exciting or as repulsive. It can be threatening, it can lead people to seek security in it or away from it, or it can just be dull. For some people, mathematics is a big blank, or something to be avoided in the future if at all possible .
>
> As a school subject it is needed to pass examinations. It often has high status because of the intellect it seems to require; failure in mathematics may lead to a loss of self-esteem.

G. Howson and J.-P. Kahane also pointed out a few surveys that demonstrated that the public image of mathematics among youth was globally poor. If such is the case, what can we expect of the mathematicians from the next decades? This question was also raised in France during the *Mathématiques à venir* colloquium that was held in December 1987 at the École Polytechnique and for once received unexpected press coverage [2].

"How do we know that we have a problem?" It was enough to hear what two young girls, just finishing high school, had to say about mathematics during the second session of the round table last July. Their never-ending complaints about the lack of intellectual purpose of mathematics and its total lack of any sensory context provoked a good deal of turmoil among mathematicians who were there. The way these students set mathematics in opposition to philosophy on the grounds that the latter was more "human" and more connected to our perceptions of the world, incited several reactions. According to Jean Brette, director of the mathematics "department" of the *Palais de la Découverte* in Paris, "the teaching of philosophy is closer to the actual work of the philosopher than the so-called school mathematics is to the work of the mathematician".

Jean-Michel Kantor and Philippe Boulanger (Editor of the French journal *Pour la science*) expressed this lifelessness in a similar fashion in a preparatory text for the European Congress of Mathematics:

What is the image of mathematics in the general public? It cannot be said that it is bad. Rather, it is evanescent. The general public, outside of the professional mathematical milieu, does not have a clue of what contemporary mathematical research is about. Mathematics retains its image as a completed science that is learned without enjoyment in high school. It is a challenge to evoke the cultural aspect of mathematics. It is fashionable to be proud of being ignorant of it, as Bouvard or Pécuchet would have put it: such ignorance is almost a sign of sanity.

The fetishistic example of Bourbakism also gave mathematics the image of a science consisting of abstract structures, in which general synthesis came before examples and formalism concealed meaning.

As early as the 1960s several reactions against this tendency had appeared, some of them related to a change in mathematical research itself. Mathematics sprouted outside of established structures and "abstract algebra" and theories that were more easily illustrated such as fractals or knots came into view. At the same time, computers brought computing aid and the visualization of results to mathematics. There were parallel undertakings to publicize the dynamics of mathematics through films, articles in multidisciplinary popularization magazines, exhibits such as the minimal surfaces exhibit at the *Palais de la Découverte* or at the *Cité des Sciences et de l'Industrie de la Villette*, specialized journals for high school students (*L'ouvert, Tangente, Quadrature, Maths et Malices, Le nouvel Archimède*), activities for youth such as *Maths en jeans* and competitions such as *Kangourou.*

These French undertakings nonetheless suffer from a lack of coordination, from the regrettable absence of mathematics in the most widespread medium, television, and from a lack of interest on the part of the mathematical community, which has better things to do, or so it seems. The torch brilliantly held up by Poincaré and Borel was only taken up by Ivar Ekeland, Marcel Berger and few others. Popularization is more alive in Russia and the Anglo-Saxon countries.

Hasn't the time come to eliminate the sophism that thinks that information delivered to the public never deals with mathematics because mathematics is incomprehensible and that this silence of the media endlessly reinforces this incomprehensibility?

A consequence of this is found in the *Frankfurter Allgemeine Zeitung* by Bernhelm Boos-Bavnbek and Philip Davis:

Given the internal difficulties, it is not surprising, that, generally speaking, the media avoids mathematics. As a measure of this avoidance, consider a few statistics. The *Frankfurter Allgemeine Zeitung*, considered one of the most intellectual newspapers in Europe, has a weekly science section and runs perhaps 20 stories per issue. An informal scan of 225 such stories revealed that not one of

> them was related to mathematics as such. If mathematics underlies a statement, it is left hidden, or at best is sloughed off with the sentence "a mathematical model shows that...".

2. Obstacles

The most common reaction is that of the two Parisian high school students. Media people and a good part of the public have the same reaction, although books and conferences capture their curiosity more than they used to. Everybody has the impression of understanding the main evolutions of astrophysics or medicine or the impact of progress in biotechnology. Ethical or civic problems that arise from the use of biotechnology rarely meet with indifference. But in mathematics how can one be anything other than a passive witness to an incomprehensible discourse?

Bernhelm Boos-Bavnbek and Philip Davis go even further in referring to intrinsic characteristics of the discipline:

> Another difficulty of the communication of mathematics is that people have fundamentally different concepts of clarity or lucidity. While the mathematical community is aware that it is tongue-tied with respect to the general public and often works hard to achieve reformulations that have a higher degree of popular comprehensibility (in some instances, it has taken centuries to do this), the fact remains that the professional mathematician has a concept of lucidity that is different from that of the nonprofessional. To the mathematician, what is lucid is that which proceeds by logical or computational steps from sharply formulated assumptions to conclusions. This procedure supports an awareness of the intricacies of human thought and allows a dramatically growing complexity in its treatment of reality. Its explication demands a kind of patience, concentration and attention to detail that is poison in a media presentation.
>
> The public wants a story; lucidity is located in the narrative mode. The story should be simple, understandable, definite, resolved, interesting. And in today's world, interesting often means sensational. A story should tell of a great accomplishment or of a great failure, should have a hero or a villain, should carry some kind of emotional charge. The public wants the concrete; it does not like the abstract, and mathematics is surely one of the most abstract of the human constructions. These considerations rule out most mathematical developments and applications.

Pursuing their argument, the two authors ask: "What is mathematics?"

> What is mathematics? Mathematics is the science and the craft of quantity, space and pattern. The first two are reasonably clear, but by pattern is meant not only visual pattern, but also any mode or organization that is distinguished or imposed and occurs repeatedly. A short definition such as this is inadequate to convey the

full extent and richness of the field. One might be inclined to add that it is the deductive study of these three notions and of the symbolisms through which this study is pursued.

Every special field from horseracing to banking to medicine has its special terms; likewise mathematics, and the special terms and ideas of mathematics are standardized, abbreviated and written with special typographical symbols. The ten digits themselves, the mathematical signs plus, minus, etc., the square root sign are met in the elementary grades. In advanced mathematics there are literally hundreds and hundreds of special signs and symbols employed, hundreds and hundreds of ordinary words given special and precise meaning, and this meaning is often located only in combinations of previously agreed upon terms and symbols. Thus mathematics is like a language, with some features of a natural language such as English or Swahili, but with many differences. The general public is at a loss as to how to pronounce the complicated formulas that often occur in mathematical statements. This is the least of the difficulties. English may be translated into Swahili and vice versa, granting that certain subleties may be lost either way, but it has become increasingly difficult to translate mathematical statements into the English of ordinary experience even when the mathematics itself is expressed in what seems to be English.

Take the following mathematical sentence (theorem) : "The zeros of successive real orthogonal polynomials separate one another', which occurs in an area of advanced mathematics with which one of the authors is familiar. If a lay person asked him for the meaning of this sentence, he might reply that certain curves interlace. He might even draw a picture. But he would be hard pressed in lay terms to state precisely what curves, how the interlacing occurs, how he knows this is to be the case, what the connection is between this mathematical sentence and others, or why anyone should trouble his head over the whole matter.

Mathematical terms, signs, formulas have a precision and an abstraction lacking in ordinary language. They usually point to themselves. This lies at the heart of the difficulty. Non-mathematical signs, even difficult ones, point to something in the world of physical existence or event. The meaning of mathematical signs is usually to be found only in the way in which they are used in accordance with some conventions of our brain.

Scientific signs are bound to one scientific context : "Hepatitis B", "Amorphous semi-conductors", "Ozone O_3". Such terms are media-friendly. Mathematical signs, on the contrary, being abstract, are movable from one context to another. They can tell too many stories at one time. The stories are too unspecific. They often lack a visual or kinesthetic interpretation, which, if present, makes for easy understanding. They make no immediate appeal to our allegorical or mythical natures as do the statements of the biological, medical, physical, astronomical sciences. Accordingly, they are not media friendly.

> According to Wittgenstein (and to every kindergarten teacher as well), showing and telling are independent modes of communicating knowledge. But there is also a kind of mathematical knowledge, located in the comprehension of the way its component parts fit together and operate, that transcends these standard modes and is not easily communicated.
>
> Mathematics, when it is applied to the physical or social worlds, stands "infra " to some topic of more primary interest to the general populace. When mathematics is self-referrent, it often displays a self-indulgent triviality of motive that can baffle all but the most passionate of its votaries. Perhaps it is the fate of all infrastructures, whether of language, of finance, of technological availability, to be taken for granted or ignored. What captures attention is the superstructure; like electricity in a blackout, what is "infra" is noticed only in a crisis. As Tolstoy once put it, "The leaves of a tree delight us more than the roots".

Still, is the question itself legitimate? According to Catherine Goldstein, "To choose to formulate the entire problem of the popularization of mathematics in terms of the 'definition of the subject' is to accept in advance a referential framework devised for privileged westerners formed in a literary culture".

We must add to this debate the even more provoking contribution by Marie-Jeanne Husset, a scientific journalist, chairperson of the second session of the round table, "Do we really need to popularize mathematics? Or can people do without an area of knowledge that is difficult and abstract?"

Will these obstacles constitute a hindrance, when some members of the mathematical communities have already engaged in a reflection on the means needed to facilitate relations with their scientific colleagues or simply with their fellow citizens?

3. Assets

In effect, every obstacle reveals new reasons and new advantages for the popularization of mathematics. In particular, the object of mathematics is not to exploit an area of the physical or social world but to devise non specific concepts and general methods. The value of mathematics lies in its availability and accessibility to the broadest and most varied public possible through education, documents or personal conversations.

There exist several documents which, upon careful reading, reveal a few key difficulties and needs one encounters outside of the mathematical community. One of these is the 1989 report on circumstances of the *Centre National de la Recherche Scientifique* (CNRS) [3]:

> Pasteur thought that there is no good applied science without a good theory. Although a theory does not necessarily lend itself to a mathematical formulation (*cf.* in contrast to

the current trend of using the vocabulary and some of the concepts of biology in computer science), many theories in fact allow it. In order to be able to use such a formulation, it is necessary to identify the obstacles that may arise.

If mathematics has returned in recent years to more open relationships with other sciences, this has not always been the case. H. Weyl (1919) is supposed to have said that mathematics and physics are like a couple who quarrel in the daytime and nourish each other in the dark of the night. According to this view, physicists dream of a mathematical supermarket where all the tools they need are available when they need them, and mathematicians get exasperated or even irritated when their concepts and tools are ignored or badly used.

This is by no means the end of the list of difficulties that arise in the interaction with other sciences. In fact, there are numerous mathematical problems coming from other mathematical problems, themselves coming from other sciences, that are currently unresolved due to lack of proper tools. Actually, mathematicians are not always able to answer the questions that are asked at the time they are posed in a satisfactory fashion.

There are several reasons for this situation. Let us assume that a mathematical model has been built to describe a phenomenon but that the analysis of this model leads to problems that are out of the reach of contemporary mathematical techniques. If mathematicians deem such a problem interesting, they take hold of it, expose its structures and isolate the difficulties and, eventually, solve it. However, the solution may require an amount of time that has no common measure with the delays that are customarily accepted by the end users. In effect a theorem is always true and a mathematical concept always useful, even 2000 years after its discovery or proof. This is rarely the case in other sciences and certainly never in the economic world. Nonetheless, it often happens that advances in mathematics made for the solution of this problem can help solve other problems.

It may also happen that, because the relevant parameters and fundamental interactions of a phenomenon fail to be identified, no reasonable and efficient model can yet be obtained. This is often still the case in biology and economics. However, even in such a case mathematics is useful in pointing out the absurdity or the limits of accepted models because it can be used to derive all their consequences.

In the meantime, the field of interaction has considerably broadened, but communication difficulties are still numerous. Results that are obtained in different languages and with different problematics often require years of mutual adaptation in order to be understood by both sides until they finally merge into an embryo of common culture. For a given theory, the important concepts for the mathematician are often not the same as for the other specialist, hence the difficulty in accepting the other's point of view and reading his publications. It is thus necessary to compile "dictionaries" to convince everyone that they are actually talking about the same concept (a recent example of this is given by the various approaches to the renormalization group in physics and in mathematics).

Since the number of professional mathematicians is limited and since their ability to grasp radically new fields is somewhat held up by the demand of the mathematical milieu to publish only really profound results, every new area of interaction almost always necessitates a new generation of mathematicians. Given the current organization of their community, mathematicians thus tend to react rather slowly.

Interaction is not encouraged by the educational system, especially in secondary education. For example, it is regrettable that high school teachers do not have more contacts with the utilizations of mathematics, due notably to the lack of appropriate documentation.

A possibly more irremediable hindrance to interaction comes from different aims and frames of mind. Mathematicians are happy with the harmonious unfolding of a theory. Due to their training, they are less sensitive to the notion of relevant parameter and to the limits of modeling. Mathematical theories have a very long life span—mathematics is somehow timeless. Mathematicians are perfectionists and are more likely to appreciate the difficulty of a given result than its usefulness. An elegant new proof of an old result is well appreciated by mathematicians since it shows a deeper understanding of the nature of the result. The distinction between pure and applied mathematics then is more a question of frame of mind than of the actual applicability of the results or of the qualification of the tools used. Applied mathematicians are in effect forced to include a dimension of utility in their work, which can lead them to reorient their research direction if their mathematical field is no longer in touch with a potential domain of application.

The same report quotes several areas where obstacles to dialogue have been overcome and research has benefitted from interaction between mathematics and other sciences. As we have already noted, mathematics does not have a privileged grasp on reality, but it is versatile enough that it can become involved in a large variety of domains. Coming back to the report, one interaction mode is rapidly gaining importance—modeling.

Mathematical modeling is not a recent creation. However, due to the current computing capabilities of computers, it is being applied to new areas. Because modeling gives structure to the phenomena under study, it is an efficient tool in pointing out causality effects and significant parameters. Furthermore, by suggesting similarities between isomorphic models, modeling can help transfer intuitions and problematics from one domain to another domain (for example, the recently discovered surprising similarities between arithmetic geometry and quantum field theory). Due to internal classifications that are made available through mathematical modeling, a certain number of qualitative behaviors can be sensed *a priori* even though current mathematical techniques are insufficient to completely solve the resulting problems (*e.g.* all the models that pertain to a given class of differential or partial differential equations, such as equations of elliptic, hyperbolic or parabolic type, when specialists in classical mechanics are led to use equations whose type may vary throughout the domain). Mathematical modeling can also, at least in the best of cases, be a great thought-saving device by providing systematic procedures for studying and solving problems.

However, mathematical modeling can only make use of the available tools. Models can be linear or nonlinear, discrete or continuous, in one or several variables, stochastic or deterministic, stationary or time-dependent, and so on. The mathematical background of the mathematician who does the modeling is often the determining factor in choosing the nature of the model. This fact can become more important than the phenomenon itself. For example, the recent advances in probability theory and their broad diffusion can be seen to result in more and more models that take chance into account. Good modeling therefore requires very experienced mathematicians who are able to overcome the technical problems that arise during the construction of their model.

In any case, if mathematics can help in understanding a phenomenon, it cannot under any circumstance be a substitute for reflection on the phenomenon and on the relevance of the model. A model only contains what has been put into it. The role of mathematicians is to extract every bit of information from the model. In doing so, mathematicians may have to make accessible information from other disciplines by using similarities that the mathematical processing of the model may suggest.

It is certain that the use of mathematics can become a danger when its only role is to throw dust in your eyes by concealing a lack of thinking concerning the phenomenon under study through a brilliant mathematical development.

A good model and a good theory must be predictive and must be able to interface with experimental reality..., even if modeling is sometimes used in those very situations where experiments are difficult or even impossible to realize (space and nuclear industries, economics, medicine and so on).

According to these remarks, if defining mathematics is still difficult, explaining its areas of applications is comparatively easy. For example, differential equations and dynamical systems are related to the description of the evolution of phenomena, probability theory to deriving certainty from uncertainty, harmonic analysis to decomposition and reconstruction by means of elementary vibrations. Mathematical concepts that arose from telecommunications such as information theory or coding, are now influencing molecular biology. The current vogue of wavelets is due to the fact that wavelets provide a unifying and generalizing framework for a host of preexisting engineering, physical as well as mathematical practices, making them at the same time more powerful and more easily communicable. Finally, what can we say about fractals except that one meets them everywhere? It thus appears that while mathematics is not just a discipline at the service of other disciplines, it is also not disconnected from other sciences. It is the very generality of mathematics that makes it necessary to integrate it into the common cultural patrimony.

There is thus not one unique way of engaging in mathematics, but a multiplicity of ways, as demonstrated in several recently published works (see for example the book by Mauduit and Tchamitchian, which is aimed at a young readership, [4]).

This multiplicity is based on the variety of subjects and their applications and also on the variety of conceptions of what it means to do mathematics.

A universal language, an intellectual game, sport or exercise? After Miguel de Guzman ("Mathematics is a generator of true beauty, intellectual beauty, profound beauty"), Jacques Mandelbrojt, a physicist and an artist, draws a parallel between art and mathematics by observing as a jest that you more often hear a mathematician speaking of a beautiful theorem than an artist speaking of a beautiful painting. Mandelbrojt also notes the analogy between the role played by mathematics in physics and the role played by abstract painting in relationship with figurative painting. To link the mathematician's methodology to that of other scientists, to make the "interactions" understood, as the CNRS report puts it, are undertakings that can reach beyond the limits of the scientific community and concern a broader public.

Lacking a global view of the profusion of mathematics, we can try and make use of the variety of access that will all shed a different light on mathematics. We must allow every subject, use all media and reach the entire public, and once more transform all obstacles into assets. It is remarkable that the mathematicians who were interviewed before the Congress placed their trust in the written word, which is the traditional way of transmitting abstract knowledge, on the one hand and on television on the other, which provides a sort of acknowledgment of the credibility of any information or know-how that is now necessary. It is as though there were no intermediate public between the colleagues who are determined to read a paper from beginning to end, even if it is slightly technical, and a vague population for whom the transmission of knowledge has to be mediated through images, through a few documents summarily shown on the screen between images of some conflict somewhere on the planet and the last electoral speeches.

On the contrary, according to Catherine Goldstein, "popularization can be a practice as well and not only diffusion of knowledge. There are several undertakings in this direction: *Maths en Jeans* in France tries to initiate high school students to research; on a completely different scale *Paulus Gerdes* in Mozambique demonstrates the mathematics necessary for agricultural techniques. More generally, ethno-mathematicians try to build on those practices of mathematics that can be put to work based on the culture of a given group—like typography, weaving, knitting or cooking". Besides, several films, radio programmes or competitions such as *Kangourou* in France have widened the spectrum of popularization activities. We will come back to this in the next section.

The discussion on the trends, what is at stake and even the nature of mathematics is also, and even primarily, a popularization means. It concerns all levels and the entire public. The question is not to teach but to bring awareness to an area of human activity. Let us again give a few examples. A debate on "Mathematical thought and reality" held the attention of a non scientific audience for a whole afternoon at the University of Orsay. The people who buy (and who read) books signed by Changeux and Connes or by Penrose are also primarily interested in the philosophical aspect, the nature of mathematical objects or the distinctions between thinking man and a sophisticated software. Once the public's attitude has been changed, new areas of interest may open up. There is no doubt that the discussion on mathematics and philosophy also belongs to the popularization of mathematics.

Of course, all these ways of approaching mathematics should not be thought of as equal. A meeting between youth and professional researchers or a lecture with slides and overhead projector will not have the same impact and will not command the same attention as a radio or television production of a few anecdotes on the somewhat chaotic adventures of nonstandard analysis. Portraits, which present the advantage of making the profile of a professional mathematician more familiar and in principle less caricatural, are more catchy when traits are exaggerated. Here is another obstacle, and potential asset, namely, the general public's ignorance of contemporary mathematicians, at least in some countries like France.

The rigidity of school mathematics is often opposed to any reflection on raising consciousness concerning great mathematical problems. School mathematics is more a discipline than a science, more a finished and linear product than a field of open questions. Its discriminatory role in curricula cannot be dismissed either. However, we must not underestimate the classical aspects of mathematics by replacing them with an immersion in which showing how to do it replaces actually doing it. On the other hand, mathematics teachers have an enormous potential for diffusion and communication. They can be involved in journals, magazines, brainstorming sessions and all sorts of competitions. It is out of the question to confine mathematics teachers to the lowly task of evaluating ossified knowledge. They stand on the front line of the diffusion of mathematical culture. Finally there exists a long tradition of great teaching treatises starting with Euclid up to Bourbaki that recalls this privileged link between research and the transmission of its results.

To sum up the spirit in which mathematicians should approach "efficient popularization", we first quote a general principle of Miguel de Guzman:

> To pay attention to the place where the persons are to which we want to pass on our message and to examine what society and the general public are ready to receive.

We then take the "recipes" proposed by Philippe Boulanger and Jean-Michel Kantor in a humorous fashion in the document presented to the Congress:

"Popularization requires efforts from the mathematical community and cannot be left to journalists, editors and producers by themselves. The latter, with the help of the mathematical community, could incite some of its members to take up the pen. Such an ideal cooperation has not yet proven very efficient.

In order for mathematics to be interesting, it has to be:
(a) appetizing,
(b) palatable,
(c) and profitable for the persons making the effort to swallow it.

(a) *Appetizing mathematics*

The primary cause of the lack of appetite for mathematics is the vocabulary, semantics and careful language that rigor requires. An elementary precaution should be to insist on rigor of vocabulary only when such affectation is important for the specific problem under consideration.

1. The subject is more appetizing if one starts from an example that is at the same time striking and as simple as possible in order to get to the heart of the subject. An example is interesting when it is unexpected, either if the result could not be presumed in advance, or if it has to do with the mathematization of an ordinary problem.

2. An historical approach can embark the reader or editor on a stream of thought or a traditional form of research in which new results have recently been obtained.

3. A mathematical result that has applications in other sciences or in solving concrete problems has more impact.

4. Game theory applied to simple games stimulates the reader's analytical faculties and gives him or her a feeling of confidence if the game is complex enough. Games that are expressly constructed to be mathematizable are not very interesting.

5. Advances in computer graphics prick curiosity and arouse the sense of beauty. They can incite readers to redo on their own computers what was shown to them and get them interested in the theory, *e.g.* cellular automata.

6. Using computers can be a means of avoiding tedious or even unending computations that in the past were a cause of repulsion. Computers are also helpful in handling mathematical objects and in mastering their study.

7. The purity and truth of a mathematical result detach the reader from a dull day-to-day life.

8. Readers like novelties. They like to feel the movement of ideas. With an appropriate effort, they can understand the evolution of modern research.

(b) *Palatable mathematics*

1. There is a broad discussion on the use of formulas in texts for the popularization of mathematics. Formulas have the drawback of greatly slowing down reading speed. They also frighten readers who are afraid they might not grasp the entire meaning of the symbolism. However, a formula can have an intrinsic beauty. In this case it can be gently introduced, if such is the goal of the presentation. We must remember that letters can also cause apprehension as they recall school memories while numbers are reassuring by their self-evidence.

2. The mental gymnastics that have to be performed in order to solve a problem can be explained. It is the author's responsibility to make clear what a proof is and how it differs from a mere presentation. The author must also indicate why such and such a method of proof was chosen (by contradiction, by induction, by analogy with a geometrical problem and so on).

3. The reading speed for a mathematical text must not be too different from that of any other text. It is up to the author to smooth out the difficulties.

4. It is important to relate the elements of the action (*i.e.*, here the proof) with points that are known to the reader. The reader feels more secure if handholding from his familiar universe is available.

5. The classical presentation is not the only possible way to explain a mathematical result. The author can use any form of dialogue and storytelling, all the resources of paradox and drama.

6. The reader must be made to understand that mathematical subjects cannot be grasped in one fell swoop, but by successive approximations and only exceptionally through flashes of global understanding.

7. Mathematics is an activity and popularizers must be concerned with making the reader a participant.

(c) *Beneficial mathematics*

1. Is that which lingers in the memory as a pleasurable moment.

2. Is mathematics that is usable in other sciences or that sheds light on some social activities. Probability theory, electoral mathematics, the laws of chance and optimization principles are very much appreciated in this respect.

3. Just like other arts, mathematics is a means of communication within social or familial groups.

4. Mathematical culture helps in mastering the omnipresence of computers. Geometry enriches the mechanistic view of the world and logic and probability theory enrich the analysis of decision making processes.

5. Mathematics can be rejuvenating for those who were familiar with it in their youth.

It would be interesting if an exchange between the various attempts at the popularization of mathematics led to more professionalism on the part of popularizers and to more interaction between mathematicians and media professionals. This can only happen if mathematicians and popularizers feel that they are working together.

The important point is that the author must not despise his public: the art of popularization is difficult and strewn with traps. There are however remarkable achievements such as the books by Penrose, Ekeland or Mandelbrot. These are not easy books and their qualities may deserve further analysis.

Mathematics helps creativity by developing a certain kind of intuition and an aptitude for problem solving. Mathematics does not claim to explain everything (no mathematician ever forced Lacan to explain the unconscious with knot theory). Mathematics, the better one knows it, however helps in discerning the moving boundaries of knowledge. It thus seems important to popularize mathematics if only to assess its power and current limits. But are all means equally good?"

The above text gives some possibilities. Let us stress one more aspect—public expectations. The public is often ahead of the media and professionals in its desire for mathematical knowledge or activities. An example, in the Sciences and Industry Museum in Chicago the mathematics room is very poor in every respect. However, when you leave the museum to enter the museum's bookstore, half of the books that are on sale and three quarters of the objects and games have a mathematical content. The public is ready to pay for mathematics.

Mathematics has political, economic, social and cultural stakes which can appeal to various elements of the public. For example, the history of mathematics, in relationship to the history of ideas, answers a widespread need. The presentation of its recent evolution, as in the CNRS report quoted at length above, can capture the attention of persons who might otherwise be repelled by detailed reasoning. Mathematics is justly considered to be general and powerful and is unjustly considered to be inhuman. Given its importance in today's world, the aim of popularization is to correct the injustice of this vision and to ensure that no one facing mathematics ever again is left with the impression of facing a closed door marked "no trespassing".

4. Evolution

Let us take as our starting point the Leeds colloquium of September 1989. As we have already said, the emphasis was on the problems, the bad image of mathematics and mathematicians, their lack of experience at the time and the intrinsic difficulties in popularizing mathematics. However, some directions were already indicated: mathematics in various countries and cultures (Czechoslovakia, Hungary, China), possible subjects, mathematical problems and games, the various levels and the targets (public at large, youth, teachers, professionals, retirees), the role of exhibits and various media (newspapers, television, radio), the transformation of what is "new" into "news" and the responsibility of mathematical societies. Starting from one of the main goals of popularization, which is to change the public's attitude, the colloquium led to the working hypothesis that the change of attitude of the public would open a new field and new needs in mathematical popularization.

The evolution since 1989 seems to confirm this hypothesis. There is a flourishing of initiatives in Europe and throughout the world that seem to meet the expectations of part of the public.

Let us first recall the importance of what the British called in Leeds "the event", the "pop.maths.roadshow". This event was made up of several exhibits (posters, montages, experiments, on-screen graphics, games, sculptures) which drew some 20,000 visitors to Leeds. There was also a series of films and audio or video cassettes as well as popular lectures for youth of different age groups, with such different speakers as Johnny Ball (a British TV star), Christopher Zeeman and Dr. Ruth Lawrence, then age 17. On the whole 20 lectures were given during the week in packed enthusiastic amphitheaters. Another main feature of the event was the conclusion of the "Leeds competition" between high schools, each team working on a real research project (three subjects on coding, tilings and dynamics) and presenting their results on posters, software or film. Add to this the excellent recreative practical exercises such as the math.night club "chez Angélique".

This event had a measurable impact in Great Britain. The roadshow, in a reduced form of exhibits, toured 21 cities in Great Britain and Ireland between September 1989 and October 1990. The total number of visitors was 250,000.

Geoffrey Wain from the University of Leeds wrote a report for round table participants ("Reflections on the Pop Maths Roadshow") that was distributed as part of the "Situation in Europe" brochure. This report contains an exhaustive description and some conclusions of general interest. We quote the following one:

> ... if popularization is to be taken seriously, performances need to be developed that convey to lay audiences the essence of mathematics without requiring of them knowledge of the technicalities. There is an attitude problem here. Many mathematicians may suspect that communication of this kind is demeaning, in some sense, of mathematics itself. The evidence of the Roadshow suggests that the opposite is the case; that many people, when they come to appreciate some aspect of the subject, may well gain a new respect for it and interest in it. It must be likely that, if people appreciate and like the subject at the popular level, more of them will want to study it for its own sake. It is important though that any study they pursue does not destroy their enthusiasm.

Leeds also had a considerable, if difficult to ascertain, impact on the persons who attended the meeting due to its exemplary impression. Let us now review the recent salient initiatives in a few European countries.

In France, following the book by Changeux and Connes already mentioned, the years 1990–1992 saw the appearance of several good books that popularize mathematics. Let us mention the beautiful one by Christian Mauduit and Philippe Tchamitchian, *Mathématiques*, aimed at high school students but touching many extracurricular subjects; the saga by Ivar Ekeland, *Au hasard*, for which its author was awarded the d'Alembert prize of the Société Mathématique de France; the very personal book by David Ruelle, *Le hasard et le chaos*, mixing the mathematical and physical aspects of deterministic chaos; *Géométrie*, by Marcel Berger, at the same time teaching treatise and popularization for the informed audience; *Le pouvoir des mathématiques*, by Moshe Flato, for the cultivated reader. In 1991 and 1992, the *Palais de la Découverte*, whose director Michel Demazure is a mathematician, held two exhibits on mathematical or physico-mathematical subjects, fractals and minimal surfaces. Furthermore, there was a festival of mathematical films there in July 1992. Mathematicians and mathematical societies have increased their communication efforts since the *Mathématiques à venir* colloquium in 1987 and the "Mathecom" exhibition in 1989. Let us also mention between September 1991 and May 1992 the Cerisy-la-Salle colloquium *Mathématiques et art*, the beginnings of an afternoon in Orsay with Alain Connes and Pierre Cartier on "mathematical thought and reality", the one day meeting *Mathématiques indiscrètes* in Marseille (exhibit, films, debates, music), the special meeting with journalists of the Société Mathématique de France on the theme "mathematics as an historical adventure", and interesting personal initiatives such as that of Josette Adda at the center for scientific information in Boulogne-sur-Mer or the creation of the summer mathematics university by Daniel Loeb at Bordeaux. TV films and radio programmes have also been produced in the same period.

Following the *Mathématiques à venir* colloquium and the *Cinquante lycées* ("fifty high schools") undertaking, a survey on the image of mathematics among high school students compiled by the Strasbourg IREM (Research Institute on Mathematical Education) in 1989, to which the Leeds colloquium had already alluded, initiatives reaching out to youth have multiplied. The most important ones are *Math en Jeans*, *Kangourou* and the *Congrès Mathématique Junior* (La Villette, July 1992). The principle of *Math en Jeans* is to have two teams from two high schools work on a given subject with a researcher for several months. The resulting presentations (*Palais de la Découverte*, April 1992) were reminiscent of the Leeds competition. *Kangourou* is an adaptation of the Australian mathematical competition that has 500,000 competitors a year. Its success was rapid: more than 100,000 competitors in 1991, more than 200,000 in 1992, with participants from several European countries. Finally, the *Congrès Mathématique Junior* held during ECM was a successful meeting between motivated youth and mathematicians. *France-Culture* radio station gave it some coverage and Alexandre Sossinsky, ex-editor of *Kvant*, gave a very appreciated report of it during the second session of the round table.

The questionnaire prepared by J.-M. Kantor and sent to several European mathematical societies showed the diversity and importance of recent initiatives in most European countries. Here is a small sample of the information received, see the appendices for more details.

In Germany, the centennial of the Deutsche Mathematiker Vereinigung was marked by a big exhibit in Bremen, *Mathematik sehen*, including, in particular, examples of dynamical systems and deterministic chaos, which was visited by 30 classrooms and was quoted in 25 newpaper articles.

In Norway, the 150th anniversary in 1992 of the birth of Sophus Lie was the occasion for an exhibit on his life and work and for the production of a "Sophus Lie" video, on which mathematicians and video professionals collaborated.

In Spain, there is a vigorous increase in publications in and around mathematics. Particularly noticeable is Miguel de Guzmán's last work *Para pensar mejor* [5].

In Sweden, there is a programme directed by Dan Laksov to collect mathematical problems that can be proposed as new activities in Swedish high schools (ages 15–18). A large number of Swedish mathematicians participated in this programme. The resulting book is currently under testing, with good reactions. Discrete mathematics (combinatorics, coding and so on) is strongly represented, together with other subjects (iterations, electoral theory, Möbius transformations, Bernoulli polynomials, Gauss integers, history of mathematics, etc.)

In Czechoslovakia, in the framework of the small, one volume, "general encyclopaedia" prepared by the Academy of Sciences, a new conception of mathematical articles has been proposed and accepted. Aside from classical subjects (circle, sine, gradient) there will also be articles on algebraic topology, Galois theory, cardinals and ordinals, the monster group, exotic spaces in dimension 4 and so on.

We finish this quick glance with information received during or after the Congress.

In Denmark, a volume called *Udsagn* (statements) was published in 1992 on the occasion of the 60th anniversary of the foundation of the association of mathematics teachers. Writers, composers, journalists, engineers, politicians, scientists and mathematicians express their feelings regarding mathematics. The book, illustrated with beautiful photographs, is a great success [6].

In Italy, Michèle Emmer, who was already well-known for his films and popularization articles, has just published three books, predominantly geometrical, [7] [8] [9], and a sumptious special issue of the arts and culture review *Leonardo* on visual mathematics [10]. He declared at the round table:

> Until a few years ago it was extremely rare to read articles on mathematics in newspapers, at least in Italian newspapers, or to see mathematical books aimed at the general public. Now the situation has changed. A field has opened in the publishing business for the publishing of such books.

The "Arts and Mathematics" theme that was discussed at the Cerisy-la-Salle colloquium of 1991 [11], was also discussed in Spain in a series of brief surveys on the cultural and social dimension of mathematics aimed at students and teachers, notably through lectures by Capri Corralès (University of Madrid) interweaving pictural developments (from Pompeï to the cubists) and mathematics (perspective and conceptions of space).

We can assess the popularization and teaching of mathematics in Poland in the special issue of "Mathematics" published on the occasion of the 7th international congress of mathematics teaching in Québec (August 1992). "The goal of the journal is to improve the professional knowledge of teachers, both mathematical and educational. We publish articles on mathematics and on mathematicians, and especially articles popularizing mathematics and its history", says the editorial. A good example of this is an article that starts from a proof of the irrationality of $\sqrt{2}$ to explain what a formalized mathematical theory is.

There was an exhibit *Mathematik in der industriellen Praxis* shown during the Hannover fair (April 1–8, 1992) that was prepared by the Center for Arts and Science of North Westphalia (Wissenschaftszentrum Nordrhein-Westfalen). The exhibit met with remarkable success, witnessed by almost 60 pages of press coverage. A few titles from these articles were *Mathematik ist überall!, Mathematik als ökonomische Ressource, Mathematik in der industriellen Anwendung, Gesichtserkennung mit dynamischen neuronalen Netzen, Erkenntnisse der Forschung für neue Produkte genutzt, Den Puls der Wissenschaft fühlen, NRW zählt auf die Mathematik.*

Let us go back to England to mention the enormous success of Roger Penrose's beautiful book *The Emperor's New Mind* [12]. The book has sold 200,000 copies and has been translated into several languages. This is not an easy book and its success reveals a certain demand on the part of the public.

To sum up, since 1989 the evolution shows an increase in initiatives in all domains, exhibits, debates, books, films and TV and in reaching out to youth. The public is more interested than in the past. Popularization of mathematics is becoming popular.

5. Dangers and Recommendations

It seems that a goodly number of mathematicians have accepted that the concepts of culture and patrimony should include a certain amount of mathematical knowledge. Such knowledge does not have to be sophisticated but it must be linked to the development, history and impact of mathematics. All responses to the questionnaire demonstrated, if unequally, the desire that at least an awareness on the part of the citizen (of the honest man) be achieved.

This is not simple and mathematicians cannot escape from this task. Besides, some of them do not want to leave the job to another intermediary: lack of confidence or timidity in expressing themselves?

If we agree that mathematics and the general public are bound to grow closer to each other, we should nonetheless outline a few recommendations in order to avoid certain pitfalls.

After the dangers caused by the gap between mathematics and the general public, we now have to face the dangers of intimity, to which neither mathematicians nor the media are well prepared.

1. The spontaneous personal attitude of most mathematicians is adverse to communication. Mathematicians are often simultaneously extremely proud of their discipline and modest and timid concerning themselves. We have to overcome this mix of reserve and arrogance and learn how to talk, in a different context than teaching, to write and above all to listen. There are some suggestions on this in part 3.

2. Most mathematicians suffer from narrowness. They have only a fragmentory view of mathematics and, *a fortiori*, of its history, its applications and its relationship with human society. Even the great can suffer from this defect. In fact, concern for popularization is the best cure against narrowness. To be able to extract the important points, you have to question many prejudices and habits. You have to broaden your own views on mathematics and its role in science, history and society.

3. It is tempting to follow the trend and inevitably popularization no doubt intensifies the effects of the trend. This is not too important if one is aware and watches out, especially of course in research, that dominant trends do not smother those to come. The dangers are distortion in the choice of subjects and even more the star system in the choice of speakers. How can we give an image of mathematics and mathematicians that is at the same time truthful and interesting, global and detailed? We must surely take advantage of the variety of research, which is wider in mathematics than in heavier and more targeted disciplines.

4. The profession of mathematics is little known and difficult to describe. This however is not a reason to seek shelter in incommunicability or exotism. We need a large number of young mathematicians for the future and we must be wary that our profession should appear to be at the same time open, dignified and accessible.

5. Scientific and educational deontologies evolved through time. In mathematics, we have a limited experience in popularization compared with, for example, astronomers and the ethics of the relationship between mathematics, media and the public remains to be defined. We must expect tensions and frictions among us.

6. Let us repeat that the media are ill-prepared for mathematics and mathematicians. More precisely, media directors are reticent. We need to work on them. We postulate that this rapprochement is going happen and that the problem is not *are we going to popularize mathematics?* but rather, *à la* Polya, *what are we going to popularize? Why? How? With whom?*

After having stated the dangers and problems, let us venture into some recommendations.

1. The International Mathematical Union has decided to make the year 2000 an international mathematics year. This decision can be accompanied by big regional initiatives, more media awareness, production of books and films, exhibits and public debates. According to Michele Emmer:

> Even if the period is still far away in the future, we should already start to think about a great exhibition of European mathematics for the year 2000,...

2. Specific actions for youth are possible. This is one of the lessons of the Junior Mathematical Congress. Such actions are necessary in order that the cultural aspects of mathematics become familiar to youth and also in order that young people find the mathematical profession interesting. It would be a good idea to take a census of such initiatives and to stimulate them on the European level.

3. We have mentioned in passing the special and indispensable role of mathematics teachers in the relation between mathematics and the public. Perhaps we should go back to Félix Klein and give these actual or future teachers the necessary elements needed to round out their culture, especially concerning the history of their discipline. We should at least facilitate their access to documentary sources.

4. Other scientists, engineers and many professionnals in various fields require taylor-made expositions in a style that might be very different from our customary one. Demand needs to be stimulated and offers need to be facilitated.

5. Insofar as they are in contact with authors, publishers and institutions, the European Mathematical Society and national mathematical societies can be catalysts or even the driving force in the production of works of synthesis, reference works and high quality popularizing books. They can facilitate the work of authors through institutions or universities and see to it that such works are subsequently acknowledged.

6. It is also the responsibility of mathematical societies to improve communication between researchers, the potential authors of such works, and the media, who first of all trade in information. As an example, a campaign was launched in France by sending a letter to journalists and various institutions on the life of the mathematical community, the way departments work or the points of view of various social or intellectual figures more or less involved in mathematics. But this campaign needs refueling. Who else but researchers themselves can do it best?

This is all the more true since journalists or other popularizers in the field are, we need to stress this again and again, rarely understood in their respective environments. Mathematicians who hate "wasting their time" do not imagine how hard it is for their interviewers to obtain a few lines or a few minutes on the air and evoke a subject whose austerity scares off their bosses. Perhaps mathematical societies could convey the message that, even if specialized research and rigor may lose a few points, public curiosity and intellectual stimulation may be aroused. They must also make it clear that a media product cannot make use of research language, but it can be provoking, thrilling, pleasant or even beautiful. From another aesthetic viewpoint: to talk about the subject rather than treating it precisely.

7. In several European countries, mathematical societies already take the initiative or even play an instrumental role in publishing the works of mathematicians. This is typically the case in Poland and Hungary. Other countries, like France, lag behind. In addition to publishing works, detailed notices—including obituaries—on the most important mathematicians should be developed. For Europe, the obituaries of the London Mathematical Society might provide a model.

8. Mathematicians must first of all learn to understand and communicate with each other by creating information networks on national and/or European levels.

This is one way of learning how to talk to non-mathematicians. This was already a key idea of the deceased President of the French Academy of Sciences, Jean Hamburger. The first "general public" for mathematicians is the mathematical community.

9. Still to be developed are the global aspects such as the relationship of mathematics with other disciplines concerning knowledge as well as comparative history and social and political stakes. Mathematicians must contribute to this undertaking which is in direct contact with the preoccupations of the public as well as of its decision-makers.

10. We need to think about the history of mathematics within the mathematical community. We also need to think about the reasons for the existence of research and its future directions, as well as the significance of mathematics within the intellectual hierarchies. Also of importance is the new role mathematics could play within a "mathematics and society" programme. In dealing with the relationships between mathematics and the general public, all aspects of the "mathematics and society" programme touched upon during the first European Congress of Mathematics must be taken into consideration and utilized and, in particular, the report written by Catherine Goldstein for the present volume.

References

[1] *The Popularization of Mathematics,* ICMI Study series, A.-G. Howson and J.-P. Kahane (eds.), Cambridge University Press, Cambridge, 1990 (also "The Yellow Book" : ICMI, *Papers on the Popularization of Mathematics,* Leeds University, England, 17-22 September, 1989).

[2] *Mathématiques à venir,* SMF-SMAI, Supplément au Bulletin de la Société Mathématique de France, 1987, t. 115.

[3] Rapport de conjoncture du Comité National de la Recherche Scientifique 1989, Presses du CNRS, 1990.

[4] *Mathématiques,* by Christian Mauduit and Philippe Tchamitchian, Collection La Science et les Hommes, Messidor-La Farandole, Paris, 1990.

[5] M. de Guzmán, *Para pensar major,* Editorial Labor, Barcelona, 1991.

[6] UDSAGN, *En mosaik om matematik,* redigeret af Torben Christoffersen og Flemming Clausen, 15+ tekster om personliger opleveiser med matematikken, Matematiklaererforeningen, 1992.

[7] M. Emmer, *L'occhio di Horus : itinerari nell'immaginario matematico,* Catalogo della mostra, Ist. della Enciclopedia italiana, tutto a colori, Roma, 1989, p. 262.

[8] M. Emmer, *La perfezione visibile : arte e matematica,* Theoria ed., Roma, 1992. (Un libro di riflessioni sulla matematica e sul rapporto tra matematica, creativita ed arte ; pp. 214, 16 p. a colori, illustrazioni in bianco et nero).

[9] M. Emmer, *Bolle di sapone : un viaggio tra arte, scienza e fantasia,* La Nuova Italia, Firenze, 1991, p. 144. (Un libro tutto a colori sulla storia delle bolle di sapone nella matematica fino ai giorni nostri oltre che nella pittura, scultura fotografia, architettura, chimica, biologia).

[10] Special issue of Leonardo on "*Visual mathematics*", in press, 1992. Guest editor Michèle Emmer. (Address : Leonardo, 672 south Van Ness avenue, San Francisco CA 94110, USA).

[11] M. Loi, P. Cartier, éditeurs, "*Art et mathématiques*", Colloque de Cerisy-la-Salle, 1991.

[12] Roger Penrose, *The Emperor's New Mind,* Oxford Univ. Press, Oxford, England, 1989.

APPENDIX: REPLIES TO A QUESTIONNAIRE

QUESTIONNAIRE

written and dispatched by Jean-Michel Kantor
CNRS, École Normale Supérieure, Paris (France)

In the prospective of the session mentioned, we would like to publicize recent experiences in the popularization of mathematics. We would be grateful if you could answer the following few questions.

1. Towards whom was directed the experience of popularization of mathematics on which you would like to report: age, socio-professional categories...? What was the initial purpose?

2. Which means were used? For example: newspapers, books, conferences, audiovisuals, exhibits, ...

In the case of audiovisuals and exhibits, give the technical caracteristics and the distribution procedures.

Were professors, investigators, artists, ... involved in the experience?

3. What aspects of mathematics were considered, and how? For example: history, recent discoveries, problems, proofs, conjectures, the life of mathematicians, relations with sciences, arts...

4. How did things happen? In particular was there participation from the public?

5. What are the conclusions drawn from the experience? What is your appreciation? In particular, what were the reactions of the audience, and the specific difficulties encountered?

The above questionnaire was sent to mathematical societies and individuals in most European countries, requesting that replies be sent to Jean-Pierre Kahane.

EXCERPTS OF REPLIES

Albania

Given our circumstances, especially in this difficult transition phase of our society, we would like the popularization of mathematics to be directed toward pupils and students. We have a certain experience in this regard. The initial goal was to make the history of mathematics and the evolution of ideas up to the present known to students and teachers in connection with a change in the mathematics curriculum in school. This work was started several years ago and is reaching its completion.

Concerning socio-professional categories, we do not have much experience. Our work consisted of trying to present mathematics through its applications in our country with the help of the media. Given the level of development of the country, this was essentially popularized by means of and through the achievements of computer science.

The popularization of mathematics is necessary for our future. This is all the more the case when we see that the best high school students do not want to pursue their higher studies in mathematics or in the natural or technical sciences. They prefer medicine, social, political or juridical sciences, and lately, economics.

We need motivated people to achieve this popularization. The mathematical community which has the capability for popularization is small. Some of them do not have the time to take care of it, or do not take any pleasure in it. Popularization is generally and on all levels a spontaneous job. There is no global policy involving several participants.

In the near future we will announce the foundation of a mathematical society that will contribute to this endeavor.

Along with the transformation going on in society, we need to rethink the content and means of our work.

The lack of audiovisual materials is a great handicap.

As a general rule, scientific culture and consequently mathematical culture does not belong to the culture of the masses. This is possibly even more true for mathematics.

A country like ours, with limited technical and financial means, needs a certain amount of foreign support.

Finally, it might be a good idea to devise a European project in the area of popularization of mathematics as a necessity for our common future.

Llukan Puka
Tirana, Albania
14.11.1991

Czechoslovakia

The Mathematical Olympiad

This nation-wide problem-solving competition has been organized since 1951. Originally designated for the students of the secondary schools (15–18 years), it was gradually extended to involve the pupils of the second-level basic schools (12–15 years). Until recently, the annual number of participants has been about 9000 students from the secondary schools and about 60 000 pupils from the basic schools. (There is a project to involve pupils 10–11 years old). Most of the work in the "field" has been done by teacher-volunteers in their free time and without any financial compensation.

The demands for the competition have been growing all the time. This was encouraged by the variety of supporting activities such as weekly seminars for the pupils, correspondence seminars, students camps with the mathematical programme organized during the summer vacation, the special edition "School of young mathematicians" with more than 60 titles published (some of them have been translated in Germany). Since 1959, Czechoslovakia has been participating in the International Mathematical Olympiad, sometimes as the organizing country.

The most successful Czechoslovak winner of several international mathematical olympiads, Jan Nekovář (born 1963) is now one of the invited speakers of the ECM in Paris.

After the fall of the communist regime in 1989, the future of the Mathematical Olympiad has been broadly discussed. The total number of participants has decreased dramatically because the students and pupils are no longer under pressure of the school authorities to take part in the competition. Fortunately, the number of successful participants (who advanced to the second round organized in the regional centers) did not decrease as much. Thus we still believe that the competition is helping to discover the "world of mathematics" to the pupils with inborn mathematical capacities but with insufficient outside motivation and an unfavourable intellectual or social environment.

The TV "Matematika Převážně Vážně" (Mathematics largely in earnest)

This TV serial took place in the years 1979–83. It was composed of 160 lessons, each lasting 40 minutes. The 16 main topics, or combinations of topics (each presented in 10 lessons) were the following: 1) Some elaboration of basic school mathematics; 2) The set-theoretical language of mathematics; 3) Problems in elementary algebra; 4) Planimetry and Stereometry; 5) Rational functions; 6) Exponential, logarithmic and trigonometric functions; 7) Transformations in geometry; 8) Vectors in geometry, Trigonometry; 9) Linear objects in the space; 10) Quadratic objects in the space; 11) Infinite sequences and series, Combinatorics; 12) Probability and Statistics; 13) Differential calculus; 14) Integral calculus; 15) Domains of numbers, Algebraic structures; 16) Methods of work in mathematics.

One of the aims of the serial was to show mathematics as a tool in science, technology and practical life. As the title suggests, some diversion was also a part of the programme.

During each lesson, the teacher (or moderator) was discussing the topic with a little group of students (= a little model class); he often introduced some prominent guest from a research institute or from industry. At the end of each lesson, the teacher proposed homework for the general audience (two problems for a sponsored competition and one additional problem).

The serial was prepared with the cooperation of several regional studios. The central management evaluated it as very successful and as having broad impact.

An experiment with the encyclopedia

The Encyclopedic Institute of the Czechoslovak Academy of Sciences is preparing a *little general* encyclopedia (probably in one volume), with some mathematical items. Our group of mathematicians has proposed a new approach which was accepted by the editors. In a general dictionary of this type, the mathematics is usually reduced to basic school mathematics and to most frequent items in engineering. We decided to use a part of the capacity available for the publicity of modern (or advanced) mathematics. We have included items which are sophisticated from the mathematical point of view but (as we believe) can be still somehow presented to a layman (in the sense that he or she will wonder that something like this exists and may become eager to learn more). Besides the standard items, like circles, sine and cosine, gradient and so on, we have included new items such as algebraic topology, Galois theory, cardinal and ordinal numbers, Monster Group, Exotic 4-space and others.

Our philosophy behind this is that (a) the general audience should see that mathematics is still alive; (b) not all people are just pragmatic and many of them need just something "for the soul" — in which case an exotic 4-space may be a good choice.

Acknowledgements. I am grateful to Leo Boček and Oldřich Odvárko for some information used in this text.

Oldřich Kowalski
Prague, Czechoslovakia
1991

Germany

In response to your questionnaire, I would like to inform you about public relations activities which have been carried out in cooperation with the Deutsche Mathematiker Vereinigung (DMV). I will try to describe these activities in the order of the questions on your form.

During the celebrations of the 100th anniversary of the DMV in Bremen in September 1990 there was an exhibition "Mathematik sehen" and a series of press releases.

1. They were addressed to the general public, schools, the media and politicians, to acquaint them with the usefulness, beauty and scientific development of Mathematics.

2. The exhibition was arranged by mathematicians of the University of Bremen, using mathematical models, computer generated objects, prints, objects of mathematical art, movies and experiments (on computers, soap bubbles), that could be influenced by the observer.

3. Some of the more prominent aspects were "Dynamical systems" and "Chaos theory".

4. The public was invited. There was excellent participation, in particular of about 30 school classes.

5. The reaction was positive. There were about 25 publications by the press media.

The DMV has continuously published press reports.

1. They were directed towards the media and the readers (listeners, viewers).

2. The press reports were for some years mainly produced by myself and distributed to journalists who showed interest in reports about mathematics.

3. News about conferences, prizes for mathematicians, celebrations of anniversaries of old mathematicians (e.g. 100 years), mathematics in society (the changes in patent laws with respect to mathematical algorithms, new applications).

4. Many of the press releases were more or less verbally taken over by the media.

5. The reaction by the journalists was mixed, some were very interested, some others didn't want to be "bothered" and wanted to be removed from the distribution list.

Invited address to a group of journalists.

1. The audience consisted of technical journalists and PR managers of the industry. The initial purpose was to show that Mathematics can be misleading if used in the wrong way and should be applied by persons with a strong mathematical education.

2. The talk was given by a Mathematics professor.

3. Use and misuse of statistical material and other mathematical tools.

4. The address was given for the society of technical journalists (TELI) in Munich, Germany with a participation of about 40 persons.

5. The reaction was very favorable (this very critical group tends to leave talks, when they get too long; these talks lasted two hours and nobody left!). As a follow up, several interviews on the broadcasting systems and other invited addresses and invitations to discussions followed. The propagation in the media was exceptionally good (for mathematical standards).

Bodo Pareignis
Munich, Germany
5.01.1992

United Kingdom

For the past ten years the London Mathematical Society has organised Popular Lectures. These take place at the end of July, when the academic term has finished and school examinations have been completed. I have coordinated the organisation of the Popular Lectures since 1985. I enclose a list of the speakers and titles.

The series of lectures started when, following a well-meaning announcement in the national press by a reporter who had seen the notice, large numbers of the general but misguided public attended one of the regular LMS meetings at which Professor Bombieri was giving a research lecture. The Society realised that there was a potential audience for the exposition of university level mathematics, but that it had to be presented in a form comprehensible to a much wider public.

For the first few years the lectures were held only in London. Beginning in 1986, we introduced a second evening of lectures at a northern university. These have been given for two years each at Leeds, Liverpool and Sheffield.

The lectures are aimed at people with a general interest in mathematics. A large proportion of the audience are school teachers and teachers in Colleges of Further Education. Many of the school teachers bring their advanced level pupils with them, and we particularly encourage this. The Society's Administrator has built up an extensive list of contacts. The lectures are advertised very fully. Since 1986 a video recording has been made under studio conditions, of each lecture. Except for 1988, the recordings were made separately from the public meeting. The LMS has, therefore, collected a library of videos, which are available for sale or hire.

The majority of the Popular Lectures have attempted to communicate the essential features of some aspects of current mathematical research. There have been two lectures (Ledermann, Wilson) that were primarily historical. The audiences in London attained the size of 300 in 1984 and have fluctuated just above that figure ever since. Our experience is that the Northern universities have been more variable, with audiences between 50 and 150.

The audiences have always responded enthusiastically. In London at least, we know that many people return every year. It does seem very important to have extensive long-term relations between the academic mathematicians and the local schools.

Philip Holgate
London, United Kingdom
18.11.1991

Speakers at Popular Lectures

1982	J. Lighthill	The Mathematics of weather forecasting
	K. Devlin	Infinity
1983	E. C. Zeeman	The Discovery of perspective in the renaissance

	D. Williams	Hopping mad: the mathematics of probability
1984	R. Penrose	Cosmic geometry
	R. Brown	The Fascination of Knots
1985	R. F. Churchhouse	Discovering Mathematics with the Computer
	D. V. Lindley	Mathematics and the Law
1986	W. Ledermann	The Rise and fall of Matrices
	J. Maynard-Smith	Games animals play
1987	F. Piper	Codes and Cyphers
	W. Hodges	Games that solve Problems
1988	M. V. Berry	Chaology
	D. G. Kendall	How should a mathematician think about shape?
1989	F. G. Leppington	Asymptotic expansions
	R. Wilson	Stamping through Mathematics
1990	P. Giblin	Geometry and Computers
	R. A. Bailey	Designing experiments in the presence of interfering neighbours
1991	I. N. Stewart	Four encounters with Sierpinski's Gasket
	A. Beardon	Wallpaper Patterns in different Geometries

Lithuania

1. Popularization of mathematics has been directed towards:

a. School children to stimulate their interest in mathematics, their willingness to participate in competitions of young mathematicians as well as their decision to study mathematics in higher schools.

b. Intellectual community to help understand the role played by mathematics in the development of sciences.

c. The public of Lithuania in general to acquaint it with the achievements of Lithuania's mathematicians in the context of the achievements in mathematics attained by the mathematicians of the world, thus boosting national confidence in the appropriate standard of abilities.

2. Popularization has been mostly carried out by means of the press: the popular scientific magazines "Science and Engineering", "Science and Life", "National School" as well as the newpapers for school children and teachers. Once in a while some of the leading mathematicians of Lithuania address the general public by radio or TV. Mathematicians are encouraged to submit papers for publication by the board of the association of Lithuanian mathematicians. One of the members of the board is responsible for the popularization of mathematics. Assistant Prof. A. Baltrunas, Prof. V. Liutikas and some other mathematicians are among those who comparatively more often tackle the problem in the press.

3. The questions considered mostly included history of mathematics, relations with sciences, putting mathematics to man's practical use.

4. Participation from the public consisted in repeated requests expressed in letters to the authors of the papers published to further dwell on certain topics.

5. At present the situation is like this:

a. no decline of the interest taken in choosing studies of mathematics by the young people is noticed;

b. the public is aware of the international status of Lithuania's mathematicians and it is duly appraised by the public;

c. the interest taken in mathematical simulation by the specialists of other fields of sciences is consistently rising.

The general public regards popularization of mathematics with favour.

Jonas Kubilius
Vilnius, Lithuania
1991

Norway

Popularization of mathematics

For a long time there did not seem to be much interest for popularization among Norwegian mathematicians, but the situation has improved noticeably in the last few years. Most of the popularization is channelled through one of the following two organizations:

"Norsk Matematisk Forening" (NMF) is the Norwegian Mathematical Society; i.e. the professional mathematicians membership organization and the Norwegian partner in the European Mathematical Society.

"Norsk Matematikkrad" (NMR) is an advisory board for the government in questions relating to mathematics, and in addition to professional mathematicians from the universities, there are members from all levels of the educational system, from the teachers' organizations, from industry, and from the research councils.

A short list of the main activities (many of them in the initial stages) would look as follows:

1. Normat is a popular mathematics journal published jointly by the mathematical societies and the teachers' organizations in the Nordic countries Denmark, Finland, Iceland, Norway, and Sweden. The journal publishes four issues per year and has approximately 1100 subscribers. Mainly because the editorial staff is situated in Oslo, the journal contains an unproportionately large amount of Norwegian material.

2. The Abel Competition is NMF's annual competition in mathematics for high school students. There are two stages in this competition; for the first round, (identical) problem sets are mailed to high schools around the country, the students answers are collected and graded, and the 20 best students are invited to the final round in Oslo. The six best students in the final make up the Norwegian team to the Mathematics Olympiad. More than 2000 students participate in the Abel competition every year, and care is taken that the first round problems contain something for everybody — we do not want give the impression that mathematics is something for the few exceptionally gifted. Problems from previous Abel competitions have been collected in a book distributed by the NMF, and this collection has become quite popular among teachers who want some variation from the routine problems of many textbooks.

3. NMR's collection of popular mathematical essays. NMR has collected a number of popular essays from Norwegian mathematicians and plans to publish them as a book. The collection is aimed primarily at senior high school students and beginning university students, but the essays should also be of interest to teachers and to the public in general. Some emphasis has been put on mathematical topics which are far from the school syllabus (there are essays on cryptography, complexity theory, and fractals as well as on bootstrapping methods in statistics and mathematical methods in surveying), and there are also essays on the history of mathematics and on the importance of mathematics in education and society.

4. NMF's booklets on mathematics. In collaboration with a commercial publisher (Gyldendal), NMF is starting a series of booklets (32–48 pages) on mathematical topics. One booklet has already appeared, and the plan is to publish two booklets per year.

5. NMR's survey of Norwegian mathematics. Earlier this year, NMR received a grant from the Norwegian Research Council for Science and the Humanities to study the state of Norwegian mathematics today and to identify important issues for the future. As part of the final report is planned a popular essay on the importance of mathematics in a modern society aimed primarily at politicians, administrators, teachers and journalists. The essay is imagined somewhat along the lines of the American report "Everybody Counts", but on a smaller scale.

6. Sophus Lie exhibition and video. In connection with the 150th anniversary of the birth of Sophus Lie in 1992, an exhibition based on his life and work is being prepared. Work has also started on a "Sophus Lie video" aimed primarily at high school students. The video will be a collaboration between mathematicians and a professional video production company.

Although it does not fit exactly into the popularization label, it may be worth pointing out that in recent years several steps have been taken to strengthen the ties between the universities and the high school system; most (if not all) universities each year organize a special day with discussions and lectures on new scientific developments for high school teachers, and the universities and professional societies have been pushing to get representatives on the curriculum committees on all levels in the educational system (as an example we can mention that the NMR in the last years has had representatives on the committees revising the mathematics syllabus in the high schools and in the teachers'colleges.) For a number of years, the NMR has been administering a mathematics test to new students at different university level institutions, and this test has been quite useful in documenting, e.g., that most primary school teachers have an insufficient knowledge and mastery of even very basic mathematics.

As you will see, most of the projects above are still in their initial stages, and it is, unfortunately, too early to report on public reactions.

Tom Lindstrom
Oslo, Norway
13.11.1991

Spain

Activities concerning popularization of mathematics

Mathematics in the newspapers. Some of the newspapers with the greatest popular impact, *El País, La Vanguardia, ABC,...*, publish a weekly science supplement. These supplements usually have some mathematical content. Often this is in the form of interviews with representative mathematicians, occasionally reviews of recent mathematics books aimed at the general public and most frequently by means of mathematical recreations. These contributions are, in some cases, written by professional mathematicians and in others by more or less able journalists with some mathematical background. We can sometimes say that they have done a wonderful job. Subjects such as chaos, fractals, the impact and use of computers and so on have been much publicized in recent years and it can be said that the impact on the general public has been significant.

Mathematics in the popular scientific magazines. There are four scientific magazines with a respectable circulation. *Investigaćion y Ciencia*, the Spanish version of Scientific American has a circulation of about 50 000. These magazines have a significant influence on a wide segment of the population.

In Andalucía, some energetic teachers have experimented with the publication of a 4-page bulletin which has a wide distribution, containing mathematical news, stories and mainly problems. This bulletin has generated a lot of interest among children and their families. It also contains excellent material for the classroom which in turn serves as a means of communication among different schools.

The recent upsurge of publications related to the popularization of mathematics. In Spain, during the past ten years, there has been a great increase in the number of publications fostering the popularization of mathematics, in one way or another. The large success of the books by Martin Gardner, most of which have been translated in Spanish and some of which have even been published in Spain simultaneously with the American edition, and similar works have stimulated some publishers to issue a large number of publications in the field of mathematical recreation.

On the other hand, works such as *The Mathematical Experience* by Davis and Hersh have reached a fairly broad audience and have had a healthy influence in correcting the understanding of mathematics among educated people. It is interesting to note that other works on the history of mathematics, such as that by Boyer, have been published in Spanish by well-known publishers. These works have been receiving attention by many people outside of the mathematical community. The work of M. Kline *Mathematical Thought from Ancient to Modern Times* has been translated and will be published in the near future by one of the most prestigious publishers of the past decade.

Mathematics on TV. In the 1980s, public television conducted a very interesting experiment. Spanish and foreign mathematicians were invited on prime time to recount their experience in their particular fields of research and their history. The intent was to present an account of what is mathematics. The series was directed at the general public and some of the programmes were quite successful. The series of eight pieces of 30 minutes each has been shown several times in the last few years. In general we can say that the results have been satisfactory.

Mathematics on radio. Again in the 1980s, some radio networks, particularly in Andalućia, have been relatively active in the area of mathematics. This is due above all to the interest and leadership of some high school teachers. They presented a series of programmes related to the history of mathematics in the form of tales for children. The tales were followed by mathematical questions that tried to involve the listeners. The success went far beyond what might have been expected.

Mathematics on the road. A few years ago, a French exhibit, *Horizons Mathématiques*, toured many cities of Spain with great success, especially among high school teachers and students. It lent a strong impetus to many educational groups to create similar exhibits for special purposes. It is now rather usual to see exhibits in some centers focused on geometry, elementary topological concepts and so on. These exhibits stimulate and deepen a certain mathematical perception among neighboring schools.

Miguel de Guzmán
Madrid, Spain
April 1992

Sweden

I have recently been active in three projects on the popularization of mathematics.

1. Book project. I was advisor to the author, K. Dahl, and author of one chapter, of a recent book in Swedish on popularization of modern mathematics for a general reader.

2. "Mathematical pictures of fractals and chaos" for students in upper secondary school. The material consists of 23 colour photos (slides) on fractals and chaos, a text in Swedish explaining how the pictures are constructed mathematically, computer programmes, and a poster. The material is now available in most Swedish schools and is shown in classes by the teachers..... The material has also been published in Nordisk Matematisk Tidskrift 38, 1, 1990, 18–31.

3. TV-programmes. Three programmes of thirty minutes each were made by the Swedish educational TV and me on the topics Fractals, Chaos, and Dimension, respectively. The programmes were directed towards a general audience and have been sent in Swedish TV some times, and they are also available to schools.

The purpose with projects 2 and 3 was to inform on some current research in mathematics. My conclusion is that all three projects have been successful — we have received a great deal of reaction from schools and individuals.

Hans Wallin
Umeå Sweden
1.12.91

Collecting mathematical problems for the high schools

All my answers below pertain to the project: *collecting mathematical problems for the high schools in Sweden.*

1. The project was directed towards Swedish high school students, ages 15–18. Its original purpose was to collect problems for *special projects in mathematics* where the mathematically interested students write a small thesis about a topic in mathematics. This project takes a sizeable part of the last year they are in high school, as much as 1/3 of the time can be spent on the *special project.*

2. The project was presented as a book consisting of 46 different topics and distributed to all the high schools in Sweden. A considerable part of the researchers in mathematics in Sweden contributed to the book. The setting, lay-out,..., we did to a large extent at the Mittag-Leffler Institute and was done in TEX, but at some point professional lay-out experts were consulted.

3. Many different aspects of mathematics were presented in the 46 topics mentioned above. Most of the material consisted of topics related to mathematics of a *discrete* character like combinatorics, coding theory,... However, I can mention topics iteration and dynamical systems, election theory (Arrow theorem), Möbius transformations, Bernoulli polynomials, history of mathematics, Gaussian integers,...

4. The collection was sent to all schools in Sweden. It has also been tested on high school teachers on several occasions. On these occasions the participation has been enthusiastic.

5. The conclusion is that there is a great need for collections of the above mentioned kind and that the participation of university researchers in the teaching in high schools is essential for stimulating the interest for mathematics. We have found that the reactions of the students have been enthusiastic. The main difficulty is that the teachers are afraid to use the collection because they think the topics are so difficult that they will not be able to solve the problems themselves. Hence it is mainly well educated and interested teachers that use the collection for their students, or teachers that have sufficient confidence in the students, and themselves, and let the students work on their own.

Dan Laksov
Djursholm, Sweden
22.10.1991

There is not much to add to my previous letter except perhaps that the idea of writing collections of mathematical topics for high school students has caught on in all the nordic countries. Both in Denmark and in Norway there will appear collections of mathematical problems at the high school level and these collections will be used not only to have the students at high school prepare small projects in mathematics, but will also be used to improve the high school curriculum in the nordic countries.

Dan Laksov
Djursholm, Sweden
25.2.1992

Round Table B: July 7, 1992

WOMEN AND MATHEMATICS

by Eva Bayer-Fluckiger

CNRS, Université de Franche-Comté, Besançon (France)

TABLE OF CONTENTS

Introduction 37

1. The Round Table 38

2. Statistical Data 41

3. Tables 43

3. Women Mathematicians in Germany 47

4. Women Mathematicians in France 51

5. Women Mathematicians in Switzerland 52

6. Women in the Mathematical Community 53

6.1 Opinions of colleagues 53

6.2 Attitudes of women 55

6.3 The only one 57

6.4 Role models 58

6.5 Family and career 58

6.6 Women participants at conferences 59

6.7 "Dear Sir" postcards 60

6.8 What would be different if there were more women in the mathematics departments? 60

7. Committee on Women and Mathematics 61

8. Groups of Women Mathematicians 62

8.1 European Women in Mathematics (EWM) 62

8.2 Femmes et Mathématiques 62

9. Concrete Propositions 63

10. Bibliography 65

Appendix: List of coordinators of EWM 71

Introduction

The proportion of women among mathematicians varies considerably from one European country to another. In some countries, the proportion of women among students in mathematics is much higher than among professional mathematicians. The data in Table 1 gives some specifics on these points.

The aim of the round table was to analyse these facts. Of course, in a relatively short text such as this one, some choices had to be made. It appeared that in several European countries, women are seriously underrepresented among mathematicians. Therefore, special attention has been paid to these countries. This is done in some detail in the case of Germany (see Section 3). The next section, dealing with the everyday life of women in the mathematical community, is also mainly concerned with these countries. This is also the case in the final section, which contains some concrete propositions aimed at improving the situation.

Several groups of people played an important role in the preparation of the round table. First of all, the EMS committee on Women and Mathematics (see Section 5), whose work is partially reflected in this round table. The group "European Women in Mathematics" (EWM, see Section 6) was founded in 1986 and has already held 5 congresses. The present text owes very much to the women who are active in this group. Finally, the information in Table 1 could not have been obtained without the help of Mathematical Societies. This text is not only a report on the talks and discussions that took place during the round table, but it also tries to take into account the experience and reflection of the groups of people listed below.

Acknowledgments: I would like to thank the many colleagues who have helped by collecting statistical data, taking part in preliminary discussions, and giving information and ideas for this text. I also thank my colleagues of the Université de Franche-Comté for their interest and support, and Catherine Pagani for the excellent job she did typing this manuscript.

I thank The Commission of the European Communities, Equal Opportunities Unit of DG V, for the support they gave to this round table.

1. The Round Table: Women and Mathematics

Tuesday, July 7, 1992

Organising Committee: Michèle Audin (Strasbourg), Eva Bayer-Fluckiger (Besançon), Bodil Branner (Copenhagen), Capi Corrales (Madrid), Catherine Goldstein (Orsay), Annette Grabosch (Tübingen), Ragni Piene (Oslo), Marie-Françoise Roy (Rennes), Laura Tedeschini Lalli (Rome).

The Round Table consisted of 5 short talks (10–15 minutes each), and a general discussion. The following are abstracts of these talks, with some slight modifications, of those that were distributed during the Round Table.

Moderator: Capi Corrales, Universidad Complutense, Madrid (Spain).
International Coordinator of "European Women in Mathematics".

Speakers:

- **Christine Bessenrodt**, Institut für Experimentelle Mathematik, Essen (Germany).
  National Coordinator of "European Women in Mathematics", Germany.
- **Ina Kersten**, Universität Bielefeld (Germany).
  Member of the EMS-committee "Women and Mathematics".
  Member of the Praesidium of the German Mathematical Society.

The German situation, compared with other European countries, is quite deplorable. Approximately one third of the students in mathematics are female, which appears not unreasonable. However, only 9% of dissertations and 7% of habilitations are written by women, and only 2% of the professors in mathematics are female.

What are the reasons for this discrepancy? One reason might be that, with the absence of female professors, students would think that mathematics cannot be a field for women. There is still a lot of resentment towards intelligent women who think abstractly. In addition, it is difficult for a woman with children to work; school lasts only half a day and kindergartens are rare, particularly in West Germany. Other discouraging factors will be discussed later in the talk.

- **Laura Tedeschini Lalli**, Università "La Sapienza", Rome (Italy).
  Member of the EMS-committee "Women and Mathematics".

30% of the permanent faculty of Italian departments of mathematics are women. An even higher percentage have permanent positions in research institutions as mathematicians. Women in Italy feel free to choose mathematics as an elective field of study, a choice not extended to other scientific or technical fields. As a consequence, a lot of them end up "making it" to the upper academic level. This seems not to be the case in Northern Europe. Is this telling us something about mathematics, about Italy, about women? As we all know, it takes a great deal of verbal approximation before sensible variables are defined, and a model can start making sense of scattered numbers. "European Women in Mathematics" proudly undertook this lengthy task, and is drawing people together to name some of the ghosts that lurk behind personal choices and social figures. We think the ongoing work is of interest to all mathematicians.

• **Mary Gray**, American University, Washington (U.S.A.).
First president of "Association for Women in Mathematics".

American female mathematicians have had a rocky road to travel ever since Johns Hopkins University withheld a Ph.D from Christine Ladd for nearly fifty years. In the twenties and thirties women achieved a certain degree of security with positions in women's colleges, but even the best had difficulty being accepted in research institutions — witness the fact that only Bryn Mawr welcomed Emmy Noether while her less distinguished colleagues went to more prestigious places. Women were accepted as temporary fill-ins during World War II, but were hustled out of math positions as they were from other traditional male occupations when the emergency was over. In 1964 the percentage of Ph.Ds in math going to women was at the same 6% as thirty years earlier. Since then the percentage has increased to 20% and, due largely to the efforts of the Association for Women in Mathematics, women are highly visible at nearly all conferences and in professional societies. However, only 4 in 303 of the tenured positions in the top ten math departments are filled by women; more depressing is the fact that only 1 of the 88 untenured junior faculty is a woman.

• **Barbara Roszkowska**, Technical University of Warsaw (Poland).
Member of the EMS-committee "Women and Mathematics".

The situation of women mathematicians in a country cannot be considered separately from the general situation in that country. There are two crucial differences between the situation of women in Eastern and Central European countries and most West European ones. The first is connected with the prosperity of a country and its lifestyle, the second with the general attitude of society towards working women. In the case of Poland, the economic crisis has made women's lives more difficult. This has influenced the level of professional activities. On the other hand, since the last war, the majority of adult women in Poland have been working. This situation is now considered ordinary and normal by most people in Poland so that many women would not like to give up the possibility of developing their own talents and making their own professional career.

The next congress of "European Women in Mathematics" has already been held in Warsaw (Poland), June 4–7, 1993. It provided a good opportunity to discuss the general situation of women mathematicans in different countries (especially in Eastern and Central Europe). The program of the meeting included expository mathematical lectures, talks and discussions on the theme "creativity" (both in mathematics and other domains) and meetings were held with mathematics students in Warsaw.

• **Marie-Françoise Roy**, Université de Rennes I (France).
First president of *Femmes et Mathématiques*.

Quantitative information about France can be found in Table 1. No progress has occurred in the last 15 years regarding the presence of women. There was rapid progress in the 70s, then a regression in the 80s, followed by slow progress in the last years.

From a qualitative point of view, it is now possible to be a woman (a wife, a mother, a lover) and a researcher. In previous generations, women mathematicians had to be single. The presence of women is accepted in the mathematical community (collective responsibilities about teaching...). But there are not many women who have scientific power, as for example, on the committees of journals.

Mathematics may very well be a woman's love. Mathematics may very well be a woman's work. The reasons why women or men love mathematics may be partly different. Different approaches to mathematics must be allowed and encouraged. It is a good thing for the whole community to be more open to women.

There is no automatic continuous progress in the proportion of women in the math community or more generally in scientific studies. The underrepresentation of women is a global and complicated problem, very deeply rooted in family, education and the social organization of labour. There is a need for studies in various social sciences to better understand what goes on: sociology, psychology, history... The role of the educational system for reproducing (or changing) the gender roles inside society, the reality of coeducation appears now in our country as an important social issue.

Important sociologists like Baudelot and Establet have studied this problem; *Le Monde de l'Education*, the most important magazine about education in France has published a special issue. In the 60s and 70s, the importance of the educational system for the reproduction of the class structure of the society was discovered; in the 80s and hopefully in the 90s, the importance of the educational system for reproducing gender differences will begin to be understood.

The evolution of the mathematical community is part of this general social process. Internal efforts have to go on and contacts with social science researchers have to be encouraged.

These five talks were followed by a lively discussion. Most of the topics that were mentioned during the round table are developed in some detail in this article.

2. Statistical Data

The Committee on Women and Mathematics decided to make an inquiry into the number of women mathematicians in Europe. In February 1991, the committee wrote to all the Mathematical Societies of Europe requesting the following data:

I. The number of mathematicians holding a permanent position in a mathematics department of a university or a research institution in the country in question.

II. The number of full professors ("full professor" being the highest academic position).

III. The number of Ph.Ds delivered in 1990.

For each of the points I-III, the total number and then the number of women among them was asked.

IV. Percentage of women among the students in mathematics.

The results of this inquiry are represented in Table 1. Almost all the Mathematical Societies did send this data. In a few cases, for a variety of reasons, including lost mail or wrong addresses, no answer came. The group "European Women in Mathematics" was able to obtain the desired information for some of these countries. One of the aims of the Committee on Women and Mathematics is to collect the data from I-IV every four years—in time for every European Congress.

Three Mathematical Societies gave additional information:

Bulgaria

Virginia Kiryakova, on behalf of the Bulgarian Mathematical Society, obtained detailed information about mathematicians working in Bulgaria. This is presented in Table 2. In this table, Total 1 gives the numbers concerning the mathematics departments of universities and research institutions.

Italy

With the help of the Italian Mathematical Society, Laura Tedeschini Lalli has collected the information presented in Table 3. In Italy, mathematics are subdivided into 4 areas: algebra and geometry, analysis, mathematical physics and numerical analysis. The second part of Table 3 shows the division of women mathematicians in these areas.

United Kingdom

Amanda Chetwynd collected, for the London Mathematical Society, data concerning all universities in the United Kingdom (see Table 3).

TABLE 1

Country	Mathematicians women/total	Mathematicians % of women	Full Professors women/total	Ph D. 1990 women/total	Students in Maths % of women
Austria (1989)	*54/762*	*7%*	*0/73*	*3/21*	*39%*
Belgium	30/219	14%	8/134	12/32	54%
Bulgaria	254/855	30%	1/42	4/23	40%
C.S.F.R.	60/500	12%	2/65	1/10	42%
Denmark	4/121	3%	1/19	1/5	30%
Estonia	32/109	29%	0/8	0/4	60%
Finland	3/127	2%	1/34	8%	41%
France		20-25%	8%	20%	30%Univ.-17%Prep.
WGermany (1987)	40/1500	3%	4/490	19/208	33%
Greece	*67/283*	*24%*	*3/56*		
Georgia	486/1217	40%	7/72	1/28	70%
Hungary	32/281	11%	4/103	5/38	15%
Italy	609/1727	35%	84/646		70%
Ireland	7/135	5%	0/9	2/3	30%
Iceland	0/10	0%	0/4	0/0	33%
Luxemburg	1/7	14%	1/5		24%
Netherlands	19/437	4%	1/88	2/41	22%
Norway	9/103	9%	3/45	0/5	20%
Poland (from 5 univ.)	107/359	30%	7/44	3/16	60%
Portugal		*40-50%*	*5%*		*50-60%*
Romania	345/1594	22%	12/177	25/81	60%
Spain	*168/1075*	*16%*	*12/279*		
Sweden	7/150	5%	0/21	1/10	30%
Switzerland	3/141	2%	0/91	4/42	25%
United Kingdom	97/1379	7%	3/267	46/266	31%
Ukraine	31/145	21%	1/33	1/3	50%
Yugoslavia	91/530	17%	12/132	4/24	60%

Statistical data on Women Mathematicians in Europe

Sources : Entries in roman : European Mathematical Societies (Questionnaire from E.M.S.)

TABLE 2

Institute, University	Total number Mathematicians	Women number, %	Asso. prof. Total	Women number. %	Full professors
Inst. of Math. Bulg. Acad. Sciences	198	45 22,72%	64	11 17,18%	18 no women
Sofia Univ. Depart. Math. and Informatics	185	58 31,35%	47	10 21,27%	18 no women
Plovdiv Univ. Depart. Math.	65	20 30.77%	30	11 36.67%	2 no women
Technical Univ. Sofia	93	42 45,16%	38	10 26.31%	4 incl. 1 woman
Technical Univ. Rousse	35	18 51.42%	7	0 0%	
Technical Univ. Gabrovo	24	9 37.5%	5	2 40%	
Plovdiv Polytechnics	15	6 40%	3	0 0%	
Rousse Section of Union Bulg. Mathematicians	240	56 23.33%	9	0 0%	
Total 1	**855**	**254** **30%**	**203**	**44** **21,67%**	**42** **incl. 1 woman** **2,38%**
Agrarian Univ. Plovdiv	4	3 75%	2	1 50%	
Higher Institute for Food Industry Plovdiv	12	7 58.33%	5	3 60%	
Teachers' Section Sofia (6° Section)	72	54 75%			
Teachers' Section Gabrovo	185	123 66.47%			
Computer Center Bulg. Acad. Sciences	42	36 85.71%			
Computer Center Gabrovo	68	40 58.82%			
Total number investigated	**1238**	**517** **41,76%**	**210**	**48** **22,86%**	**42** **incl. 1 woman** **2,38%**

TABLE 3

Academic career

Level	% of women
Students	70
Researchers	55
Associate professors	33
Full professors	11
Principal investigators	5

Women/total by area

Research area/faculty position

Colonna 1	Researchers number, %	Associate Prof. number, %	Full Professor number, %	Total number, %
Algebra, Geometry	120/181 66%	77/107 72%	20/209 13%	217/497 44%
Analysis	63/138 45%	153/366 42%	37/281 13%	253/785 32%
Mathematical Physics	26/60 45%	17/101 17%	8/102 8%	52/263 20%
Numerical Analysis	43/77 56%	27/51 50%	11/54 20%	79/182 43%
Total	**252/456**	**274/625**	**76/646**	**602/1727**

ITALY

TABLE 4

University	Mathematicians	Women	Full professors	Women
Aberdeen	14.3	0	1	0
Aber	16	1	4	0
Aston	5	0	0	0
Bangor	16	0	1	0
Bath	27	1	5.2	0
Belf	31	1	8	0
Birkbeck	8	1	2	0
Birmingham	28	2	6	0
Bradford	13	0	1	0
Bristol	38	2	8	0
Brunel	21	2	2	0
Buckingham	2	1	0	0
Cambridge	72	1	15	0
Cardiff	26	2	5	0
City	12	1	3	0
Cranfield	4	2	1	0
Dundee	27	0	7	0
Durham	24	1	3	0
East Anglia	14	2	3	0
Edinburgh	30	2	4	0
Essex	13	1	3	0
Exeter	18	0	1	0
Glasgow	35	2	7	0
Goldsmiths	10	2	0	1
Heriot Watt	23	1	5	0
Hull	14	2	5	0
Imperial	39	3	12	0
Keele	12	2	2	0
Kent	21	1	3	0
Kings	26	3	6	0
Lancaster	12	1	3	0
Leeds	21.5	1	5	0
Leicester	8	1	0	0
Liverpool	35	2	4	0
Loughborough	31	2	3	0
LSE	35	5	7	0
Manchester	42	2	7.5	1
Newcastle	38	3	7	0
Nottingham	24	0	3	0
Open	34	8	5	0
Oxford	64	9	12	0
QMC	42	2	15	0
Eading	15	2	5	0
RHBNC	20	3	4	0
St Andrews	24	1	6	0
Salford	15	1	2	0
Sheffield	50	5	9	0
Southampton	38	2	7	0
Stirling	5	0	1	0
Strathclyde	23	0	5	0
Surrey	23	1	3	0
Sussex	21	1.5	3	0
Swansea	17	0	4	0
Ulster	14	1	0	0
UMIST	26	1	5	0
UCL	21	1	5	1
Warwick	28	1	9	0
York	13	1	4	0
	1379	97	267	3

UNITED KINGDOM

3. Women Mathematicians in Germany

Anyone who has visited a mathematics department in Germany has realized that professional women mathematicians are very, very rare. The data in Table 1 confirms this.

Christine Bessenrodt started her talk at the round table as follows: "Investigating the status quo and compiling a list of obstacles that women in mathematics have to face in Germany was a very discouraging task; considering all these difficulties, it is very surprising that there are women mathematicians at all in our country!".

Before giving additional (depressing) details on these difficulties, I would like to point out that there is also reason to hope: more and more German mathematicians think about this question, wish to understand the reasons for the under-representation of women among mathematicians, and would like to help improve the situation. The German Mathematical Society organised a panel on women mathematicians at the 1991 annual meeting in Bielefeld ([49]), and a discussion on the same subject was held at the 1992 meeting in Berlin. For the first time in its hundred-year-old existence, the German Mathematical Society has a woman (Ina Kersten) in its Praesidium. Recently, two excellent young women mathematicians became full professors: Hélène Esnault in Essen, Ursula Hamenstädt in Bonn (two nominations may not seem very many, but they represent a 50% increase, from 4 to 6 women full professors!).

Studies and academic positions in Germany

In order to enter a German university, the students have to have an "Abitur" (high school degree). This degree is normally obtained in 13 years, so the students getting an "Abitur" are about 19–20 years old. If a student decides to study mathematics, he (or she) has the choice between "Staatsexamen", preparing for teaching in a high school, and a "Diplom", preparing for a job in industry or an academic career. The statistics about the proportion of women among the students in mathematics (Table 1) concern both types of studies.

The highest academic level in Germany is full professor (called C4-Professor). This is a respected and highly-paid position. Usually a full professor has several assistants and a secretary. According to the Table, the number of full professors of mathematics in Germany was 490 in 1987. There are also "associate professors" (C3-Professors). Most permanent positions are either full or associate professors. In addition, there are also some "Akademische Räte", as well as a few other types of permanent positions. These types of positions are taken into account in Table 1. No evaluation was made on the non-permanent positions. Most of these are post-doctoral positions, limited to 6 years.

Statistics

Whereas the percentage of women among the professors is low in all subjects, there is an important difference between mathematics and other subjects as far as women Ph.D's are concerned (see [63]). The explanation is probably that in most other subjects, a Ph.D leads to careers other than academic ones.

All the statistical data concerns the former West Germany. Since reunification, many changes have occurred in the eastern part of the country. It appears that women mathematicians who had jobs before reunification are losing them at a higher rate than their male colleagues; this is also very alarming.

Why are there so few women mathematicians in Germany?

In her talk at the Round Table, Christine Bessenrodt gave a list of about 12 obstacles that prevent women from becoming mathematicians. As she points out, these can be separated into two groups: those coming from the general public, and those coming from the university. I would just like to present here the difficulties that seem to be the most important, and those that are most specific to Germany. As in many other countries, there is a preconception in the general public that mathematics is not a feminine subject. This seems to be especially strong in Germany: the fact that until quite recently the textbooks of mathematics and physics were different for boys and girls seems to indicate this.

In spite of this, a large number of females are interested in mathematics. Among students in mathematics, about 33% are girls. However during their years of study, females find that whereas they can complete their studies and then become high school teachers or perhaps find a job in industry, they have almost no chance of becoming professional mathematicians at a university. Very often, they have the impression that nobody expects them to pursue such a career.

In a 1960 poll of professors, 64% were against female students at the universities, and 79% were against female university teachers [5]. Hopefully the situation is better now—but I doubt that the present figures would be 0%! There are few female professors in any domain, but the situation is especially bad in mathematics (see [63]). It is also rare to see a woman invited to speak at seminars and colloquia. Girls don't find any role models, and more often than not, they give up the idea of having an academic career.

Some women persist in spite of all these difficulties, and get a Ph.D and a habilitation. In spite of international recognition of their work, many do not find permanent positions for many years (if ever). The only possibilities open to them are temporary jobs or emigration. When I discuss this with German colleagues, and point out that this or that excellent woman mathematician does not have a job in Germany, invariably the reply is to the effect that men have the same difficulties. However, it is truly questionable as to whether the examples given of such men awaiting jobs were in any way "better mathematicians" than the women I listed. It seems that in the minds of some colleagues women should wait until all men of comparable level have jobs, and only afterwards could they be considered. Obviously, with such a policy, it is no wonder that there are few women mathematicians in Germany. Unfortunately, it is not to be expected that the situation will change spontaneously in the near future. This indicates the need to introduce affirmative action to increase the proportion of women among the mathematicians in Germany.

Affirmative action: facts and suggestions

Many German universities have some sort of affirmative action aimed at increasing the number of women on the scientific staff. These programs are recent and often of a tentative nature: first versions are publicised and then later different versions are put into effect. Therefore it is difficult to give a precise picture of the situation. However, the following aspects seem to be important:

1. Scholarships for completing a habilitation. In order to become a professor in Germany, one has to have a habilitation. Therefore it is natural to want to increase the number of women holding such a degree. The problem with scholarships is that they are low-paid and (even more importantly) they are very limited in time: one or two years at most. This is too short to finish an habilitation, especially if the scholar is just getting back to research (these programs mainly seem to be intended for women who interrupted an academic career in order to take care of their children). I have also heard complaints that the women holding such scholarships are not well integrated into the universities where they work. This starts with very basic things like having a library card and an office.

Suggestions. Maintain this type of scholarship, but extend it with some sort of "career plan": for instance, give a scholarship for two years with possibility of renewal for two more periods of two years, if the work of the scholar is successful. Make sure that the scholars are integrated and respected in the institutions where they work; this involves office space, access to the library, the possibility of working with advanced students in a seminar. It may also involve increased financial support. Moreover, it would be a good thing not to reserve these scholarships for women. Ideally, a young mathematician preparing a habilitation should have the option of a research scholarship or a job as an assistant, paying slightly more but involving more teaching. Of course, one should pay attention to ensure that a sufficient percentage of these young mathematicians are women!

2. *Women commissioners.* Many universities have recently elected a "woman commissioner". This is usually a woman professor in the university. She tries to make sure that women get equal treatment when they apply for a position and when working at the university.

Taking over the task of a woman commissioner means a significant extra burden for the few women at the universities, in particular if this comes with hardly any reduction of other duties and with hardly any support (like office staff). Moreover, it appears that often the male colleagues are not helpful, and on the contrary they make jokes about the whole issue and try to ridicule anybody involved, making it very explicit that this is only some sort of gimmick and not to be taken seriously.

Suggestions. An obvious suggestion is to give more support to the women commissioners: office staff, and a reduction of her other duties. This is mainly a financial question. It is more difficult, but even more important, to get a more constructive attitude from the colleagues in the department.

3. Equal treatment for women candidates. Some universities even advertise positions with an accompaning text saying something to the effect that "in the case of equal qualifications, women candidates should be preferred"—as long as women are underrepresented. I have heard that, in the case of more junior positions such as assistantships, this can play a positive role for women. As long as "qualifications" means having a diploma, having good grades or having a Ph.D, it is something that can be checked and compared easily. In the case of applicants for a professorship, "equal qualifications" does not seem to have any meaning, as there is no uncontestably objective way to compare the candidates at this level. The woman commissioner can make sure that the women candidates for a professorship are invited to give a talk. However, this seems to be totally useless: if the professors are not interested in hiring the woman in question, they will just not come to the talk. This is usually what happens. The woman commissioner can recommend changing the order of the candidates on the list. But this has a very negative effect, in fact it decreases the number of women who appear on lists of candidates at all.

Suggestions. For junior positions, the strategy of preferring women in the case of "equal qualifications" seems to have good results, so it is a good idea to maintain it.

It is much more difficult to suggest an efficient course of action to ensure equal treatment for women candidates for professorships. The heart of the problem is in the attitude of the professors in the mathematics departments. If they are really open to the idea of hiring women, one can make some suggestions. If not, then I really do not know what measure could help, except imposing quotas for women. So, here are some suggestions: when a position is advertised, usually the area of mathematical expertise is specified. It is not at all clear that there will be women candidates in just that subject. However, highly qualified women candidates may be interested in the position. The department should consider them as well, even if they work in a different area. Another suggestion is to think ahead, and encourage young women mathematicians with a Ph.D to work on a habilitation in the department.

4. *Visiting professorships for women.* The Sofia Kovalevskaia guest professorship in applied mathematics at the University of Kaiserslautern has existed since 1991/92. It is internationally advertised and awarded.

5. *Permanent professorships for women.* It has been suggested to start a program similar to the "Fiebiger-program": professorships for women in those fields where they are seriously underrepresented (this of course includes mathematics).

Conclusions

Women are very strongly underrepresented among the mathematicians in Germany. This situation will not change spontaneously. The affirmative actions attempted so far have not accomplished much, but they provide some women with the opportunity to work towards a habilitation. The Sofia Kovalevskaia visiting professorship is certainly a very good initiative as well. It would be important to:

(1) introduce a control of effectiveness in the affirmative action programs;

(2) find an effective way of increasing the number of women professors, with and without special programs. The latter is possible if the basic attitude towards women in mathematics becomes more positive. There is certainly room to believe that there is hope for such a change!

This section is essentially based on Christine Bessenrodt's talk, and on the preliminary work of several women mathematicians working in Germany (Annette Grabosch, Sandra Hayes, Ina Kersten, Roswitha März). The article of Irene Pieper-Seier [63] was also an important source of information and ideas. As Christine said in her talk: "we were all very glad to meet so many good women mathematicians in Germany, and also several male colleagues who wish to see the number of women mathematicians increase."

4. Women Mathematicians in France

It would be very interesting to have a detailed analysis of the situation of women mathematicians in France. Because of lack of time and space, it was not possible to do this here. I hope that a detailed text on France will appear in the Proceedings of the next European Congress.

I would like to make some preliminary observations. First of all, some estimations concerning the proportion of women among mathematicians in France (communicated by *Femmes et Mathématiques* and by the Société Mathématique de France):

30 % in scientific classes at the end of secondary school (Terminale C)

15 % in "Classes préparatoires" (2 years of preparation for the scientific Grandes Ecoles)

30 % in "Premier cycle scientifique" (first two years of the University)

20 % in Engineering schools

50 % in "licence et maîtrise de mathématiques" (third and fourth year of the University)

25 % of Ph.Ds

50 % of secondary school teachers (much less than in other subjects)

25 % of "maîtres de conférence"

8 % of University professors.

Some comments:

It is clear that women mathematicians are no exceptions in France. This is shown by the statistical data, which is also reflected in the everyday experience. As Marie-Françoise Roy said in her round table presentation: "It is now possible to be a woman (a wife, a mother, a lover) and a researcher. In former generations, women mathematicians had to be single. The presence of women is accepted in the mathematical community". Women are present at every academic level. They are also increasingly present in the administrative experiences of academic life in France: in the various decision-making committees in mathematical departments, as well as in the Council of the SMF (Société Mathématique de France) and the SMAI (Société des Mathématiques Appliquées et Industrielles), in the Committee of the CNRS, and so on. The work of women mathematicians is also occasionally recognised by prizes. For instance, Claire Voisin received several French prizes, as well as one of the European prizes of the Congress.

Still, the proportion of women is very low among editors of mathematical journals, main speakers of important congresses or seminars. For instance, it is hard to understand why there are so few talks by women at the Bourbaki seminars (about 1,5 %).

It would be a dangerous mistake to think that the proportion of women mathematicians in France will increase spontaneously. The evolution of the situation in the last 25 years was the following: fast increase in the 70s, then a slight decrease in the 80s and again a slight increase in the early 90s. The merging of the two Ecoles Normales Supérieures will certainly have a negative effect on the proportion of women mathematicians in the near future. These problems are currently studied by sociologists, for instance Baudelot and Establet.

5. Women Mathematicians in Switzerland

The highest political power in Switzerland is held by the Federal Council. The composition of this 7 member council follows the "magic formula": a quota system regulating the representation of political parties and of regions.

In Switzerland, there are very few women holding positions of responsability or power. In particular, there are very few women university professors. This is true in any subject, and is especially striking in mathematics. More and more people would now like to change this situation. Most universities express the desire to increase the number of women professors.

One way to obtain this would be to use a quota system. This has been suggested, but most university professors are very much against it. If one thinks about it, this is really surprising. Why not do in Universities what works so well in the Federal Council?

For the time being, only very mild measures are taken. They are similar to those introduced in Germany. For instance, when a position is available, the announcement contains a sentence saying, "the University wishes to encourage women to apply", or something similar. But even this irritates many professors.

Most Swiss Universities and Polytechnic Schools now have a "woman commissioner" whose role is to see that women are not discriminated against. This is a very good thing, but of limited efficiency since the woman commissioner only has a consultative role.

6. Women in the Mathematical Community

6.1. Opinions of colleagues

The reactions of men (and women) to their women colleagues and students are quite varied. They depend a great deal on the country and of course on the personality of the colleague.

(1) Colleagues who think that women can be as good as men in mathematics: More and more colleagues do think so, not only in the countries where there is a high percentage of women mathematicians but even in those where there are hardly any. In the second case, they also realise that the underrepresentation of women among the mathematicians in their country is not natural. Many of them are ready to help in one way or another. They join associations like AWM or *femmes et Mathématiques*, or take part in round table discussions on this issue. More importantly, they are encouraging their women colleagues and women students.

(2) Colleagues who think that some women can be mathematicians, but not too many: Some men think that some women can be good mathematicians, and have great respect for the work of some of their women colleagues. However, the majority think that this happens only very rarely and that women are gifted for mathematics much less frequently than men.

(3) Encounter with colleagues who don't think that women should be mathematicians.: As already pointed out, there are many types of reactions. I will give a non-exhaustive list here, and try to understand the reasons for each of them.

The weight of tradition

Some people, even mathematicians, find it very strange and unusual that women are mathematicians, or that mathematicians can be women. In many countries, this combination is very rare. As a result, many people rarely meet any women mathematicians. It is a frequent reaction to feel uneasy in the presence of unusual phenomena. Perhaps surprisingly, it is even more difficult to accept things or beings that resemble closely known ones (for instance, mathematicians) but are different in one important aspect (they are women). This is because two types of reactions are involved: one tends to have a certain type of behaviour towards mathematicians, and another (quite different) towards women. To a mathematician, one would talk about theorems, conferences, job openings. One would open the door for a woman, and ask her about the children. Being in presence of somebody who is both a mathematician and a woman may cause problems. People (usually) prefer to avoid problems, and this one can be easily avoided by not having women colleagues in the department, not inviting women to conferences, and so on.

Illustrations:

(1) This is a story from the special issue on women and mathematics in the *Notices of the AMS* [3], p. 732. "A graduate student at Warwick gave a ride to a visiting star in her field. She started to talk to him about his lecture and saw him chuckling to himself. She asked him what was funny and he bent over double, laughing, "A women, talking mathematics, and foliations, it's too much!" ".

(2) This story has nothing to do with women; it just illustrates the difficulty of relating to someone having two attributes that are traditionally incompatible. A young German mathematician was nominated "assistant professor" (a new type of temporary position) in a German university. He took the elevator with a full professor. The full professor said to him: "These new positions are really confusing; I really think that they should not have been created. This situation is embarrassing, you are not really an equal, but still a professor somehow, so how am I to relate to you now?"

This is to show that in communities with strong hierarchical structures the relation-question is really very serious: if it is hard to figure out how to relate to an assistant professor, it is much harder to figure out how to relate to a woman professor!

Fear of loss of status

Professors at prestigious mathematics departments are usually very proud of this. They want to preserve the prestige of the status of professor, and of the department. Many of them think that this is incompatible with hiring women. Women just cannot be prestigious mathematicians. When they think of the faculty members they would like to hire, they have a profile in mind, usually of a young mathematician having a strong research record in a fashionable topic, and an excellent reputation on the international level—and a man. Consciously or unconsciously, this condition is added.

Fear of competition

In the very same countries where there are very few women mathematicians, it is currently also hard for men to get jobs as mathematicians. Therefore it is understandable that many of them don't welcome the arrival of new competition. Simple arithmetic shows that if a new category of people arrive on a job market, the competition will be harder. Moreover, they sometimes fear that this will be unfair competition. Some people think that pretty young women have an extra advantage. In the countries having affirmative action programs, being a woman is actually an advantage—that's the whole point of the program. Most European countries do not have such programs; some are just starting to have them. But in the U.S., some young mathematicians seem to be very bitter about this, and tend to say very negative things about the programs and also about women mathematicians, individually or collectively. I have frequently heard the comment, "It is no longer possible to get a job in mathematics for a white male". The data about new appointments in the U.S. ([4]) shows that this is not true, but the comments are still there.

Women should be happy

Some people (even mathematicians) think that real happiness for a woman is getting married and having children. And they also think that this is incompatible with a career as a mathematician. This is really a variant of the "weight of tradition" reaction, but a specially nice and unselfish form—they only want the happiness of the women, in particular women graduate students or women candidates. They feel that these women make a big mistake by wanting a thesis, or wanting a job.

Conclusion

These reactions are unpleasant, but natural. Most stem from traditions and stereotypes. I think that the most effective way of dealing with this problem is to talk about it. Talk about it to colleagues or fellow students, organise panels about it. Another very efficient method is to make women mathematicians more visible by inviting them to give talks more frequently.

6.2. Attitudes of women

Girls and boys are educated differently, and this has some consequences in their attitudes towards professional activities. Girls are brought up to have "feminine qualities": they should be nice, not too independent, always ready to help others. Boys are brought up to be independent, competitive. Girls aim to be liked, boys aim to be respected. Of course, society is changing and the patterns of education will also change, but perhaps not very rapidly.

This issue is discussed in many books and articles. I would like to recommend two popular psychology books: [26] and [83]. They are both pleasant to read and can be very useful in analysing our attitudes in professional life, in particular as mathematicians. I would like to point out some attitudes that many of us have in common, often without realizing it, and that such attitudes can have a negative effect in our work as mathematicians.

1. Lack of confidence

Many sociological studies show that girls tend to have much less confidence in their capacities and scholarly performance than do boys. When questioned after a test, boys usually think that they performed better than they actually did, and girls usually think that they performed not as well.

One often sees the same attitude in professional women mathematicians. They do not dare to submit their papers to very good journals, but think that any recognition of their work is due to good luck.

Of course, such an attitude is rather nice and also effective in reducing stress; if one does not expect much, one is less often disappointed. However, this is clearly not the attitude that will lead to the greatest professional success. It may even lead to not studying mathematics at all, even if one has good grades and is interested in the subject. At all levels, a lack of confidences makes women give up their professional expectations.

2. Fear of success

This issue is closely related to the first one. Many women seem to think that having too much professional success is not "nice", that they will be less popular, that they will have less chance to find a boyfriend or a husband. For men, exactly the opposite is true. A very striking illustration of this is given in [26]: A test was conducted with American students. They were given the beginning of a story and they had to complete it. The story began as follows: a student, John (for the boys) or Anne (for the girls) finds out that he has the best results of the whole university at an exam. Most of the boys predicted a highly successful professional career and a happy life for John. Most of the girls were afraid that Anne would be lonely and nobody would like her.

3. Competition?

Most girls (women) don't like competition. Some do, to a certain extent, but there is a significant difference in boys (men). This can cause problems, for instance, when taking a competitive exam in order to enter an institution (such as the "Grandes Ecoles" in France), and in many other professional situations.

4. Finite energy?

I would like to point out a "mistake" that I have seen several women students or graduate students make. They marry early, and expect to complete their studies or Ph.D while also doing almost all the housework, making a living, and perhaps even taking on an administrative duty such as being a student delegate. In all the cases I know, this situation ended with a divorce and the woman became a professional mathematician living alone or with someone who did a reasonable share of the household duties (I probably did not meet those who gave up mathematics instead).

The same attitude also exists among professional mathematicians. We seem to think that we *have* to take on "community duties", for instance, administrative work. Laura Tedeschini Lalli explained this attitude very well during the round table discussion. She said: "We think that we can think about mathematics at night, take care of our children in the morning and run the department in the afternoon. Let me remind you—we only have finite energy!"

Conclusion

I would like to conclude with three remarks. First of all, not *all* women have these attitudes. Moreover, some men do have them. But statistically, they are more common in women then in men.

The second remark is: I do not want to say that these attitudes are wrong and that all women mathematicians are invited to think this over and change. All I am saying is that they seem to be incompatible with the highest performances in mathematics. So we have to make a choice, and it is just as well if we know about it.

Thirdly, some women mathematicians outgrow these attitudes and they tend to "overreact"; moreover, they are perceived by their colleagues as extremely aggressive. It is not easy to give up the behaviour we were brought up with, and to find a balanced attitude of non aggressive, modest self-confidence. Moreover, some people tend to find *every* successful woman too aggressive— the mere fact of being successful is often perceived as unfeminine. This brings us back to point (2).

6.3. The only one

In some countries, women mathematicians are so rare that when there is a woman in the mathematics department, she is the only one. The situation is nearly as bad at the graduate student level, as well as in conferences and mathematical research institutes. With some luck, there may be one or two other women, but usually not more. To live with this situation year after year, conference after conference, is really depressing. One starts to wonder, especially on the days when one did not prove any major theorem, whether one is really in the right place and whether the popular wisdom according to which "mathematics are not for women" is not correct.

There are now groups of women mathematicians, email networks that can help (see Section 6).

Possibly, there may be women mathematicians who enjoy being exceptions—but I have not met any so far.

6.4. Role models

Role models (professors, thesis advisors, and so on) play an important role in the development of every mathematician. It seems that gender plays a role in this process: women can find it difficult to identify with men, and vice versa. This poses a serious problem to the women students in the countries where there are almost no women mathematicians.

Role models in action

I have always been impressed by Pilar Bàyer (no relation to me). She has been a professor in Barcelona since 1981, and has built up a number theory group there. She has had 9 Ph.D students so far, 5 of whom are women. Of course there are many women mathematicians in Spain, but still, none of her male colleagues have such a high percentage of women Ph.Ds. In September 1991, at the "Journées Arithmétiques" in Geneva, 17 participants came from Spain, and 10 of them were women!

I could give several similar examples of women mathematicians in France and in the U.S. who have had many women graduate students. This seems to happen in every country with a reasonable proportion of women mathematicians (15% at least). In these countries, most women mathematicians who have Ph.D students at all, seem to have a large percentage of women students. In the countries where there are very few women mathematicians, one can observe a different but related phenomenon. There are often several women Ph.D students working with the same advisor, or with different advisors but in relatively close subjects. They talk a lot to each other, and they partially function as role models.

Psychological analysis of role models of women mathematicians

This is entirely outside the scope of this article. However, I would like to point out that a start has been made on this subject by Claudine Blanchard-Laville. She took part in the 1991 EWM congress in Luminy, and wrote a very interesting article which appeared in [39]. At the same conference, a discussion about personal experiences of role models took place. A short report on this, written by Laura Fainsilber, appeared in [39].

6.5. Family and career

Having a professional career as well as children is often difficult. The degree of difficulty depends on the country: whether or not there are easily accessible kindergartens, how long a maternity leave women can take, and so on. They also depend on the partner: how large a share he takes in the housework and the education of the children.

None of the above questions are specific to mathematicians. This is certainly not the place to start a detailed general study on the subject of family and career. The only thing that I can do here is to present in a certain number of concrete examples of how some women mathematicians deal with this problem.

1. First child just before a Ph.D: Many women mathematicians had their first child just before, or just after, defending their thesis. I know several of these, who now have 2 or 3 children, and are successful in research as well. In all of these cases, the father does about half of the housework.

2. Career first, children afterwards: Some women wait until they are very well established mathematicians before they have their first child. In the cases I know, this also works very well and the women have been able to continue their research at the same level after the birth of their child. Some of them even say that the break of a few months was an opportunity to change their mathematical subject.

3. Children first, career afterwards: This may be difficult to realise, and depends very much on the conditions. However, I know some excellent women mathematicians who had 2 or 3 children first, and only started to work towards a thesis afterwards. In general, they say that they have been lucky. There are now affirmative action programs that give scholarships to women with children to start research, or come back to research.

6.6. Women participants in conferences

Women at mathematical conferences are frequently mistaken for "wives" (which they perhaps are), that is, "accompanying persons". Women participants at mathematical conferences are still rare. This is especially true for those conferences where the number of participants is limited, so the mathematicians must be invited.

Typical examples of such conferences are those that take place every week in Oberwolfach (Germany) and Luminy (France). The number of participants is limited to 42 in both cases (with a possible extension to about 50, under favourable circumstances). The number of women at the Oberwolfach meetings usually varies between 0 and 3. In Luminy, this number is much higher: it usually varies between 4 and 8. This is certainly due to the fact that the number of women mathematicians is so much higher in France (even though the conferences are international, a large proportion of the participants comes from the host country).

Signs that times are changing: In May, there was a conference in Oberwolfach that is part of a series existing since 1978. There are usually 2 women mathematicians at these meetings. This time not only were there 5 women, but two of them came with their husbands and children. The husbands came as accompanying persons, and they looked after the children during the conference!

In September, there was a meeting in Luminy with 10 women participants, and 7 of them gave talks. Even more impressive: I just saw an announcement for an international conference of combinatorics, where half of the invited speakers were women.

6.7. "Dear Sir" postcards

Many mathematical departments have printed postcards to be used for reprint requests. They contain a text that runs about as follows:

"Dear Sir,

Please send me a reprint of your article... that appeared in...
Many thanks in advance. Sincerely,"

A card that I received recently actually had this in three languages, namely:

"Lieber Herr Professor,
Monsieur le Professeur,
Dear Sir".

So that there is no ambiguity about it... Of course, the person who is requesting a reprint does not necessarily know whether the author is a man or a woman. But actually *printing* postcards with such a text implies that only men can be authors of papers that people want to read — and this is clearly false and insulting.

6.8. What would be different if there were more women in mathematics departments?

(1) Different atmosphere in the department: The best source for this is to ask the colleagues who work in departments with a high percentage of women mathematicians. They think that this gives a more natural atmosphere to the department, that everyday work is more pleasant. A German colleague told me that he would like to have more women in his department because there would then be less tension at the department meetings. I think that he is right to expect this.

(2) Equal opportunities: It appears, for instance from the statistical data, that in certain countries, many women would like to be professional mathematicians and cannot (because they are discouraged, or not hired). This is not fair — men and women should have equal opportunities.

(3) More gifted students and colleagues, more qualified people in scientific and technical professions: If women have equal access to the mathematical profession, there will be a new arrival of gifted students in the universities— more confident, because they will have role models and will not feel isolated. This will be a gain for the mathematical community. At a more general level, this evolution will also contribute to encourage girls to choose scientific and technical professions, and this will be a gain to society.

7. Committee on Women and Mathematics

The EMS committee on women and mathematics was founded in January 1991. At present, the committee members are: Eva Bayer-Fluckiger (France), Ina Kersten (Germany), Ragni Piene (Norway), Stewart Robertson (U.K.), Barbara Roszkowska (Poland), Laura Tedeschini Lalli (Italy), Vera Trnkova (C.S.F.R.). At the time of this exposition, the committee has done the following:

(1) Collection of statistical data: With the help of the mathematical societies, data concerning the situation of women mathematicians in Europe was gathered. This is presented in Section 2.

(2) Analysis of the situation in Germany: It became evident that Germany is one of the European countries with the lowest proportion of women mathematicians. This made us decide to start a more detailed analysis of the situation of women mathematicians in Germany. This work was mainly done by the women working in Germany: Christine Bessenrodt, Annette Grabosch, Sandra Hayes, Ina Kersten, Roswitha März, Irene Pieper-Seier and others. The results of this analysis are presented in Section 3.

We plan to continue to work on these two points:

(1) More statistical data: The information gathered so far is very interesting. It is important to collect similar data every 4 years, so that we can have an idea on the evolution of the situation. Moreover, we also want to do our best so that the statistical data coming from different countries can be really comparable. It is not always clear whom to count as a "mathematician". We think that the decision, within certain limits, should be taken by the Mathematical Societies. However, we can help to obtain greater consistency.

(2) Further analysis of the situation in Germany, possibly extended to other countries: We would also like to continue analysing the German situation, especially in order to understand how effective the encouragement programs are and how the situation develops in the coming years. This project should be continued as long as there is such severe underrepresentation of women among the mathematicians in this country. It would be very interesting to extend this project to other countries, for instance the Netherlands, Switzerland, Scandinavian countries and the U.K.

(3) Study of encouragement programs, concrete propositions: We will give special attention to the affirmative action which exists, or is planned, in the above countries. We will try to evaluate the results of these programs and make some concrete recommendations. A start on this is made in Sections 3 and 7.

8. Groups of Women Mathematicians

Since 1986, several groups of women mathematicians have been formed in Europe. Some of them are associations, others just informal groups. For most of the women taking part in these groups, the main motivation is to meet other women mathematicians. It is really pleasant and encouraging to see that there are so many excellent women mathematicians. This is of course especially important for the women working in the countries where there are hardly any women mathematicians. It helps them to fight the feeling of isolation and the impression of being "strange" that often arises when they work for a long time with male colleagues only. Another motivation is to talk about mathematics in a non-competitive context.

I would like to present here two such groups: European Women in Mathematics, and the French association *Femmes et Mathématiques.*

8.1. European Women in Mathematics (EWM)

At the ICM in Berkeley in 1986, a panel discussion was held about women mathematicians. It was organised by the "Association for Women in Mathematics" (AWM), an association founded more than 20 years ago in the U.S. (see for instance [3]). Several European women took part in this panel. They decided to create a European association of women mathematicians, and that is how EWM was born.

Since then, several EWM congresses have been held (in Paris, Copenhagen, Warwick, Lisbon and Luminy). At each of these congresses, there were expository mathematical talks, and discussions about subjects of interest for women mathematicians (for instance : family and career, role models, etc).

The longest EWM-congress so far (5 days) was the fifth one, held in Luminy (France) in December 1991. It was attended by 49 participants from 19 countries. For the first time, two "satellite meetings" took place after this EWM congress: in Besançon (March 1992) and Schwerte (Germany, May 1992). A report on the congress and its satellites is available [39]. It also contains an article by Caroline Series about the history of EWM.

The communications network for EWM is through coordinators (international and regional). There is also an EWM email net. All those interested are invited to contact one of the coordinators (see Appendix).

8.2. Femmes et Mathématiques

The association *Femmes et Mathématiques* was founded in Paris, in March 1987. It has more than 100 members. In addition to giving women mathematicians the opportunity to meet each other, the association very actively cooperates with other professional associations of mathematicians or teachers of mathematics: for instance, *Femmes et Mathématiques* organised a round table at the congress "Mathématiques à venir" in 1987 (see [23]). This Association is now one of five participating in the common framework of "Mathématiques à venir" for the promotion of mathematics in France through public debates, articles in newspapers, brochures, etc... One result has been that brochures (for students and for policy makers) on mathematical careers in France includes quotations from, results by, and portraits of women mathematicians. On the other hand, *Femmes et Mathématiques* also collaborates in activities aimed at secondary schools: constructing and testing teacher programs promoting equality in schools (the European program TENET for example and units in the new Institutes of Teacher Training) and organizing conferences and workshops at various meetings of the Association of Teachers in Mathematics. The Association has also given funds for girls permitting them to participate in mathematical summer schools. Members of the association often give talks at schools, and take part in conferences about mathematics education.

9. Concrete Propositions

During and after the round table discussion, several people made interesting suggestions as to how to improve the situation of women mathematicians in Europe. On the other hand, it also became clear that there is a need for concrete propositions in this direction. This demand comes from Mathematical Societies as well as from individual colleagues who would like to help.

1. Information about positions: Many people, women and men, pointed out that they did not always get the information they needed concerning available positions. This probably does not concern all countries, but has been observed (at least) in Germany, Spain, Switzerland and the United Kingdom. Particularly in the countries where there have been very few job openings in mathematical departments in the last 10 or 15 years, it would be profitable for everybody to have an efficient network for such information. The national Mathematical Societies could help very efficiently in this issue. They could collect information about the job openings at all universities. They could establish an email net to distribute this information to all the interested members of the Society. This would also make membership more attractive to young mathematicians. A similar network could also be created for scholarships and fellowships, if there is a need for it.

2. Gender related information in statistical data: In all sociological studies about mathematicians, it would be important to take gender into account. This is also an issue in which Mathematical Societies can help.

3. Encouragement: It has been observed that women are less often encouraged than men to do postgraduate work, and that the teachers are often surprised to see how well some women students do. This is natural enough, considering that mathematics have been regarded as an unfeminine subject for such a long time. It is important to know that many of us tend to react in this way, and make the necessary adjustments.

4. Relax or suppress age limits for women: It has been pointed out that in many countries, fellowships are limited to candidates below a given age. If a woman has children, she will be disqualified on age alone from all these fellowships. But even women without children often start late because it takes them on the average longer than men to feel confident enough to consider a career in mathematical research. One solution to this problem would be to relax or suppress age limits when women are concerned, as long as women are underrepresented among mathematicians.

5. Doctoral or postdoctoral scholarships for women: Some countries (Denmark, Germany, Norway) have such scholarships. One can make the following remarks:

a) Such scholarships are useful, even if it appears that the women who get them are also eligible for other scholarships (this is for instance the case in Denmark: all the girls who got a doctoral scholarship in a special program for women were also qualified for non-specific scholarships). It was often pointed out that girls tend to have less self-confidence than boys, especially concerning subjects that were traditionally considered as reserved to men. Therefore it is very good if they can think that they have a reasonable chance of success.

b) It is important that such scholarships do not appear to be "second best", less considered or less highly-paid than others. It is also important to ensure that the women who hold such scholarships are welcome in the mathematics departments where they work, that they have access to the library, some office space and (if they wish) opportunity to work in seminars with advanced students. This way they can act as role models, and at the same time be better integrated in the department.

c) It is not enough to give a scholarship for one or two years: in the case of successful work, it should be possible to obtain further scholarships or extensions.

d) There should be monitoring of effectiveness for these scholarships, so that one can see how to modify them to make them better.

6. Change of attitude in making appointments: In the countries where the proportion of women among the mathematicians is very low, it is very important to find a way to increase this number fairly rapidly. Scholarships for women students are not enough, if they have no role models among the professors. Therefore, the first step would be to increase significantly the number of women professors. It is not difficult to find good women candidates. Hiring women mathematicians does not involve lowering standards! But it may involve some change of policy. If the department wants to hire someone in a specific subject, and perhaps insist on giving promotions to those who "have been around for a long time", then this may be incompatible with hiring women. The real change of attitude would be to change the order of priorities. That is, hiring a woman who is perhaps in a different field. The department may need a professor in a given field, but it needs a woman professor even more!

One More Suggestion: As Marie-Françoise Roy pointed out in her talk, it would be good to encourage sociologists and psychologists to work on the subject of "women and mathematics".

Bibliography

[1] ALIC Margaret, "Hypatias Töchter", Unionsverlag, Zürich, 1987.

[2] Satter Prize to Dusa Mc Duff, *Notices of the AMS* **38**: 3 (March 1991), 185–187; or *AWM Newsletter* **21**: 3 (May–June 1991), 8–10.

[3] Special issue on "Women in Mathematics", *Notices of the AMS* **38**:7 (September 1991), 701–754; 774–775.

[4] 1991 Annual AMS-MAA Survey, *Notices of the AMS* **39**: 6 (July–August 1992), 573–581.

[5] ANGER Hans, *Probleme der deutschen Universität*, Tübingen, 1960.

[6] AVELSGAARD Carol, "Women in mathematics: The Silent Minority", *The Mathematical Intelligencer* **10** (1988), 32–34.

[7] BATH Corinna, SCHÄLICKE Nicola, KNIPPING Christine (eds.), *Frauen und Mathematik*, Fachbereich Mathematik an der Freien Universität Berlin, Arnimallee 2–6, 1000 Berlin 33.

[8] BENBOW Camilla Persson, STANLEY Julian J., "Sex Differences in Mathematical Ability: Fact or Artifact?", *Science* **210** (1980), 1262–1264.

[9] BENBOW Camilla Persson, STANLEY Julian J., "Sex Differences in Mathematical Ability: More Facts", *Science* **222** (1983), 1029–1031.

[10] BENDITT John (ed.), "Women in Science, 1st Annual Survey", *Science* **255** (March 13 1992), 1325–1480.

[11] BERGHAHN Sabine, AAROE Kirsten, TAPPESER Beatrix, SCHUCHALTER-EICKE Gabriele (eds.), *Wider die Natur? Frauen in Naturwissenschaft und Technik*, Elefanten Press, Berlin, 1984.

[12] BLUM Leonore, "Women in Mathematics. An International Perspective, Eight Years Later", *The Mathematical Intelligencer* **9** (1987), 29–32.

[13] BLUM Leonore, *Assessment and Gender Differences in Mathematics*, Proceedings of the European Conference held in Birmingham, 1992, to appear.

[14] BOYKAN POU-EL Marian, "Spatial separation in family life: A Mathematician's Choice". In: *Lynn A. Steen: Mathematics Tomorrow, Issues of Equality*, Springer Verlag, Heidelberg, Berlin, 1981, pp 187–194.

[15] BREWER James W., MARTHA K. Smith (eds.), *Emmy Noether. A Tribute to her Life and Work*, Marcel Dekker, New York, Basel, 1981.

[16] BRUSH Stephen G., "Women in Science and Engineering", *American Scientist* **79** (1991), 404–419.

[17] BUCCIARELLI Louis L., DWORSKY Nancy, *Sophie Germain. An Essay in the History of the Theory of Elasticity*, D. Reidel Publishing Company, Dordrecht, Boston, London, 1980.

[18] Bund-Länder-Kommission für Bildungsplanung und Forschungsförderung: Förderung von Frauen im Bereich der Wissenschaft (Helft 19 der Materialien zur Bildungsplanung und zur Forschungsförderung). Bonn, 1989.

[19] CHOWDHURY M.R., "Koblitz, Klein and Kovalevskaia", *The Mathematical Intelligencer* **8** (1986), 68–71.

[20] CLARK KENSCHAFT Patricia, "Charlotte Angas Scott (1858-1931)", In *A Century of Mathematics in America, Part III*, Duren P. (ed.), pp. 241–252; AMS, *History of Mathematics*, vol. 3, 1989.

[21] COOKE Roger, *The Mathematics of Sonya Kovalevskaya*, Springer Verlag, New York, 1984.

[22] COSTE-ROY Marie-Françoise, *Les mathématiques et les femmes*, Université Paris-Nord, IREM, Collection Philosophie et Mathématiques, Juin 1981.

[23] COSTE-ROY Marie-Françoise et al, "La place des femmes en mathématiques: problèmes actuels, perspectives d'avenir". In: *Mathématiques à venir* (SMF-SMAI). Supplément au Bulletin de la Société Mathématique de France, t. 115, 1987, pp 320–342.

[24] CRANNELL Annalisa, "Applying for Jobs: Advice from the Front", *Notices of the AMS* **39**: 6 (July/August 1992), 560–563.

[25] DICK Auguste, *Emmy Noether. 1882–1935*, Birkhäuser Verlag, Boston, Basel, Stuttgart, 1981.

[26] DOWLING Colette, *The Cinderella-complex: Women's Hidden Fear of Independence*, 1981.

[27] ECKELT Irmgard, *Mathematik - nix für Frauen ?!*, Albatros Verlag, Frankfurt, 1981.

[28] ETH-Frauen, *Bulletin der Eidgenössischen Technischen Hochschule Zürich*, **237** (Oktober 1991).

[29] FEINGOLD Alan, "Cognitive Gender Differences are Disappearing", *American Psychologist* **43** (1988), 95–103.

[30] FENNEMA Elizabeth, "Sex-related Differences in Mathematics Achievement: Where and Why?" In: *Women and the Mathematical Mystique*, L.H. Fox et al. (eds.), The Johns Hopkins University Press, Baltimore, 1980, pp 76–90.

[31] FEYL Renate, *Der lautlose Aufbruch. Frauen in den Wissenschaften*, Sammlung Luchterhand, Darmstadt, 3. Auflage 1989, 1. Auflage 1983.

[32] FLAD-SCHNORRENBERG Beatrice, *Der wahre Unterschied: Frau sein - angeboren oder erlernt?*, Herder Verlag, Freiburg, 1978.

[33] FOX Lynn H., BRODY Linda, TOBIN Dianne, *Woman and the Mathematical Mystique*, The John Hopkins University Press, Baltimore, London, 1976.

[34] Frauenförderung, "Frau, mach Dich schlau", In : *vom Ministerium für Wissenschaft und Kunst Baden-Würtemberg*, Prisma-Neues aus Wissenschaft und Kunst, (eds.).

[35] FROUGNY Christiane, PEIFFER Jeanne, "Der mathematische Formalismus - eine Maschine, die Wahres aussondert". In: *Naturwissenschaftlerinnen: Einmischung statt Ausgrenzung. Feministische Studien*, **4** (1985), 61–77. Beltz GmbH and Co.KG, Weinheim, 1985.

[36] FUCHS Margot (ed.), *Naturwissenschaft und Technik-doch Frauensache?*, Dokumentationsband. flv - Frauenliteraturvertrieb, Wiesbaden, 1986.

[37] GERHARDT Uta, "Frauenrolle und Rollenanalyse". In: *Frauensituation. Suhrkamp Taschenbuch Wissenschaft*, U. Gerhardt, Y. Schütze (eds.), 1988.

[38] GOLDSTEIN Catherine, "La situation des mathématiciennes en France et en Europe", Communication à l'Université Libre de Bruxelles, Mai 1992.

[39] GRABOSCH Annette, PAYS Isabelle, "Report on the fifth annual EWM meeting", *Femmes et Mathématiques*, Laboratoire de Mathématiques Fondamentales (UFR 921), Univ. Pierre et Marie Curie, 75252 Paris Cedex 05.

[40] GRABOSCH Annette, ZWÖLFER Almut (eds.), *Frauen und Mathematik - Die allmähliche Rückeroberung der Normalität*, Attempto Verlag Tübingen, Studium generale, 1992.

[41] GRABRUCKER Marianne, *Typisch Mädchen (...)", Prägung in den ersten drei Lebensjahren*, ein Tagebuch. Die Frau in der Gesellschaft. Fischer Taschenbuch Verlag, Frankfurt A.M, 1985.

[42] GREEN Judy, LADUKE Jeanne, "Women in American Mathematics: A Century of Contributions". In: *A Century of Mathematics in America, Part II*, Duren P. (ed.), pp 379–398. AMS, *History of Mathematics, Vol. 2*, 1989.

[43] GRINSTEIN Louise S., CAMPBELL Paul J., "Anna Johnson Pell Wheeler (1883-1966". In: *A Century of Mathematics in America, Part III*, Duren P. (ed.), pp 293–297. AMS, *History of Mathematics, Vol. 3*, 1989.

[44] GRINSTEIN Louise S., CAMPBELL Paul J. (eds.), *Women in Mathematics. A Bibliographic Sourcebook*, Greenwood Press, Inc., Westport, Conneticut, 1987.

[45] HORSTKEMPER Marianne, *Schule, Geschlecht und Selbstvertrauen. Eine Längsschnittstudie über Mädchen-sozialisation in der Schule*, Juventa-Verlag, Weinheim, München, 1987.

[46] INHETVEEN Heide, "Frau und Mathematik". In: *Kritische Stichwörter zum Mathematikunterricht*, Dieter Volk (ed.), Fink-Verlag, München, 1979, pp 67–85.

[47] JACKSON Allyn, "Encouraging Women in Math and Science", *Notices of the AMS* **37** (1990), 5–6.

[48] KAISER-MESSMER Gabriele, "Frauen und Mathematik - ein verdrängtes Thema der Mathematikdidaktik", *Zentralblatt der Didaktik der Mathematik,* (1989), 56–66.

[49] KERSTEN Ina, Forum "Frauen in der Mathematik". In: *DMV-Mitteilungen* (1/1992).

[50] KIEN Jenny, "Die unsichtbaren Frauen in den Naturwissenschaften". In: *Frauen in Naturwissenschaft und Technik. Natur und Technik - Geist und Gesellschaft*, Scherer Irene, Thunecke Inka (eds.) , Tüte Sonderheft, Tübinger Termine Verlags GmbH, 1989, pp 16–20.

[51] KIEN Jenny, "Gibt es 'weibliche' und 'männliche' Naturwissenschaft? Analyse von Beispielen aus der Hirnforschung". In: *Frauen in Naturwissenschaft und Technik. Natur und Technik - Geist und Gesellschaft,* Scherer Irene, Thunecke Inka (eds.), pp 43-46. Tüte Sonderheft, Tübinger Termine Verlags GmbH, 1989.

[52] KOBLITZ Ann Hibner, *A Convergence of Lives. Sofia Kovalevskaia: Scientist, Writer, Revolutionary*, Birkhäuser Verlag, Boston, Basel, Stuttgart, 1983.

[53] KOBLITZ Ann Hibner, "Sofia Kovalevskaia and the mathematical community", *The Mathematical Intelligencer* **6** (1984), 20–29.

[54] KOBLITZ Ann Hibner, "A Reply to Chowdhury", *The Mathematical Intelligencer* **8** (1986), 71–72.

[55] KOECHER Max (eds.), *Hel Braun. Eine Frau und die Mathematik 1933-1940. Der Beginn einer wissenschaftlichen Karriere*, Springer Verlag, Berlin, Heidelberg, 1990.

[56] LAFORTUNE Louis (ed.), *Femmes et mathématiques*, Les éditions du revenue-ménage, Montréal, 1986.

[57] Mc CLURE Donald E., "1991 Annual AMS-MAA Survey" (Second Report), *Notices of the AMS* **39**: 6 (July/August 1992), 573–581.

[58] MAURER Magarete, "Die Verdrängung der Frauen aus Naturwissenschaften und Technik". In: *Lila Schwarzbuch: zur Diskriminierung von Frauen in der Wissenschaft*, Schlüter Anne, Kuhn Annette (eds.), Schwann Verlag, Düsseldorf, 1986, pp 234–256.

[59] MOHR Wilma, *Frauen in der Wissenschaft, Ein Bericht zur sozialen Lage von Studentinnen und Wissenschaftlerinnen im Hochschulbereich*, GEW-Texte, Dreisam Verlag, Freiburg i.Br., 1987 .

[60] NOLAN Deborah, "Women in Statistics", *Statistical Science* **7**: 2 (May 1992).

[61] OSEN Lynn M., *Women in Mathematics*, MIT Press, 9th printing, Cambridge, MA, 1974, 1988.

[62] PERL Teri, *Math Equals. Biographies of Women Mathematicians + Related Activities*, Addison-Wesley Publishing Company, Philippines, 1978.

[63] PIEPER-SEIER Irene, "Studien- und Arbeitsbedingungen von Mathematikerinnen in westdeutschen Hochschulen". In: *Frauen und Mathematik – Die allmähliche Rückeroberung der Normalität?*, Grabosch Annette, Zwölfer Almut (eds.), Attempto Verlag Tübingen, Studium generale, 1992, pp 30–47.

[64] POIANI Eileen L., "The Real Energy Crisis". In: *Lynn A. Steen: Mathematics Tomorrow, Issues of Equality*, Springer Verlag, Heidelberg, Berlin, 1981, pp 155–163.

[65] REES Mina, interviewed by DANA R., HILTON P.J. In: *Mathematical People. Profiles and Interviews*, D.J. Albers, G.L. Alexanderson (eds.), Birkhäuser, Boston, Basel, Stuttgart, 1985, pp 255–267.

[66] REID Constance, interviewed by ALEXANDERSON G.L.. In: *Mathematical People. Profiles and Interviews*, D.J. Albers, G.L. Alexanderson (eds.), Birkhäuser, Boston, Basel, Stuttgart, 1985, pp 269–280.

[67] REID Constance, ROBINSON Raphael M., "Julia Bowman Robinson (1919-1985)". In: *A Century of Mathematics in America, Part III*, Duren P. (ed.), pp 405–413. AMS, *History of Mathematics, Vol. 3* , 1989.

[68] RUNG Donald C., "Highlights of the 1990 CBMS Survey", *Notices of the AMS* **39**:6 (July/August 1992), 570–572.

[69] SARGES Heidrun, "Frau und Mathematik - Mädchen und Mathematikunterricht". In: *Louise F. Pusch: Feminismus. Inspektion einer Herrenkultur*, Suhrkamp Verlag, Frankfurt A.M., 1983.

[70] SCHAFER Alice T., "Women and Mathematics", In: *Lynn A. Steen: Mathematics Tomorrow, Issues of Equality*, Springer Verlag, Heidelberg, Berlin, 1981, pp 165–185.

[71] SCHERER Irene, THUNECKE Inka (eds.), *Frauen in Naturwissenschaft und Technik. Natur und Technik - Geist und Gesellschaft*, Tüte Sonderheft, Tübinger Termine Verlags GmbH, 1989.

[72] SCHILDKAMP-KÜNDIGER Erika, *Frauenrolle und Mathematikleistung*, Pädagogischer Verlag Schwann, Düsseldorf, 1974.

[73] SCHILDKAMP-KÜNDIGER Erika, "Special Problems of Women in Mathematics". In: *ZDM* **13** (1981), 12–17.

[74] SCHMIDT Willi, SCRIBA Christoph J. (eds.), *Frauen in den exakten Naturwissenschaften*, Franz Steiner Verlag, Stuttgart, 1990.

[75] SCHOCH Annemarie, "Frauen in Naturwissenschaft und Technik. Ein Überblick über die neuere Literatur", Diplomarbeit Bibliothekswesen, Stuttgart, 1989.

[76] SCHÜLTER Anne, KUHN Annette (eds.), *Lila Schwarzbuch: zur Diskriminierung von Frauen in der Wissenschaft*, Schwann Verlag, Düsseldorf, 1986.

[77] SELVIN Paul, "Does the Harrison Case Reveal Sexism in Math?", *Science*

252 (28 June 1991), 1781–1783.

[78] SERIES Caroline, Review of the "Problem of Women and mathematics" by L.H. Fox, *The Mathematical Intelligencer* **5** (1983), 52–53.

[79] SIEGEL Patricia J., KAY TH. FINLEY, *Women in the Scientific Search. An American Bio-bibliography 1724-1979*, The Scarecrow Press, Metuchen, N.J., London, 1985.

[80] SMORYŃSKI C., "Julia Robinson. In Memoriam", *The Mathematical Intelligencer* **8** (1986), 77–79.

[81] SRINIVASAN Bhama, "Ruth Moufang, 1905-1977", *The Mathematical Intelligencer* **6** (1984), 51–55.

[82] SROCKE Bettina, *Mädchen und Mathematik*, DUV, Wiesbaden, 1989.

[83] TANNEN Deborah, *You Just Don't Understand - Men and Women in Conversation*, William Morrow and Co., Inc., New York, 1991.

[84] TAUSSKY-TODD Olga, "An Autobiographical Essay", In: *Mathematical People. Profiles and Interviews.* D.J. Albers, G.L. Alexanderson (eds.), Birkhäuser Verlag, Boston, Basel, Stuttgart 1985, pp 309-336.

[85] TEDESCHINI LALLI Laura, *Fatti, numeri e femine*, Sapere, dic. 1992.

[86] TEE G.J., "The pioneering women mathematicians", *The Mathematical Intelligencer* **5** (1983), 27–36.

[87] VOUILLOT F., "Sciences et techniques, pourquoi si peu de filles", L'orientation Scolaire et Professionnelle, *revue de l'INETOP*, special issue, **20**:3 (1991).

[88] ZIMMER Dieter E., "Sexualhormone und die Mathematik", *Die Zeit* **49** (Dezember 1988).

[89] ZITELMANN Arnulf, *Hypatia*, Beltz-Verlag, Weinheim, 1988, und Deutscher Taschenbuchverlag GmbH and Co. KG, München, 1990.

Eva Bayer-Fluckiger
Faculté des Sciences
URA CNRS 741, Mathématiques
16, route de Gray
25030 Besançon Cedex
France

Appendix: List of Coordinators of European Women in Mathematics

- International coordinators:

Marketa NOVAK
Dept. of Computer Science
Chalmers University of Technology
41296 Göteborg, Sweden
tel. 31-721090 (from April 1992: 7721090)
email : NOVAK@CS.CHALMERS.SE

Capi CORRALES
Departemento de Algebra y Fundamentos
Facultad de Ciencias
Universidad Complutense de Madrid
28040 Madrid, Spain
tel. 1-3520696
email : CAPI@EMDUCM11.BITNET or CAPI@EMDUCM11.EARN

Marie DEMLOVÁ
Dept. of Math. Faculty of Electrical Eng.
Czech Technical University
Technicka 2
16627 Praha 6, Czechoslovakia
tel. 42-2-3323587
email : DEML@CSEARN.BITNET

REGION/COUNTRY	COORDINATOR
BELGIUM	Isabelle PAYS Université de Mons-Hairaut Section de Mathématiques 15, Avenue Maistriau 7000 Mons, Belgium spays@bmsuem11.bitnet
CZECHOSLOVAKIA	Marie DEMLOVA Dept. of Math. Faculty of Electrical Eng. Czech Technical University Technicka 2 16627 Praha 6, Czechoslovakia deml@csearn.bitnet

DENMARK

Bodil BRANNER
Mathematical Institute
Building 303
The Technical University of Denmark
2800 Lyngby, Denmark
maibb@vm.uni-c.dk

FINLAND

Marjatta NÄÄTÄNEN
Dept. Math.
University of Helsinki
Hallituskatu 15
00100 Helsinki, Finland

FRANCE

Catherine GOLDSTEIN
Université de Paris-Sud
Mathématiques
Bâtiment 425
91405 Orsay Cedex, France
cgolds@matups.matups.fr

GERMANY

Christine BESSENRODT
Institute for Experimental Mathematics
University of Essen
Ellernstr. 29
45326 Essen, Germany
mat440@de0hrzla.bitnet

Sandra HAYES
TU München-Arcisstr. 21
80333 München, Germany
sandra.hayes@mathematik.
tu-muenchen.dbp.de

GREECE

Vassiliki FARMAKI
University of Athens
Department of Mathematics
Panepistimiopolis-Ilisia
157 84 Athens, Greece

HUNGARY

Hajnal ANDRÉKA
Mathematics Institute of the Hungarian
Academy of Sciences
Réaltanoda u. 13-15
Budapest 1053, Hungary

ITALY

Valentina BARUCCI
Dipartimento di Matematica
Università di Roma “La Sapienza”
Piazzale Aldo Moro, 2
00185 Roma, Italy
barca@itcaspur.bitnet

NETHERLANDS	Coby GEIJSEL University of Amsterdam Faculty of Mathematics and Computer Science Plantage Muidergracht 24 1018 TV Amsterdam, The Netherlands coby@fwi.uva.nl
NORWAY	Ragni PIENE Department of Mathematics University of Oslo P.O. Box 1053 - Blindern 0316 Oslo 3, Norway ragnip@math.uio.no, ragnip@ma-mail.uio.no
POLAND	Anna ROMANOWSKA Mathematical Institute Warsaw Technical University Plac Jednosci Robotniczej 1 00661 Warsaw, Poland aroman@plwatu21.bitnet
	Barbara ROSZKOWSKA Mathematical Institute Warsaw Technical University Plac Politechniki 1 00661 Warsaw, Poland barosz@plwatu21.bitnet
PORTUGAL	Isabel LABOURIAU Grupo de Matemática Aplicada FCUP Rua das Taidas, 135 4.000 Porto, Portugal islabour@ncc.up.pt
RUSSIA and BULGARIA	Roza GALEEVA JINR, LCTA P.O. Box 79 101000 Moscow, Russia roza@lcta.jinrc.dubna.su galeeva@main2.jinr.dubna.su
SPAIN	Capi CORRALES Departemento de Algebra y Fundamentos Facultad de Ciencios Universidad Complutense de Madrid 28040 Madrid, Spain capi@emducm11.bitnet,capi@emducm11.earn

SWEDEN

Marketa NOVAK
Dept. of Computer Sciences
Chalmers University of Technology
41296 Göteborg, Sweden
novak@cs.chalmers.se

SWITZERLAND

Eva BAYER-FLUCKIGER
Equipe de Mathématiques
U.F.R. des Sciences et Techniques
16, Route de Gray
25030 Besançon Cedex, France
bayer@grenet.fr

TURKEY

Sahnur YAPRAK
Faculteit Wetenschappen
Departement Wiskunde
Celestijnenlaan 200 B
3001 Leuven, Belgium
sahnur=yaprak%gst%wis@cc3.kuleuven.ac.be

UKRAINE

Polyna AGRANOVICH
Institute for Low Temperature Physics
and Engineering
47 Lenin. Ave.
Kharkov-164
310164, Ukraine
agranovich%ilt.kharkov.ua@relay.ussr.eu.net

UNITED KINGDOM

Amanda CHETWYND
Mathematics Department
Lancaster University U.K.
Lancaster LA1 4YF, Great Britain
maa003@central1.lancaster.ac.uk

- Coordinator of EWM email net:

Laura Tedeschini Lalli
Università degli Studi di Roma "La Sapienza"
Dipartimento di Matematica
"Guido Castelnuovo", Piazzale Aldo Moro, 2
00185 Rome, Italy

Round Table C: July 8, 1992

MATHEMATICS AND EDUCATIONAL POLICY

organized by Jacques Camus
Université de Rennes I, Rennes (France)

Introduction

Why mathematics and educational policy as a subject for a round table? What does the title mean? Mathematics as a subject taught in schools and universities has a place within the social and educational politics of a country. In part, the place of mathematics in society is indicated by the structure of the educational system and the importance of mathematics within this system, the numbers of students taking the subject, and so on. Due to the greater mobility of the student population in Europe and the increasing need for a mathematically well-educated population, comparisons between the systems in the different European countries are being made more frequently than before. However, many of these comparisons are based only on personal impressions, often even on misinformation. Therefore, the results are usually not wholly accurate or complete. Decisions are made based on such comparisons and unless mathematicians at all levels have access to reasonably accurate information, the risk of perpetuating or repeating errors is enormous.

The purpose of the round table was thus to try and give a limited but more accurate picture of the place of mathematics within the educational systems of a small number of European countries. In fact, it became obvious as we began preparing the round table that our own knowledge of the general educational systems in countries other than our own was incomplete. Thus the report does not restrict itself to mathematics alone, but also gives a description of part of this general educational structure.

The task was huge, but was made more manageable by the publication of Geoffrey Howson's book *National Curricula in Mathematics*, [GH]. This 235 page book attempts to map out the structure and content of the national curricula of the twelve E.C. countries plus Hungary and Japan. The book is divided into two parts, the first of which contains an analysis of the points of divergence in the various syllabi and a discussion as to the way in which the models of the learning process influence the content and the context of the national curriculum. He notes (p. 8): "It is at the secondary level, however, that differences in the structure of the various systems become most pronounced. This reflects not only different assumptions as to the desirability of streaming children by ability and attainment, but also different beliefs and traditions, considering the degree to which the student's curriculum should be tailored to ability, aptitude and aspirations." He discusses various structures used for differentiating between students of various levels of ability and, where possible, looks at the reasoning behind these structures. The real aims of education are not always overt.

In his analysis of the content of the various curricula, Howson presents a table showing the relative stage of a pupil's school life in which certain concepts are introduced in a selection of the different systems (p. 19). This is very instructive since for some topics there is very little spread in the age of introduction, but in others the spread can be six or eight years.

The divergence in course content and structure of both the curricula and the assessment procedures are such that it is clear that there is no simple comparison to be made between these structures. However, the transfer of ideas from one system to another might be possible with care. There are other solutions to the problems inherent in the teaching of mathematics than those found in our own indigenous traditions—and this is the great value of the book.

The second part of the book presents a reasonably detailed summary of the national curricula of the fourteen countries, each in a separate section. Each section starts with a diagram indicating the structure of school education in the country followed by a general discussion of this structure together with an annotated summary of the national curriculum in mathematics. The extent to which the system has been in full evolution nearly everywhere is striking. This is clearly evident in Spain, for which two structure diagrams are given since the system was in the process of changing when the book was being written.

Howson states in his introduction that he has not discussed many features of the educational systems that are relevant to mathematics, but rather has limited himself to topics directly related to national curricula. Problems of teacher training and numbers of teachers are thus not mentioned in detail (see the following report for a description of the recent complete renewal of the teachers educational system in France). The questions of the school/university transition, the university system, the structure of higher education qualifications and so on, are also not discussed. In our report we will try to make a start on providing some of this information, at least for a small group of E.C. countries (Belgium, France, Germany, Italy and the United Kingdom). Thus we hope to continue and expand the task that [GH] set out to do. The space available here is not sufficient to do more than describe and summarize the information, but we hope to produce a more lengthy version in the not too distant future.

Reference

[GH] A. G Howson, *National Curricula in Mathematics,* The Mathematical Association, 1991, (259 London Road, Leicester LE2 3BE, England).

Note: In view of this round table, a questionnaire was sent to the various mathematics associations in Europe in December 1991. The aim of this questionnaire was to gather information concerning the role of mathematics in the European educational systems. We received answers from fourteen countries, but not all the information could be fully exploited. These documents were handed out during the round table and still available from J. Camus (Department of Mathematics, University of Rennes I, France). The countries were: Belgium (W. Dewilde, N. Rouche), Czechoslovakia (J. Bures), Denmark (H.J. Munkholm), Finland (I. Laine), France (J. Camus), Germany (P. Bungartz, W. Haag, R. Strässer), Hungary (T. Nemetz), Italy (V. Villani), Luxembourg (J.P. Pier), the Netherlands (M. Bakker, P.C. Baayen, D. Sierma), Poland (Z. Bobinski, A. Rozkosz), Spain (J.D. Godino, M.C. Batanero), Sweden (T. Tambour) and the United Kingdom (A.C. McBride, G. Howson, T. Porter, J.C. Robson).

The Teaching of Mathematics

I. Pedagogical structures

The pedagogical structures are generally decided by central governments. There are a few interesting exceptions such as

– Germany. Each of the sixteen Länder has its own state organization and there is an attempt at harmonization through the Council of Ministers of Education.

– Belgium. The Belgian system with its three French, German and Dutch speaking communities is becoming more and more similar to the German system, with three Ministries of Education.

– The United Kingdom. The four regions (England, Scotland, Ulster and Wales) possess an educational network, the LEA (Local Education Authorities).

– Spain. Much like the UK, each region of Spain has an Education Council. Secondary education is generally divided into two cycles whose importance is quite varied depending on the country. For example, the first cycle is sometimes included in primary education as in Denmark, or nonexistent as in Germany. The French situation is unique.

II. The curricula

We can distinguish between two different kinds of situations. There is a first category, in which compulsory national (as in France) or regional (as in Germany) curricula exist that spell out contents and objectives and contain detailed instructions. The second category consists of countries where the national curricula only define national standards and where there is a certain amount of initiative left to local authorities or teachers, as is the case in the United Kingdom, Denmark and the Netherlands.

In the first case, the emphasis is rather on the contents of each discipline, whereas in the second case, it is more the overall coherence in view of objectives to be attained that is stressed.

Several countries went through the same problems as France concerning the so-called "modern mathematics" reform. What is the current state of affairs? In France, the last curriculum modification has just been applied in Terminale (last grade of high school). However, the pedagogical renewal in high schools that started this year in Seconde will certainly slightly affect the syllabi of Première and Terminale. It is worth noticing that since 1989 the curricula are under the responsibility of the Comité National des Programmes (National Curriculum Committee).

Concerning the contents, we refer the reader to the excellent book by G. Howson, *National Curricula in Mathematics*, published in 1991 by the Mathematical Association. This books gives in particular detailed information on the mathematical curricula in secondary education in fourteen countries. Without going into much detail, we can say that statistics and probability theory take a non-negligible space in British secondary education, that these disciplines are becoming more present in France and Italy and that the notions of structures, linear algebra and vector spaces remain in the Gymnasium teaching in Germany but have completely disappeared from secondary education syllabi in France.

III. Mathematics

Is mathematics a service discipline or a cultural discipline? In the scientific streams, the learning of mathematics is more cultural in the sense that emphasis is put on reasoning, the sense of proof and imagination. In other streams, mathematics is taught more as a service discipline. In short term higher education, mathematics is essentially an element of culture with a precise goal, mathematics as a tool. This can nonetheless involve relatively specialized knowledge as in certain sections of IUT (Instituts Universitaires de Technologie), for example for a DUT in statistics.

In long term higher education, mathematics is taught in every aspect depending on the stream considered, for example in France the Classes Préparatoires aux Grandes Ecoles/Grandes Ecoles stream or the pure or applied university streams.

Is mathematics playing a training role or a selection role? It is difficult to give an objective answer to this question, in spite of its importance for the layman. A partial answer may reside in the relative importance of mathematics in education, and more specifically in secondary education, and the weight of mathematics in the various modes of evaluation and orientation, notably in the transition between secondary and higher education.

The number of hours devoted to mathematics in secondary education is extremely varied, depending on the country. For example, the number of hours allocated to mathematics in the last year of the scientific stream of secondary education is, according to the answers to the questionnaire sent on the occasion of the European Congress of Mathematics:

– Belgium 8 h

– Denmark 5 h

– France 8 h
– Germany 5 h
– Italy 3 h
– Luxemburg 8 h
– the Netherlands 8 h
– the United Kingdom 4.5 h

Note: In this table, the h unit was not always made precise and should be understood as corresponding to 45 to 50 minutes.

IV. Orientation, evaluation and certification

The transition between the first and second cycles of secondary education varies with the countries, depending on their educational structure but also sometimes on the passing or failure of an exam. This is precisely the case in the UK with the General Certificate of Secondary Education (CGSE). In France, the *Brevet des Collèges* has no influence on the continuation and orientation of studies.

The end of secondary education is marked by a diploma or certification that falls into one of three categories:

–a national diploma, passed before a jury of teachers and possibly taking into account the results obtained by the candidates during their studies. This is the case in Denmark, France (Baccalauréat), Germany (Abitur) and Italy (Maturità);

– a diploma delivered by the teaching insitutions according to certain standards. This is the case in Belgium and the Netherlands;

– a diploma delivered by private examination centers. This is the case in the UK (A-levels). These differences imply that the end of secondary education diploma does not guarantee access to higher education in every country. There is a case by case selection in Germany and a very strong selection in the UK for certain universities. In Belgium, a validation procedure is applied that allows the candidate to pursue academic or non academic studies.

Mathematics Teachers in Secondary Education

I. Recruiting

The recruiting of permanent teachers in secondary education for state owned institutions always involve a compulsory level of qualification that can depend on the cycle of secondary education in which the teacher is being recruited.

Recruiting can be national as in Belgium, France and Italy, regional as in Germany and the UK or local (e.g. by municipalities) as in Denmark and the Netherlands. In Belgium, as well as in the Netherlands, these remarks only apply to the state network, which is relatively modest compared to private schools who rercuit directly. In the UK, regional authorities generally ratify the proposals of the institutions. All teachers thus recruited are civil servants, except in Belgium. In principle, mobility is generally possible, although in fact in Germany and in the UK cross-regional mobility is very difficult.

II. Training

Depending on the country, the training of first cycle teachers may or may not be distinct from that of second cycle teachers. This difference is directly related to the structural differences in educational systems:

– a unique body for 7 to 16 year old pupils and specialized teachers for the second cycle, 17 to 19 year olds, in Denmark.

– specialized teachers in the first cycle as well as in the second cycle, in France and in the UK.

– specialized teachers for levels or streams of studies associated with the institutions, in Germany, Belgium, Italy and the Netherlands.

For secondary education teachers, France presents the particularity of a univalent specialization, which is also the case in the second cycle in the Netherlands. Otherwise, teachers are generally bivalent as in Denmark (2nd cycle) and Germany, trivalent in Belgium and Italy or even polyvalent in Denmark (1st cycle).

The length of training varies according to the kind of teachers and the country. Training is usually performed within the university (Germany, France, Italy) or sometimes in specialized institutions (UK). In principle, a compulsory level of qualification is required to teach mathematics. However, in those systems where teachers are bivalent, certain teachers may have to teach a discipline for which they have not been trained.

Outside of France, or even the UK, few countries face a shortage crisis for teachers in mathematics, such as the one we are now facing. On the contrary, in Belgium, Italy, Germany and Luxembourg there is a surfeit in the opposite direction and there are many graduates looking for a position.

Mathematics in the School Curricula in Flemish Speaking Belgium

by W. Dewilde

Mathematics Inspector Secondary Education

I. The Teaching of Mathematics in Secondary and Higher Education

I.1 Structures

In Flanders, schooling is compulsory until the age of 18. From the age of 16 on, a student can choose betweeen part-time and full-time secondary education. It

is a three-tier educational system:
— Basic schooling which includes pre-school (2.5- to 6-year-olds) and primary (6- to 12-year-olds) education.
— Secondary education which consists of a 6 year study (12-to 18-year-olds) divided into 3 levels of 2 years each.
— Higher education.

Secondary education is organized into four streams:
— general secondary education;
— technical secondary education;
— vocational secondary education;
— artistic secondary education.

Higher education is organized in two streams:
— university education
— non-university education which comprises undergraduate courses (2-3 years) and graduate courses (4 years).

To gain admission to an undergraduate course only a certified Secondary School Graduation Diploma is needed. To go to university or enroll in a graduate course of higher education an ability certificate awarded by secondary teachers or a state board of examiners is required too.

Note that for civil engineering courses the applicant has to take an admission exam organized by the university or a state examining board. For a global view of the organization of education in Flanders, see Appendix 1.

I.2 The place of mathematics in secondary education

I.2.1 Number of hours per week

In order to interpret correctly the following figures giving the number of hours per week devoted to teaching mathematics in secondary education one must note that

— vocational education has been left out;

— the figures concern only the so-called "unified structure", which in any case will be effective throughout the country from 1994 on;

— in a total of 32 hours a week for general education and 36 hours a week for technical education, the Education Secretary determines the basic number of hours (e.g. 27 in the first year) while the education networks are free to allot the rest.

age	year	minimum	maximum
13	1st	4	5
14	2nd	4	5
15	3rd	3	5
16	4th	3	5
17	5th	2	8
18	6th	2	8

Apart from the fact that, on the whole, mathematics is given a major share in the curriculum of secondary education, it must be noted that:
— after the 3rd year, minimal hours are to be found in language-orientated sections;
— the 8 hours repeated in the 5th and 6th years are in intensive mathematics sections;
— the 5th and 6th year "science" sections offer only 4 or 6 hours twice;
— both theses sections have the same number of hours in physics and chemistry and the absence of mathematics hours are devoted to biology and geography;
— some schools have a 7th year almost completely devoted to mathematics (20 hours a week) to enable students who may be lagging behind to catch up. This is especially to prepare students from non-mathematics sections for entrance examinations for civil engineering schools.

I.2.2 The programmes

In 1968, like many other countries, Belgium made a clean sweep of its previous mathematics curricula. It was the reform called "modern mathematics". Practice has shown that students, especially the younger, were not always able to cope with an often too abstract and linear approach.

In the eighties, mathematics curricula were totally reconstructed in four directions:
— reduction of the theory of sets and relations;
— postponement of the study of structures;
— reevaluation of geometry;
— increased attention to algorithms.

Teachers welcomed these new, more balanced curricula, which are a synthesis of traditional and modern mathematics, even though they are still felt to be too extensive. Concerning the curricula themselves, in the near future the following concepts could be introduced:
— mathematics as a tool to solve problems;
— mathematics as a means of communication;
— mathematics as a source of reasoning and abstraction;
— mathematics as a link between subjects.

I.2.3 A few figures

In the first years of secondary education, the student population ratio between general education and all other types of secondary education is 40/60. These figures, due to the "cascade system", are reversed towards the end of secondary education.

If we admit that 1/4 of the student body in general secondary education choose science or mathematics sections, we can infer the approximate percentage of students to be taken into account for possible scientific studies in higher education. These figures, which depend considerably on the schools and regions concerned, must be viewed in perspective.

I.3 The place of mathematics in higher education

I.3.1 Non-university higher education

Among the eight categories which this type of education comprises, it is in economics and mainly in graduate technical studies, e.g., industrial engineers, that mathematics is given a prevailing place. The same applies to teacher training for those who will teach in the first three years of secondary education (*régents*).

In all the streams of secondary education leading to these studies, a minimum of 4 hours of mathematics per week is required.

I.3.2 University education

It is of course in science and applied science faculties, e.g., training of industrial engineers, that the place of mathematics is decisive. The best preparation for such studies implies a minimum of 6 hours of mathematics per week, and preferably 8, for mathematics, physics and civil engineering studies.

A few figures about the student population of science faculties in all the universities in Flanders are shown in Appendix 2.

II. Mathematics and Engineers

II.1 Civil engineers

In Belgium, civil engineers are trained in universities specifically in applied science faculties. The studies last 5 years: 2 qualifying years, and three years of specialisation. In the first 2 years mathematics is essentially a means of selection, while in the last 3 years it is mainly a tool.

It must be noted that the student applicants have to take an entrance examination wholly based on mathematics. Besides, it is this examination combined with a reputation of high standards in these studies that explain why the best secondary students are recruited, and also the high rate of success at the end of the first year. The difficulty of the studies results in a relative stability in the number of students.

II.2 Industrial engineers

In Belgium, industrial engineers are trained in higher education colleges of technology (graduate studies). The studies last 4 years. There is no entrance examination and as a result the students' levels are heterogeneous, some of them having studied mathematics for only 4, or 6 or as much as 8 hours per week in secondary education.

Mathematics is more used as a means of selection than as a tool.

III. Recruiting and Training of Mathematics Teachers

III.1 Initial training

Secondary education mathematics teachers can be divided into two groups:

— Regents, who are trained in higher education schools of education (called training colleges/schools of education), teach in the first stage of secondary education;

— university graduates-associate professors, who are trained in universities, teach in the third stage of secondary education and at the undergraduate level of higher education.

Note that both groups can teach in the second stage of secondary education. The training of "Regents" takes 3 years and is centered rather on practice, half of the last year being devoted to teaching three different subjects.

University graduates-associate professors are trained at a university for the last two years. Their training is rather theoretical and training periods are limited. As part of the reevaluation of teaching professions it is increasingly envisaged to give academic training to all teachers (from nursery school up to the undergraduate level of higher education).

III.2 Continuing education

Although demand for continuing education is very high, there are few opportunities and moreover these are poorly coordinated. They come from the inspectorate, from university departments or teachers associations.

III.3 Recruiting

Recruiting and appointing of teachers is one of the prerogatives of the education networks that pass through local authorities. Since the decentralization, this does not prevent heads of schools from recruiting supplementary teachers and proposing applicants.

There is a more serious problem in the near future: the relief. The number of training college and university students choosing to study mathematics has sharply dropped in the past few years. When we add to this the fact that a significant number of mathematics graduates are choosing the private sector, there is likely to be a shortage of mathematics teachers resulting in a foreseeable drop in standards since mathematics will be taught by teachers who will not have the appropriate qualifications. Higher education teachers (graduate level) are recruited among people with a doctorate degree. Due to a shortage of applicants, graduates with several years of experience in higher education are appointed.

IV. Some Aspects of Research on the Teaching of Mathematics

Mathematics curricula for primary and secondary education are prepared by separate commissions with scarcely any efforts at harmonization. These two stages of education are separate, not only on the level of organization, but also at the research level. All Belgian research on mathematics at the nursery and primary school level is carried out in faculties in psycho-pedagogy, and therefore without the collaboration—except occasionally—of mathematicians.

As for research on mathematics at the secondary education level, it is carried out in all mathematics departments. Assessed in terms of doctorates, scientific achievement at this level remains limited. A research commission, composed of teachers from all levels and having the necessary time and means could produce well thought out, explicit documents from which the curricula commissions could draw their inspiration.

V. Conclusions and Recommendations

The following recommendations come from the first report of the Scientific Commission on the Teaching of Mathematics and Science entitled "Mathematics teaching prospects in the Belgian French-speaking community." The fact that this extract can easily be extrapolated to the Flemish-speaking community amply demonstrates that both communities, if we except a few minor points, have the same problems to solve.

Also note that most of the following conclusions and recommendations have a much wider scope than just that of the teaching of mathematics.

- It is urgent to improve the moral and material status of teachers.
- The initial training of the teachers of the lower and upper levels of education must be harmonized.
- Continuing teacher training programmes should be seen as a duty and as a right.
- Research groups should be set up and should provide simultaneously a coordinated global view of nursery, primary, secondary, and higher education, as well as continuing education offered to teachers.
- It seems indispensable for all categories of teachers to benefit from part-time secondments and to be associated with continuing education.
- Any curriculum commission should associate representatives from other commissions that have a natural link with them as well as representatives of research groups.

- It would be useful to draw up an inventory of the equipment and books considered necessary for the teaching of mathematics at the various levels and to inquire at the universities about the part devoted to problem solving and, in a more general sense, about training in mathematical reflection.
- It would be desirable to suggest to science faculties that they should rethink the concept of *agrégation* studies trying to stress and develop the link between mathematics at the university and mathematics in secondary education.

FORMATION DES ENSEIGNANTS

TEACHER TRAINING IN BELGIUM

EN BELGIQUE

(Secteur néerlandophone)

(Flemish-speaking part)

1. QUALIFICATION REQUISE POUR ENSEIGNER

Required qualification to teach

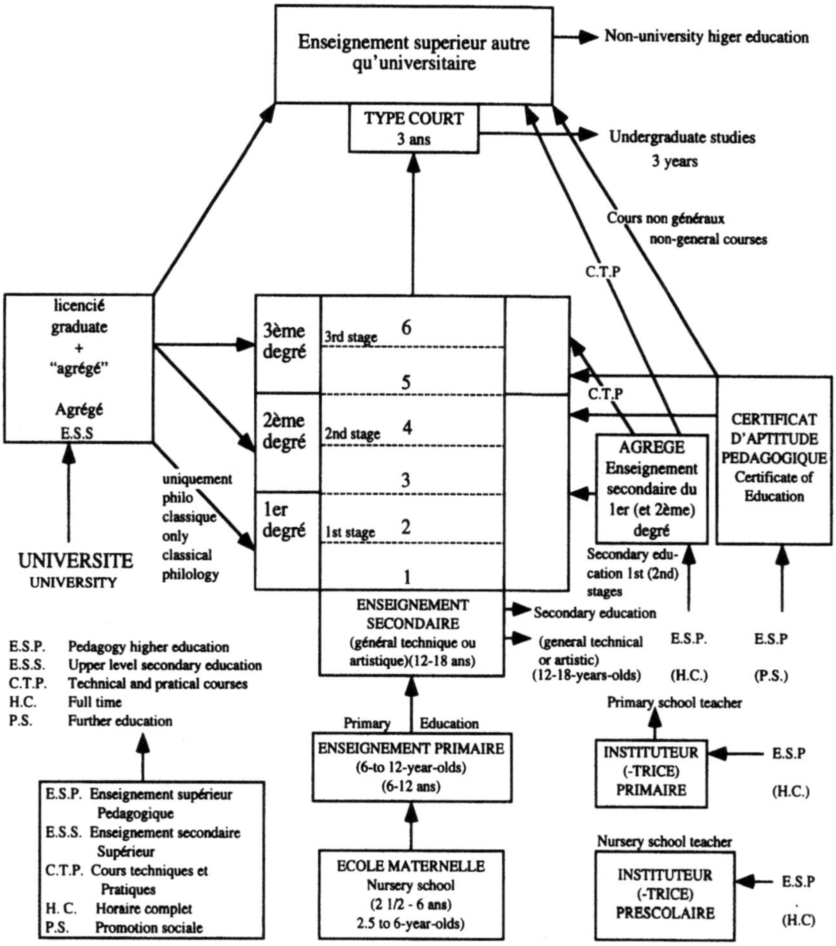

APPENDIX 1 (Required qualification to teach), organigramme

APPENDIX 2

TABEL II. 4 Evolutie van het aantal hoofdinschrijvingen per studierichting uitgesplitst naar geslacht en nationaliteit (Belg/buitenlander) - totaal over alle Nederlandstalige instellingen (année 1991-92)

Wetenschappen

	1984-1985				1985-1986				1986-1987				1987-1988			
	B		BTL		B		BTL		B		BTL		B		BTL	
	M	V	M	V	M	V	M	V	M	V	M	V	M	V	M	V
Combinaties	106	54	63	17	121	60	74	12	134	67	61	15	131	69	78	24
Wisk.-Natuurkunde	80	45	3	0	112	75	4	1	74	51	2	0	42	34	2	0
Wiskunde	486	425	9	4	459	411	9	2	437	400	11	2	410	396	13	3
Informatica													305	136	19	2
Natuurkunde	329	146	12	3	303	129	18	2	318	137	17	1	15	2	0	0
Scheik.-Biol.	86	67	2	1	106	84	2	4	79	67	1	2	57	62	1	0
Scheikunde	570	423	32	3	542	443	31	7	560	496	34	9	580	505	33	16
Biologie	552	537	24	23	501	533	29	15	476	456	39	14	429	440	54	26
Aard- en delfstofk	126	46	61	9	92	37	66	14	72	38	80	14	67	32	79	12
Aardrijkskunde	182	172	10	0	142	141	9	1	101	114	10	0	79	86	10	1
Totaal	2517	1915	216	60	2378	1913	242	58	2251	1826	255	57	2115	1762	289	84

	1988-1989				1989-1990				1990-1991				1991-1992			
	B		BTL		B		BTL		B		BTL		B		BTL	
	M	V	M	V	M	V	M	V	M	V	M	V	M	V	M	V
Combinaties	139	71	85	36	189	131	101	51	276	193	96	50	340	255	144	66
Wiskunde	419	398	17	5	431	363	26	10	403	301	23	5	315	249	20	7
Natuurkunde	317	144	21	1	287	137	23	3	317	128	22	2	318	129	28	3
Scheikunde	635	589	32	18	614	532	34	18	632	553	38	13	643	501	38	16
Biologie	442	440	61	30	444	435	74	37	377	405	101	42	375	383	102	49
Aard- en delfstofk	56	32	86	13	50	34	79	14	55	41	78	15	62	48	72	17
Aardrijkskunde	54	74	8	1	53	67	6	2	48	77	9	2	76	93	14	2
Totaal	2062	1748	310	104	2068	1699	343	135	2108	1698	367	129	2129	1658	418	160

Mathematics in the French Educational System

by Jacques Camus
University of Rennes I

I. Secondary Education

A. The French school system

The French school system is highly centralized: structures, regulations, curricula and syllabi are laid down at the national level. National competitive exams are organized to recruit most secondary school teachers who are appointed by a central authority. The private sector represents just under 20%. Most of it is subsidized and controlled by the state. The percentage of children attending school is extremely high: 99% of the children go to school as soon as they are three years old and the proportion of 16 to 18 year olds has nearly doubled in the past 20 years. This effort has resulted in larger classes and a shortage of teachers, especially in mathematics.

B. The organization of secondary education

Secondary education follows nursery education (2 to 6 year olds) and primary education (6 to 11 year olds). There is no examination at the end of primary education. Secondary education is divided into two cycles and lasts 7 years.

— The first cycle concerns 11 to 15 year olds who are taught at a *collège*. It consists of the 6th, 5th, 4th and 3rd forms in ascending age order.

— The second cycle concerns 15 to 18 year olds who are taught at a *lycée*. It consists of the 2nd, 1st and *Terminale* forms.

All pupils in the first cycle receive the same education. At age 13, selection and guidance for general, technological or vocational streams takes place. This first cycle leads to an examination, the *brevet*, which has no influence on further studies or selection. The pupil is guided, often by his teachers, towards general, technological or vocational education, the last one being separate from the others. Compulsory schooling ends at the age of 16, following the end of the first form of the second cycle. At this point, general education is divided into diverse streams (A, B, S until 1992; L, ES, S after). Some of these streams are subdivided either at the age of 16 or 17. At the end of their secondary education at the age of 18, students take a national exam, the *Baccalauréat*. A vocational stream, which also leads to a *Baccalauréat*, has been developing rapidly in parallel with the general and technological streams.

C. Mathematics in secondary education

I. Mathematics at the *collège*

In the 6th and 5th forms, 3 hours are allocated to the teaching of mathematics in a global average of 25 hours a week. A fourth hour is however provided, either for all the pupils or for supportive or workshop activities. In the 4th and 3rd forms, pupils spend 4 hours a week on mathematics. New maths curricula were gradually introduced between 1986 and 1987, starting in the 6th form. The teaching is centered on tutorials in geometry, numerical analysis, data organization and processing and functions. Literal calculus and geometry are gradually introduced but are actually taught in the 4th form. An introduction to statistics is given and the use of pocket calculators is systematic.

II. Mathematics at the *lycée*

A complete pedagogical reorganization is currently under way, but will take effect only for the 2nd form at the beginning of 1992-1993. The 2nd form asserts itself as a guidance form for all. The notions studied are mainly notions which are likely to be useful to all pupils, especially with more practice in statistics. The weekly math timetable of 4 hours and 15 minutes includes a three-quarter hour period for modules. These modules are a novelty and aim to give the best possible individual help to each pupil in need, or in any case to help the pupil develop his or her abilities as much as possible. As these modules have no specific programme, they can give way to teaching initiatives while using as a basis the math content of the curriculum.

At present, pupils spend 2 hours a week in mathematics in the "literature and arts" and "literature and languages" sections, 5 hours in the "economics" and "literature-mathematics" section, while in the "science" section, they spend 6 hours a week in math in the 1st form and 6 and 9 hours respectively in the "math-biology" and "math-physics" *Terminale* sections.

After the excesses of the "modern mathematics" reform, new curricula were set up for the second cycle in 1981. New reorganized curricula were introduced in 1985 for the 5th forms and in 1986 for *Terminales*. To follow up the 1986-1990 curriculum reorganization in the first stage, new adjustments were necessary in the 2nd and 1st forms at the beginning of 1991 and 1992 respectively. The present mathematics curricula are not very ambitious concerning their contents. They no longer include linear algebra or studies of algebraic structures as such, and develop only a pragmatic and intuitive approach to the questions of limits and continuity. However, on the other hand, the basic geometry of configurations is quite developed with an emphasis on the "practical function of geometric applications." In addition, the study of statistics is globally reinforced as well as that of probability theory. They are both taught in the 1st form in an elementary and pragmatic way, without being centered primarily on combinatorial analysis as it used to be.

D. Recent modifications

The Ministry of Education has recently launched several major reforms:

— the creation of a National Curriculum Council in charge of proposing the broad orientations of the education programmes;

— the creation of the Instituts Universitaires de Formation des Maîtres (IUFM). The IUFM are an amalgamation of the former institutions for the training of primary, general secondary, and vocational education teachers which were previously separate;

— the setting up of yearly national assessments of pupils at the ages of 8, 11 and now 15 (CE2, 6th and 2nd forms);

— a pedagogical reorganization of the lycées.

The last measure will concern 15 year old students in 1992 (2nd form), 16 year olds in 1993, and 17 year olds in 1994. As mentioned above, in the 2nd form modules have been introduced in four core subjects, one of which is mathematics. For each pupil this represents about 27 hours a year in each of the four subjects, during which he will work in a more flexible way and in a class reduced in size. This will be organized slightly differently in the following two forms, lst and *Terminale*, where in each section a complex system of optional courses, offering either further study of a subject already taught or a new subject will be increasingly provided.

It is too early to pass a judgement on the impact of these measures.

II. Higher Education

A. The organization of higher education

The French higher education system is characterized by the coexistence of a multiplicity of institutions with different goals, structures and admission requirements. This diversity is reinforced on the level of the organization of studies and of the diplomas awarded.

One of the particularities of the French system is its heterogeneous structure in undergraduate higher education and the dualism of two types of graduate higher education institutions.

The *baccalauréat* is required and sufficient to undertake higher education studies. With the *baccalauréat*, students can choose between two systems: a system where entry is not selective at university and another one where entry is selective (school records or competitive exam) either at university in an *Institut Universitaire de Technologie* (IUT) or in a *Section de Techniciens Supérieurs* (STS) for short technological courses, or in a *Classe Préparatoire aux Grandes Ecoles* (CPGE) in a lycée or a *Grande Ecole* for graduate courses.

The student population in these four streams has considerably increased over the last ten years. The number of university students has increased by 3.3% a year on average, with a 7.8% increase in selective streams, particularly in the STSs where it trebled between 1980 and 1990.

Undergraduate higher education

Undergraduate higher education essentially concerns industry and the tertiary sector. Training generally lasts two years (sometimes three), at the end of which a vocational diploma is awarded. The courses are generally provided by IUTs within universities and lead to a *Diplôme Universitaire de Technologie* (DUT), or within lycées in STSs, where they lead to a *Brevet de Technicien Supérieur* (BTS) and where the training is more specialized than in IUTs.

Graduate higher education

Graduate higher education is provided by universities and *Grandes Ecoles*, whose syllabi are organized differently.

I. The universities

Universities, which are state-run in their vast majority, have the monopoly of awarding diplomas. They have huge numbers of students and provide basic and practical courses augmented by research.

Universities are self-governing and multidisciplinary-oriented institutions. This autonomy also appears in the studies: universities determine their own curricula, which sometimes must be validated by the Ministry, the organization of their courses and how their diplomas are awarded.

Globally speaking, university studies can be divided into three cycles:

— a first cycle of multidisciplinary studies for students with the *baccalauréat*. It lasts two years and leads to an undergraduate degree, the *Diplôme d'Etudes Universitaires Générales* (DEUG) with possibly mentions of a specific discipline. The first-choice stream for mathematics is the "DEUG A SSM" which generally combines physics or computer sciences with mathematics. Other streams increasingly combine mathematics with social sciences (DEUG MASS), often with economics and statistics courses. The DEUG is essentially intended to serve as a qualifying cycle for the second university cycle;

— a second cycle of high-standards in-depth studies preparing students for professional responsibilities or for a third university cycle. The second cycle leads to the *Licence in one year*, the Maîtrise in two years and the *Magistère* in three years;

— a third cycle, lasting one or several years, which is highly specialized and oriented toward research. Two types of training are provided:

  - a one-year course with emphasis on vocational training and a compulsory training period in a firm, leading to a *Diplôme d'Etudes Supérieures Spécialisées* (DESS),

- training for and by research leading to a post-master qualification, a *Diplôme d'Etudes Approfondies* (DEA) and then in principle in three years time to a doctorate.

II. The *Grandes Ecoles*

The *Grandes Ecoles* are the second type of graduate higher education institutions. They are specific to the French system.

The *Grandes Ecoles*, which have a smaller intake of students than the universities, provide high-standards training with vocational goals: training of teachers and researchers, of the vast majority of engineers and business executives. Access to the *Grandes Ecoles* involves a two or three years preparation in a CPGE within some *lycées* and success at competitive exams. It is generally the choice of the brightest students. Mathematics is relatively important in the scientific CPGEs. There are two types of *Grandes Ecoles*: on the one hand, the specialist engineering schools, which train engineers and civil and defense executives: *Ecoles Polytechnique, des Mines, des Ponts et Chaussées, Centrales, Navale, ENI, ENSAM, Instituts Nationaux des Sciences Appliquées* (INSA = university colleges of applied science), and on the other hand, colleges of education, such as the four *Ecoles Normales Supérieures* (ENS), "the *Ecole Normale Supérieure* (Ulm-Sèvres) and the ENS of Fontenay-Saint Cloud, Lyon and Cachan (ENSET), which train teachers and researchers. Studies in the *Grandes Ecoles* last three or four years. The specialist engineering schools award an Engineer's Diploma recognized by the Council for Engineers Awards, while business schools award a diploma of higher business studies.

The ENS prepare their students for national university diplomas and also to take competitive examinations for the recruiting of teachers (CAPES, and especially *Agrégation*).

B. Mathematics in higher education

I. At the university

About 50 of the 70 universities devote a fairly important share of their activity to the teaching of mathematics. The student population in science, without taking into account medicine, has risen very sharply to reach 215,000 in 1989-1990. This situation has disrupted the teaching practices and methods in universities entirely. This is even more so in mathematics since we are confronted with both a shortfall in the recruiting of secondary school teachers and in the future replacement of university professors, as well as with the need of taking charge of new teaching duties and new, more vocational courses that were set up these last few years.

Combined with the changes in secondary education curricula, this rapid increase of the student population has prompted universities to alter the content of their courses, especially DEUG courses, as well as their teaching methods, particularly in the first cycle.

In addition, in order to attract more students to research, an active policy of incitement to research has been—and still is—developed for DEAs and doctoral courses. This policy is supported by granting students more facilities to carry on third cycle studies, especially through grants awarded by the Ministry of Research and Technology.

There has been a remarkable increase in the number of students who are awarded mathematics diplomas at all levels. However, the number of graduates is still definitely insufficient to meet the recruiting crisis of secondary education mathematics teachers. Similarly, the number of completed doctorates is insufficient to ensure that it will be possible to replace all higher education mathematics teachers when they retire. In addition, a genuine policy to make courses more vocational is being set up in universities. Mathematics has not escaped this change and several very vocationally-oriented mathematics courses have been designed: *Maîtrises de Sciences et Techniques* (MST), *Magistères, Diplômes d'Etudes Supérieures Spécialisées* (DESS) and a few mathematics engineering courses.

I.1 Mathematics in the first cycle of the university

It is in the DEUG A-SSM that mathematics has the most important position. But even in this section, the importance of mathematics and the type of training provided in it varies according to the streams (such as mathematics / physics, physics / mathematics, mathematics / physics / chemistry, mathematics / computer science, and so on).

Organization of the courses. Most of the first year is common to the various streams, or has a common core syllabus for about one semester. In nearly all cases, students have to pass an exam in order to proceed from the first to the second year. The DEUG diploma is awarded at the end of the second year after passing another examination. Generally speaking, continual assessment is given a great deal of weight: midterm examinations, written and practical work, "dissertations" or projects. In the first year students spend 9 to 10 hours a week in mathematics. In the second year, the number of hours varies notably according to the section considered, from about 12 hours a week in sections where mathematics prevails to 4 to 6 hours where physics and chemistry prevail, plus two to three hours a week in computer science at some point in either year.

Teaching methods. The courses consist largely of lectures. However, in some universities and in some sections there is a mixture of lectures and "problem classes and tutorials". These initiatives are becoming increasingly difficult to maintain, due to the increase in the student population which has not always been accompanied by the appointment of more teachers. Some universities have also introduced practical study of mathematical applications using desktop computers.

Mathematics curricula. The curricula have been significantly altered over the past few years, in particular because they needed to be adapted to the new secondary education curricula which are far less abstract than before. Basic notions of linear algebra are now taught in the first cycle of higher education. Probability theory is generally taught in the second year.

I.2 Mathematics in the second cycle of the university

I.2.1 The *Licence* (Bachelor's degree)

The mathematics Licence involves a minimum of 500 hours, 300 of which are devoted to the following topics: differentiation, complex analysis, differential equations, Fourier analysis, integration and probability theory and numerical analysis. In addition to this, the mathematics Licence must include a course in computer science on the structure of computers, algorithmics and the learning of one programming language.

The mathematics *Licence* is not meant to be a final examination but should enable the student to acquire the necessary basic knowledge to carry on with more specialized studies like an M.Sc or vocational studies to prepare for a CAPES, which is a competitive exam for the recruiting of teachers.

While the number of students who obtained a mathematics degree had decreased in the 1980s, especially because of the development of computer science and of the low number of positions offered in competitive exams for the recruiting of teachers such as the CAPES and the *agrégation*, we now see a dramatic increase in the number of mathematics graduates: 2,600 in 1989-1990, that is to say, a 160% increase since 1986, whereas the number of computer science graduates has remained constant at about 1,300 since 1985.

I.2.2 The *Maîtrise* (Master's degree).

The mathematics *Maîtrise* involves of a minimum of 500 hours, including computer science training, and can be obtained in a wide range of options:

— The "pure mathematics" option must include 250 hours in the following topics: algebra, geometry and analysis;

— The "mathematics and basic science applications" option (MASF) must include a minimum of 250 hours in mechanical analysis, numerical analysis, probability theory and statistics and basic physics;

— The "mathematical engineering" option (MIM) must include a minimum of 300 hours in the following topics: numerical analysis, scientific computation, partial differential equations and distributions, probability theory and statistics, elements of mechanics and physics.

Again, the organization and content of *Maîtrises* degrees vary considerably. However, the pure mathematics *Maîtrise* primarily leads to teaching and basic research, the mathematics and basic science applications Maîtrise is designed for applications of mathematics to physics and mechanics and the mathematical engineering *Maîtrise* for scientific computing and industrial applications.

There also exist other *Maîtrises* where mathematics holds a dominant position, such as the "mathematical and computer science structures" *Maîtrise*, which is closer to computer science, the "discrete mathematics" *Maîtrises* with emphasis on logic, symbolic computation and cryptography and a few "science and technology" *Maîtrises* (MST).

The choice of a *Maîtrise* determines the choice of third cycle studies, either short, vocational (DESS), or long (DEA, doctorate). Moreover, with a *Maîtrise* a student can sit for the agrégation in mathematics.

I.3 Mathematics in the third cycle of university

The third university cycle is characterized by several diplomas at the level Baccalauréat + 5 years of studies, the Diplôme d'Etudes Approfondies (DEA), the Diplôme d'Etudes Supérieures Spécialisées (DESS) and the Magistère. After obtaining a DEA, a Magistère or a DESS, students can carry on their studies and prepare for a doctorate.

I.3.1 The DEA

The preparation of a DEA takes another year after a *Maîtrise*. It is the first year of doctoral studies at the end of which the student submits a doctoral thesis. The aim of a DEA is to consolidate in depth the knowledge acquired in a *Maîtrise* and to acquire more specialized knowledge. A DEA mainly consists of lectures (about 100 to 250 hours depending on the type of DEA), rarely combined with tutorials and practical work. Initiation to research is always a strong characteristic of the DEA, which provides training for research thanks to a training period. The DEA training period lasts several months and consists either in the study of recent research papers or in applied research work within a firm. Each training period is supervised by a professor and results in a written dissertation and an oral presentation. The final natural purpose of a DEA is to start preparing for a doctorate. However, because of their variety, DEAs also play the role of final examination which is appreciated in the work market.

I.3.2 The DESS

Unlike the DEA, the DESS is a final diploma of university studies. It provides highly specialized training and prepares students to directly take a job in mathematical engineering. It does not however entitle its holder to an engineer qualification, which is reserved to graduates from schools whose diploma has been validated by the "Council for academic awards". Its structure is comparable to that of a DEA (lectures + training period) but more vocational. The nature of job opportunities implies a strong emphasis on numerical analysis, scientific computing, probability theory and statistics and even computer algebra. There now exist some university units that are entitled to award engineer qualifications, which proves that universities are becoming more vocational.

I.3.3 The Magistère

The Magistères are top-level university courses that combine basic knowledge, training periods, vocational qualification and initiation to research. The Magistères, which were introduced recently, are university diplomas validated by the Ministry of Education. Studies last 3 years and concern small groups of students selected by a competitive exam taking place after the DEUG or a classe préparatoire aux *Grandes Ecoles.* The possible openings are positions as engineers in a firm and/or researcher. At present, there exist in France six Magistères in mathematics, three of which are associated with the Ecoles Normales Supérieures. The three others are offered by large provincial universities that have important departments of mathematics, applied mathematics, and computer science. In addition to the traditional courses corresponding to national diplomas, students are offered further courses that are more vocationally oriented: applied mathematics, modeling and scientific computing. A common characteristic of all Magistères is the place reserved for personal projects and training periods in firms or laboratories.

I.3.4 The doctorate

The doctorate is the normal continuation of a DEA. It is generally prepared within a team or a laboratory of research in mathematics. The system of doctorates was recently changed in France: now there is only one doctorate. The preparation of a doctorate takes about three years. The doctorate must consist of original research work, conducted at university or in a firm, or in a mixed university-firm environment. A doctorate holder can then apply for a teacher-researcher or a senior lecturer position, or continue his career as a researcher or a mathematical engineer in a firm.

II. The CPGE-*Grandes Ecoles* stream

Except for a few institutions, especially the Instituts Nationaux des Sciences Appliquées (INSA), studies begin in the Classes Préparatoires aux *Grandes Ecoles* cycle in order to gain admission to the *Grandes Ecoles*, Ecoles Normales Supérieures (ENS) or engineering schools. The aim of this cycle is to intensively prepare the students for the competitive entrance exams to the *Grandes Ecoles.* This is achieved through hard and steady work, which also has a formative effect on the working methods of the students. The main CPGEs in which mathematics has a significant share are: science and technology, agronomy and biology, economics and business classes préparatoires.

The classes préparatoires to the *science and technology* Grandes Ecoles

The scientific *classes préparatoires* are the first cycle of engineering studies, but they also contribute a lot to the training of mathematics and physics researchers as well as secondary and higher education teachers. They prepare students in two years for the competitive entrance exams to the various science and technology *Grandes Ecoles*, which generally train engineers as well as some researchers and teachers.

The schooling

It lasts two years. The first year, called *Mathématiques Supérieures*, cannot be repeated. At the end of the second year, called *Mathématiques Spéciales*, students take several competitive entrance exams to the schools they have chosen. Many of these exams are common to several schools. On the average, 90% of the students accepted in *Mathématiques Spéciales* manage to enter a school after one or two years.

The courses

Courses are mainly scientific: mathematics, physics and chemistry represent 60% to 80% of the syllabus. They also include technology, which can even be important in some sections, a foreign language and French. Students have a busy timetable: from 31 to 38 hours a week of lectures and "class tutorials", to which practical computer science work, oral and written tests, not included in the timetable, must be added. The courses follow a national curriculum and the share allocated to mathematics depends on the section considered (from 12 to 14 hours a week).

The competitive exams

The 90 schools or so that belong to the *Conférence des Grandes Ecoles* organize competitive exams to recruit their students, most of them at the level of *Mathématiques Spéciales* and a few at the level of *Mathématiques Supérieures.* Apart from the three scientific *Ecoles Normales Supérieures* (Ulm-Sèvres-Lyon-Cachan), which essentially train teachers and researchers, these schools lead to an engineer's diploma. The competitive exams generally consist of written and oral exams that take place at the end of the academic year. Mathematics, through a system of coefficients that is specific to each exam, plays a significant role in the results of the exam.

III. Mathematics and Engineers

Short IUT and STS courses do not confer an engineering qualification. Some IUT sections have sections in which mathematics is important (DUT in statistics). In those two types of courses, mathematics is more likely to be a service discipline. The role of engineering schools is not to train mathematicians, but an engineer's training must include substantial mathematical training. In the case of CPGE-engineering schools, training generally lasts three years after two (or three) years in a *classe préparatoire* during which students have studied a lot of math mainly in view of the competitive exams and in which mathematics seems to be more a selection tool than a real culture. The mathematics taught in engineering schools is essentially applied mathematics and few schools provide a specialization in the field.

IV. Training and Recruiting Mathematics Teachers

A. Secondary education

For several years, the recruiting of teachers in science, especially in mathematics, has been in a state of crisis. This crisis is due to several main factors: retirements of teachers, increase in the student population in the *lycées* and university first cycles, low number of graduates and a job that has become less attractive on the job market.

This crisis was made worse by the development of post-baccalauréat education, especially that of computer science. In addition, the number of positions offered through the national mathematics CAPES and *agrégation* competitive exams had dropped sharply between 1975 and 1980. It has been increased again in the past few years. This policy resulted in a sharp drop in the number of candidates. However, their numbers are now comparable to those of the mid-seventies. In the same period, the number of mathematics graduates increased considerably (from 1,052 in 1986 to 2,899 in 1991).

To curb this crisis, the Ministry of Education sponsored reflection groups and made a certain number of decisions concerning the recruitment competitive exams and the training of future secondary education teachers: more positions at the CAPES/*agrégation* recruitment exams, revaluation of teachers salaries, grants and allowances for students who want to embark on teaching careers. The global answer is however the setting up of the *Instituts Universitaires de Formation des Maîtres* (IUFM) in all regions since the beginning of the academic year 1991-1992.

The IUFMs replace previous structures: *Ecoles Normales d'Instituteurs* (ENI), *Ecoles Nationales Normales d'Apprentissage* (ENNA), which trained the teachers of vocational secondary schools, *Centres de Formation des Personnels de l'Enseignement Technique* (CFPET) and the *Centres Pédagogiques Régionaux* (CPR) which were designed to prepare those who had passed the theoretical CAPES exam for the practical teaching part of the same exam. IUFM students are recruited among Licence graduates, on the basis of school records and after an interview. IUFM training lasts two years. For future *collège* and *lycée* teachers, the first year is devoted to the preparation of the CAPES competitive exam, a preparation which necessarily includes practical teaching experience. The second year, which is more vocational, is centered on a training period when students are in charge of a class. During that year their status is that of "trainee" teacher.

B. Higher education

To accommodate the foreseeable need for educators and professors caused by the increase in student population and retirements, the Ministry of Education has significantly increased the number of fellowships awarded since 1989. The same year, *Centres d'Initiation à l'Enseignement Supérieur* (CIES) were set up. These centres have been designed for post-graduate tutors and post-graduate-Normalien-tutors recruited among students who have been awarded research fellowships. A tutor's position is not meant as a pre-recruiting of teacher-researcher, but gives the student an extra grant (an incentive) for higher education training. For the first time this year, the first batch of such "tutors" took the 1992 higher education recruitment competitive exams.

V. Some Aspects of Research into the Teaching of Mathematics

Research into the teaching of mathematics began within university teams and *Instituts de Recherche sur l'Enseignement des Mathématiques* (IREM) in the early seventies and gained momentum in the wake of the modern mathematics reform. Gradually, the didactics of mathematics has developed around mathematics teaching but also around teachers' training. From now on, with the setting up of the new IUFM institutions, teacher training will systematically include a didactic course and the CAPES exam a more vocational test taking this new aspect into account.

Note: A longer version of this report with more detailed figures is available from the author.

Acknowledgements and disclaimer: This paper has been prepared with documents and information provided by many colleagues, teachers associations and departments of the Ministry of Education to whom I am very grateful: M. Artigue, P. Attali, D. Barsky, C. Burgaud, M.-F. Coste-Roy, J. Dablanc, G. Debeaumarché, G. Demengel, J. Deshayes, J. Giraud, P. Legrand, C. Leroy, J.-Y. Merindol, D. Monasse, L. Moret-Bailly, J.-L. Ovaert, J. Pichaud, J.-P. Raoult, G. Renault, M. Rogalski, J. Rolland, L. Schwartz, F. Virtel and the departments of the Ministry of Education, Inspection générale, DPDU, DEP, DRED and MRT.

The Role of Mathematics in Educational Policies in Germany

by Paul Bungartz[1] and Rudolf Strasser[2]
Bonn (Germany)[1], IDM, Bielefeld (Germany)[2]

Before presenting any detailed description concerning the Federal Republic of Germany, we must mention the drastic changes that took place in Germany from 1989–1990. After decades of separation between East- and West-Germany, Germany became politically united. As a consequence, German authorities are still struggling to develop a uniform, or at least compatible, educational structure for the whole nation. Education continues to be under regional control with the now 16 *Länder* as the decision making bodies. Presenting a national picture implies overlooking important regional differences in West Germany and emerging structures in East Germany.

I. Secondary and Higher Mathematics Education in Germany

I.1 Structure of general education

For secondary education in Germany—especially West Germany: the new *Länder* tend to imitate West German structures—there are two most important borderlines. The first one separates primary from secondary education after 4 years of schooling (for the majority of *Länder*. A few place this borderline between grades 6 and 7). At this point, the pupils normally decide which type of secondary school they wish to attend. This decision already implies the choice students have at the second borderline. After 10 years of schooling, at the age of 16, full-time compulsory education is over. Students then either attend a continuation of general education (in most cases a Gymnasium, a grammar school, about 30% of the population) with the objective of obtaining an entrance qualification for the university (higher education), or the majority of 16 to 17-year olds enter vocational education and/or the work force with additional part-time college attendance until the age of 18 (see Figure I next page). Special schools exist for the physically and/or mentally handicapped, but will not be described in the following.

Table 1 (adapted from Weidig) offers some statistical data on the respective importance of the schools and their development over the years.

Table 1. Students in Secondary Schools.

All percentages only apply to the 11 *Länder* forming the Federal Republic of Germany before 1990.

School (without special schools)	1931	1966	1988	1988 (from state to state)
Volksschule/Hauptschule	78%	62%	36%	16%-45%
Realschule	6%	18%	29%	21%-35%
Gymnasium	16%	20%	30%	27%-35%
Gesamtschule (integrated type)			5%	1%-28%

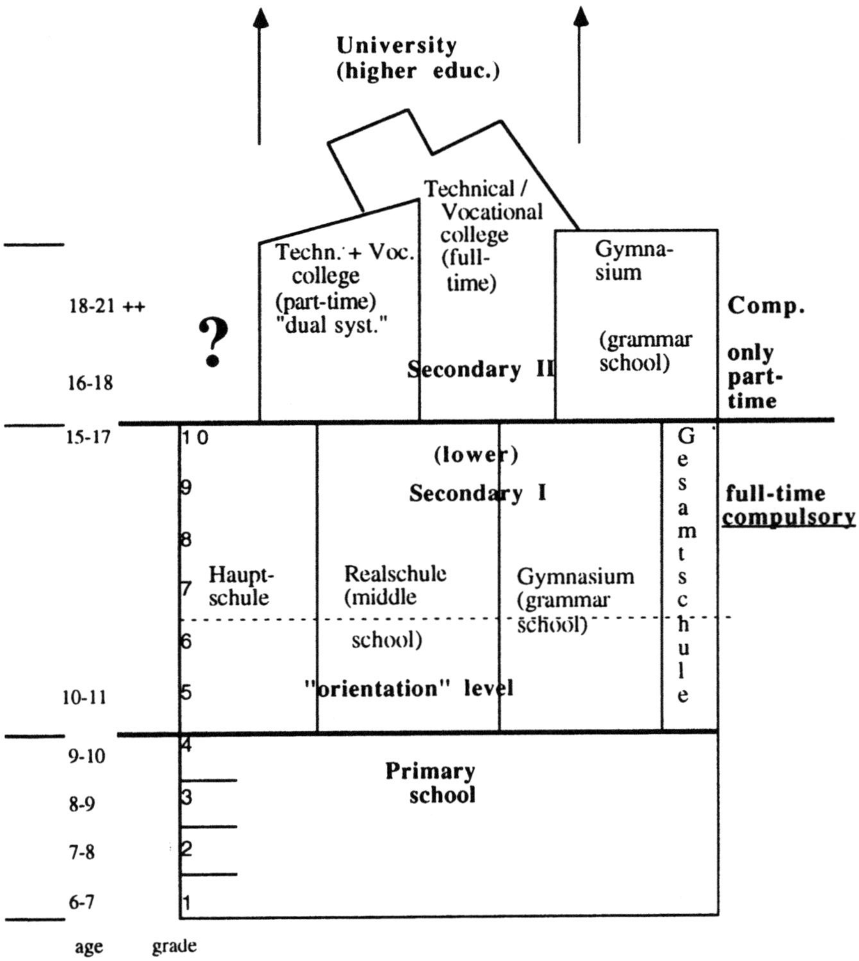

Figure 1. Educational system of Germany

I.2 Mathematics education in secondary general education

All students have to attend mathematics courses till the end of the lower secondary level. At the age of 13, the student normally has four lessons in mathematics per week. In most *Länder*, mathematics is compulsory until the end of *Gymnasium* and most of the technical/vocational colleges have compulsory mathematics until the final examination. For the last two years of *Gymnasium*, two different mathematics courses are offered: About 75% of the students take a basic course in mathematics (normally 3 lessons per week), while about a quarter of *Gymnasium* students take an extended mathematics course (5 or 6 lessons per week). For them, mathematics is a subject with a written examination for the "Abitur" (the final examination in *Gymnasium*). There are large differences between male and female students concerning the choice of mathematics courses. At the university level, 18,3 % of the male students take sciences and/or mathematics, while 13,8% of female students take sciences and/or mathematics.

Mathematics is taught in all schools and colleges as a separate subject, with the rare exception of some minor and highly specialised types of vocational colleges in grades 11 and higher (for details see Blum and Strasser). Nevertheless, it is somewhat difficult to briefly describe the mathematics curricula because of two bones of contention:

(1) There are 16 *Länder* with 16 different curricula; one *Länd* cannot relate to the another; each one fights for the regional right to make decisions affecting the particular curricula.

(2) With the tripartition of the educational system at the secondary-l-level, for most of the *Länder*, we find three different curricula for mathematics. Normally, the syllabus for *Hauptschule* would be more oriented toward arithmetic and practical aspects of the application of mathematics, while the syllabus for *Gymnasium* would be oriented toward preparing for university entrance, especially in mathematics. With the role of *Realschule* differing from *Länd* to *Länd* (and some new *Länder* do not even have *Realschule*), the mathematics syllabus of *Realschule* is in some cases more oriented to Hauptschule, in others more to *Gymnasium*.

Table 2 (adapted from Weidig) presents an overview on the subjects taught, identifying topics usually taught in *Gymnasium*. Syllabi in *Hauptschule/ Realschule* normally present a reduced table of contents.

It is in the teaching of geometry that the differences from *Länd* to *Länd* are the most important. In some *Länder*, transformation geometry was reduced to the study of the symmetry of some figures in grades 5 or 6 (e.g. Nordrhein-Westfalen and Rheinland-Pfalz). In others, e.g. Hessen and Niedersachsen, all of transformation geometry is included in the syllabi. In Table 2, probability/statistics is mentioned only for grades 11 to 13. Nonetheless, Hessen, Niedersachsen and Nordrhein-Westfalen prescribe this subject also in *Hauptschule* (grades 7 to 10).

Table 2. Subjects taught in mathematics in secondary education (especially *Gymnasium*)

grade	subjects
grades 5 and 6 (lower secondary education)	use of variables, simple equations, transforming of magnitudes, area of rectangles, volume of blocks, fundamental geometric forms, digits and number systems, elementary number theory (prime numbers, g.c.d., l.c.m.), fractions and decimal fractions
grades 7 and 8 (lower secondary education)	relations and functions, commercial arithmetic (proportions, percentages), congruence transformations, congruent triangles, quadrilaterals, angle measurement and related theorems, area of polygons, volume of prisms, transformation of terms, sentential forms, linear equations, algebraic structures, integers and rational numbers
grades 9 and 10 (lower secondary education)	the real numbers, quadratic functions and equations, similarity, theorems on right triangles and circles, powers, power and exponential functions, area of circle, volume of pyramid, cyclinder, cone, sphere, trigonometry
grades 11, 12 and 13 (*Gymnasium* only, upper secondary education)	calculus, vector spaces, linear algebra, analytic geometry, geometric transformations, conics, probability and statistics

On the whole, there may be about 10% of the curriculum devoted to probability and statistics at the lower secondary level, while at the upper secondary level of *Gymnasium*, stochastics may occupy 20% of the curriculum. Algorithmics is not a separate discipline in mathematics, whereas computer science is offered as a separate course in *Gymnasium* (upper secondary level, 3 lessons per week) and as an extension course in most schools of the lower secondary level. Because of the organisational and curricular problems, mathematics teaching has nearly no explicit links with other school and college disciplines. However, some teachers try to connect their math teaching with problems from other disciplines.

As for the teaching method, most of the time is devoted to an activity characterized by questions from the teacher followed by answers from the students (*fragend-entwickelnder Unterricht*, see Maier and Voigt). Lecture-type lessons (similar to those in the university) seem to be rare, below 10%. More time is reserved for exercises and problem-solving by individuals or groups of students. It is impossible to give an indication on the importance of projects, reports and so on, because the relevance of this type of activity differs widely even from school to school. A certain focus on project work can be identified in the pedagogy of comprehensive schools.

As for evaluation and tests, most of the tasks assigned seem to be of the same problem type as is common in homework. Apart from a large number of tasks involving variations of parameters in problems already solved in the classroom (mere exercises), word problems seem to play a role in evaluating and grading the students. Some teachers also use multiple choice tests to evaluate student progress (this is not widespread) and with higher grades, some reasoning may be required in tests. Real problem solving seems to be restricted to project work.

I.3 Mathematics teaching in university and higher education

As each individual *Länd* has the right to decide on cultural and educational matters, we have several types of higher education: traditional universities, technological universities (*Technische Hochschule*), comprehensive universities (*Gesamthochschule*) and correspondence courses (also at university level). In the following, we will speak of "university" whenever we want to describe features of higher education that are common to all these types of institutions. All universities are autonomous in teaching and research. For the training of future teachers, each *Länd* has its own regulations concerning the studies and examinations.

All universities (of whichever type) have a department of mathematics, which in most cases is part of a faculty of (natural) sciences, mathematics and computer sciences. Teaching mathematics at university level is comparable to math teaching all over the world at the same level. Students who study mathematics usually have to take calculus, linear algebra and applied mathematics during the first 4 semesters and pass a first examination (*Vordiplom*). For the final examination, the *Diplom*, they have to major in at least two mathematical subjects (e.g. topology, algebra, number theory, geometry, applied mathematics, statistics and so on), write a paper and pass the mostly oral final examination. Each university has its own curriculum and regulations for the diploma. Nevertheless, a diploma awarded by any German university will be accepted everywhere in Germany.

If the candidate passes his (her) diploma very successfully, he or she is invited to write a Ph.D thesis and to do research in a special field. After being awarded the title of *Doktor* of mathematics/natural sciences (which is comparable to a Ph.D) he/she normally finds a position as assistant-professor and prepares for the *habilitation*, an outstanding mathematical research work. After passing this examination, the person gets the so-called *venia legendi*, the privilege of teaching and doing research mathematics at university level. The title then is *Privat-Dozent*. One now can apply for a full professorship position everywhere.

Estimated time spent at the university :

Math diploma: 6 to 7 years;
Doktor (Ph.D): an additional 2 to 3 years;
Habilitation: an additional 2 to 4 years.

Total duration of studies: 10 to 14 years!

We have a *hiring crisis* for university teachers caused by the restrictive policy of the *Länd* governments. A young university teacher (*Privat-Dozent*) has little chance of being appointed in his or her profession at a university.

II. Training Teachers: Mathematics as a Service Discipline

Mathematics plays a role in many disciplines and studies undertaken at *Fachhochschul-* and university-level. Consequently, it is taught to a large number of students not enrolled in mathematics as a service discipline. It is rather difficult to obtain a clear picture of how this type of mathematics is taught, who teaches it, and what role it actually plays in the studies of the students. In fact, service mathematics may be taught either by staff from the faculty of mathematics or by (non-)specialists from the discipline mathematics is serving. In smaller universities, there may even be lectures for mathematics students (teachers and/or diploma) which are also attended by students studying the subject as a service discipline.

Depending on the subject to be served, it may or may not be compulsory to take mathematics for a specific diploma, while in some areas students may be able to avoid studying mathematics altogether. Usually, at least students in physics, chemistry, biology and engineering will have to take some mathematics. For these disciplines, the lectures usually cover calculus, linear algebra and (for physics and engineering) differential equations and numerical analysis. Students in social sciences are forced to attend some lectures in probability/statistics. It is known, but not researched in depth, that mathematics is one of the subjects that increases dropout rates in a variety of disciplines.

III. Training Mathematics Teachers

In relationship to the structure of the educational system in Germany, we have different kinds of teachers for general education. In this paper, we will not comment in detail on the training of teachers for vocational/technical colleges.

III.1 Primary and secondary-I teachers

For the primary level (*Grundschule*) and parts of the lower secondary level (*Hauptschule, Realschule*), we have pedagogical academies (*Pädagogische Hochschulen*) which were intended to train teachers in an elementary fashion separated from the university. Some *Länder* integrated the *Pädagogische Hochschulen* into universities, while in other *Länder* the *Pädagogische Hochschulen* still exist. Pedagogical high schools or faculties of education are heavily overcrowded.

Every future *Grundschule, Hauptschule* or *Realschule* teacher has to attend some lectures in elementary mathematics and has to pass an examination in mathematics (elementary algebra, calculus, geometry, number theory, stochastics and so on). The reason for this is that every primary level teacher should be able to teach mathematics.

The training of these teachers proceeds in two phases:

1st phase (theoretical):

Studies at the *Pädagogische Hochschule* or at the university in two subjects (e.g. mathematics and ...), with only a few hours of practical teaching in schools during this phase of study. The 1st examination includes the writing of a paper in one subject and an examination by an examining board set up by the *Länd* government.

Estimated duration of studies: 6–8 semesters.

2nd phase (practical):

Work as trainee teacher with part-time teaching under the guidance of experts in various types of schools, additional lectures and seminars at a teacher training institution (Seminar) on didactics, methodology of teaching and psychology. The 2nd examination includes the writing of a second paper, in most cases a report about a teaching unit, the teaching of two test-lessons and an oral examination at the *Seminar*.

Time spent at the Seminar: 2 years.

The total training time for *Grundschule, Hauptschule* and *Realschule* teachers is about 5 years.

There is a hiring crisis: many students pass both examinations but only a few of them are employed by the governments.

III.2 Lower and upper secondary level: grammar school *(gymnasium)*

1st phase (theoretical):

The future teacher has to study two subjects (e.g. mathematics and ...) and pedagogy including didactics at a university or comprehensive university. The government of each *Länd* determines regulations. Each university has its own curriculum. It is impossible to give a detailed overview.

Normally, the 1st phase is divided into two blocks:

"basic studies" in calculus, linear algebra, applied mathematics, algebra, complex analysis, stochastics for teachers and so on, with an examination at the end of about 2 years (4 semesters).

"main studies" in at least two different fields of mathematics: algebra, geometry, topology, probability, number theory, applied mathematics and so on, for 3 to 4 years (6 to 8 semesters).

The final examination at university level (*Erstes Staatsexamen*) includes a paper (theoretical work on...) in one of the candidate's subjects (e.g. mathematics or ...), two written examinations and a one hour oral examination in each subject plus half an hour in pedagogy and didactics. It is controlled by an examining board of the *Länd*.

2nd phase (practical):

The teacher training institutions (Seminars) are totally separated from the universities (and from the Seminars for primary school teachers). They are institutions under government control with experienced teachers of grammar schools as trainers. Under the guidance of these expert teachers (*Fachleiter*), the teacher trainees are expected to learn how to teach (e.g. mathematics). In addition to practical work, they receive further didactical, pedagogical and psychological instruction. This phase ends with an examination (*Zweites Staatsexamen*), organised by the expert teachers (*Fachleiter*) and a delegate from the government. The examination is made up of a test of the teaching qualifications in each subject, a written paper (often a report about a teaching unit) in one subject and an oral examination before a committee.

The total time necessary to train a teacher for Gymnasium *is about 9 years!*

There is a hiring crisis again. Although many schools sadly lack teachers, especially in mathematics, there are restrictions on hiring.

IV. Research into Mathematics Teaching

Didactics of mathematics

In Germany, a lot of research on the teaching of mathematics is done in the field called "didactics of mathematics". Nearly every university has some staff specialising in this area, which to a large extent is organised in a research association called *Gesellschaft für Didaktik der Mathematik* (GDM)–Society for the Didactics of Mathematics. The GDM has its own research journal (*Journal für Mathematikdidaktik*–JMD). There is one national institute for research on didactics called the *Institut für Didaktik der Mathematik* (IDM) in Bielefeld, which also houses a documentation centre and library on the teaching and learning of mathematics.

In order to give you an idea of the research in the field, we adapt Burscheid, Struve and Walther and present a non-exhaustive list of research-foci in Germany:

— Elementarisation of mathematics (*Stoffdidaktik*) analyses pieces of mathematics in order to develop teaching units on the topic. Detailed studies on the role of proof in mathematics are undertaken.

— Applications of mathematics are studied and transformed into teaching units to enrich the teaching of mathematics and boost motivation.

— Methodological aspects of mathematics education are studied with a special

attention to the construction of mathematical knowledge in the classroom.
— Historical and philosophical investigations are used to better understand the epistemological dimensions of mathematics education and optimise mathematics teaching.

V. Concluding Remarks

Rather bluntly, we present four conclusions on the role of mathematics in German education:
— In Germany, the educational system seems to be not as comprehensive but more selective than in other European countries. Comprehensive schools only play a minor role in Germany; partly because of the unification of Germany in 1989, partly due to the growing number of students in *Gymnasium*, there is a political and educational debate on the future structure of the German educational system ("comprehensive" vs. "tripartite", the role of *Gymnasium*).
— In Germany, we have a hiring crisis for mathematics teachers and researchers partly because of financial pressure on education as a whole.
— From a historical perspective, the curriculum in schools proved to be rather stable. This is illustrated by the fading influence of the "modern math movement" on long-term curriculum development.
— Judging from the topics taught, mathematics seems to play a major role in the selection of the work force, contradicting the widespread legitimizing of school mathematics by the importance of mathematics for the competence of the work force.

References

[1] Blum, W. and Sträßer, R., "Mathematics Teaching in Technical and Vocational Colleges Professional Training versus General Education". In: *Mathematics Education in Germany*, H. Schupp (ed.), ZDM, Karlsruhe, 1992, pp 242–247,
[2] BMBW, O.J. (ed.), *Grund- und Strukturdaten - Ausgabe 1991/92.* Karl Heinrich Bock, Bad Honnef.
[3] Burscheid, H.-J., Struve, H., and Walther, G., "A Survey of Research". In: *Mathematics Education in Germany*, H. Schupp (ed.), ZDM, Karlsruhe, 1992, pp 296–302.
[4] Maier, H. and Voigt, J. (eds.), *Interpretative Unterrichtsforschung*, Aulis Verlag, Köln, 1991.
[5] Schupp, H. (ed.), *Mathematics Education in Germany*, ZDM (special issue no. 7/92 of ZDM), Karlsruhe, 1992.
[6] Weidig, 1., "On the School System in Germany and the Regulations of Mathematics Teaching". In *Mathematics Education in Germany*, H. Schupp (ed.), ZDM, Karlsruhe, 1992, pp 214–219.

Italy: Mathematics in Higher Education

by Vinicio Villani
University of Pisa, Pisa (Italy)

I. Pre-University Education

The Italian school system is centrally governed by the *Ministero della Pubblica Istruzione*. Education is compulsory between 6 and 14 years of age, i.e., grades 1 to 8. It is divided into primary and junior secondary level.

The primary school (*scuola elementare*) lasts five years (grades 1 to 5).

The junior secondary school, also called middle school, (*scuola media*) lasts three years (grades 6 to 8).

After grade 8, approximately 80% of the corresponding age group enroll in some kind of upper secondary school. Up to grade 8, there is only one type of school (with mixed ability classes); hence differentiation starts at this moment. The different types of upper secondary schools may be grouped into four categories:

1. *lyceums* (classical or scientific);
2. technical schools (many options);
3. specific institutes for the training of elementary school teachers;
4. vocational schools (many options).

The *lyceums* and technical institutes last five years (grades 9 to 13); the institutes for elementary school teachers last four years, while vocational schools may last from a minimum of two to a maximum of five years. At the end of the five year courses and at the end of some of the four year courses, there is a nation-wide final examination, called *maturità*. The following table shows the most important quantitative data (updated to the year 1990):

School type	% of age group enrolling in grade 9	Number of students who passed the maturità in 1990
Classic *lyceum*	5	40,000
Scientific *lyceum*	12	90,000
Technical schools	36	170,000
Elem. teachers inst.	5	40,000
Vocational schools	22	70,000
Total	80	410,000

With the only exceptions being some vocational schools, mathematics is compulsory for all grades, but the time devoted to the subject is insufficient: for instance, in each of the last three years of the classic *lyceum* only 2 hours per week and in the scientific *lyceum* only 3 hours per week (out of a total of 30) are devoted to mathematics.

The syllabi for all school types include algebra, geometry (both Euclidean and Cartesian) and the study of a few standard functions (polynomial, exponential, trigonometric and so on), while calculus is only taught in the scientific *lyceum* and in some types of technical institutes.

A reform of the upper secondary school system is planned. The main points of the reform will be: two more years of compulsory education (until grade 10), teaching computer science (including practical computer activities) within the mathematics courses and an increase of one or two hours per week in mathematics for all school types. Many upper secondary schools have already adopted new (experimental) curricula and syllabi in accordance with the planned reform. The assessment is usually performed both on the basis of written and oral examinations. The use of multiple-choice tests is not popular at all.

II. Universities

In Italy, universities are centrally governed by the *Ministero dell'Università e della Ricerca Scientifica e Tecnologica*. The number of universities is growing every year. At present there are about 40 universities and 2 polytechnical schools, i.e., university-level institutions for engineering and architecture. Mathematics degrees are awarded by almost all universities. Computer Science degrees are awarded by an increasing number of universities (10 at present). Engineering degrees are awarded both by the main universities and by the two polytechnical schools.

With few exceptions (medicine and computer science), there is no *numerus clausus*, nor are there admittance examinations. Every student who obtains any kind of *maturità* at the end of a five year secondary school is free to enroll in any university. Students who obtain a *maturità* after a four year secondary school must spend one more year in school before entering the university. Approximately 30% of Italian 19-years olds enroll in the university, but the academic success rate is very low: in most faculties, only 1 student in 3 completes his studies successfully.

Italian universities award two kinds of degrees in mathematics:

1. *Laurea* (roughly equivalent to a Master's degree): The studies officially last four years. In practice for most students, they last five years or even more.

2. *Dottorato di ricerca* (roughly equivalent to a Ph.D). Access is limited and based on selection. To be admitted students must have a *laurea* in mathematics or in some related field such as physics or computer science. Those who are admitted receive a small grant from the government. The studies last four years after the *laurea*.

Note: From 1992 on, a third type of degree called *Diploma* (roughly equivalent to a Bachelor's degree) will be available. The studies will consist of two years in mathematics and three years in other subjects such as computer science or engineering. Although the curriculum for the *diploma* shows significant differences with the curriculum for the first years of the corresponding *laurea*, it will be possible, with some penalties, to continue studying after a *diploma* to obtain a *laurea*.

Quantitative data

In Italy, each year approximately 4,000 students enroll for the *laurea* in matematica but only 1,300 complete their studies successfully. In percentages, these 4,000 students correspond to 0.5% of their age group and the 1,300 students to less than 0.2% of their age group.

The low academic success rate, i.e., the fact that only one student in three reaches graduation, is very unsatisfactory but, as already mentioned, this is not a specific feature of mathematics: 4,000 freshmen in mathematics represent approximately 1.7% of the total number of Italian freshmen in all disciplines, and the 1,300 students who obtain their *laurea* in mathematics represent approximately 1.7% of the total number of Italian students who obtain their *laurea* in all disciplines.

It is hoped that the introduction of the *diploma* will increase significantly the number of students who complete their university studies with a degree—although at a lower level than now—in mathematics as well as in other disciplines.

The figures concerning mathematics have been more or less steady during the last ten years, with a slight tendency to decrease. This fact may be due to a strong increase in the number of students (now nearly 6,000) who enroll each year for the *laurea* in Informatica (computer science), as well as to a modest increase in the number of students (now nearly 35,000) who enroll each year for a *laurea* in faculties of Engineering and Architecture. In these faculties, the academic success rate is approximately the same as in mathematics—between 30 and 35%. Each year, approximately 160 students enroll for the *Dottorato di ricerca in matematica* and 100 complete their studies successfully. Since the *Dottorato di ricerca* was only introduced 10 years ago, these figures are likely to increase in the next few years.

Organization and curriculum

In order to obtain the *laurea in matematica*, students must attend a total of 15 whole year courses during the four years of their university studies (4 courses during each of the first three years and 3 courses during the last year). Typically, a course consists of 60 to 70 hours of lectures for large audiences (up to 300 students or even more for some first year courses) and 60 to 70 hours of exercises, practical work, and so on for the same class, usually split into two or three smaller groups.

There is a specific examination at the end of each of these courses. Several courses are in physics. Some whole year courses may be divided into two one semester courses.

At the end of their studies, students must also write and discuss a dissertation (*Tesi*), which is supposed to contain original results but in most cases remains below the level of a research publication.

The curriculum is, at least formally, the same all over the country and is quite rigid. The *laurea* in mathematics is split into three branches: "General mathematics", "Applied mathematics" and "Didactics of mathematics". Of the 15 courses each student has to attend, 10 are compulsory and are common to all these branches. Typically these are:

3 courses in calculus (including differential equations);
3 courses in geometry (including topology);
1 course in algebra;
3 courses in physics.

Two more courses are compulsory but differ from one branch to another. Therefore, after having enrolled in a specific branch students can only take three one year courses, or equivalently six one semester courses, of their own choice. There are courses in computer science and in numerical calculus for mathematics students in all universities. They are not compulsory, except for those who are enrolled in the "Applied mathematics" branch.

The assessment for most mathematics courses is based both on written and oral examinations, with more emphasis than in other countries on the oral part. This procedure is rather time consuming, but has the advantage that students become accustomed to interact with examining teachers. Also, oral examinations seem more appropriate than written exercises to evaluate the students understanding of theoretical aspects of the matter. Students who fail to pass the examination for a course at their first attempt can try again up to 3 times during the same year.

The curricula for the *Dottorato di ricerca* are less rigid than those for the *laurea* . They include courses, workshops and personal research work. At the end of their studies, students must write an original research dissertation.

III. Teachers Training

Up to now, elementary school teachers completed their training at the secondary level. In the future, in accordance with a new law, there will be a specific university degree for these teachers too, including mathematics and science courses.

The training of secondary school mathematics teachers takes place within universities in the branch "Didactics of mathematics". Usually, the same teacher also has to teach other subjects such as physics and computer science (in upper secondary schools) or natural sciences (in middle schools). More than 50% of the mathematics students who obtain their *laurea* in mathematics become secondary school teachers. Generally speaking, there is no shortage of teachers in Italy. However, the new generations of mathematicians are barely sufficient to ensure the turnover in upper secondary schools, while the majority of those who teach mathematics in Italian middle schools have a degree in natural sciences and not in mathematics.

The training of Italian secondary school teachers is good in the specific topics they have studied at the university, but these topics may not cover all the subjects they have to teach. Moreover, their training lacks pedagogical and sociological aspects.

The Italian university teachers are divided into "Full Professors", "Associate Professors" and "Research Fellows (Assistant Professors)". At present, there are approximately 800 mathematicians working in each one of these three categories. This corresponds to about 5% of the total number of Italian academics in all disciplines. Although these figures may seem high with respect to the rather low number of mathematics students, it should be noticed that mathematics professors do not only teach in their own faculty, but also in several others (the main bulk being in engineering and computer science) as specified in paragraph 4. In particular, the number of research fellows is not at all sufficient in order to meet adequately the needs for teaching and research in mathematics. Recruiting teachers at all levels, from elementary school to university, is organized nationally. For primary and secondary schools, it is based on selective exams, whereas at the university level, the scientific work of the candidates is the primary criterion.

IV. Mathematics as a Service Discipline

University students of many faculties must attend one or more mathematics courses. Here are some typical examples: computer science (5 one-year math. courses, some of them split into two semesters), physics (4 math. courses), chemistry (3 math. courses), biology (2 math. courses), engineering (4 math. courses), economics (3 math. courses), medicine (1 math. semester only). Most of these courses are taught by mathematicians in a rather formal and theoretical fashion, but physicists, chemists and engineers tend to involve their own staff in the teaching of some of the more specific courses, such as "mathematical methods for physics, for chemistry, for engineering".

V. Research into Mathematics Teaching

There is a growing interest in this field. In several universities there are active groups (called *Nuclei di ricerca didattica*) of university and school teachers working together on innovative projects, but the centralized structure of the Italian school system allows only for a limited freedom with respect to the national syllabi. Nevertheless, these groups have had a positive influence on the updating of the ministerial curricula and syllabi for elementary and middle schools as well as on the (experimental) curricula and syllabi for the upper secondary schools.

One of the main concerns is the role of calculators and computers within the traditional mathematics classes at all levels, from elementary school to university. Also the teaching of geometry at secondary level is under debate: the prevailing opinion is that, in spite of an introduction of coordinate and vector geometry, the synthetic point of view is still very important and should therefore maintain its traditional place in the new syllabi. The problem is when and how to start with an introduction of axiomatics.

A major difficulty in the Italian school system is the re-training of in-service teachers: most of them had no university preparation in computer science, nor in other non-traditional subjects only recently included in the new syllabi, such as probability theory or transformation geometry. In the past five years there was a national project for the re-training of upper secondary teachers in view of the introduction of computers (mainly in grades 9 and 10).

The project had some positive effect, since now most teachers are capable of using computers and writing at least simple programmes (mainly in Pascal). However, the project was limited to the more technical aspects of computer science. Hence, it did not contribute significantly to an improvement in the teaching style in the classroom, as would have been possible and desirable thanks to the new technology, nor did the project cover the other new topics mentioned above.

Mathematics and Education Policy in the United Kingdom

by Timothy Porter

School of Mathematics, University of Wales at Bangor, Bangor (United Kingdom)

As Howson points out in [5], p.216, "There is no single educational system for the U.K.; each of England, Northern Ireland, Scotland and Wales has its own education office (or ministry) and its own tradition." Although there is now a "national curriculum", this only applies to England and Wales and it only handles the 5–16 year age range. Even within England itself, different local areas may organise their school systems differently; thus although the majority of English counties have a co-educational comprehensive school system at secondary level, some have single sex, selective schools (usually called "grammar schools") existing in parallel with the comprehensive system. A further complication has been that although most state-run schools are funded via local education authorities, recent government reforms have introduced the possibility of a school "opting out" of local education control and obtaining direct funding from central government. The national curriculum applies to all state schools in England and Wales, but "independent" schools (that is the private sector of education) do not have to follow it, although as the national curriculum influences the content of the end of school examinations, set by independent examination boards, the private schools often do use the national curriculum as a basis for their syllabus structure. (For details of the English national curriculum in mathematics, see [5], pp 216–235).

This heterogeneous structure in secondary education is also present at the level of Higher Education Institutions (H.E.I.'s). The structure of higher education has been changing rapidly for the last ten years. Until now, there has been a "binary" system consisting of Universities and Polytechnics. The degrees in a university are awarded directly by the institution, whilst those in Polytechnics, and other non-university H.E.I.'s, have been validated by a central agency, the C.N.A.A. (Council for National Academic Awards). At the time of writing (April, 1992) this council is in its last year of existence, and the distinction between Polytechnics and Universities will shortly disappear. Polytechnics have a tradition of handling more vocationally inclined students, training them for work in industry and commerce. The position of mathematics in this sector of the higher education system has not always been well defined, nor has it been particularly strong politically. In addition to these conventional H.E.I.'s, there is also the Open University, which will be considered separately.

The curricular autonomy of universities has been almost complete. The degrees awarded are classified (First, Upper second, Lower second, Third and Pass for Honours degrees, as well as Ordinary degrees). The decision on what grade to award to a particular student is taken by an examining board, which includes external examiners whose report is sent to the university authorities. This report will include comments on the academic level of the courses, the organisation of the examinations and so on as well as on the fair conduct of the examination. Apart from this, there has been little or no external influence on what is taught, nor on how it is taught. This level of academic autonomy is changing as more verification of the quality of the teaching, examining and so on performed in the universities is being required by central government.

Acknowledgements and Disclaimer: I would like to thank Sue Burns, Paul Coackley, Geoff Howson, Adam McBride, John Mason, Chris, Robson and Derek Woodrow for providing, or feeding me, valuable information. They have, of course, no responsibility for the way in which I have interpreted that information.

I. Mathematics Teaching in Secondary Schools and Higher Education

Overview

The education system in the United Kingdom is organised roughly as follows:

ENGLAND AND WALES: From 5 to 11, primary school; from 11 to 16, secondary school with the G.C.S.E. exam as the culmination of this period. From 16 to 18, either sixth form within a secondary school or a "sixth form college". At the end of the sixth form, the typical qualification is up to three A level or an equivalent number of AS level exams. University education normally lasts from the age of 18 to that of 21, ending with the award of a B.A. or B.Sc.

NORTHERN IRELAND: This is approximately as above, but primary education starts at 4 and secondary schools are selective.

SCOTLAND: Somewhat as above; 99% of local education authority schools are comprehensive with multi-ability intake and 90% of these have a six year course, so 12-18 is more usual as the age range in secondary school in Scotland than 11-18. The end of school qualifications are differently organised and have different names than in England and Wales.

I.1 Secondary education

In all areas schooling is compulsory until the age of 16, (but see Howson's comment [5] p.7). The usual age at the end of secondary education is 18. At the age of 13, mathematics is a compulsory subject in school and a "typical" pupil (in England) would spend about 3.5 hours a week on it. At age 18, it is no longer compulsory, and at that age a pupil who is taking it, would be spending about 4.5 hours a week on mathematics. In Scotland, the comparable figures are a minimum of 3.5 hours in the third year of secondary education, whilst in their last year in secondary school, the course is designed to last 160 hours per year, which means about 4 hours a week on average. This, of course, depends on whether the pupil is following advanced or lower level mathematics courses at that stage.

The major qualifications taken at secondary school in England and Wales are the G.C.S.E. (General Certificate of Secondary Education), usually at the age of 16+, although some pupils take it at age 15 or less, and A (Advanced) or AS (Advanced Supplementary) level, taken at 17 or 18. The G.C.S.E. was introduced in 1988, but no review of A-levels occurred at the same time. Shortly after, AS-level was introduced. One of its aims has been to provide more options at a level below that of A-level, (it has approximately half the content of an A-level), in an attempt to encourage a broader academic profile for the end of school, its style though is still more compatible with the A-level than with the G.C.S.E.

Typically pupils take three A-levels or two A's and two AS's. There is worry about the falling numbers of candidates taking A-levels in mathematics, physics and chemistry. There is wide agreement that there is a crisis in as much as the supply of mathematically qualified candidates for university courses in science, engineering, finance, etc. and eventually of "numerate" graduates to take jobs in industry and commerce, is insufficient. The existence of a crisis is agreed; what is not is a detailed analysis of the causes, and more importantly of the possible solutions, especially if the solution involves expenditure of large amounts of public money. The number of those taking A-level Mathematics is declining. In 1991, the 18 plus age-group was 79% of that of 1984, but the number of pupils taking A-level Mathematics was 70% of the corresponding figure for 1984. Attempts to explain the decrease in the entry for A-level Mathematics point to the recent reforms in the pre-16 syllabus. The G.C.S.E. introduced a different approach including open-ended coursework, projects and "investigations" as well as more traditional "dogmatic" teaching with each type having the corresponding type of assessment. No such reform was done of A-level, which retained its emphasis on formal algebraic manipulation, the beginnings of the differential and integral calculus and a start on mathematical rigour.

The Scottish system is different and unfortunately cannot be described in detail in the limited space available. A detailed discussion of the syllabus both in that part of secondary education (11-16) during which the national curriculum applies in England and Wales, and post-16, can be found in [5] pp. 216-235 and there is no room to repeat it in detail here.

I.2 Higher education

The structure of higher educational studies is typically:

First cycle: a 3-year course leading to a Bachelor's degree.

Second cycle: a taught 1-year course possibly with a short dissertation, leading to a Master's degree.

Third cycle: research leading to a Doctorate, after at least 3 years. Students may have omitted the second cycle.

I.2.1 First cycle. Entry to University or Polytechnic is selective on the basis of the A-level results and students do not tend to go to their local H.E.I. Students themselves select a limited number of degree courses for which they wish to be considered, (e.g. Southampton: Mathematics with Actuarial Studies ; Bangor: Mathematics with Insurance, etc.) based on a course profile in the relevant University or Polytechnic prospectus. The requests for admission are sent to the H.E.I.s and typically certain students, whose school study-profile looks promising, will be asked for interview. An offer of a place may be given dependent on achieving given grades at A or AS-level. The student compares the offers received and short-lists two courses. When the examination results are available (in August), offers are confirmed for those who have obtained the needed grades, whilst those who have not, are handled through "clearing" and are helped to find a place that suits them.

A certain understanding of this quite complex selection system is needed to comprehend the problems of mathematical education in the initial years of a degree course in mathematics in the U.K. Not only may there be quite a large mix of A-level results—some selected early with good grades, some arriving through clearing, etc., but also there is the effect of the spread of geographical origins of the students already mentioned above. In the recent report "The Future for Honours Degree Courses in Mathematics and Statistics", ([8], p. 5), when discussing the present position one finds: "The background of students is very varied. Different schools may have taught the core syllabus in different styles and with differing levels of thoroughness, and students entering higher education with a qualification in mathematics may have studied different branches of the subject. They may, for example, have chosen to specialize in mechanics or in statistics. In these topics first-year higher education courses therefore need to cope with students having very different levels of knowledge." This poses problems, in particular, for first year teaching. At present the number of students taking degree level courses in mathematics and statistics is of the order of 3,000 in each year. Thus within a present age group at 18 years, of approximately 700,000, of which just under 20% enter higher education, there are about 3,000 students taking mathematics, possibly combined with another subject, that is about 0.5% of the age group and approximately 5% of those taking an A-level in mathematics. The low number is causing an increasing financial crisis for departments of mathematics in the U.K., exacerbated by the extremely low "guide price" paid by the government to H.E.I.'s per mathematics student. This "guide price" is considerably lower than all other sciences and is even lower than the average for all "humanities".

The majority of mathematics degree courses at the University or Polytechnic last three years. A few courses where the emphasis is more vocational, or more concentrated towards industrial applications of mathematics, have a period working in industry and a subsequent four year duration. Some courses recently introduced have a foreign language component and students spend their third year studying mathematics at another European university, after which there is then a fourth and final year back in the home university. (For the detailed structure of degrees in the ex-Polytechnics, see [3])

Joint degrees and combined studies. Students following courses that are described in this way, take their other subject for half of the time, and the subject matter within the mathematics part of their course will probably be restricted so as to provide a selection of those courses judged most suitable for the interaction with the second subject. For instance, students doing a joint degree in mathematics and economics might have to specialise in courses such as operational research, optimisation, statistics and probability, etc. in their third year.

Course content. The typical mathematics degree at a U.K. university has a wide mix of both pure and applied mathematics together with probability and statistics.

The applied mathematics is likely to include applied aspects of differential equations, examples of various modelling situations including mechanics, and perhaps biological modelling, fluid mechanics, and applications of stochastic processes, operational research methods, and numerical methods for solutions of problems, together with related aspects of computing. The pure mathematics, during the first two years at least, would typically be foundations of calculus and analysis, linear algebra, some discrete methods and some abstract algebra. Third year courses often provide students with a range of options. Computer algebra and graphical packages are increasingly being used. The statistics and probability content may vary considerably, but statistical data analysis, using statistical computing packages such as MINITAB, is quite common, leading on to the more theoretical areas later. Also students are increasingly being expected to be "computer literate", being asked to use word processors, spreadsheets, statistical packages, graphical packages, etc. in the production of assessed material.

Teaching methods. The predominate teaching method has traditionally been the hour long lecture, usually augmented by problem classes and"tutorials". The tutorial can take many different forms. At Oxford and Cambridge, in a tutorial there would normally be one tutor and two students. The hour-long session would consist of a discussion of recent material from lectures, based on completed assignments that the students should have handed to the tutor in advance. Other universities with fewer resources use tutorials more sparingly and tutorial groups are larger. "Class tutorials" or problem classes are also used. Ideas such as "student-centred learning", "transferrable skills" and "self-paced learning" have increasingly been applied in an effort to handle the increasing number of students, the general weakening in their level of algebraic manipulatory ability and, on the positive side, to try to encourage and develop more independence and maturity in students.

Assessment. Higher education institutions tend to use "traditional" three hour written examinations on each course or group of courses, as the dominant method of determining whether or not a student should proceed from one year to the next. These examinations are usually taken at the end of the year. Increasingly, however, course work, continual assessment, project work and variants on the examination method, (open-book examinations, take-home papers, etc.) are being introduced.

The Open University and mathematics. Set up in 1969 with its first student intake in 1971, the O.U. has had a considerable influence on mathematics teaching in universities and polytechnics. The O.U. student is typically mature and in full time work. The teaching is done by course units and carefully chosen texts. Motivation, extra background, etc., is provided by radio and television programmes. These latter have given a very positive image of mathematics to the large number of people, not only students, that have watched them. There are also tutorial sessions organised in local centres and, on many courses, summer schools.

The O.U. degree is based on units and accumulated credits. A student can take a wide range of different units from different subject areas and so might perhaps combine mathematics, say, with technology, or humanities, in many different proportions. Because of this the figures of O.U. students taking mathematics have to be interpreted slightly differently from those of more conventional systems. At present there are about 17,000 students studying mathematics courses with the O.U. This figure does not represent the number who will get a degree in mathematics (as such a degree does not exist), but does indicate the considerable contribution to the mathematical training of adults made by the O.U.

I.2.2 Second cycle. The next level of degree is the M.Sc. (or M.A.), that is, the Master of Science (or Arts) degree. Little public funding is available for students to follow such courses and fees are too high for many students to finance taking such a course themselves.

The number of students awarded an M.Sc. (or Post.Grad.Diploma.) is approximately 300 a year, including overseas postgraduates. The numbers of funded places to study pure mathematics and traditional areas of applied mathematics at M.Sc. level is small, but they are also widely sought after by postgraduates from other countries as the first step towards a doctorate. In addition the Open University has some taught M.Sc courses. The number of students registered for these courses have been in the order 200 for each of the last 3 years (1989–1991).

I.2.3 Third cycle. The doctorate (Ph.D. or D.Phil.) nominally takes 3 years. In mathematics, a limited number (about 160 each year) of postgraduate studentships are available from the Science and Engineering Research Council (S.E.R.C.). These provide funds for 3 years. The S.E.R.C., in an attempt to ensure good supervisory practice, is applying pressure on institutions at which students are taking more than 4 years to complete their research programme. Such pressure is, of course, handed on to the student! The number of students involved in Ph.D. programmes is of the order of 350 to 400 in any one year. Of course these include a high percentage of students with non-U.K. backgrounds, who will return to their home countries upon completion of their studies.

II. Mathematics and Engineers

There are no specialist engineering schools in the U.K., all engineering schools being a part of a university or polytechnic. The prestige of the engineering profession in the U.K. is not that great at the moment. Manufacturing industry would seem to be less well considered by the general public than are the financial sector and service industries, and in the last ten years many of the best mathematically qualified of the 18+ age group have opted for degree courses in accountancy, business studies, etc., rather than engineering or computer science. Once in an engineering course at a University or Polytechnic, the mathematical demands placed on the student vary greatly from course to course. The professional engineering bodies have been putting pressure on departments of engineering to ensure that students spend as much time as possible on engineering rather than on mathematics, physics, etc., and this together with the decrease in the mathematical and physical backgrounds of incoming students has led to complex problems. Including both an increased amount of engineering and the necessary areas of mathematics and physics, is increasing the risk of student "overload" in first degree courses in engineering. The report [4] (quoted in [8]), for example, includes a suggestion to "Teach only the mathematics... applicable to their chosen kind of engineering degree" and a proposal to "Reduce "analytical theory ". Such developments are a cause of worry for many mathematics departments, as too much integration of the mathematics content in engineering degrees, into purely engineering courses means that (i) the importance of mathematics as an independent discipline in the global view of the governing authorities of the H.E.I.'s is decreased, (ii) there is a risk that too narrow a mathematical education may lead the engineer to apply inappropriate mathematics to practical problems. Another more positive view is emerging in some quarters. "Where mathematics is not covered in engineering degrees, it will need to be supplied by mathematicians, but by mathematicians who know how to communicate with technologists. In fact, there is already a need in industry for well-trained mathematicians with a sufficiently deep and wide range of knowledge to enable them to develop the applications of mathematics to major technical problems." ([8], p.8)

Thus the position of mathematics in engineering courses is very variable. Where mathematics is taught as such and by mathematicians, the level of mathematical knowledge of the students is even more varied than in the specialist mathematics degree courses. The numbers involved are often large and given the attitude of the engineering profession, there are big problems of motivation and consequently often poor attendance rates at lectures. Some experiments with using "student paced learning", Computer Aided Learning, and variants of these have been tried with varying success. This is an area in which more research into the learning of service subjects is urgently needed if the mathematical knowledge of engineering students is going to be adequate and appropriate to their needs.

III. Recruitment and Training of Mathematics Teachers

The majority of the teacher training in the U.K. is at present through courses leading to a Postgraduate Certificate in Education (P.G.C.E.). A student taking a P.G.C.E.course in mathematics need not have a degree in mathematics. For instance in 1982, Woodrow, [12], found that recruitment to university P.G.C.E. courses contained about 60% mathematics graduates. More recently Straker, [11] found that at the University of Newcastle, the percentages of 1st degree mathematicians in the P.G.C.E. class decreased from a peak of 86% in 1983 to 50% in 1987.

The Cockcroft report, [2] gave national figures of 61% in 1978, 63% in 1979 and 58% in 1980. Thus the problem is not new. To have the nation's children trained in mathematics by teachers who do not have a degree in mathematics seems a bit like getting dentists to act as heart surgeons! This suggests there is a "hidden" shortage of qualified mathematicians in British schools. "The proportion of mathematics graduated opting for a career in teaching has been in decline for the past 40 years and even in recent times, the proportion has declined from 17% in 1974 to 8% in 1984" [11]. The average percentage for 1989 and 1990 is 5.6%. [8].

There is no direct link between the number of mathematics posts in secondary schools and the number of places for students on P.G.C.E. courses. It is therefore also difficult to estimate the extent to which there is a shortage of supply of teachers of mathematics.

In an attempt to curb this decline, in January 1986, the D.E.S. announced that it was introducing a tax-free bursary for intending teachers of shortage subjects such as mathematics and physics. This was an effective overall grant increase of about 60%. In 1986, the figures for P.G.C.E. were an intake of 703, by 1988 they were up to 867, but by 1990 some of the effect seems to have worn off (779). As these years correspond to a down-turn in the job market, it is not certain that this increase in entry was due to the bursary or to the general economic climate. How many of these students mean to try teaching as a career is difficult to tell. Some may be intending the P.G.C.E. to be an additional qualification which will stand them in good stead for future job hunting inside or outside education. A recent survey [13] for the Joint Mathematical Council (conducted by Derek Woodrow of Manchester Polytechnic) indicated that only about 70% of P.G.C.E. students in 1991 were known to have taken up teaching posts.

There are thus several problems: (i) a low percentage of mathematics graduates amongst the Mathematics teachers; (ii) a shortfall in recruitment to training courses (except for this year); and (iii) the low percentage of qualified teachers entering the teaching profession, which combine to give a very worrying profile of the future of mathematics teaching in the U.K. In addition to these, one has two factors not specific to mathematics. The first is the large number of skilled teachers (and mathematics and physics teachers in particular) who have left the profession in the last 15 years. The other factor is that experienced, mature teachers have been more likely to leave the profession than young relatively inexperienced ones. Their posts are either not being filled at all or are being taken by much younger (and hence cheaper) teachers. Given the falling percentage of mathematically qualified teachers, etc., this effective removal of experienced mature colleagues to whom the younger teacher might turn for advice and guidance, is causing a lot of worry. The shortage of highly qualified mathematics teachers in the U.K. is not new. It was already apparent in the early 1960's. Officially there is no crisis as a crisis might imply that someone was to blame, but officials have admitted to there being a problem—but have not come up with any solution as yet. The problems hinted at earlier of the jump in difficulty from G.C.S.E. to A-level Mathematics may further cut the number of students taking A-level Mathematics and hence even if total student numbers increase in higher education as planned by the present government, it looks unlikely that there will be a corresponding increase in the number of mathematics graduates. If that is what happens, it is difficult to see how sufficient mathematics graduates can be attracted into teaching to fill the admitted shortfall, let alone the hidden one that is not admitted to exist.

Content of a P.G.C.E. course in mathematics. As mentioned at the beginning, Universities are autonomous and so they can design their own P.G.C.E. courses within some general guidelines from the D.E.S. All such courses seem to be different, so here we will only look at the main features of the courses on which I have details (see [1], [6], [7]).

In the last twenty years, there has been a swing away from formal courses on, say, the philosopy and history of education and more assessment is based on coursework and practical teaching experience, rather than on examinations. There is also a tendency to make the P.G.C.E. more school based. At present the government requires that at least 12 weeks are spent on teaching practice, usually some in primary schools and the rest in secondary. This is likely to increase with government pressure for more school based or in the extreme case, wholly school based training. The Open University is preparing a radical training package, starting in 1994, and using a mix of tutoring, distance learning and radio/television, that may provide a way in which this may be done with reasonable effectiveness.

As an example of what P.G.C.E. training entails, we will look at the course at Bangor. The School of Education at Bangor runs a small P.G.C.E. course in mathematics. (Most of what follows is adapted from the Course Handbook, as supplied to me by Paul Coackley.) The course is largely seminar/workshop based. As well as the periods of teaching practice, regular visits are made to schools at other times. These visits sometimes have a specific focus or may merely be opportunities to gain experience of as many different schools as possible. The content of the workshops/seminars (see below) is derived from the topics the student will be asked to teach or from parts of the National Curriculum, G.C.S.E. or A-level. The students might, for instance, be required to consider the resources necessary, previous knowledge, or methods of assessment suitable, for a given subject area or to look at relevant documentation on assessment methods.

The content of the course can perhaps best be gained from the workshop topics listed below: (the list is not complete).

The Nature of Mathematics Problem Solving and Investigational Work in School Mathematics; Watching yourself teach!; Generic Software—Logo and its possibilities; Open and Closed Activities; Key Documents in Mathematics Education; The National Curriculum—its structure and terminology; Lesson planning—

nature and purpose; Setting, marking and recording work; Some teaching approaches to Mathematics topics—area, volume and fractions; Mathematics and language—an introduction to cross-curricular themes; An introduction to Theories of Learning; Teaching practice review—lessons and targets; Generic Software—Spreadsheets; GCSE Mathematics—structure, variety and assessment; GCSE coursework—what is it and how do you mark it? The use of calculators in the classroom —good practice or bad?

Unlike in many other countries, teachers in the U.K. are not civil-servants. They are recruited by the school. This is usually done by placing an advertisement in the national press. This explains the lack of data available on the number of teaching posts in mathematics that are filled by non-mathematics specialists. The selection committee in a school may have very little competence in mathematics, but may call in local mathematics advisors, inspectors, etc., if it is felt to be necessary. With increasing independence of schools and the lessening of the resources of the local education authorities, it is not clear if there soon will be enough specialists to assess the candidates' mathematical ability. Certain aspects of this increased independence are therefore quite worrying, especially in shortage subjects such as mathematics.

Higher education teachers. Jobs in Universities and Polytechnics are also advertised in the national press. A selection committee is appointed to examine the candidates' applications. This is done totally internally to the institution with no recourse to any body outside. The selection process will often involve a presentation of the applicant's research in a short seminar, however no other indication of teaching ability is usually possible. As the monetary resources of a department depend in part on its research profile, the decisions as to whom to appoint are usually based on research potential and the compatibility of the candidates research area with those already in place in the department, although with the appointment to more senior posts some considerations of an educational nature may be involved.

There is no requirement that a beginning teacher in higher education have any training in teaching methods, or even any knowledge of relevant educational issues. In any case it is difficult to see how an institution based scheme can provide subject specific training given the small number of people involved and their disparate needs.

IV. Some Aspects of Research into Mathematics Teaching

Research into mathematical education takes place in most of the institutions involved in the training of mathematics teachers. Some of these have centres for science and/or mathematics education within a school or department of Education, in others the "mathematical educators" are attached to a School of Mathematics. Some of the centres are relatively well funded with money from industry augmenting the governmental funding (e.g. the Shell Centre for Mathematics Education at the University of Nottingham), but this is the exception rather than the rule.

Research into, and development of, new teaching material has recently received a lot of impetus and importance, given the recent changes in the structure of teaching in secondary schools. Some centres such as the Nuffield Advanced Level Project at the University of London, are designing new syllabi for schools, others such as a small group within the School of Mathematics and of Education at Bangor, have been designing and experimenting with new material for structured investigations at G.C.S.E. level (13 to 16 age group), whilst the Centre for Mathematical Education at the Open University has been developing resource packs for professional development throughout the age range (cf. [9]).

The Mathematical Association and the Association of Teachers of Mathematics are two organisations which provide a useful service to teachers, encouraging the sharing of research findings at a practical level through their journals. (For instance Mathematical Association publishes reports including that by [5] and summaries of research findings.)

Finally it is worth mentioning the amount of innovation in the teaching of mathematics at degree level. This innovation brought about by the changing academic profile of the incoming students has occurred at a time when ideas relating to "quality control", "enterprise", and "public accountability" are entering into the language that is being employed in higher education institutions and are forcing a re-evaluation of the traditional aims and objectives of a degree in mathematics.

Note: a longer version of this report with more detailed figures is available from the author.

References

[1] *PGCE Mathematics*, Handbook, UCNW, Bangor, School of Education, 1991-92.

[2] W.H. Cockcroft, "Mathematics Counts", Report of the Committee of Inquiry into the teaching of mathematics in school, H.M.S.O., 1982.

[3] Review of subject and course development in mathematical sciences, Council for National Academic Awards, January, 1992.

[4] "The Future Pattern of 1st Degree Courses in Engineering", Discussion document prepared by a working party of the Engineering Professors' Conference, February 1991.

[5] A. G. Howson, "National Curricular in Mathematics", The Mathematical Association 1991, (259 London Road, Leicester LE2 3BE, England).

[6] "P.G.C.E., Mathematics Programme", Centre for Educational Studies, King's College, London, 1991.

[7] Information for Secondary Mathematics Students, Graduate Certificate In Education, Centre for Studies in Science and Mathematics Education, The University of Leeds, 1991-92.

[8] "The Future for Honours Degrees in Mathematics and Statistics", Final report of a group working under the auspices of the London Mathematical Society, London, Feb., 1992.

[9] "Supporting Teachers of Mathematics", Information booklet, CME: GA/2.91, Centre for Mathematics Education, The Open University.

[10] A Strategy for Support of Core Science, Science and Engineering Research Council, 1991.

[11] N. Straker, "Recruitment trends in PGCE Mathematics", Math. Education for Teaching, vol. 5, 1988.

[12] D.Woodrow, The Supply of Mathematics Teachers in England and Wales, Journal of Education Studies, vol. 8, No. 1, pp 26-33.

[13] D.Woodrow, "The JMC Survey of Secondary Maths Output in 1991", The Joint Mathematical Council, 1992.

Round Table D: July 9, 1992

LET'S CULTIVATE MATHEMATICS!

organized by Y. Chevallard[1] and A. Rouchier[2]
IUFM, d'Aix-Marseille (France)[1], IUFM, Orléans-Tours (France)[2]

Introductory Remarks

by Yves Chevallard

Let me begin with a few remarks in order to outline the topic of our round table discussion. The title — *Let's Cultivate Mathematics!* — gives a good idea of our main concerns. However, I would like to stress some hidden aspects of the situation to which it implicitly refers.

(1) In today's societies mathematics has come to be one of the main constituents of the societal machinery. This point should be carefully appraised. For instance, mathematics as a driving force had not as yet become central in 18th century European countries. It certainly played a key role in some branches of economic activity, such as commerce, but it had not achieved the vital importance that it has gained today. In present-day societies, mathematics is indeed an all-pervasive ingredient which has lodged itself in every nook and cranny of the social structure. If we could cut off mathematics just as we can cut off electricity, very little would survive for long. For our societies run on mathematics as they run on electricity. I shall not address the question of why this is so, or whether it could be otherwise. I take reality as it is, whether we like it or not. And I shall emphasize rather an apparent paradox, which will lead us to the main problem we are currently confronting.

(2) The part played by mathematics in yesterday's societies was not impressive, but nevertheless mathematics had achieved *cultural visibility*. What I mean here is that people generally knew that mathematics was underpinning at least *some* aspects of social activity. All European languages have apparently recorded this state of affairs, using different expressions, e.g., in English, "I don't understand the mathematics of it".

(3) The gap between the man in the street's vision of society and the reality of social functioning was brought about by two overriding factors, the first of which is the industrialisation and the subsequent rationalisation of our societies. In this long-term process, the traditional mathematics of commerce, accounting and banking was overshadowed by the much less visible mathematics of industry and technology, which went deeper and deeper into the inner recesses of the social machinery — the division of labour, which conceals from us the true structure of society, and the inability of culture to provide us with an image of social reality that could compensate for this loss of lucidity. This made the man in the street unaware of the growing role of mathematics as a tool of production for society and, consequently, of the growing mass of mathematics "crystallised" in the whole gamut of objects that surround us in daily life.

A second factor has had equally significant consequences: compulsory schooling. This of course is a major fact of our times. However I would like to single out one aspect which seems most relevant to our topic. School was thought to be a vantage point from which society could be revealed through the study of a number of fields of knowledge. From then on, generations of budding citizens met head-on with mathematics, but this encounter took place under very special conditions. Mathematics was locked away in school; school mathematics, as we may call it, was cut off from its social roots, and its social significance and utility gradually disappeared. Mathematics thus became a school-bound reality.

(4) The intention behind mathematics at school was twofold. On the one hand, school was supposed to provide society with the necessary scientific workers, i.e., mathematics for the few. In this respect it succeeded rather well, although in the long run its effectiveness has slowly but steadily declined, in relative terms. On the other hand, it was also the task of school to help culture keep pace with the evolution of society. That was mathematics for the many, and this is where the problem arises. The mission assigned to school was to generate and spread a vision of society more faithful to the changing nature of our societies. In this respect it failed. The gap between the two cultures, one supposedly "humanistic", the other "scientific", gave way to a permanent culture lag, a fact almost unknown, at least in some circles, for instance in the Age of Enlightenment.

(5) We have now arrived at a turning-point in the history of our science-based societies. These depend heavily on technology, science, and therefore mathematics. But science, mathematics and technology have not yet gained the recognition that would allow them to live a peaceful life in the realm of culture. While their *social relevance* is high, their cultural attractiveness remains very low. In the long run, this conflict is sure to lead to the situation that some of our countries already experience — a situation marked not only by the shortage of mathematics teachers, but also by a shortage of "scientific workers". The current exclusion of mathematics from the realm of culture (regarded both as a body of attitudes, values, representations and knowledge, and as a selection of these) makes it difficult for anyone, especially among the youth, to be spontaneously attracted to mathematics.

(6) To sum up: mathematics remains secluded, first at school, then in scientific communities. Now the problem is that mathematics, as a body of knowledge and a tool for action, cannot live long in such a splendid isolation. It has to be supported from outside, that is, it has to receive cultural support by being reinstated into culture. Society as a complex whole has *needs*, which include mathematical needs. Cultural unawareness of these needs is a blatant fact which endangers the very development of present-day societies. Let me add here that such a problem arises in every hamlet of the "Global Village" in so far as, willy nilly, the Western model of development is being extended to the whole planet.

(7) The *mathematics awareness* we should try to spread meets with a difficulty of which we will not easily rid ourselves. There is no evident relation between society's mathematics needs and those of the individual. Mathematics needs can be felt at the level of society as a complex machinery. But, as a consequence of the division of labour, they usually remain foreign to the individual. This is why the man in the street can so easily ignore mathematics.

(8) However, the process of separation between the needs of society and those of the individual can never be totally complete. More and more people, as individuals, and especially as professionals, are personally affected by the growing presence of mathematics around us. Many are very likely to be confronted with personal mathematical needs at different levels of sophistication. Now a new problem arises. Cultural ignorance of mathematics impedes the mere recognition of these mathematical needs. Thus, we are taken back to the central question of our round table: how can we help to increase cultural awareness of mathematics as a social reality — not only for the few, but indeed for the many?

Institutional Dimension in the Definition of Cultural Strategies in Mathematics

by André Rouchier

A round table is a place where it is possible to intervene on two levels, opinion on the one hand and technique on the other. It is sometimes necessary to go beyond the greater or lesser competence of the expert to produce the elements needed for a debate in which each intervention represents a term. I am expressing myself in French because that is the language in which I can best express those notions to which I make reference.

(1) Mathematics is extraordinarily present in numerous aspects of society, in technical and industrial activities, for example. It is, in combination with physics, chemistry and computer sciences, embedded in a large number of objects in daily life, both in their material as well as functional aspects. Sometimes it is so embedded that we do not know how to identify it without a specific search. As for extracting it, most of the time that could only be done by destroying the object or by concealing the technique. We agree with Yves Chevallard in saying that mathematics has great *social relevance*, closely linked to a no less greater *social legitimacy*.

(2) We must admit, however, that mathematics does not enjoy the cultural acknowledgment which is its due, given its importance. Again following Yves Chevallard, we can characterize this situation by saying that mathematics does not enjoy *cultural relevance*. Everybody knows this whole set of attitudes that consists of distancing oneself from any positive demonstration regarding mathematics. To say the least, a lot of people have ambivalent attitudes and it is infrequent to find someone who claims to have mathematical ability, apart from the professionals. What manifests itself through negative reactions and the frame of mind they reveal could be studied on several levels. The sociological level could consist of studying the relationships between what is said and the reality of social positions, notably through the cultural signs that serve as beacons. The psychological level, which is more interior, could consist of describing the patterns of these reactions in relation to the structure of causal elements. This is not our subject here, and we will consider the lack of cultural relevance as a recognized fact.

(3) Outside of professional areas where the presence of mathematics is evident, either directly through the use that is made of it, or indirectly because material or symbolic tools loaded with mathematics are involved, the encounters of a given individual with mathematics almost exclusively occur at school. To go beyond this educational framework, to show mathematics under another angle and to communicate through carefully organized practical procedures are the basic ingredients of what is classically called popularization. Traces of it can be already found in the Middle Ages. It is an ongoing attempt to fit mathematics into the cultural mould of "civilian" society, to lay the foundations for mathematical taste which could also be a taste for mathematics. That mathematics should be made to look commonplace could be seen in various seemingly ordinary situations. "How do we call the number that, multiplied by itself, gives a given number?" "What is the theorem that says that, in a right-angled triangle, the square of the hypotenuse equals the sum of the squares of the other sides?" Here are, for example, two questions asked in the *mathematics* question set of a French televised game show. That mathematics should be present on this level of social life is still quite uncommon, and the answers do not always require a cultural background that goes beyond vague memories. There is undoubtably a definite reach toward certain categories of people, notably school students, in order to achieve a different relationship with mathematics. These initiatives will be discussed in another round table of the European Congress of Mathematics. The *Horizons mathématiques* exhibit is on display during the Congress. The *Rallyes mathématiques* or *Maths en jeans* initiatives are also in the process of receiving some recognition among the community of mathematicians. However, do these initiatives have both the form and the contents which could convince us that mathematics has been made more ordinary through them?

(4) If we push our questioning further, we may also reflect, based on the assumed lack of cultural relevance, on the objects on which and from which the aforementioned fashionable disdain appear. At the same time, we may also reflect on those mathematical objects and concepts that are not a customary part of an "ordinary" mathematical culture. Common terminology, on the contrary, has a tendency to forget what has been learned in school, or is not able to incorporate what mathematics has to say. It is thus that a fraction always stays a fraction and never appears as a rational number, a designation confined within the mathematical world and that shows as well the foundations of a rational number. Even more so, all that has to do with probability theory and statistics never emerged in or contributed to the constitution of a culture.

(5) It is not our intention to brand as simplistic the argument according to which the specificity of mathematics is the main reason for its lack of cultural expansion. Mathematics demands that a specialised vocabulary based on abstract and complex concepts be manipulated in written form. Mathematics very rarely presents the statements it produces in the light of the phenomenal realm of ordinary life. Mathematics does not deal with the physical world in the same fashion as physics, chemistry and the natural sciences, which have the possibility of placing themselves from the start in the immediate vicinity of the manifestations of the real world, the one that imposes itself massively on us through our senses. Which order of reality, familiar enough so as to appear to be concrete, could be dealt with and within a culture that is connected with mathematics? To speak in mathematical terms of a fact of everyday life seems to require involving complex symbolic systems from which the usual language and recourse to imagery and metaphor seem to be excluded. All this would soon sound excessive and pedantic. In mathematics, writing seems to be indispensable, because the definitely more familiar registers of oral and gestural expression are no longer sufficient. To make the underlying idea of the motto *Let's cultivate mathematics!* alive and instrumental, it appears important to take into account the specific difficulties related to the original or non original semiotic elements that will have to be introduced. This is an unwritten law that is well-known to all popularizers.

(6) There is another factor that contributes to the fact that such "limited strike" initiatives as an exhibit or a "rally" competition, do not receive all due credit as carriers of a mathematical culture. These activities lack temporal dimension and are not inscribed into a duration that could accommodate itself to the laws of this culture within its own time scale. They cannot instigate the establishment of a relationship that could refer to a specific universe. The autonomy of such a universe must be clearly distinguished from the institutional universe in which we are put in prolonged contact with mathematics, namely the educational universe. Is it possible to let two different operational patterns referring to the same area of knowledge live together or coexist on the individual level as well as on the collective level? If we exclude the various professional contexts in which individuals can be mathematics users without considering any cultural problems, the question posed may be reformulated in terms of *social devices* as well as in terms of patterns of *manipulations of mathematics*. It thus seems to me that we must include into the terms of the *Let's cultivate mathematics!* challenge the involvement of a creative level that reaches not only up to social configurations, but also to the symbolic and/or cultural objects that could live and be exchanged in adapted ways in such configurations.

(7) Naturally, mathematicians are not the only ones involved in this. Mathematicians are supposed to attend to, I am sure you will forgive the jest, the conservation and development of mathematics as a body of knowledge. They have no particular authority to control relationships other than those for which they are in charge. The problem we consider here does not befall entirely to mathematicians. During this presentation, we were able to go a little beyond the question posed, but this displacement was not gratuitous. Although it was proposed and elicited in a form that is not necessarily familiar to all, its aim was to point out the constraints we have to take into consideration.

Television Mathematics: Cultivation or Distortion?

by Celia Hoyles

Institute of Education, University of London, London (United Kingdom)

What is mathematics? Without entering into a debate as to the epistemological nature of the subject, it is possible to at least agree about what mathematics is not: namely, rows of sums, a mindless following of rules. Why is it that so many people regard mathematics as having just these characteristics? Mathematics is not seen to touch everyday life — it is certainly avoided but more fundamentally is not recognized as a structuring influence on work or social processes. Mathematics is something that is confined within schools — it becomes "school mathematics" — and is strongly associated with anxiety and boredom. There is a pervasive ignorance about what mathematics is, what mathematicians do, and what mathematical culture involves. So how is it possible to cultivate mathematics?

First I suggest some aims for cultivating mathematics:
— to enhance the visibility of mathematics;
— to break down the isolation of mathematics;
— to improve public awareness of mathematics;
— to transcend the school-boundedness of mathematics.

Second, a decision has to be made as to how to achieve these aims, that is, within the education system or outside it? The advantages of the former approach, within education, are that there is a "captive" audience, so all pupils will at least come into contact with some kind of mathematics, and the teaching situation provides means for clarification, support and suggestions for fruitful lines of inquiry.

The main disadvantage of working to cultivate mathematics in schools, however, is that schools exist to select as well as to teach, and this tension can militate against the fostering of genuine mathematical activity. But, if we avoid this tension by working outside the education system, what medium can best be exploited and how? I will now briefly describe an experiment to cultivate mathematics through popular television. The aims were modest — to enhance awareness of mathematics and to begin to change attitudes. We did not seek to improve competence in the subject or raise performance.

The program was called *Fun and Games* and was transmitted at prime-time (7.00 pm) throughout the United Kingdom every year between 1987 and 1990. *Fun and Games* was formed around puzzles in order to capitalize on people's interest in puzzles by using them as vehicles to think about the embedded mathematical ideas; after all, recreational mathematics has been the source of a great deal of mainstream mathematics. The programme was set in the studio with an "ordinary" participating audience — they were not selected nor rehearsed. There were no prizes. We also strove to include within one evening session at least one puzzle that involved some non-trivial mathematics — a situation which was counter- intuitive, intriguing and conceptually challenging. There were two presenters, a host and a mathematics professor. The host, Rob Buckman, accompanied the puzzlers through the games, set them at ease, explained the rules, was witty and friendly and acted as liaison between audience, puzzlers and the professor. The professor, myself, was supposed to be knowledgeable about mathematics, demonstrate strategies, and exude enthusiasm. I aimed to provide a mathematical perspective on a puzzle, to give a hint part way through the action leading to a useful strategy by highlighting a particular structure or relationship. I tried to give the puzzlers, and the audience, a mathematical lens, through which they could look at the puzzle and then, with this new view, see ways to a solution.

So, was the program successful? As far as viewing figures were concerned, surprisingly so; we had up to 10 million viewers on some occasions. Some reporters expressed amazement at this success; for example, *Northants Evening*

Telegraph under the heading, "Calling all Clever Cloggs," wrote on 27 April 1989: "Television doctor and clever cloggs Celia Hoyles team up again tonight to present another program packed with puzzles, conundrums, brain-teasers and games. *Fun and Games* is the show that aims to shake the cobwebs out of mathematics. But probably the biggest puzzle is how it has managed to capture the imagination of the public."

Some people maintained that they thought differently about mathematics and mathematicians as a result of watching the program. In a questionnaire survey after the 1989 series, respondents were asked, "if having seen *Fun and Games* they had changed their view of mathematics". Yorkshire Television were a little disappointed that only 20% said 'yes.' I was absolutely delighted — if we had made an impression on say 20% of nine million people, we were not doing badly in cultivating mathematics!

So by some measures the program achieved its aims. But how far did we really cultivate mathematics? Working in television subjects the presentation of mathematics to severe constraints which possibly distort the very thing we are seeking to cultivate. What are these constraints? The first is *time pressure*. Each *Fun and Games* program was twenty-five minutes five seconds long and, on the average, we presented five puzzles per show: so, roughly, each puzzle took about four minutes. In this short space of time, the puzzle had to be stated clearly, the mathematical hint given, and then hopefully the puzzle brought to a triumphant conclusion. Why is all this pace necessary? From a perspective of encouraging mathematical problem solving, a longer time on fewer puzzles is self-evidently preferable — but television producers have other agendas and I was assured we would lose our audience. Prime-time television needs to be entertaining and action packed!

The second constraint was the avoidance of *teaching* or giving any "whiff" of school. Sometimes it would seem to me to be obvious to use a diagram or, dare I say, a blackboard. The rationale against doing this was that it would "put off" the audiences (unpleasant memories of the classroom?). It also though has undoubtedly to do with the politics of television — we were leisure TV, not educational TV — and higher status!

A third constraint concerned the selection of puzzles. We tried to use a variety of types of puzzle: topological, number manipulation, geometric, but all of them had to involve the puzzlers in *doing something active*. It is obviously bad television to see people just standing thinking; but perhaps this is just what doing mathematics is all about — it goes on in the head!

The final constraint was that *formal mathematical language* of any sort was not allowed, the argument being that it would be too mysterious for the average viewer. This naturally restricts the generality of any solution and thus runs the risk of failing to convey the power of the subject, which is exactly why it is important within the fabric of society.

So what can be learned from this experiment to cultivate mathematics through television? Firstly, the four constraints identified above all have positive as well as negative sides: we had no time for irrelevancies and we had to spend considerable effort discussing *how* to draw out mathematical ideas; nothing was obvious. From a personal point of view, I was forced to think about ways to suggest a mathematical perspective without resorting to mathematical "jargon" or the formal ways that are deeply familiar yet frequently obscure. I had to search for *intuitive* approaches which could be made accessible through everyday language and straightforward strategies. I became more familiar with the views of the public about mathematics — by listening to the reactions of the studio audience and reading the letters written to the TV company. I came to be rather critical of mathematics education as a member of another community who saw our endeavours in a rather sceptical light. I learned to respect the power of the media to reach an enormous number of people — albeit in a superficial way. I am certainly more convinced of the influence of role models — the fact I was a *female* mathematics professor was undoubtedly of importance to many viewers.

But by trying to popularize mathematics did we distort it? Did we show something as playful and entertaining which is in fact serious and intellectually challenging? In our defense, we did not expect viewers to grapple at length there and then with all the mathematical ideas thrown up during the puzzling. All we hoped for was to leave the viewers with an intriguing idea that they could struggle with later at their leisure — and some certainly did. Learning mathematics requires hard study, considerable thought and persistence. But if we want to extend the mathematical culture, first of all surely we must open people's eyes to what mathematics could involve and why it is so influential in today's society? As Gattegno has so cogently put it, "one can only educate awareness".

It is however important to reflect upon this experience in order to identify the limitations as well as the potential of this particular approach to cultivating mathematics. Television has an immediate and considerable impact — people remember the faces, the incidents — but I wonder if the images are too fleeting to have any lasting impact or sustained message? The aims of the television program need to be reinforced by schools. But there is little evidence that cultivating mathematics outside the system has any effect on the system. Certainly the developments in the UK mathematics curriculum since 1990 exhibit a rather different agenda than *Fun and Games*! If we aim for more lasting change we need to turn our attention to the *transitions* between outside and inside school. Can these be achieved without artificially making home and leisure pedagogical or without returning mathematics to school mathematics? Theoretically this would seem to be impossible — particularly in the short term — yet, in practice, we might make small steps towards the long-term goal of an evolution in mathematical culture.

Mathematicians and the Need for Critical Self-Reflection

by Christine Keitel
Technical University of Berlin, Berlin (Germany)

The title of our round table — Let's Cultivate Mathematics! — reminds me that the majority of our colleagues sometimes feel that they have to be interested in cultural issues because they see themselves as merely offering cultural items to a world which rejects them. In my contribution I would like to take up one of the questions passed from the floor and to speak about this question. The question: is there sufficient interest within the community of professional mathematicians concerning general cultural questions? If not, how should this be rectified?[1] In my interpretation this question matches very nicely what I generally criticize when considering the social (societies' and mathematicians') view of the relationship between mathematics and culture.

If we look at the role of mathematics in our society — which is generally very rarely done by mathematicians themselves — it is bewildering that we find a big disjunction between the generally accepted technical necessity of mathematics for society and a broad cultural indifference of society toward mathematics. Everybody agrees that society needs more and more mathematics because there is an increasing mathematization of our societies. On the other side, however, mathematics is not acknowledged or appreciated as a worthwhile cultural heritage that has enriched our cultural acitivities, that frees us from some of our daily problems and is part of our enjoyment. To be provocative, I would like to state that until recently mathematicians were very happy with this disjunction. It ensured more or less uncontrolled working conditions and allowed them to remain undisturbed in their ivory tower, and nevertheless supplied adequate funding while leaving mathematicians completely free to pursue their interests. It is not rare that the social necessity of mathematics is used to make subjective interests and intentions of mathematicians appear as objective needs.

This has changed a little today, but will change more rapidly. This change has

[1] The question was brought up by Allan Muir from The City University of London. The written version of my contribution to the panel is based on a summary of my talk he kindly sent me afterwards, together with a newsletter on *How Mathematicians Work*, which he started recently. This newsletter helped me to clarify my ideas, and it shows that there is already a group of mathematicians concerned with self-reflection and willing to address cultural issues in mathematics and mathematics education in a broader sense. Actually, the round table has become for me the starting point of an interesting correspondence as I have received some reactions to my talk which encouraged me to write down even provocative statements.

come about through the advent of computers and the computer sciences. Computers are now perceived by politicians and the general public as making mathematicians less necessary. Computers have not only influenced our social life in general, but have entered into nearly every social activity including all spheres of production, labour and service. In addition, it has started also to change the paradigms of mathematics itself. [2] Mathematics is still the source of the production of mathematical models and theories as descriptive, as well as predictive means. In this function the scope of mathematics is almost unlimited. The level of abstraction can vary, one can prefer an aesthetic point of view instead of a utilitarian one; one can model structures of which the existence is only guaranteed in a mathematical sense, and so on. Certainly, models and theories were and are invented for very specific reasons raised outside of mathematics. Through the computer as materialized mathematics the subject becomes a kind of metatechnology, a technology of technology or a "thinking tool". Mathematics as materialized in the computer — as a technological problem-solver — is just as universal as a concept-producer. Mathematical models serve not only the purpose of description and prediction, but also, primarily and increasingly, of prescription. So the process of decision-making is coming more and more under the influence of mathematical models, which today can be realized by the computer.

Formalization of language and formalization of actions are closely connected types of formalization which presuppose and foster the use of mathematics. We find transmissions and mediations between the two types of formalization. The application of a formalized language makes us "see" formal structures in reality, and by doing so, reality is adapted to our image of reality. We highlight some things and ignore others. If our object is the work process and our intention to systematize and "taylorize" production processes, a formal description of what is going on could facilitate our next step. The formal description makes it easier — because it is decontextualized and depersonalized — to lay out new algorithms for the working processes.

By means of mathematics, conceived as a formal language and formalized routines, we are able to make mathematical models of reality. On the side of the piece of reality and its context, which has become the object of modelling, the agreement with a "matching" mathematical model presupposes a theoretical conceptualization of that part of reality in question. Theories about objects are intermediate steps of formalization by which the relevant elements or properties of the object are selected and their relationship is determined.

---

[2] Parts of the following ideas were developed in cooperation with E. Kotzmann (Austria) and Ole Skovsmose (Denmark) and will be included in an extended version in *Beyond the Tunnel Vision — Mathematics, Society, and Technology*, (to appear as chapter 8 of Keitel/Ruthven (eds.): *Learning from the computer. Mathematics Education and Technology* (NATO-ASW- series, Springer, Heidelberg, 1993).

Normally, a mathematical model is not a model of a segment of reality as such, but a model of the conceptual system, created by a specific interpretation based on a more or less elaborated and more or less explicit theoretical framework. The selection of the theoretical framework as well as of the model is guided by interests and purposes. To understand and evaluate the model we have to analyze the premises and assumptions of these intentions and purposes.

As concerns the mathematical aspect of the model, the coincidence of both the instrumental (tool) and the systematic (theoretical) aspects are constitutive, as mathematical concepts are always means and ends, and mathematical activity presupposes an awareness of that fact. Comparatively, in the concretisation of a structure outside of mathematics by creating an algorithm of real actions or a computer program, this balance tends to get lost. It is in order to enable us to do without the systematic background that the construction is undertaken. The mathematical modelling process is encapsulated in the technological structure — the application is cut off from the requirements of explicit mathematical knowledge; it has no regard for an *a posteriori* deduction or understanding.

The mathematical model provides a morphism to a conceptual system, and conceptualizing reality reveals possibilities of applying mathematical techniques or establishing algorithms which on being operated offer results. In dealing with them we follow the well-established idea that by a mathematical model we are in fact modelling "something". The model describes an object to a certain degree of approximation. We talk about the model as an image of reality, and in numerous introductions to modelling it is explained that modelling must start with reality. We have to model something!

This is the phenomenon of "objectification" connected with formal languages and formal routines. Of course, objectivation is not always a problem. To identify the shape of the earth with a sphere will become a reasonable description with lots of purposes in mind (although in some situations an ellipsoid will give a better description). To use a mathematical model as an image of "something" does not seem to be a mistake in this connection. However, the problem is that sometimes a model is not a model of "something", although the mathematical and formal language easily gives a descriptive interpretation. If the descriptive interpretation cannot find anything in reality which fits with the description, it is possible to change reality in such a way that the language in fact describes something. It is possible to invent an object for the formal description, and this invention may be based on a formalization of routines. By this the "world between" becomes part of our world. The use of mathematics becomes prescriptive, and formal structures are established as parts of our social reality.

The step of inventing an object for the formal language is the kernel of the transformation of thinking abstractions into reality, to make them into realized abstractions. While thinking abstractly could influence our way of interpreting what we see, realized abstractions influence reality as such. The realized abstractions become real objects; they can interact with reality as well as with other realized abstractions. Realized abstractions exist in the same sense as geographical or industrial infrastructures. Realized abstractions could be useful, or dangerous, or formidable. We have created a second nature made up of realized abstractions, within which the social development now takes place.

The trend towards systematization and formalization which was initiated and continuously fostered by the use of formal thinking is now reinforced by the ubiquitous use of the computer. The computer transforms the organizational pattern of the most influential social institutions in our society into universally applied and valid formal structures. In considering the social role of the computer we want to focus on some aspects which are not usually addressed: the aspects of interests, purposes, capital and investment, values transported with and created by computers in this process of transformation. The development and the first and main usage of computers are closely connected with management problems in economy, in dealing with the problem of directing and controlling production processes, i.e., applications made easy because the applied underlying (mathematical) model is simple and the structures to which it applies are highly processed by economical quantification and strict algorithmisation.

The computer reinforces the increasing formalization and mechanization of that domain in particular. The complete area of book-keeping, storing, calculation and investment planning with all detailed activities formerly done by people is now run by the computer. It is there that we find by far the most extensive use of computers, and by this use, a maximum of control is gained. Statistics for all purposes could be continuously and quickly prepared in any time to provide information for management decisions. The computer presupposes formalization but also produces formalization at a higher degree: in order to use computers efficiently enough they must be used extensively and everywhere, with the effects that formalization takes over larger fields of society. The fact that the computerization was already prepared by the formal structures of many social activities and by the need to maintain or improve these "social algorithms", supported and facilitated this process. In addition, decisions based on computerized information and calculations have a kind of logical stringency inherited from the discipline which brought it alive, i.e., from the rigor and logical constraints of mathematics.

The organisation of larger groups of people in huge social institutions and industrial companies needs algorithmic principles for collective communication and cooperation. In addition to the management process another obvious and important example is the formalization and later computerization (automatization) of the production process as the foundation of industrial society, especially in the case of assembly-line production, which is characterized by an extreme segmentation and standardization. Mass production would have been impossible without algorithmic principles (the same input produces the same output). Every worker has a well identified and limited sphere of activity and clear instructions about what to do. All abilities and individual capacities of a worker which are not necessary are excluded from the process, i.e., for the process the worker's identity is only his/her qualification to do the work. The human individual is thus reduced to the "human factor", and leaves out an "uninteresting rest of individuality". In many cases, these particularized segments of the production process as well as activities in the administration can be fulfilled by computers or computer directed machines.

Before returning to our general question we want to comment on the impact of the computer on one of its own main sources, mathematics. Some pure mathematicians use the computer as a comfortable tool. There is, however, no necessity to do so. Until now the computer did not revolutionize pure mathematics — e.g. the discussion about the computer-aided proof of the four-colour problem already calmed down. If the main goal of pure mathematics is to get insight, then tools for gaining comprehension like a computer have to be able to serve three purposes: avoiding boring calculations, saving time (first of all, if the human life span is not long enough to carry out the procedure), and producing illustrative materials like graphics and numerical data. But as simplicity and elegance still deserve the highest value, there is a tendency to eliminate the laborious elements of mathematical reasoning as far as possible. Reasoning without the use of the computer is by far more understandable and more elegant, therefore the computer is often used as a kind of unofficial tool, i.e., for solving the problem in any way. But afterwards every trace of the heuristic work with the computer has to be removed from the final version of the solution by a deductive proof. In this way, the discovery of some remarkable pieces of pure mathematics heavily depend on the use of the computer, although mathematicians still believe that they use only their mind, and cling to the view that they are not dependent on complicated instruments. So the computer comes to belong to the heuristics of mathematics, for instance to support inductive reasoning and probing of certain cases to get insight.

The microelectronic revolution caused an immense demand for specialists who could implement their theoretical know-how on the computer. This had, of course, a big influence on the economical support which is granted to research and university institutions. So called "Applied fields of mathematics" (numerical analysis, statistics, and appropriate parts of other disciplines) received more resources than the so far dominating pure mathematics. Applied fields of mathematics grew rapidly. Starting as a quantitative change, the computerization of mathematics turned into a qualitative change, that of the scientific paradigms of mathematics.

By the increasing use of mathematical model building for mathematical problems on the computer, too, we approach a new view of mathematics as an "empirical science". This is supported by the growing role played by quantification and discretization processes for all kind of problems as a prerequisite of the use of the computer for mathematical models. It is not enough to dispose of a mathematical model; between the mathematical model and the final computer output there are two more stages, namely the quantification and discretization of the model (algorithm) and the computer program. The quantification and discretisation of the model are partly necessary for modelling without the computer too, simply to find numerical methods. But in this case, the numerical analysis does not offer such a big variety of trials, of possible models and results as when playing with computer technology. While the traditional restrictions of numerical methods were diminished by improvements of the mathematical modelling, the importance of mathematical modelling steadily increased. Today, instead of focusing on the underlying theory, one just counts on the brutal computing force of the machine. This procedure contradicts the demands of traditional mathematics in searching for simplicity and theoretical progress. But for the user the solution may be the most important. Information technology started this development of empirical mathematics.

One of the most difficult problems of computer science is the verification of computer programs. The verification is often more complicated than the program itself. The reason to believe in computer outputs lies in the empirical fact that the program has run successfully on different machines very often, that many of the single steps of the algorithms and subroutines have been tested, and — last but not least — that the products of modelling have been considered to be correct. What a difference with the understanding of the verification of the classical models of mechanics, not to speak of the kind of proof which logical deduction offers, when mathematicians claim for experimental methods and the acceptance of unproved hypothesis!

Since the ancient Greeks, the postulation of a unified theory was considered an achievement of mathematics as a science, in contradistinction to the view of mathematics as just a heterogeneous accumulation of recipe-like solutions, familiar to the old oriental cultures of Egypt and Mesopotamia. The recent claim for mathematics to serve more and more for efficient practical solutions can be interpreted as a tendency back to an non-homogeneous development lacking rigorous uniform guidelines for structuring mathematics. It is not much exaggerated to say: What is accepted as a solution of a problem, is defined by the practical testing depending on many — often mutually contradicting — criteria; solutions have priority over integrating insight.

Undoubtedly, these new trends towards empiricism and heterogeneity cause an interesting dispute. Pure mathematics has to put up with the blame to be of scholastic character. The new practitioners of applied mathematics heap reproaches on traditional theorists, depicted as concerned only with hair-splitting concepts for obtaining knowledge and problem solving, as falling back on unimportant questions, in isolation from the real problems. This dispute has also consequences for the organisation of mathematical research institutions and their financial support. Mathematics, in this view, is essentially driven by problems raised outside of mathematics, the original impulse of most of the abstract theories coming from reality.

Computer aided, applied mathematics changed the social image and the manifestation of mathematics in society. The new paradigm of empiricism and heterogeneity towards which mathematics has been forced to move by social constraints of utility can be summarized as follows: 1) the focus clearly is problem orientation; 2) modelling (with preference given to discrete methods) deals with a much greater complexity (more variables, nonlinear problems); 3) experimental aspects are emphasized (algorithmic variations and concrete comparison of algorithms, heuristics, complexity theory); 4) interdisciplinary representations, problems of visualization, communication with non-mathematicians get more consideration.

The consequences of the very existence of the computer as a competitor of the mathematician finally has frightened mathematicians into lobbying for popular appreciation and this has been done by starting to address cultural issues. First of all, one has started renewals of movements towards the popularization of mathematics. Slogans like "Mathematics is fun, mathematics is nice, mathematics can be loved" have been incorporated into activities and become a message for the general public. But the message reached only a very few people; it has not solved the disjunction between the social acceptance and the cultural neglect of mathematics.

Popularization programs, in my view, are the wrong measures to overcome the aforesaid disjunction. We have to recognise, too, that the mathematisation of society, reinforced by the computer, constitutes a greater part of what is regarded as mathematics while it makes what is experienced as school mathematics look obsolete or at least obsolescent. What people have learned in school as mathematical skills and knowledge in greater part is no longer appropriate nor useful for tackling today's problems in work or social life. One can speak of an increasing demathematisation[3] of the individual, while an increasing mathematisation of society takes place. This is bewildering people, too, and reinforces a negative image of mathematics.

This also applies to an increasing number of study courses directed to future teachers, which still are mainly determined by mathematicians. These courses of study are used as means to ensure certain teaching capacities, and a number of chair for mathematicians. The amount of mathematics to be taught and learned by future teachers is determined without any regard to the possible use — in a general sense, not a utilitarian one only — they can make of it. And our teachers are left alone

---

[3] This term has been used first by Yves Chevallard in his talk at ICME 6 in Budapest (1988); see Burkhardt/Keitel/Malone (eds.): *Curriculum for the year 2000* (Curtain University Press, Perth, 1989).

with the problem of how to create cultural acceptance of mathematics, a problem about which they never get any help from mathematicians during their training. Meta-knowledge of mathematics, not only skills and procedures, self-reflection about the what and the how are not ingredients of study courses for teachers.

This question is rarely addressed by mathematicians. They still behave as if they were just there to invent something for which they bear no responsibility. Mathematics and culture, in my view, means first of all that mathematicians should address issues of culture but in quite another sense, namely by paying more attention to the pernicious uses of their craft, by reflecting on the social use of mathematics as outlined above.

Looking at mathematicians we see little attempt at critical self-reflection. Mathematicians are generally quite staggeringly uninterested in the human processes of doing mathematics as opposed to recounting the product. The dislocation between process and product seems to me deeper than in any other human activity. It seems to be a typical result of the socialization by and the increasing specialization of mathematicians and those people educated by them. I have often wondered how one might provoke mathematicians into self-reflection on the methods and reasons of their work and how to make these considerations a part of the normal discourse of the mathematical community. At least in the education process it should be an essential constituent.

It is astonishing that in contrast to the vast outpouring of mathematical results, very little research indeed has been done into processes which lead to results: How mathematicians communicate and collaborate, how mathematicians vary in their behaviour, personality and experiences, how mathematics is created and why a certain problem is chosen to be studied, what is the role of aesthetics and intuition in mathematics, what is the impact of tools like computers on the way mathematicians work and perceive their work. Of course, this is essentially a question of a multidisciplinary nature, one which can be attacked only by an interdisciplinary community which recognizes mathematics as a rich but puzzling cultural phenomenon and which believes that research can unravel at least some of the mystery. But it has to be done by mathematicians, too!

Mathematics in its Historical Context

by Lucia Grugnetti
University of Cagliari, Cagliari (Italy)

We must have more education for scientists in the humanities and in history. We cannot afford to be ignorant technicians. We must have less rigidity in our thinking. The solution lies in mixing science and technology with the rest of life in proper proportions (Davis, Hersch, 1986). This gives us the new demarcation line between traditional areas of culture, science and technology. This new border also goes inside mathematics. In fact, we have to deal with two different approaches to mathematics, and two kinds of social (and cultural) demands for it. When seen from the point of view of the super-technologies, as well as of the needs of information society, mathematics is the set of methods and models which are useful in two main areas, both connected with decision-making: data processing and problem solving. Let us now turn our attention to the second, pure part of mathematics. Its future role is to help elaborate the new picture of the world. Efforts to create this picture will go very deeply. They can involve all main categories of rational reasoning. The fundamental character of incoming changes opens new opportunities for mathematics, assuming it is able to be used appropriately to this task. And it means coming back to the old Platonic and Pythagorean visions of the world and the role of mathematics in it (Waszkiewics, Wojciechowska, 1989). Through the universal concept of mass education in a fast changing world, "mathematics for all" reaches an unprecedented dimension as a social endeavour. It makes it urgent to question, in a much deeper and broader way than ever before, its socio-cultural roots, and, indeed, the place of mathematics education in society as a whole. Being in such a privileged position in western thought, mathematics may be, at the same time, an essential instrument in building up modern societies and a strong disrupting factor in cultural dynamics (D'Ambrosio, 1984).

If we take into account the previous statement, what does it mean to cultivate mathematics at the secondary school level? P. Ernest (1989) has distinguished three philosophies concerning mathematics and mathematics education.

1) *The instrumentalist view*, in which mathematics is an accumulation of facts, rules and skills to be used in the pursuance of some external end. Thus mathematics is a set of unrelated, but utilitarian rules and facts, to be known as separate entities.

2) *The Platonist view*, in which mathematics is a static but unified body of a certain kind of knowledge, whose elements are discovered, not created. This view involves *a global understanding* of mathematics as a consistent, connected and objective structure.

3) *The problem-solving view*, in which mathematics is a dynamic, continually expanding field of human creation and invention. Mathematics is a process of inquiry and coming to know, not a finished product, for its results remain open to revision. This view involves a *dynamically organised structure* located in a social and cultural *context*.

The third view brings to the fore that mathematics becomes a cultural product when it is connected with the dynamism of scientific knowledge. This dynamism depends on two main aspects: firstly, the acquisition of ever new data, provided by more and more complex and refined techniques of observation; secondly, the investigation of the results previously obtained. My contribution to this round table will focus on the importance of investigation of the results previously obtained, as a component of the cultural aspect of mathematics, in particular at the lower secondary school level.

Richard Courant once wrote:

> Differential and Integral Calculus and Mathematical Analysis in general is one of the great achievements of the human mind. Its place between the natural and humanistic sciences should make it a singularly fruitful medium of higher education. Unfortunately, the mechanical way in which calculus sometimes is taught fails to present the subject as the outcome of a dramatic intellectual struggle which has lasted for twenty-five hundred years or more, which is deeply rooted in many phases of human endeavor and which will continue as long as man strives to understand himself as well as nature. Teachers, students, and scholars who really want to comprehend the forces and appearances of science must have some understanding of the present aspect of knowledge as a result of historical evolution.

The teaching of calculus in the new programs for Italian higher secondary schools is presented thus:

> It will be possible to follow the historical itinerary which, starting from Archimedes, by the contributions of Cavalieri, Torricelli and other precursors of calculus, leads to Newton and Leibniz.

On the contrary, in the textbooks for higher secondary schools, the presentation of calculus does not follow the historical development. In fact it goes from the concept of function to the concept of limit (one of the most complex mathematical concepts), and then from the concept of derivative to the concept of integral. If *culture* is not only an accumulation of items of knowledge, but a set of critical attitudes about knowledge, then the question is not about the quantity of transmitted knowledge, but about its quality. Why did this concept arise? Under which historical conditions? The traditional idea that the development of mathematics is purely cumulative is largely out-of-date. Teachers must remain aware of the inherent relativity of knowledge, and of the fact that, in the long run, providing students with an adequate view of how science builds up knowledge is more valuable than the mere acquisition of facts (von Glasersfeld, 1991).

The discovery of calculus is usually attributed to Newton and Leibniz, and associated with the period 1665-75. There is no reason to dispute this attribution, but it would be wrong to imagine that these two men invented calculus more or less "out of the blue". In fact, "infinitesimal methods" were in the air, and virtually every mathematician of note in the seventeenth century contributed to this development. Concerning the period, it makes even less sense to argue about who discovered what. The question is rather, why did such a tremendous advance occur in the seventeenth century, after such a long period of stagnation? It was undoubtedly part of the new spirit in art and science associated with the Renaissance. Strong stimuli for the development of theoretical mechanics, and thus for the elaboration of computational methods and the application of infinitesimal methods to the study of motion and change, came from sources such as the increasing use of machines in early forms of industry, or the "new astronomy" associated with Copernicus, Tycho Brahe and Kepler. All this raised hopes that the science of mechanics might account for celestial as well as earthly phenomena, and so on.

So, to answer the previous question, "why did this concept arise?", we and our students must leave the narrow boundaries in which school mathematics is usually confined. If we think that history is an essential means of making mathematics meaningful as a subject matter, instead of an accumulation of unconnected and inconsistent formal procedures (Jahnke, 1990), we can indroduce calculus by starting to show (as in a noteworthy Italian textbook by L. Mancini Proia and L. Lombardo Radice) the way in which Archimedes calculated the area between a parabola and one of its chords. Then we can illustrate the methods of Indivisibles used by Luca Valerio to calculate the volume of a sphere, by Kepler and Torricelli (in different ways) to calculate the area of the circle, by Cavalieri to find the area of the ellipse. The problems of velocity and of the tangent in the way they were treated by Galilei, Newton and Leibniz introduce modern differential calculus. In doing so, we would give meaning to a previous statement: "Mathematics is a process of enquiry and coming to know, not a finished product, for its results remain open to revision". Indeed, mathematics is a growing, living fabric; a changing experience. Each attempt to view mathematics as existing outside of time and human society strips away a layer of meaning and exposes a dessicated kernel. Detemporalization leads to a naive faith that formal manipulation may be productively and authoritatively invoked in any situation. Detemporalized mathematics cannot tell us what mathematics is, why mathematics is true, why it is beautiful, nor how it comes to exist. On the contrary, if one places mathematics squarely within human time and experience, it becomes a warm and rich source of possible meanings and actions (Davis, Hersh, 1986).

The historical approach offers the opportunity to present mathematics as a continuous process of improvement, instead of a collection of truths that stays unchanged in the course of time. Following this line of thought, with its insecurities and also its progress, we make the concepts more significant. The point of the question at this stage is, how to develop methodologically an itinerary of the historical type? In this sense the history of mathematics implies a passage from disciplinary to interdisciplinary teaching, in the wi(l)dest sense of that term. Mathematics has been influenced by agriculture, commerce and industry, by war, philosophy, physics, astronomy and every type of human activity and, conversely, the study of the transmission of mathematical knowledge and astronomy is, as O. Neugebauer once remarked, one of the most efficient of instruments to enhance the existence of connections between different civilisations.

So, why do we hide from our students this truly humanizing and deeply cultural aspect of mathematics? I think that, developed in this way, mathematics can finally lose all the aridity that, according to our students, should be regarded as the essential reason for their failure. At the methodological level, almost everything remains to be constructed. But the history of mathematics as an introduction to a critical and cultural study of mathematics could constitute one of the most important challenges for mathematics educators and didacticians.

References

[1] Boyer, C. B., *The History of the Calculus and its Conceptual Development*, Dover Publications, New York, 1959.

[2] Courant, R., *Differential and Integral Calculus*, Translated by E. J. McShane, Vol. II, 2d ed., London and Glasgow, 1937.

[3] D'Ambrosio, U., "Socio-Cultural Bases for Mathematical Education". In: *Proceedings of the Fifth International Congress on Mathematics Education*, Majorie Carss (ed.), Adelaide, Australia, 1984, Birkhäuser, Boston, 1986, pp 1–6.

[4] Davis, P. J. & Hersh, R., *Descartes' Dream: The World According to Mathematics*, Harcourt Brace Jovanovich, San Diego, 1986.

[5] Ernest, P. "The Impact of Belief on the Teaching of Mathematics". In: *Mathematics, Education and Society*, Christine Keitel et al. (eds.), UNESCO, 1989, pp 99–100.

[6] Glasersfeld, E. von, *A Constructivist's View of Learning and Teaching*, International Conference on the Psychology of Learning and Mathematics Learning, 1991, preprint in English; also published in *L'Educazione Matematica*, April 1992, 27–39).

[7] Jahnke, H. N., *History of Mathematics and Mathematics Education: A General View of the Problem*, preprint.

[8] Menghini, M., "Some Remarks on the Didactic Use of the History of Mathematics". In: *Proceedings of the First Italian-German Bilateral Symposium on the Didactics of Mathematics*, Luciana Bazzini and Hans-Georg Steiner (eds.), Pavia, October 4–9, 1988, 1989, pp 51–58.

[9] Sondheimer, E. & Rogerson, A., *Numbers and Infinity (An Historical Account of Mathematical Concepts)*, Cambridge University Press, Cambridge, 1981.

[10] Waszkiewcz, J. & Wojciehowska, A., "The Future Cultural Role of Mathematics and its Impact on Maths Education". In: *Education and Society*, Christine Keitel and others (eds.), UNESCO, Paris, 1989, pp 187–189.

Ecology of Mathematical Knowledge: An Alternative Vision of the Popularization of Mathematics

by Juan D. Godino
University of Granada, Granda (Spain)

The title of this round table, *Let's Cultivate Mathematics!*, leads us in a natural way to the use of an implicit metaphor: mathematics can be considered to be like a plant that grows and develops in certain habitats and which needs care to maintain its vitality. We believe that the analysis of the problem of the use of mathematics in different institutions and their relation to other sciences can be facilitated by viewing such a problem in the light of the ecological paradigm, i.e., "the scientific discipline that is interested in the relations between the organisms and their past, present and future environments. These relations include the individual's physiological responses, the structure and dynamic of the populations, the interactions between species, the organization of the biological communities and the processing of energy and matter in the ecosystems " (*Ecology*, Journal of the American Ecological Society).

Since Lakoff and Johnson's work (1980) on the relevant role of metaphorical concepts in structuring the human conceptual system, the use of metaphors has been justified as a means to understand and experiment one reality in terms of another. We think that the ecological metaphor proposed by Chevallard (1989) in didactic analysis constitutes an important tool in order to understand the genesis, the development and the functions of mathematical knowledge in human institutions. The analysis of the institutional ecology of a piece of knowledge leads us to study its habitats, i.e., the places where we can find the objects with which it is associated, the supporting structures and the functions of these interrelations, that is, the ecological niches of the different aspects of mathematical knowledge.

In this presentation we are interested in the problem of the use of mathematics, its characteristics and conditions of development in present-day culture and society, making use of the ecological metaphor and, to be more precise, of the concept of econiche. A modern approach to this concept, based on the general theory of systems (Patten and Auble, 1980) enables us to apply it to inanimate objects, by replacing the criteria of viability, persistence or indefinite existence, by the notions of utility, availability or fitting.

Ecology of Mathematical Knowledge

The application of the ecological metaphor to the study of the evolution of knowledge implies that we consider it as a system of "organisms" or "objects" that interact and play a role within the institutions where its cultural existence is recognised and which have become its habitat. It seems clear that it is not possible to think of knowledge regardless of the people who think and use it. However the identification of the existence of a piece of knowledge needs collective recognition. Institutions are thus the habitat of knowledge (Berger and Luckmann, 1968). One of the possibilities that the ecological paradigm offers consists in its capacity to give meaning to new questions that otherwise would seem obvious or uninteresting. Likewise, it leads to focus our attention on contextual aspects and interactions that frequently go by unnoticed. As an example of this we indicate some of these questions.

(a) Which institutional habitats does mathematical knowledge occupy at present? What are the different uses of mathematics in these habitats?

(b) Are there any institutions in which mathematics could be used more intensely and adequately?

(c) What types of environmental constraints (limiting factors) make it difficult for mathematics to occupy empty ecological niches?

(d) How are mathematics related to the mathematical knowledge existing in different institutions?

(e) Is it possible to identify subspecies (subknowledge) as a result of the phenomenon of adapting to the environment?

(f) Are there any special relations of competition, symbiosis and dominance and control between knowledge and subknowledge that condition the diffusion of mathematics?

(g) In general, in society, mathematics is not sufficiently appreciated and so leads a precarious existence: What are the factors that determine this "mathophobia"?

We will try to offer some partial answers to these questions. It is interesting to identify, in the first place, three types of mathematical "subspecies", as a result of adaptation to different institutions, i.e., pure, applied and school mathematics. The coexistence between these sub-species is not free of problems which make an optimal diffusion of mathematics difficult. The didactic transposition (Chevallard 1985) can be regarded as the phenomenon of adapting pure and applied mathematics to the school environment. Also between the latter two we would have to differentiate phenomena of mutual adaptations that we could call "modelling transpositions". The subspecies sometimes coexist in the same institutions. At the university level, for example, the normal thing is for teachers to be researchers involved, at the same time, in the production of new knowledge. A problematic aspect is the phenomenon of "dominance and control" which the PM institution (pure mathematics) exercises on AM (applied mathematics) and EM (educational mathematics), which has negative consequences. Applied mathematics is frequently seen by "pure mathematicians" as something of an inferior category. The doses of creativity required in the process of modelling real-world problems and in the necessary educational contextualization is frequently ignored and underestimated.

Within the different habitats of mathematics, the first in order of importance with respect to extension is Mathematics Education, understood as a system that includes not only the teaching and learning of mathematics but also curricular design, the elaboration of didactic material, and didactic research. At the different educational and professional specializations levels, mathematics is found to be omnipresent in education, although the phenomena of dominance and control of theory over applications, to which we have referred, is reproduced. But, regardless of the tool-object dialectic that is characteristic of mathematics (Douady, 1986), mathematics education frequently emphasises the conceptual aspects (objects) as opposed to the applications.

In teaching as well as in research or the applications, mathematics coexist and interact with other knowledge, giving rise to new fields or "species", as for example econometry, psychometry, etc. (Benzecri, 1982). New problems of competition of the sciences over mathematics, i.e., usually professional mathematics, arise. Another conflicting aspect is the difficulty of communication between the mathematician and the specialist in other sciences, due to the use of scientific languages that are different in each specialty. The users of mathematics are those who pose the problems, but it is the mathematician who has the tools to solve them. In the process of posing the problem to the mathematician by the user and of communication of the solutions achieved by the mathematician, a double process of didactic transposition appears from one discipline to another. In this process there can be disagreements that perturb the adequate use of mathematical tools.

It is expected on other occasions that the response to a mathematical problem is immediate, without any sort of creative reflection (Barnett 1988). In school practice, each problem has a solution, which is frequently unique, and in every case the teacher knows this solution. Society does not appreciate the professional mathematician because it is understood that mathematics teaching, from school to university, should make citizens and different professionals capable of solving their own mathematical problems. This is unreal and prevents an adequate cultivation of mathematics. There are usually different mathematical techniques adapted to a given problem. Moreover, each one of them is based on a series of hypotheses of a theoretical nature which are really satisfied. The professional mathematician should assess, among the available methods, the degree of fitness between hypothesis and data. Mathematical modelling is frequently highly complex and requires sophisticated technical skills as well as a level of creativity. This can only be achieved by people with a certain level of specialization and professional dedication.

On the other hand, the application of many routine mathematical techniques is today a laborious task, due to the great quantity of information that it is necessary to process in these applications. The computer is necessary. This poses the problem of the cooperation with yet another science, i.e., computer science. The mutual influence between both disciplines is full of potential but also of difficulties.

Problems of the Popularization of Mathematics

Recently the term popularization has been coined to refer to a set of activities whose specific objective is to try to fill the gap between science and the public's understanding of the same (Howson and Kahane, 1990). It deals with trying to share mathematics with the widest possible public, to encourage people to be "mathematically active", to induce a recognition and a favourable attitude towards mathematical activity. The aforementioned authors distinguish the concept of popularization from that of mathematics teaching by its specific characteristics:

- — it deals with freely made decisions that are not subject to the compulsory character of teaching: they do not suppose hard work, but freedom and pleasure;
- — it is proposed for a wider audience; it should affect all groups of people, from the researchers themselves to retired people;
- — it tries to use all modalities of communication;
- — it affects all mathematical topics.

The necessity of carrying out activities of popularization arises from the present reality of rejection, reluctance and the bad image that mathematics has among the public in general. "The popular image of mathematics is that it is difficult, cold, ultra-rational, important and strongly masculine" (Ernest, 1992). This situation is negative, from a personal as well as from a collective point of view. In every country there is a need to increase the number of scientific professionals and a bad image of mathematics implies a handicap in this respect.

The activities of popularization are interpretable in the perspective of the ecology of knowledge, as efforts to use mathematics in a wide variety of institutions, and to a certain extent in a way adapted to them. So, it has to do with the promotion of the use of mathematics in all the potential ecological niches, by using the necessary selection and adaptations. To each group should be proposed some mathematical activities adequate to their interests and possibilities. This implies recognition of a greater variety of types of relations to mathematical knowledge, constructed by the subjects themselves by using a very wide range of mathematical activities. The popularization thus arises as an emblem of a new epistemological formation, of a new "knowledge" that competes to a certain extent for the same space as the didactics of mathematics.

In general terms the objectives of popularization seem valuable and necessary. But the ecological perspective leads us, before proposing indiscriminated actions, to reflect on the "biotic" and "abiotic" factors that determine mathophobia. The heart of the problem that the popularization is trying to remedy, i.e., the social rejection of mathematics, cannot be solved with the type of activities that are currently proposed. To understand mathematics is something so complex, that includes so many nuances and levels (Sierpinska, 1989), that it seems ingenuous to try to achieve it merely with the activities of popularization. The analysis of the conditions on which a favourable attitude towards mathematics depends, is the object of study of one scientific discipline: the didactics of mathematics. The difficulty to fulfill these objectives is obvious if we observe the embryonic state in which this discipline is found in a majority of countries.

It seems clear that the aforesaid failure should be attributed to the particular type of existence that the "educational noosphere" (Chevallard, 1985) has given to date to school mathematics, through the processes of didactic transposition. However, the actions that are proposed as a counterpoint to achieve the objective of popularization do not seem clear. To learn to see the mathematics crystallized in products, to illustrate textbooks, to propose passtimes and crosswords in newspapers, is this enough to make mathematics activity accessible and pleasant for the many?

In our opinion, the actions of popularization do not always have the adequate orientation. It is not always possible to make mathematics pleasant since they do not necessarily have to be so. It is not a pleasant task to transpose a matrix or develop a function in power series. We must create situations in which the use of these objects is the most reasonable. It is not necessary to make a hammer popular for someone to use it to put in a nail. Nobody likes to have a vaccination; however, responsible mothers periodically have their children vaccinated because they know that medical science has demonstrated the efficiency of these vaccinations in the face of illnesses with serious consequences. In the case of medicine, the popularization of the same has not consisted in making it pleasant, but in putting it within the reach of everyone through health centres and professional doctors. For mathematics, the key problem lies in the fact that the pupil (the citizen) receives complex, incomprehensible and, accordingly, useless knowledge at school.

Some Proposals for Action

The teaching institutions (schools, universities, etc.) can be seen in this perspective as special habitats of knowledge, as "nurseries" where these "organisms" grow, since the users of mathematics in the different fields of society receive their education in these centres. Mathematics should be contextualized, adapted to the conditions of particular habitats. The contrary would generate risks of a general rejection.

The development of a common language which makes mutual understanding and communication possible is necessary for the development of a "symbiotic coexistence". This requires several types of actions among which we point out:

— A formation of all the citizens and professionals relating mathematics to the problems of their own environment and interests, and enabling them to distinguish the situations in which the collaboration of the expert mathematician is necessary. Mathematical instruction should provide each citizen with the capacity to identify mathematical situations and to discern when the collaboration of a mathematical professional proves necessary.
— The creation of mathematics consultancies in the universities, in symbiosis with the computer consultancies, and perhaps in the secondary school institutions, that permits the creation of habits that facilitate the integration and the cooperative use of different fields of knowledge.
— The setting up of interdisciplinary teams in research and development units is likewise a key action for mathematics to be used intensely and adequately.

Finally we consider that it is fundamental to support the development of didactic studies, since these analyze and identify the conditions of the "support structures" (Alley, 1985) of these "organisms" in the different institutions where they can survive. Didacticians, the group of people who critically and systematically reflect on the production and communication of knowledge, play the role of "fertilizers" for the knowledge to fully develop its potential.

References

[1] Alley, Th. R., "Organism-environment mutuality epistemics, and the concept of an ecological niche", *Synthesis* **65** (1985), 411–444.
[2] Alsina, C. *et al.*, Hacia unas matemáticas populares, *Suma* **4** (1989), 83–120.
[3] Barnet, V., "Statistical consultancy. A basis for teaching and research". In: *Proceedings of the Second International Conference on the Teaching of Statistics*, R. Davidson and J. Swift (eds.), University of Victoria, 1988, pp 303–307.
[4] Benzécry, J.-P., *Histoire et préhistoire de l'analyse des données*, Dunod, Paris, 1982.
[5] Berger, P. and Luckmann, T., *La construcción social de la realidad*, Amorrortu, Buenos Aires, 1968.
[6] Chevallard, Y., *La transposition didactique*, La pensée sauvage, Grenoble, 1985.
[7] Chevallard, Y., "Le concept de rapport au savoir. Rapport personnel, rapport institutionnel, rapport officiel", Séminaire de Didactique des Mathématiques et de l'Informatique, Université Joseph Fourier-Grenoble I, 1989.
[8] Douady, R., Jeux de cadres et dialectique outil-objet, *Recherches en Didactiques des Mathématiques* **7**: 2 (1986), 5–31.
[9] Ernest, P., "The Popular Image of Mathematics", *Philosophy of Mathematics Education Newsletter*, 4 & 5, 1992, p.19.

[10] Howson, A. G. and Kahane, J. P., *The Popularization of Mathematics: A study overview*, ICMI Study Series, Cambridge University Press, 1990.

[11] Lakoff, G. and Johnson, M., *Metaphors We Live By*, University of Chicago, Chicago, 1980.

[12] Patten, B.C. and Auble, G. T., "Systems approach to the concept of niche", *Synthesis* **43** (1980), 155–181.

[13] Sierpinska, A., "Some remarks on understanding in mathematics", *For the Learning of Mathematics* **10**:3 (1990), 24–36.

Round Table E: July 6, 1992

MATHEMATICAL EUROPE, MYTH OR HISTORICAL REALITY?

coordinated by Catherine Goldstein[1] and James Ritter[2]

CNRS, Université Paris-Sud, Orsay (France)[1], Université Paris VIII, Saint Denis (France)[2]

The founding of a European Mathematical Society and the holding of its first Congress would seem an ideal time to reflect on what has been and on what might yet be a mathematical Europe. It was for this reason that, practically from the very beginning of the project, a Round Table was constituted on the History of Mathematics.

Yet such a review partakes of something very like a paradox right from the outset. For what sense does it make to talk of European mathematics at a time when mathematics makes, and largely justifies, its claims to universality? The close network of researchers and teachers, seminars and colloquia, publications and publishing houses seems to belie any attempt to fit it into a restricted geographic mould such as "Europe".

The constitution of mathematical Europe is not, in general, attributed to political contingencies but rather to an argument cultural identity: Mathematics as a discipline, many would say, was born in Europe, and developed there into its present state. That is, the privileged link between mathematics and Europe is to be found precisely in the history of the former.

Indeed the standard version of the genealogy of the field, which is held by many working mathematicians, runs something like this: Mathematics was born in classical Greece, the purest product of the Greek invention of reason. It lay submerged and forgotten for over a thousand years, until the Western European Renaissance of the sixteenth and seventeenth centuries laid the foundations for the rediscovery and then the transcendence of the Greek legacy. Finally the spectacular development of mathematics and its applications, particularly in the nineteenth century, created an irresistible momentum, carrying mathematics beyond the limits of Europe into the world as a whole.

This version of the history of mathematics, often taught as an adjunct to school mathematical programmes and reinforced by the popularity of such hagiographic writings as E. T. Bell's *Men of Mathematics*, or even the more scholarly book by Morris Kline (*Mathematical Thought from Ancient to Modern Times*), is, like most foundational histories, largely mythical. History, like mathematics, advances in its methodology and its results and contemporary historians of mathematics have for some time now been looking quite closely at these questions. As we shall see, the picture that emerges is in striking contrast to the established myth.

What is the interest of such a revision? There is, of course, the simple regard for truth — or at least for the best truth that we can know at a given time — and this aim ought to be especially dear to mathematicians whose professional practice is so intimately linked to questions of truth. But more practically, there is the fact that much of the interactions between the mathematical community and others is mediated by various perceptions of the history of mathematics. We are seen, as it were, by society as a whole through the screen of their main access to our practice: school mathematics. Made up of our past mathematics, its understanding is intimately linked to the understanding of our past as a whole. Moreover, the history of mathematics is often called to the rescue when presenting mathematics to a public composed of non-mathematicians (and, especially, young mathematicians-to-be). It appears to be a privileged means of importing flesh and blood into an otherwise rather abstract domain. But the manner of presentation, inevitably developed with the level of knowledge of a certain historical period, is often now inadequate for present-day use and serves merely to propagate an out-of-date image of current mathematical practice.

Even the undoubted universality of mathematics takes on a deeper, and more complex, sense when viewed carefully in an historical perspective. All civilizations possessing writing have developed a mathematical activity, but their aims, objects, practices and results have varied profoundly. This variety not only separates one civilization from another but also exists *within* Ancient Greece and European countries. The priorities and ranking of achievements have been selected through general and not merely mathematical processes of transmission and abandon.

Our first aim in what follows is to show concretely how some standard myths about European mathematics were created, involving the truncating and distorting of both the contours of Europe and of mathematical activity. In other words, the very definition of mathematics, like that of Europe, has undergone important shifts throughout its history. Their boundaries, what counts as important or central and, as such, is worth transmitting, reworking and recording in historical accounts, has depended on complex and multileveled priorities of the various actors involved. We shall concentrate on the following specific aspects, each being illustrated by concrete examples: why and how was Greek mathematics perceived, since the sixteenth century, as the (European) source of our (European) mathematics? Are founding figures like Descartes objective precursors of modern points of view on mathematics? Was the mathematics developed inside Europe the only possible one, and had it nothing to learn from other traditions? Are pure mathematics more universal than applied, less affected by national or regional priorities? Does the diffusion of mathematical knowledge as well as its universality merely translate internal criteria of excellence?

Each of these themes will reveal rich and interesting realities and our second objective is to present briefly the elements for a survey, taking into account recent advances in the history of mathematics in Europe. We shall mention the importance of mathematical research programmes and how working mathematicians have implicitly proposed a view of their own history. We shall see how the field of mathematics has been tightly and diversely related to society as a whole and specifically to such areas as economics, law, religion and education. We shall also examine the links between the development of mathematics and the institutions which condition it, as well as the question of the relative status of pure and applied mathematics.

A large part of the information included in this report comes from the Satellite-Colloquium: "Mythes et Réalités historiques de l'Europe mathématique / Myths and Historical Realities of Mathematical Europe", held in Paris in April 1992 and soon to be published. References to the communications presented at this Colloquium will be indicated in what follows by " (*Proceedings*)".

This report was prepared with the help of the chairperson of the Round Table, Herbert Mehrtens (Berlin, Germany) and the participants: Henk Bos (University of Utrecht, Netherlands), Karine Chemla (CNRS, France), Giovanna Cifoletti (Milan, Italy), Hélène Gispert (University of Paris-Sud, France), Eduardo Ortiz (Imperial College, UK). Details will be found at the end of this article.

The word "Europe" itself is problematic; the Greek term originally signified simply the Ægean portion of the continent without any particular cultural value attached to its use. The first mention of the term in reference to a political unit was by a chronicler of the Battle of Poitiers in 732, for those forces in the field opposed to the Muslims.

Another connotation of the term "Europe" concerns the idea of political unity; realized during the brief period of Carolingian domination, it has remained a dream ever since. But the nature of that dream changed, from a region united under single rule to the notion of the balance of centralized powers, such as that between England and France in the nineteenth century. Though these two nations began to acquire a certain importance in the seventeenth century, it was two centuries later that, expanding to new territories all over the world, conflicts of dominance were generated which posed problems for those countries excluded from the main path of this development.

The history of mathematics is closely tied to all these events; a century after the Reformation, when the concept of Christian Europe was falling apart, the Church of Rome no longer able to impose a central unity, mathematics sometimes served as an important constituent in other — rational — models of unity for European thought, science and philosophy. It is this development that has been designated as the First Scientific Revolution.

The establishment of new states and their new needs in the areas of administration, economy and teaching, determined new forms of institutionalization for mathematics, implying changes in the latter's practices and resources. After the

French Revolution, the creation of a bourgeois state and of civil liberties, the Industrial Revolution ushered in what is often called the Second Scientific Revolution; mathematics and science became professions in more or less the modern sense, that is, in relation with teaching and research.

The history of the world order is also partially reflected in the history of mathematics (and vice-versa), a history of centers and peripheries, or, more exactly, of peripheries of different types, their roles changing as respective positions of leadership changed in one domain or another.

Here, for obvious reasons of space, we will privilege two moments in time, each important for both the myths and realities of our contemporary mathematics: the sixteenth and seventeenth centuries, considered decisive for the intellectual origins of the discipline, and the nineteenth century, crucial for the institutional foundations of our field, when mathematics became a full profession in universities and research organizations. Of course, important as these two moments are for us, they represent only a minute fraction of the mathematical activity of the last five millennia.

The first European mathematicians long predated Descartes' generation. Several universities, such as that at Bologne, were founded before 1200; Peterhouse, the oldest Cambridge college was created in 1284, while the University of Prague dates from 1348. Scholars at Oxford and Paris were using mathematics — and graphs — to study the laws of motion in the thirteenth and fourteenth centuries.

The economic changes of the fourteenth century, particularly the increasing complexity of commercial links among the burgeoning cities of the period, favored the development of specialized calculation techniques and accounting skills. Abacists, professional mathematicians who lived by offering consulting and training services to merchants on questions involving banking and rates of interest and exchange, used and perfected algebraic techniques as tools, marks of their professional competence. Their methods circulated, first in Italy, then the Rhineland and throughout Europe. A century later, this tradition provided the first printed mathematical "best-sellers". The abacists drew their knowledge from many sources: Indian, Greek, Byzantine, Arab.... The precise details are not yet clear, but what is certain is that the Islamic channel was essential, operating either directly through travel and contacts with Arab mathematicians or through the intermediary of Arabic works, original and translated. Thus, for example, decimal notation, once adopted in Western Europe, spectacularly increased mathematicians' capacity to handle complex calculations and approximations. It is in this same tradition, mixing commercial practice and theoretical algebra, that one must locate, for example, important fifteenth century mathematicians such as Luca Pacioli and Nicolas Chuquet.

With the Renaissance we see a shift and a new selective sorting of material coming from the earlier period. Of course, there was now the diffusion of works coming directly from the Greek, but the crucial questions are why such a shift was then judged useful, by whom, and in what context — after all, the first Humanists were not at all interested in mathematical works. As to how this change was propagated in the subject matter of mathematics itself, there is a hint, for example, in Rafael Bombelli's *Algebra*, published in 1572, where a part of the traditional practical, commercial preoccupations is set aside while an attempt is made to integrate problems stemming directly from Greek sources.

The story of the development of algebra is particularly revealing for our purposes. Now taught at an elementary level, its "formulæ" remain for many the distinctive mark of mathematics. But at the beginning of the seventeenth century algebraic calculation represented an astonishingly powerful technique, strong enough to generate for certain mathematicians of the period hopes that were analogous to those raised fifty years later by the infinitesimal calculus, or closer to our own times, by approaches using categories or schemes. Its history underwent a major change at the end of the sixteenth century; a new community, interested in mathematics for reasons quite different from the abacists, undertook to rewrite the chronology of this science.

The Creation of (the History of) Algebra

based on the presentation of Giovanna Cifoletti

University of Milan, Milan (Italy)

One of the first myths about Europe and mathematics tells the story of the birth of algebra. According to this story, symbolic algebra was created by the Greeks: the discoveries on the one hand of the text of Diophantus (*ca.* third century C.E.), providing an early form of algebra, and on the other of those of the classical geometers Apollonius (*ca.* 220 B.C.E.) and Pappus (*ca.* 320 C.E.), providing the "Greek method of discovery in mathematics", would have come at a time when, after nearly 2000 years of ignorance, early modern European mathematicians had at last regained the intellectual heights last reached by the Greeks and were thus able to make use of it. Furthermore, they would have rapidly improved it and, in the hands of those giants, Viète and Descartes, perfected it. Viète, for instance, went so far as to criticize the Euclidean synthetic presentation, claiming at the same time that he had rediscovered and pushed to perfection the real Greek method, obscured by Euclid: this was the Art of Equations, the analytic method (as opposed to synthetic) whose efficient tool was algebra. He even stressed the incompetence of the Arabs in the so-called transmission of the ancient wisdom: they have, he said, introduced a lot of improper terminology, so that the Art has become "vile" and the great potential of the art had yet to manifest itself since "they had used it to find so little".

Algebra was new in this story, born in a rising European culture, flourishing in the midst of religious wars. Sixteenth century Europe, and France in particular, had been celebrated as the authentic heir of classical Greece, in the form of a new civilization in which the reference to the Greek legacy stressed more a prestigious lineage than a direct intellectual debt. A larger French and European identity was in the process of being forged, in contrast with the more recent medieval past, while the prestige of the new discipline was guaranteed by a differentiation with respect to more recent traditions in mathematics.

One of the most interesting things about this story is that it first starts being put in place by those very early modern mathematicians who were involved in it: principally Descartes and Viète but also, before them, Ramus and others. To these have been added recently some lesser-known figures (G. Cifoletti, "Mathematics and Rhetoric", Thesis, Princeton University, 1992): the sixteenth century French algebraists connected with the Collège Royal who were trying to elevate algebra from the role of practical art, as it was taught in the abacus schools, to a noble theory: Jacques Peletier, Jean Borrel and Guillaume Gosselin. They well knew, as we well know now, that there was a double source for much of the algebra that was developing at this period: the work of many Arab mathematicians, such al-Khwarizmi and Omar Khayyam, and the medieval school of abacists of which we have spoken above. Yet both these groups of actors are missing from the story.

How was this myth created? The origins of algebra had in fact been discussed for about a century before Viète. The fifteenth century scholar, Regiomontanus, had announced the existence of Diophantus' text in 1463 and had claimed that it contained algebra. Regiomontanus had divided the mathematical sciences into theoretical and practical, and while he often attributed the practical to the Arabs, the theoretical are attributed exclusively to Greek authors. Regiomontanus' position, mentioned over and over by sixteenth century French algebrists, illustrates a very important aspect of the myth, according to which Diophantus would have provided algebra with its theoretical underpinnings, for the Greeks had founded all theoretical disciplines, while the Arabs would have amalgamated theory and practice in their arts, developing the mixed sciences. Here we see one of the main purposes of sixteenth century authors; to differentiate their enterprise from the Arabic tradition, crucial in the Middle Ages, both for showing the theoretical status of their work and for denying a heritage or a dependence which for political and cultural reasons had become an anathema. Since the Arabs themselves had attributed authority to the Greeks, it was easy to justify this oblivion within a traditional classification of sciences, which divided them into the theoretical and the merely practical. This accounts for the surprising fact that algebra's main founders, such as al-Kwarizmi and Omar Khayyam, could be dissolved in the dozens of transcribers and commentators of the Greek heritage; they had "no mathematical originality".

As to their other main accusation leveled against the Arabs, the fact that they had introduced a vulgar language, it is a topic that brings us back in the middle of early discussions on the new scientific culture in sixteenth century France. We shall focus on Peletier's case for whom the project of an algebra in French is combined with a radical criticism of the rhetorical and scientific skills of the Arabs. He wrote:

> "Our mathematics were never better than they are at present, nor in a better position to be heard in their perfection. And because their truth is manifest, infallible and constant, think what immortality they might bring to a tongue were they, the good and true methods, to be written up in it. Now look at the Arabs; though they are far from us, almost as if in another world, nonetheless there be those in our Europe who have wished to learn their language, principally for astrology and other secret things which they have dealt with in their common tongue but oh how badly. For one knows what sophistry they have mixed into medicine and even mathematics. And nonetheless they have made their language a necessary acquisition to that end. Let us then consult together in how we might act so that we may do, not so much, but incomparably more with ours."

This project of writing the sciences in the national language and in a national language codified by a new classical style was typical of many sciences and many national languages of the period. Peletier founded his project on an Erasmian theory of the imitation of Cicero: Cicero had translated from Greek into Latin, hence we should not write in Latin, but translate from classical languages into our national language. It is thus only fitting that Peletier used this theory for a discipline which had never existed in Latin and took as an example to imitate, not a classical authority, but the Arabs.

But Peletier was able to recognize that the Arabs, at least as a people, had contributed to the invention of algebra. In this attribution he followed Herodotus' criterion for the invention of the arts, according to which they should be ascribed to a people not to an individual to avoid mythical figures.

The other two French algebraists we have mentioned, Borrel and Gosselin, did not even go so far. Borrel excluded the possibility of any Arabic influence on algebra and, indeed, any mathematical authority whatsoever for them by means of a strong critique of the Arabic "contamination" of Euclid's *Elements*. Gosselin was more aware of Peletier's work and of contemporary debate; he came out in favor of a completely classicized discipline. Writing at a time in which algebraic manuals were printed in France but in Latin, algebra was, for him, a speculative science, *subtilior arithmetica*.

The main mathematical innovations by these authors concern research on notation, an extended notion of algebraic questions, the treatment of problems in several unknowns, the expression of roots in terms of coefficients and above all a vision of algebra as a theory of equations. If we add all these results to the ones already contained in Cardano, Bombelli, which themselves included the achievements of the Arabic tradition, we find most of the raw materials for symbolic algebra in place, showing a slowly growing product of many contributions from many sources. In particular, efficient notation and a theory of equations had reached a good level even before the entrance of Viète. Diophantus' *Arithmetic* certainly did not contain these tools: to call it "the" privileged source of inspiration for algebra could not refer to the mathematical content. What this work really contained was a collection of problems and, above all, authority, theoretical status as opposed to the practical art of algebra.

To summarize, the French algebraists of the sixteenth century detached algebra from its Arabic origins and from the Mediterranean tradition of the abacus schools. Yet they might have taken seriously the previous Italian tradition coming out of the abacus school; Cardano, whom they used directly from the algebraic point of view, admitted very clearly the Arabic origin of algebra. The separation was accomplished in two related ways: by the development of a tradition of algebra conceived for a milieu of learned culture and by the writing of a new history of algebra. The understanding of this requires us to connect their scientific and their humanistic work, including their specific commitment to national language, culture and history, typical of learned juridical circles of the time. This complex context illustrates how the theoretical status of a discipline can be connected to a search for identity in the culture to which it belongs.

What thus appears as essential and will reappear as a leitmotiv in what follows is the concrete process of the transmission of knowledge — from one civilization to another, one place to another, one generation to another. This is not a spontaneous activity, a simple, natural expansion of the "best" mathematics, but rather a selective transmission, depending on extremely varied criteria. Nor is such a phenomenon particular to Europe; the mathematicians of the Islamic world (which included, one must recall, many of the lands once occupied by Greek mathematicians) also considered themselves the legitimate intellectual heirs of the latter. They not only developed Greek mathematics while integrating into it material of their own and other cultures, but they operated a process of selection on Greek sources following their own priorities.

We must stress the falseness of the widely held idea that mathematics was at first an amateur affair, before being recognized as useful and thus necessitating a professional community. What is true, however, is that the appearance during the Renaissance of a type of "amateur" mathematician, often coming from or near to the nobility and less financially dependent on his mathematical practice, was often accompanied by a reevaluation of the objectives of the discipline and by a rewriting of its genealogy. This introduced new criteria of selection into the available, disparate corpus, criteria which privileged Greek origins and a vision of mathematics as theory which tended to devalue the professional practitioners. In a significant fashion, this moment corresponds also to that in which Mersenne and others launched a call, via their correspondence networks, for exchanges which would help constitute a European mathematical community. Though it was not the first such, it would be the first to unite mathematicians at this social level. Among those involved figure a number of 'heros' of modern science: Galileo, Wallis, Descartes and Huyghens among others.

The Perenniality of Programmes

based on the presentation of Henk Bos

University of Utrecht, Utrecht (Netherlands)

Descartes is often mentioned as the founding father of modern European mathematics. But there were some others at that time who thought about mathematics in as visionary and programmatic a way as Descartes. One of these was Viète and there are strong similarities between Descartes' programme and the one Viète formulated some forty years earlier. This view is somewhat unorthodox because the current view stresses the differences rather than the similarities in the achievements of these two mathematicians. According to this view, Viète's great achievement was to have introduced into algebra the use of letters both for unknowns and for indeterminates; but he bogged down in unwieldy notation and superfluous adherence to the homogeneity of equations. The current view about Descartes is that he introduced "our" notation in algebra, disregarded homogeneity and, most importantly, applied algebra to the geometry of curves, for which he created coordinate geometry and the method of studying curves by means of their equations.

These views are not completely wrong if you look at what lasted of Viète's and Descartes' achievements, but they are strongly misleading if you want to know what they really did and especially why they did it.

Thus, there is a tension between what is now considered important in the work of Viète and Descartes and what they themselves felt to be important. That tension has to do with realities and myths, to which we shall return, but first their programmes.

Both Viète and Descartes wanted to develop algebra as a tool for problem solving according to the following schema:

> translation of problem into algebra ⟶ equation ⟶ (algebraic solution of equation, if possible) ⟶ solution by retranslation of answer (the equation or its algebraic solution) back into original context

The context decided what would count as a solution; when the problem is about numbers the solution is a number arrived at through calculation, but when the problem is geometrical, a solution is a configuration effected by means of a construction. The latter case requires the *geometrical construction of the roots of equations*. And since such roots often cannot be constructed with ruler and compass alone, decisions have to be made about what will count as acceptable means over and beyond the ruler and compass.

The problems which Viète's and Descartes' general algebraic method were meant to solve are not related to modern ideas of algebraic structures; they aimed their new methods at problems largely inherited from classical mathematics. Thus, there was a rich tradition of geometrical problem solving, dealing with such questions as:

- the elementary Euclidean constructions,
- mean proportionals (including duplication of the cube),
- angular sections (dividing an angle in a given number of parts or following a given proportion)
- addition of similar areas or volumes,
- placing a given section between two given curves and directed towards a given point (the *neusis* problem),
- division problems (dividing an area or volume by a line or plane according to a given ratio)
- triangle problems (constructing a triangle from three given elements)
- varia (e.g., find the normal to a parabola from a given external point)

These were distinguished, in a classification recorded by Pappus, into *plane* (solvable by ruler and compass), *solid* (non-plane but solvable by the intersection of conics) and *line-like* problems (all the rest).

Descartes worked out the programme further than Viète had done. Indeed his main achievements in his *Géométrie* of 1637 was, formulated in seventeenth century terms, that he gave a complete theory of geometrical construction of roots of equations. To do so he explored the use of higher order curves as a means of construction. He thus instituted a research programme that lasted for about a century, then died, and is now completely forgotten. Coordinate geometry, curves and equations also figured in the *Géométrie*, but primarily in the service of the geometrical construction of roots of equations. They were taken up later, enthusiastically, but almost *despite* the programme of the book. What Descartes really aimed at was soon lost; what people took over were the techniques, not the motivation. In that sense one may see Descartes as a tragic hero.

One could of course dismiss an interest in such abandoned programmes as those of Descartes and Viète as irrelevant for real mathematics and, therefore, as uninteresting history. As someone recently wrote about a study which endeavored to make a serious analysis of Weierstrass' concept of number:

> "Why did Weierstrass cling to such a clumsy and badly presented method as late as 1886, when much more satisfactory ones were available?"

Clearly such dismissals hinge on value judgements about past events — and that is where myths are born!

Obviously one might think that this interest is valuable as well, for instance, because the story of Descartes' and Viète's programmes shows how mathematicians struggle to interpret the aims of their activity and to decide what should count as exact or proper mathematics and what should not.

But in defending this interest in past mathematics which is no longer considered alive, one is up against a strong ideology which is at the root of much confusion, misunderstanding and even contempt for the history of mathematics.

That ideology of mathematics starts from the conviction that the only value in mathematics resides in what is in its top level, in brilliant and lasting mathematical results or proofs or arguments; and that consequently, the only relevant selection criteria for thinking about mathematics are the brilliance and permanence of the results. Applied to history, that conviction leads to a teleological view of the development of mathematics; the essence (if not the logic) of the development of mathematics lies in how the brilliant results we know now as those essentially valuable in mathematics were arrived at.

For instance, in a recent review of a historical study, the reviewer complained that what he called "the essential and therefore the perennial" had not been made clear and proceeded to make it clear — in mid-twentieth century symbols and notation.

With respect to an historical understanding of mathematics, this ideology, and the resulting teleological view of the history, is inadequate, if not harmful. It is so because it is intellectually unsatisfactory, unserious about people and mathematics, and not contributive to self-understanding (it also produces boring history). By concentrating on the "great" results, one gets a depersonalized history; the people, the mathematicians, are no longer the active agents but rather the mathematical ideas or concepts which "emerge" in the course of history. In this view Descartes is graced with the distinction of having helped coordinate geometry to "unfold". But others may be blamed for "holding up" this development. Such a misfortune happened even to Newton, about whom a standard history of the calculus has this to say:

> "Had Newton devoted more of his time to clarifying the elements of thought in his demonstration by ultimate ratios, the calculus might have been established upon the concept of the derivative a century before the time of Cauchy."

Such a point of view takes seriously neither the people involved nor their mathematics. The mathematicians of the past worked at their mathematics, not at those of the twentieth century. Nor do mathematicians now experience their work in terms of what will be recognized as valuable in the twenty-third century. Looking back at history while keeping to the values dictated by this ideology creates myths; it does not bring an understanding of mathematics as a human activity nor can it contribute to self-understanding — and that is what history is about.

The example we have just studied concerns someone considered as an important mathematician. It shows the role of ambitious research programmes in the establishment of mathematical communities, the manner in which they have served, even when forgotten and unrealized, to transmit techniques by focussing attention and mobilizing personnel around them. An algebraic treatment of geometric problems existed before the times of Descartes and Fermat; it can be seen, for example, in the work of Nicolas Chuquet, or, among non-European mathematicians, the Chinese. It is suggestive, just at this time when the new European network was being set up through a series of exchanges — correspondence, books and manuscripts — and proto-academies were being privately constituted on a local level, to take a look at the extensions and the limits of European mathematics. The interaction between two different and independent mathematical traditions will give us a privileged position from which to view this along with the mathematical priorities accorded by Europeans. One of the questions to be addressed is then: what might the expression "European mathematics" mean when we look at it from an extra-European point of view: the Far East will be our example.

Mathematical Europe from Outside Europe

based on the presentation of Karine Chemla

CNRS, Paris (France)

How did the Chinese encounter with mathematics coming from the West work out? This occurred at the end of the sixteenth century and was mediated by European missionaries, above all by the Jesuits. They arrived at the end of the sixteenth century at the southern door of China, only to find the Chinese authorities forbidding their entrance. But the curiosity of these same Chinese about the scientific and technological knowledge of the Europeans would conquer this resistance.

The policy of the Jesuits consisted of importing into China both religion and science, using the latter to open up contacts, but presenting such scientific knowledge as having been made possible by Christian faith. The missionaries had studied mathematics in their own countries and were able to speak about Euclidean geometry and similar subjects. The Jesuit Matteo Ricci understood the potential of this situation and proposed bringing Jesuits from Europe who knew as much mathematics and astronomy as possible in order to be able to reform the calendar.

European knowledge was put forward not only as "European" but, more specifically, as "Christian" to incite those who were interested by the first to adhere to the second. For example, Euclidean geometry was presented as a Christian achievement and the translation into Chinese of the first six books of Euclid's *Elements* in 1607 carries on its cover the effigy of Christ. For people on the receiving end of this mathematics it was not so easy to dismiss this view and we will see a bit later an example of the way this connection between mathematics and religion was to influence the way mathematics developed in China in the seventeenth century.

In spite of this approach, Chinese mathematicians showed a great receptiveness to the mathematical accomplishments of Europe. If we look, for example, at the case of the indigenous geometry of China — indeed of the Far East as a whole — not only are there no axioms or definitions, but the subject was not considered to be principally concerned with statements of the kind, so common in Euclid, that such-and-such a figure has such-and-such a property. In the Far East one had a metrical geometry, treating mainly of the right-angled triangle and its algebraic properties as well as the application of this to the analysis of certain figures in space. Yet once introduced, the Euclidean style easily found its place in Chinese geometry.

It is often said that while the Chinese were at the forefront of the development of mathematics until the fourteenth century, mathematics then decayed in China in such a way that, when the Jesuits arrived, Chinese mathematics was nearly reduced to arithmetic, and the Jesuits enabled China to embark upon the mathematical research again. This account, of course, fits very well with the needs of the Jesuits in the seventeenth century. If they had accepted that there was mathematics in China it would have been impossible to say: If you want to do mathematics, you have to become Christians. So it was evidently impossible for Jesuits to recognize the existence of real mathematics in China.

Was there really no interesting mathematics for a European in China at that time? Indeed, after the thirteenth century, when mathematics had reached a high peak of development in China, mathematical knowledge had undergone a steady retreat. In the thirteenth century, Chinese mathematicians were able to use numerical methods to find roots of algebraic equations, polynomials in the search of solutions to problems, the so-called Chinese remainder theorem, etc. Many of these results had been forgotten four centuries later, but not all. Besides our main point, it is important to underline the frailty of scientific tradition; it requires a great deal of care and protection to sustain development; it is not given once and for all.

Among the mathematical techniques that were still in use when the Jesuits arrived, we will look at just one example: since the first century A.D., Chinese mathematicians had designed very clever methods to solve systems of simultaneous linear equations. These were represented as a table (of coefficients), the first equation represented by its array of coefficients in the first column of the table, etc. We find this matrix representation as early as the first century mathematical text, *Jiuzhang suanshu* (Nine Chapters on Mathematical Procedures). Equally to be found in this early text is the explanation of how to solve the system, by use of the so-called "Gauss elimination" process (see the following page): one eliminates the first coefficient of the second row at the right, and one goes on in order to get triangles of 0s. This means the introduction of positive and negative numbers and the development of rules of signs, among others. This method was still known at the end of the sixteenth century in China but, as far as we know, the Jesuit mathematicians paid no attention to it, did not want to see it or introduce it into Europe.

We know that they had access to it because Li Zhizao, a Chinese mathematician and one of the first converts, and Ricci wrote a book together, the *Tongwen suanshi* (Arithmetic Indicator Uniting the Cultures) (1612), which was designed as an introduction to European mathematics. Li Zhizao knew about this method for linear equations, available in standard books about mathematics in China. He noticed that this method was not a part of the corpus imported from Europe, but rather complementary to it and he added the method to the book. But this book of course was never translated into Latin or any European language.

Together with this activity of synthesis, Chinese mathematicians began to be interested in the history of their own mathematics; many results and methods had been lost, many sources had disappeared, and so an attempt was made to recover and understand past achievements. Among the things which were presented as having come from Europe under the title of 'European mathematics', some had in fact been known in China in the thirteenth century. Certain numerical methods, for instance, of finding solutions of equations, like the so-called Ruffini-Horner method, were understood not to be exclusively European.

Moreover, there was a gradual understanding that mathematics were not in fact so closely linked to Christian religion. The seventeenth century mathematician, Mei Wending, is a case in point. He had in his hands some 'European' theorems about polyhedra, presented without proof. So he went to his room and stayed there some weeks and proved them. In the work in which he presented his proofs of the theorems, he said (cited J.-M. Martzloff, *Historia Scientiarum* **21**, 1981, p. 34.):

> "I began to be convinced that the various geometrical constructions could be understood, although they [= the Occidentals] tell us that this is a secret teaching of the Gods and that we [Chinese] reject them as heresies."

$$\begin{array}{ccccccccc}
{}^1a_1x_1 & + & {}^2a_1x_2 & + & \dots & + & {}^na_1x_n & = & b_1 \\
{}^1a_2x_1 & + & {}^2a_2x_2 & + & \dots & + & {}^na_2x_n & = & b_2 \\
\vdots & & & & & & \vdots & & \\
{}^1a_nx_1 & + & {}^2a_nx_2 & + & \dots & + & {}^na_nx_n & = & b_n
\end{array}$$

$$\downarrow$$

$$\begin{array}{ccccc}
{}^1a_n & \dots & \dots & {}^1a_2 & {}^1a_1 \\
{}^2a_n & \dots & \dots & {}^2a_2 & {}^2a_1 \\
\vdots & & & \vdots & \vdots \\
{}^na_n & \dots & \dots & {}^na_2 & {}^na_1 \\
b_n & \dots & \dots & b_2 & b_1
\end{array}$$

Eliminating between columns...

$$\downarrow$$

$$\begin{array}{ccccc}
{}^1a_n & \dots & \dots & 0 & {}^1a_1 \\
{}^2a_n & \dots & \dots & {}^2a_2' & {}^2a_1 \\
\vdots & & & \vdots & \vdots \\
{}^na_n & \dots & \dots & {}^na_2' & {}^na_1 \\
b_n & \dots & \dots & b_2' & b_1
\end{array}$$

...until one gets a triangular matrix.

$$\downarrow$$

$$\begin{array}{ccccc}
0 & \dots & \dots & 0 & {}^1a_1 \\
\vdots & & \cdot^{\cdot^{\cdot}} & {}^2a_2' & {}^2a_1 \\
\vdots & \cdot^{\cdot^{\cdot}} & & \vdots & \vdots \\
0 & & & \vdots & \vdots \\
{}^na_n' & \dots & \dots & {}^na_2' & {}^na_1 \\
b_n' & \dots & \dots & b_2' & b_1
\end{array}$$

"GAUSSIAN" ELIMINATION

Transformations in socio-economic conditions are reflected in transformations in mathematical practice. The birth of modern states in the seventeenth century was accompanied by the institutionalization of scientific activities and the "nationalisation" of Academies; starting in 1662, the Royal Society was the recipient of royal funds, the Academy of Sciences is created in Paris in 1666, that of St. Petersburg in 1724. Periodicals begin to appear as the vehicle of scientific communication. In the eighteenth century, alongside the traditional universities, there begin to appear schools for the training of engineers or military officers; the French École Polytechnique, founded in 1794, will serve as the model for many others. New professional forms of mathematical community begin to operate during the nineteenth century, very like (at least in first approximation) those we know today.

At the Beginning of our World: The Nineteenth Century

based on the presentation of Hélène Gispert
Université de Paris-Sud, Orsay (France)

All throughout the nineteenth century, a series of mathematical notions, tools, contours of disciplinary fields and, more generally, large areas of mathematical life as a whole were constituted which are very close to ours. This period is of course very complex if for no other reason than the quantity of information available. To set some guidelines, the concepts of group, analytic function, manifold, real number, genus of a curve, integration (Cauchy, Riemann, Lebesgue), set theory, topology, began during this century. And so did algebraic geometry, differential geometry, the theory of functions of complex variables, of real variables, rigorous foundations for analysis, etc. On the other hand, the nineteenth century is also the century when "suddenly" as it seems, the number of names of mathematicians, of mathematical journals, of reference papers that a mathematician, not a specialist in history, would be able to quote, increased considerably, in large part due to the development of teaching institutions linked to national economic needs during this century.

The increase in the number of teachers of mathematics in the universities, the engineering polytechnics, the military schools, and the increasing role of mathematics in the training of engineers themselves, gave new dimensions, new bounds and new aims to the mathematical communities. Different models of mathematical communities and activities emerged, which were closely tied to a variety of national or regional priorities: the most famous and most closely studied examples are the École Polytechnique and the importance of mathematical physics in France as opposed to the Prussian universities, which favored pure mathematics and whose implantation can be connected with the importance of neo-humanistic philosophy and a less developed stage of industrialization at the beginning of the century.

This account is important because the new rich ideas at the center of contemporary mathematics have acquired their efficiency in the measure in which they have been recognized and adopted by mathematicians themselves, i.e. insofar as they have been taken over and *reinvested* by those practioners. And this reinvestment has not been due to exceptional mathematicians in correspondence among themselves but to the community as a whole, at the national and international level. Its story is really the untold story of nineteenth century mathematics. The history of these communities is a necessary part of the history of mathematics for its own sake; moreover, it participates in the identification and the constitution of the mathematical disciplines. Let us see some examples:

(1) The spectacular development of Monge's descriptive geometry at the beginning of the last century was sustained by the École Polytechnique in France (created a few years earlier after the French Revolution) and it persisted and participated in the mathematical production of the nineteenth century in the various European countries because of its role in the training of engineers. It was developed by Poncelet into projective geometry, which appeared to mathematicians and educators to be the "universal language for the engineer". It kept an essential place in France (more than elsewhere), even after it had disappeared from the forefront of mathematical research, in part because of the origin in the Polytechnique of many members of the mathematical community. In England, on the other hand, the main educational institutions were not engineering schools but universities, under the charge of clergymen; more than half of the students in Oxford and Cambridge took holy orders during the first half of the century — they were still a full third between 1850 and 1900 (See J. Jenkins & D. C. Jones, *British Journal of Sociology* **1**, 1950, p. 93–116, cited in M.-J. Durand-Richard, *Proceedings*).

(2) On the other hand, the non-Euclidean geometries found no echo for years; not, in fact, until Beltrami, in the 1860s, gave them an interpretation in the framework of classical geometry, as surfaces of constant curvature, a procedure which gave them the legitimacy they needed. This was nonetheless not at all universally accepted; it touched Germany and Italy more than France, where only Poincaré and a few others showed any interest in a subject which remained marginal.

(3) After the middle of the century, the development of the theories of Cauchy and Riemann for functions of complex variables might serve also as an example of this phenomenon of reinvestment, the cuts and surfaces of Riemann remaining completely unexploited by, if not unknown to, the French mathematical community, which restricted itself to Cauchy's legacy for thirty or forty years.

(4) A last example is that of Cantorian set theory in the late nineteenth century. It remained in a marginal situation for more than 15 years before being integrated, outside Germany — and outside the Cantorian framework — by a part of the French mathematical community interested in the theory of functions as an efficient and useful tool for analysis. Moreover, during the theory's border crossing, it was searched and the bulk of the "metaphysics", as the analysis of the transfinite was called, simply confiscated, disposed of and left to German and Italian mathematicians.

So mathematical communities could not only encourage but also filter and segment mathematics. They helped to choose and decide what would be seen as important and these choices often imply real, effective decisions in terms of jobs and publications, and finally in terms of the development of the field. One has to underline strongly the importance of local peculiarities in this general process, both in the appropriation and in the diffusion of new ideas and the influence of not only the institutional context but also the cultural and political ones. After the unification of their country in 1870, Italian mathematicians accepted and encouraged almost every branch of their discipline, while those in France rather drastically restricted those theories which were considered important and therefore to be encouraged, ignoring some of the most innovative.

In order to learn quickly about advances in mathematics and to accelerate the diffusion of these new ideas (or at least of the ideas which were recognized as useful), mathematical communities had to find new forms of organization in place of the academies and learned societies of the preceding centuries. Between 1860 and 1920 most of our national mathematical societies were born, in Europe and also in the USA and Japan: Czech 1864, British 1865, Bulgaria 1896, American 1888. These societies were of a new type, because their vocation was national and generally non-elitist: their members were teachers, engineers, soldiers, clergymen and "amateurs", depending on the country and the educational system (See H. Gispert, *Gazette des mathématiciens* (Paris), no. 53, June 1992, p. 25–31). But more and more they tended also to undergo professionalization, and many became centered around what was becoming the "hard core" of the mathematical community, the teachers. In parallel with this phenomenon and participating in it, some fields detached themselves from the common structure, constituting autonomous branches and gave birth to specific communities which had their own societies and their own journals: the new professional domains of actuarial studies, statistics, astronomy, geodesy. The main mathematical newspapers which were founded in the nineteenth century shared the field of academic mathematics, but specific communities published within their own network of publications, a system whose level could be very high but which remained marginal in the sense that it often escaped the attention of the recognized official mathematical milieu.

A shift in our attitudes towards these choices has now taken place because of the current evolution in the interests of contemporary mathematicians: the development of new problems, new possibilities, tied to the use of computers, which is at the origin of some rereadings of the mathematics of the nineteenth century. This change consists of reevaluating works and authors who were, during their lifetime, neither in the "center" of the mathematical communities, nor, above all, in the center of "our" mathematics from the beginning of this century up to about 1960. The nineteenth century has been only recently recognized as also being the century of applied mathematics. Fourier, for instance, whose work on series had been a landmark for the nineteenth century mathematical community, had also written works which were not considered important enough, around 1880, to be included in the publication of his "complete" works; for he was also a pioneer in the fields of mathematical modeling and algorithms. In his analytical theory of heat, but also in his work on the analysis of determinate and indeterminate equations, Fourier sought methods which allowed one to achieve explicit numerical solutions and granted due importance to computation with its virtues of precision and completeness.

More recently, a hitherto rather unknown French mathematician, Joseph Boussinescq, has been rediscovered because of his studies on partial non-linear differential equations describing the propagation of gravitational waves on the surface of a fluid; the equations of Boussinescq have had a renaissance of interest due to their soliton-type solutions. Moreover, he proposed procedures for approximations that are now used in numerical analysis for the modeling of such waves as well as for studies on the stability of localized solutions of infinite-dimensional Hamiltonian systems.

In this sense, "one century may hide another" or many others.

In the nineteenth century then, Europe is neither an already unified field of shared knowledge nor even well along towards such a goal. It is rather a collection of zones, ranging in size from an Institute to a State, among which exchanges operate to a greater or lesser degree according to complex criteria. In particular, educational choices are very important; if the École Polytechnique constituted a model at the beginning of the nineteenth century, it was because Napoleon's victories were partly attributed to the efficiency of such an organization. But, even so, it is a model which was rarely exported as such, without being adapted to local conditions. Ironically, in Prussia, it was technicians like August Crelle, who reproached the French physico-mathematical school for attaching too great an importance to applications and not enough to questions of foundations. The rehabilitation of textbooks based on a Euclidean model in post-independance Italy was argued for, as Gert Schubring has noted (*Proceedings*), on the basis of mathematical presentation, the importance of geometry and the refusal (partly on political grounds) to continue the use of French and Austrian textbooks. In this context, it is worthwhile to study the nature of the expansion of mathematics produced in Europe and the manner in which mathematical, cultural and political priorities interacted.

Centers and Peripheries

based on the presentation of Eduardo Ortiz
Imperial College, London (United Kingdom)

The concept of mathematical periphery is of course a relative one and it is also historically determined: the imaginary frontier between the center and the periphery has moved considerably throughout history. By the times of Adelard of Bath, at the beginning of this millennium, such a frontier left most of Europe in the periphery; the physicist Abdus Salam tells of the medieval scholar Michael the Scot arriving in Toledo to learn mathematics: he was advised to learn something useful rather than acquire mere theoretical knowledge which would not be of any use upon his return to his wild and uncultured homeland. Up until the beginning of this century, the United States of America was on the wrong side of the frontier. In the course of this century Japan and other countries have firmly entered into the group of central mathematical communities.

How did these changes take place, how and why were there contacts between center and periphery? The French Revolution and the events associated with it created conditions for the development of a number of institutions, among them early engineering professional schools where mathematics played an important role because of practical, military and civil, as well as of philosophical or ideological reasons (for instance, the belief that mathematics and science were sufficiently independent of birth and family culture to render a meritocracy possible). The success of these institutions, seen, for instance, as underlying the military victories of Napoleon, caused them to be imitated across the world. The imitations that interest us are those modelled on the École Polytechnique of Paris. Examples of such clones are distributed over an extensive geographical area: the West Point Military Academy in the USA and the Escula de Ingenieros in Madrid, both founded in 1802; Prague's Polytechnical Institute, created in 1806; the St. Petersburg Engineering School of 1809; the Instituto Matemático of Buenos Aires, Argentina, set up in 1815, were just some of them.

The textbooks used by the new French Écoles were also transmitted abroad and adopted in peripheral countries; in particular, Monge's *Géométrie descriptive* and the extensive treatise of Lacroix, covering geometry, arithmetic, algebra, trigonometry, calculus and some of their applications. In the early 1800's Monge's book was translated into Spanish, while Lacroix was translated into Italian in 1804; Spanish in 1807; Polish in 1811; and (Brazilian) Portuguese in 1813. It is interesting to remark that the English translation due to Babbage, Herschel and Peacock dates from 1816. A Greek translation was made as early as 1826. A most remarkable case is that of Portugal, where there was an early interest in mathematical analysis. Lagrange's *Théorie des fonctions* was translated in 1798, one year after its publication in Paris.

If institutional models and textbooks were exported from the central countries to the peripheral ones, the mechanism by which such a transmission took place is sometimes unusual. Favored by political instability, intercommunication between some peripheral countries was far more intense than is commonly believed. The following example relates institutions already mentioned and involves Russia, Spain and Argentina.

Two Spanish mathematicians, Agustín de Betancourt and José Maria de Lanz (the latter born in Mexico) went to Paris on scholarships and made considerable progress there. A few years later, in 1808, they wrote a famous book on the mathematical theory of mechanisms, which was published by the École Polytechnique of Paris to be used as a textbook there. Later these scientists found it advisable to leave France and unwise to return to Spain. Betancourt moved to Russia where he was the founder of the Polytechnic School of St. Petersburg, having been earlier the founder of Madrid's engineering school. His friend Lanz moved to the newly independent country of Argentina where he reorganized the Engineering School of Buenos Aires before finally returning to Paris for good.

Throughout the nineteenth century and more particularly during the second part of it, the different layers of knowledge, interest, applications and activities which make up what we call the mathematical community experienced important changes and, above all, its activities were further diversified. The expansion of international business forced the emergence of the professional as a well defined cog in the new and complex machinery of law, business, health, the armed forces, banking and engineering. This process gave new life to the schools created under the impact of the French Revolution.

A special layer of the contemporary mathematical community developed very rapidly. It was made up of those concerned with coaching students for the entrance examinations to the large and prestigious new professional civil and military engineering schools and sometimes with supplying extra tutoring to keep them there. We will refer to this level as that of *intermediate mathematics*, keeping in mind the elegant definition given by two Belgian mathematicians, Mansion and Neuberg, who contributed to its development: "The elementary part of higher mathematics and the higher part of elementary mathematics". Teaching at this intermediate level became a focal area of professionalization for mathematicians in the second part of the nineteenth century and continued to be so well into the twentieth.

A similar situation is to be found in the peripheral countries as they were integrated into the sphere of influence of international commerce. Importing the mathematics necessary to produce local versions of such professionals became a national objective for several peripheral countries, since the original product was scarce and expensive.

The demand for intermediate mathematics tutors was so intense that the Spanish mathematician, José de Echegaray, tells in his memoirs that by 1880 it was far more profitable for a mathematician in Madrid to coach students privately than to teach them officially at the engineering school.

Mathematical communication was one of the many areas affected by the expansion of the late nineteenth century. But such expansion was not confined to those dealing with research. Intermediate mathematics teachers realized, by the 1870s, that the size of their market allowed for the production of specific journals dealing with questions and solutions of the entrance examination papers to specialized schools; reviews of suitable new textbooks, and the occasional specialized paper intended to whet the appetite of possible future mathematicians.

The readership of the intermediate journals published in the leading countries, such as the French *Journal de Mathématiques Élémentaires*, crossed the frontiers of Europe and reached areas outside it. In the matter of some 15 years, peripheral countries started to imitate them and to produce journals with an identical structure but written in the local language. This wave reached Portugal in 1883; it crossed the Atlantic to the USA, where the *Bulletin of the American Mathematical Society* dates from 1888 and in 1889 it reached Argentina; turning back to Europe, a similar journal was started in Spain three years later, in 1892.

In several peripheral countries mathematical life pivoted around these new intermediate journals, much more lively than the proceedings of the academies of science. They contributed towards establishing a bridge between a few research mathematicians and a much larger and mobile audience of intermediate mathematics teachers. As in central countries, such a community included military men as well as civilians, as the main military schools relied heavily on mathematics for the selection of some of their future officers. Local mathematical heroes began to be recognized or created, mainly from among the editors or the leading contributors of such journals. Scientific correspondence of a fairly considerable size developed between mathematicians of advanced and peripheral countries in connection with these intermediate journals.

The nineteenth century pattern of transmission of pure and applied mathematics to the mathematical periphery was irregular. In the second half of the century, the almost exclusive initial influence of France began to be challenged by Germany, Italy, and to some extent, by England. The choice of topics received in the periphery often depended more on local cultural and historical circumstances than on any mathematical innovation such new research may have carried. There are also curious patterns of conservatism. The mathematics transmitted was related mainly to local needs but those were not the only incentives. Philosophy also played an important role, as it did in the central countries of Europe. In the late nineteenth century, positivism and the criticism of it had a remarkable mathematical impact in several peripheral countries. It is possible to show that, once introduced, some topics managed to acquire a high status, while others were quickly abandoned. Those which were absorbed would be reviewed and renovated by successive generations of mathematicians in a given geographical or cultural area.

Projective geometry was one of those topics which gained firm acceptance in the Hispanic countries by the late 1860s. There the geometric tradition was strong and only mildly rivaled by mathematical analysis, right up to the end of the century. The perception of the importance of projective or pure geometry was enhanced by the fact that it supported the understanding of descriptive geometry, which was then the language of engineers. Graphical statistics also arrived early to that area and along very different routes. It fit quite well with the geometric tradition and was soon appreciated as a powerful tool for geometric computation.

Local variations in the absorption by the periphery of mathematical research from the central countries also have some interest. Portugal, geographically adjacent to Spain, but more closely linked to Europe for most of the nineteenth century, shows a very distinctive pattern in which mathematical analysis dominates geometry. The theory of elliptic functions and Riemann surfaces showed up there almost 30 years before its arrival in Spain.

It is also possible to find in some peripheral countries of the nineteenth century an independent appreciation of mathematical ideas or techniques which managed to have a reciprocal effect on central countries. For example, the revival in Europe of Gräffe's method for the approximate solution of algebraic equations owes much to the efforts of Miguel Merino, a noted astronomer from the Observatory of Madrid, who publicized its advantages for numerical calculations.

The universalization of mathematics has not been always carried out on the level of what may be considered at a given time to be the most advanced sectors, nor *a fortiori* those that appear now to be so. The evolution of mathematics in Prague, studied by J. Folta (*Proceedings*), is yet another example. Although, at the beginning of the nineteenth century, such important and distinct personalities as Bernard Bolzano, F. J. Gerstner and Jacob Kulik worked on questions of analysis, rational mechanics and number theory, it was actually a school of projective geometry which became firmly established in the second half of the century, partly linked, as already mentioned, to engineering institutes. Some of its members were called to Austria and to Bulgaria to help in the setting up of a system of education.

Relations between center and periphery are not always straightforward; a simple swarming from occupied to unoccupied hives. The various attempts to "reconstruct" mathematical schools in the nineteenth and early twentieth centuries in such places as Czechoslovakia, Greece, Poland (these examples treated in *Proceedings*) — and even France — are still other examples of the complexity and conflictual nature of the choices made. The Japanese situation (studied by A. Horiuchi, (*Proceedings*)) is particularly revealing for here a very important local school of Chinese origin, *wasan*, existed before the forced "opening" of Japan in 1853. Some Japanese intellectuals attributed the success of Europe in military and economic affairs to mathematics. Debates between the traditionalists and the Occidentalists continued for a number of years after the Meiji government decided to favor the adaptation and translation of Western textbooks.

Conclusions

Having come to the end of this very rapid review of some mathematical myths and historical truths, we can hope to delineate some partial conclusions.

• The concept "Europe" has varied drastically in the last thousand years, not only in a strictly political but also mathematical context. The polyvalence of the concept "Europe" has served as a backdrop to the polyvalence of relations between it and mathematics: at various times a religious entity — synonymous with Christendom; a cultural force — the "rightful heir" of the Classical heritage; a once and future political unity — past empire and future federation; "Europe" has meant all that and more. It would be then fruitless to imagine a fixed or "natural" relation for the relationship of this protean entity that is Europe to the world of mathematics.

Each conception of mathematical Europe has been relative to a particular group of mathematicians, a public that has itself undergone evolution. The mathematical Europe of scholars, the mathematical Europe of abacists, the mathematical Europe of teachers, the mathematical Europe of missionaries have all been quite distinct, in their geographical extent as well as the mathematical culture and practices that each has presupposed.

• The "universality" of mathematics is as problematic as the term "Europe". The multiplicity of practices has engendered a multiplicity of forms and themes; what has constituted the "mainstream" of mathematics at any given time has been a product of conflict and tension into which many factors enter. Moreover, even judgements of what constitutes "good" mathematics have varied and continue to vary. This leads to constant reevaluation (or "rediscovery") of whole areas which had been previously abandoned.

The ultimate success of one or another field of mathematics has often depended not only on "intrinsic merit" but also, among others, on the channels of diffusion created and used by various fractions of the mathematical community (and other communities as well) which favored or opposed it. In this regard it is important not to underestimate the crucial role played by textbooks and their patterns of translation and transmission in the determination of the final outcome. No theory of single origins or preordained development can realistically account for our contemporary mathematics.

• History, like all other domains of knowledge, is constantly being enriched not only by newly discovered sources, but also by newly developed methods. In particular, the history of mathematics is not merely the establishment, once and for all, of a simple chronological ordering of names and facts but also the permanent reworking of older material to take into account the new perspectives opened up by methodological and conceptual innovations in the field.

• A final point resides in the intimate connection between mathematics and its history. The writing, rewriting and transmission of the "history" of mathematics has played a large part within the development of mathematics itself. The history of any domain serves, among other things, to establish, restore and guarantee filiations between communities of the past and the present. That these filiations be true or false has often been less important than the role that they might play in the (re)production of the mathematical community itself; recruitment of the next mathematical generation depends strongly on how mathematics is presented to and received by the young. What today appears to working mathematicians as the "natural" form and content of their field is often the concrete result, solidified by the system of schooling, of debates and controversies engaged in by past generations with agendas of their own.

The history of mathematics is not that of the automatic unfolding of an immutable Truth. Mathematics is a human phenomenon and as such cannot be isolated from the human actors who create and transmit it. At important moments mathematicians and others have mobilized linguistic, political and economic forces as allies in favor of one side or another in what have appeared to be internal debates. Reciprocally, mathematics has been used as an argument to enforce extra-mathematical ends. This, of course, is *not* to claim that any given political or social fact has an impact on any given mathematical result; subject matter and actors interact in subtle and dialectic ways that give all the savor — and all the difficulty — to their joint product which we call mathematics.

The Round Table "Mathematical Europe, Myth or Historical Reality?" was organized by Catherine Goldstein (Paris-Sud/CNRS, France), Jeremy Gray (Open University, UK), Jens Høyrup (University of Roskilde, Denmark), Jim Ritter (University of Paris 8, France), Norbert Schappacher (University Louis Pasteur, Strasbourg, France), Erhard Scholz (University of Wuppertal, Germany).

The authors would like to thank all those historians of mathematics whose work, as reported by them in the Satellite-Colloquium of April 3–6, 1992, in Paris under the title "Mythes et Réalités historiques de l'Europe mathématique / Myths and Historical Realities of Mathematical Europe," has been crucial for the elaboration of this report:
Harvey Becher (Northern Arizona University, USA), Henk Bos (University of Utrecht, Netherlands), Umberto Bottazzini (University of Palermo, Italy), Karine Chemla (CNRS, France), Giovanna Cifoletti (Milan, Italy), Jean Dhombres (University of Nantes, France), Roman Duda (Ministry of Education, Poland),

Marie-José Durand-Richard (Paris, France), Colin Fletcher (University College of Wales, UK), Jaroslav Folta (Historical Institute, Czechoslovakia), Hélène Gispert (University of Paris-Sud, France), Annick Horiuchi (ENS, France), Jens Høyrup (University of Roskilde, Denmark), Tony Lévy (CNRS, France), Herbert Mehrtens (Berlin, Germany), Luboš Nový (Prague, Czechoslovakia), Eduardo Ortiz (Imperial College, UK), Christine Phili (National Technical University, Greece), Kim Plofker (Brown University, USA), Zofia Pawlikowska-Brożek (Academy of Metallurgy and Mines, Poland), Roshdi Rashed (CNRS, France), Peter Schreiber (Ernst-Moritz-Arndt University, Germany), Gert Schubring (Bielefeld University, Germany), Renate Tobies (University of Leipzig, Germany), Bernard Vitrac (CNRS, France).

Of course, the use to which their material has been put does not engage them in any way and is entirely the responsibility of the authors of this report.

The Satellite-Colloquium was funded by: UNESCO (Divsion "Philosophie et Ethique"), CNRS (Sections 01 (SPM) and 35 (SHS)), DAGIC, MRT (Programme Acces) and University of Paris 8.

A good starting point for further information is, of course, the *Proceedings* of the Satellite-Colloquium mentioned above, to be published under the same title. To this can be added, among a number of excellent works, the following works, perhaps closest in spirit to the approach adopted here.

Further Readings

[1] Henk Bos, Herbert Mehrtens, Ivo Schneider (eds.), *Social History of Nineteenth Century Mathematics*, Birkhäuser, Boston, Basel, Stuttgart, 1981.

[2] John Fauvel, Jeremy Gray (eds.), *The History of Mathematics: A Reader*, Macmillan, Open University, London, 1987.

[3] *Échanges d'influences scientifiques et techniques entre pays européens de 1780 à 1830*, Éditions du C.T.H.S., Paris, 1990.

[4] Patrick Petitjean, Catherine Jami, Anne-Marie Moulin (eds.), *Science and Empires*, Kluwer, Dordrecht, 1992.

Table Ronde F : 7 juillet 1992

PHILOSOPHIE DES MATHÉMATIQUES : POURQUOI ? COMMENT ?

organisée par Hourya Sinaceur
CNRS, Paris (France)

L'idée de l'organisatrice était de susciter quelques interrogations sur le pourquoi et le comment d'une activité qui requiert une compétence double, philosophique et mathématique, et des échanges entre mathématiciens et philosophes. De ces interrogations elle n'attendait pas moins que diverses formulations et des esquisses de réponses aux questions suivantes : qu'est-ce que la philosophie des mathématiques ? Comment l'exercer ? Quels objectifs poursuit-elle ? Quelles exigences cherche-t-elle à satisfaire ? Quelle est sa portée pour les mathématiques et pour la philosophie ?

Ce programme était d'autant plus ambitieux qu'il impliquait, en fait, une réflexion sur les relations réciproques de trois disciplines : la philosophie, les mathématiques en général, la logique et l'informatique en tant qu'elles constituent un partenaire nouveau du mathématicien au travail et une cible priviligiée pour le philosophe. Il supposait sans doute quelques mises au point sur la logique, et sa mutation récente d'une position fondatrice, externe par rapport au travail du mathématicien, à une position efficiente interne. Nous avons montré ailleurs[1] que cette mutation est *antérieure* au développement informatique et technologique, qui la confirme aujourd'hui de façon pratique. Le dédoublement, qui en est résulté, en logique "pure" ou analyse des théories, des méthodes et des résultats mathématiques, et logique "appliquée" ou instrument de découverte, nous paraît déboucher sur une collaboration étroite plutôt que sur une opposition de ces deux aspects. En langage hilbertien, nous dirions que le travail *méta*mathématique conflue, sur le front de problèmes déterminés comme cette fameuse conjecture arithmétique d'Artin résolue par des méthodes modèle-théoriques, avec le travail mathématique. Bien entendu, cette confluence n'est pas la seule possible ou intéressante, il y a aussi la confluence physico-mathématique. Mais enfin elle existe. Qui plus est, elle prend des allures de plus en plus variées. Après avoir d'abord concentré ses échanges avec l'algèbre ou l'arithmétique, au temps héroïque des débuts de la théorie des modèles et de la théorie de la démonstration, voici que la logique se tourne vers la géométrie. Une "géométrie des preuves"

---

[1] H. Sinaceur, *Corps et modèles, Essai sur l'histoire de l'algèbre réelle*, Paris, Vrin, 1991, avant-propos et 4e partie.

se substitue à "l'analyse des preuves", comme dans l'entreprise de Jean-Yves Girard (brièvement évoquée par l'exposé de Giuseppe Longo). Et comme toutes les mathématiques semblent marquées aujourd'hui par un "retour" au géométrique, qui ne saurait cependant se situer *en deçà* du symbolique et du formel, ni exclure par ailleurs une avancée simultanée du calcul (dont les stratagèmes sont forcément bornés au fini et au discret), les détracteurs de la logique au nom de l'intuition du géomètre devront revoir leurs arguments.

Qu'est-ce que la philosophie des mathématiques ? Bien sûr, aucun des intervenants à la table ronde n'a attaqué de front cette question ni n'a tenté l'impossible : proposer une *définition*. Tous les cinq pourtant, comme on le lira dans leurs conférences, ont proposé des angles d'attaque, mentionnés déjà pour certains dans le premier texte de présentation de la table ronde [2] : comparer connaissance mathématique et connaissance philosophique (G.-G. Granger); rendre compte de la richesse, de l'objectivité et de l'applicabilité des mathématiques (J. Petitot); redonner à l'image et à la vision un rôle épistémique, généralement perdu depuis la fameuse promotion au début du XIXe siècle de la "rigueur analytique", et proposer en conséquence une alternative géométrico-cognitive aux points de vue du formalisme et du réductionnisme (G. Longo); expliquer la difficulté des mathématiques par une notion d'information spécifique qui combinerait deux types de complexité, computationnelle et conceptuelle (J. Dubucs); retrouver, sinon surprendre, "le geste" de la création mathématique dans les brouillards de la recherche et du pressentiment, se replacer "au cœur de l'opérativité", où diagrammes et "expériences de pensée" suscitent le *mouvement* de la pensée (G. Châtelet). Et ont été abordées les questions importantes du mode d'engendrement, d'existence et de devenir des procédures mathématiques; de la part d'auto-réflexion et d'intertraduction des théories mathématiques (par quoi se révèle la corrélation indissoluble de la forme et du sens en même temps qu'une dynamique relativement autonome du sens;[3]) de l'émergence et de la prolifération de "contenus formels"; de la référence ou non à l'espace sensible; de l'exigence de confirmer ou non l'irrésistible réalisme spontané du mathématicien; des avantages et difficultés du constructivisme; de la pertinence de préciser un concept proprement mathématique de "nécessité"; des modalités visuelle ou dynamique aussi bien que combinatoire et conceptuelle de l'imagination inventive; de l'affinité ou du contraste de la science avec l'art, etc.

---

[2] Publié dans le volume des *Résumés*, distribué aux participants du CEM.

[3] Heureusement, comme l'écrit Giuseppe Longo, même la programmation, qui dût son premier essor au formalisme, est aujourd'hui "affectée par le sens". Sur l'origine, dans les travaux de Tarski, de la prise en compte théorique du sens des théories mathématiques, voir notre ouvrage, *Corps et modèles, Essai sur l'histoire de l'algèbre réelle*, 4e partie, et en particulier les pages 350 – 355 consacrées à l'analyse épistémologique du théorème de Bott–Milnor et du principe de transfert.

On ne saurait nier la diversité, les divergences même des positions, des postures et des méthodes des conférenciers. On ne saurait manquer d'observer leur effort commun de tirer de l'expérience mathématique "actuelle", sur le mode contraignant de l'argumentation ou par simple suggestion, de quoi défaire des évidences anciennes. Peut-être la vertu cardinale de l'attitude philosophique est-elle de déjouer les pièges des idées irréfléchies, de bousculer les habitudes toutes faites, de favoriser l'audace et l'innovation. En ce sens, de grands novateurs comme Leibniz ou Galois ou Poincaré ont pratiqué l'exercice philosophique. Et en dehors de cela, si l'on en croit les réactions de plusieurs mathématiciens présents dans la salle lors de la table ronde, le mathématicien a "besoin" de philosophie, sinon pour inventer, du moins pour interpréter sa propre expérience.

Saluons donc l'initiative des organisateurs de ce premier colloque européen de mathématiques d'avoir donné la parole à la philosophie. Et gageons que le dialogue ainsi ouvert encouragera la coopération, qui jusqu'à présent a fait si cruellement défaut en France plus qu'ailleurs, de tous ceux, mathématiciens ou philosophes de profession, qui ont à cœur de faire sa place, dans la société et la culture, à l'expérience mathématique.

Formes, opérations, objets en mathématiques et en philosophie

par Gilles-Gaston Granger
Collège de France, Paris (France)

Si la philosophie est une espèce de connaissance, on lui a quelquefois reconnu une certaine parenté avec la connaissance mathématique. Sur quels points peut porter cette parenté, jusqu'où va-t-elle ?

Nous prendrons pour guides trois notions qui interviennent dans les deux disciplines. Ce sont ici des *métaconcepts*, c'est à dire non des concepts qui appartiennent à une science et pourraient avoir une définition formelle, mais appartenant à la philosophie elle-même. Ils portent, au second degré, sur des représentations de représentations. Ils sont essentiellement liés entre eux, le métaconcept d'opérations fonctionnant comme médiateur. On peut les appliquer à la philosophie comme aux mathématiques.

Formes

1. C'est d'abord un concept *empirique* et mathématique (géométrique) s'appliquant à des objets perçus et à des objets construits. Cependant, les psychologues de la *Gestalt* ont fait, dans cette perspective même d'une détermination empirique, deux découvertes qui nous orientent vers une réinterprétation plus fondamentale. La première est l'extension de la notion de forme à des perceptions non nécessairement spatialisées ; la seconde est le rôle essentiel reconnu à l'opposition forme-fond. L'une et l'autre suggèrent la possibilité de *concevoir* la forme comme déliée *en droit* de toute réalisation dans une perception spatiale, voire même de toute perception. D'autre part, l'opposition forme-fond se réduit en dernier ressort à l'opposition purement logique du *oui* et du *non*, de la présence et de l'absence, qui caractérise une situation de *frontière*, condition nécessaire de l'apparition d'une forme. Bien entendu, il n'est pas question de nier le rôle joué par la saisie intuitive, spatiale ou non, d'une forme comme *continuité*, continuité pour ainsi dire interne, mais toujours corrélative d'une rupture, d'un décrochement. En tant que pensable comme concept, la forme est justement cette corrélation, saisissable aussi bien dans les formes perçues, spatiales ou sonores, ou sans doute même relevant de façon autonome d'autres registres des sens, que dans la forme d'un discours (et non pas seulement de sa réalisation sensible), d'un raisonnement, d'un sentiment, d'une pensée.

2. C'est ce thème que retrouve la théorie des "Catastrophes" de René Thom, en donnant à cette corrélation du continu et de la discontinuité une expression mathématique savante, propre à être appliquée à la description d'objets, mais peut-être par cela même réductrice. En premier lieu, elle semble parler uniquement et explicitement un langage d'espace. Mais il ne faut pas oublier que cette spatialité, qui est en réalité celle du géomètre, n'est pas nécessairement celle du monde perçu, bien que René Thom insiste constamment sur la réalisation concrète de ses "catastrophes" dans l'espace du sens commun. En second lieu, la typologie spécifique des formes déduites par le mathématicien dépend d'hypothèses dynamiques, caractérisant un substrat qui est un champ où jouent des forces. Aussi bien cette théorie des formes est-elle directement applicable, en principe, à une description de phénomènes physiques au sens large, dont elle proposerait d'expliquer la morphogenèse. Mais il me semble qu'il n'en est plus de même lorsqu'on veut en étendre directement l'application à des phénomènes de *sens*, comme en linguistique, où il ne s'agit pas seulement de construire un modèle "énergétique", ni même "cybernétique", lequel n'atteint alors que le niveau d'un substrat, d'une infrastructure.

Or, même si l'origine spéculative en est étroitement liée à la recherche d'un métaconcept, le concept constitué par René Thom est bien un concept *scientifique* structurant des phénomènes d'une certaine espèce, et c'est ce qui en fait une innovation importante ; il n'est cependant pas forcément adéquat pour d'autres phénomènes, ni ne saurait tenir lieu du métaconcept philosophique, non directement organisateur d'objets, mais qui exprimerait ce que *signifient* les divers concepts scientifiques possibles de forme.

3. Ce métaconcept doit pouvoir s'appliquer, mais *médiatement*, aussi bien aux phénomènes saisis dans le sensible—mais alors sans doute de façon scientifiquement inefficace — qu'aux œuvres de pensée. L'exemple le plus proche en serait sans doute la *forme logique*, qui consiste alors en une corrélation parfaite d'un système d'opérations et d'une possibilité d'objet, dépouillé comme telle de tout contenu effectif. Le passage des différents concepts scientifiques de forme au métaconcept philosophique se présenterait ici comme une sorte de "montée sémantique", entendue comme neutralisation et expulsion de tout contenu empirique, et même, dans ce cas extrême, de tout contenu fût-il purement formel, transformant l'opposition forme-contenu en une corrélation duale de *contenus seulement possibles* et d'opérations organisées. Et c'est en fin de compte de cette manière que je crois pouvoir interpréter l'*ειδος* aristotélicien. Pour Aristote, la forme est détermination d'un contenu, la *υλη* matière, étant absence de détermination effective des contenus possibles. Et l'actualisation de ces contenus par la forme s'effectue comme opération.

4. La forme logique n'est cependant pas identifiable au métaconcept même dont elle est, *parmi les concepts scientifiques*, la plus abstraite des incarnations. La notion d'objet y apparaît déjà en effet, en tant qu'objet *possible* de pensée. Mais c'est dans les applications du concept de forme aux connaissances scientifiques que le pôle objectal prend toute sa force ; les formes sont nécessairement alors *formes d'objets*.

En mathématiques même les objets possibles sont déjà pourvus de qualités : les formes y *déterminent des contenus* comme corrélatifs de systèmes opératoires. Ces contenus, que je nomme "contenus formels", manifestent justement la consistance des êtres mathématiques, consistance, ou si l'on veut "réalité", qui ne doit rien en droit à l'empirie, même lorsqu'il est possible de s'en former des représentations empiriques.

5. En philosophie, les formes sont assurément inséparables de contenus, et le niveau des contenus est celui de *l'expérience vécue*; mais contrairement aux formes mathématiques, les formes philosophiques à leur propre niveau, *n'engendrent pas des contenus.* Je voudrais dénoncer sur ce point ce qui me semble être une double erreur concernant le statut des formes en philosophie.

(1) L'erreur hegelienne : des formes (essentiellement logiques ou même purement linguistiques) sont supposées engendrer des contenus intuitifs. La "dialectique" serait alors la description des règles de cet engendrement. Mais on constate que les formes ne *déterminent* pas univoquement ces contenus, qui n'apparaissent que parce qu'ils sont connus par avance.

(2) L'erreur axiomatique : on croit pouvoir décrire les formes philosophiques au moyen d'un système d'axiomes, qui ne capte que des formes logiques, certes inhérentes à l'architecture du système philosophique, mais qui ne l'épuisent pas.

Les formes en philosophie organisent des significations attribuées à l'expérience humaine, à un moment donné de son histoire, lesquelles ont pour *contenus* des formes *et* des contenus de toute espèce, en particulier empiriques. La visée de cette organisation philosophique, l'attribution d'une signification à notre expérience concrète, est la constitution de cette expérience en une totalité, non effectivement réalisée, mais virtuelle.

Objets

1. Les sciences en général ont des *objets*. Kant requérait que ces objets soient rattachés au sensible. Il suffit de reconnaître qu'ils sont les *corrélats de sytèmes d'opérations* plus ou moins précisément explicités.

Je définis la connaissance proprement scientifique comme connaissance par construction de modèles abstraits des phénomènes, ainsi transmués en *objets*. Le métaconcept d'objet serait donc constitutif de toute science, quel qu'en soit le domaine phénoménal de référence. Après une proto-histoire de visées incertaines, chacune d'elles ne commence vraiment à se déployer que lorsqu'elle détermine, en ce sens, un objet, à la fois manipulable au moyen d'opérations formelles dans un espace de représentation, et plus ou moins directement rattachable aux phénomènes par un système d'opérations matérielles. A ce niveau, la pluralité des définitions de l'objet, à l'intérieur même d'une discipline comme la physique, suggèrerait qu'on parle d'un *transconcept* d'objet. Ce transconcept d'objet, caractérisant toute visée scientifique, Kant, comme on l'a dit plus haut, avait cru pouvoir en désigner la pierre de touche par la possibilité d'une référence au sensible, ou du moins à ses formes *a priori*. On peut montrer, par la considération du statut des objets scientifiques dans la science post-kantienne, que cette condition n'est pas essentielle. C'est en revanche un trait métaconceptuel déjà souligné à propos de la notion de forme qui m'a semblé déterminer la signification la plus générale et la plus profonde de l'objet, à savoir une *corrélation* plus ou moins parfaite avec un système opératoire. De telle sorte que le métaconcept d'objet, ainsi que sa réalisation transconceptuelle dans les sciences, aussi bien que ses réalisations esthétiques dont il sera question plus loin, sont inséparables du métaconcept de forme, par la médiation du métaconcept d'opération sur lequel il nous faudra bientôt revenir. On soulignera ici, derechef, que cette codétermination est parfaitement compatible avec une absence de référence à la spatialité. Il n'y a donc d'objets, dans la science, que si la représentation des phénomènes s'effectue par le moyen d'un système d'opérations suffisamment articulé, et le degré d'articulation de ce système correspond au degré de distinction de la description des objets. On prendra garde, toutefois, de ne pas assimiler cette proposition à une thèse simplement instrumentaliste. D'une part, ce n'est pas l'opération qui *crée* l'objet ; la pensée *détermine conjointement l'opération et l'objet*, en vue d'une représentation de l'expérience. D'autre part, l'opération est tout autant — et le plus souvent bien davantage — conceptuelle que matérielle.

2. Le développement d'une visée scientifique d'objets conduit du reste à une situation très caractéristique, propre à confirmer cette corrélation de l'opération et de l'objet : la science considère fondamentalement des faits — et par conséquent des objets — *virtuels.* C'est-à-dire que les faits sont introduits par une science avec une *détermination abstraite complète* dans le système de référence adopté pour représenter les phénomènes, mais une *détermination essentiellement incomplète* comme faits empiriquement observables ou observés. Ce jeu laissé à l'imagination conceptuelle est l'une des conditions *sine qua non* de la représentation des phénomènes par des modèles abstraits. On comprend que les objets virtuels intervenant dans de tels faits ne puissent être considérés autrement que fixés par le réseau des opérations qui seules donnent pouvoir de représentation au référentiel où sont déterminés les faits virtuels. Il en résulte, par ailleurs, que les capacités de prévision d'une science ne peuvent être retenues comme critères de validité de sa connaissance des phénomènes que dans la mesure où l'on tient compte du niveau de précision qu'autorise le statut opératoire exact de ses faits virtuels : de tels statuts pouvant différer *toto cœlo* selon la période de l'histoire d'une science et selon le domaine phénoménal qu'elle veut embrasser.

3. Il est évidemment surtout ici question des objets de la science. Mais le métaconcept d'objet que nous nous efforçons d'introduire doit étendre sa domination sémantique à toute autre réalisation d'objets apparaissant assez distinctement dans l'expression de nos expériences. Dans cette perspective, il suffit en effet pour qu'il y ait objet que joue à quelque niveau et à quelque degré de clarté *l'opposition* que nous reconnaissons comme catégorie fondamentale de la pensée, de l'opératoire et de l'objectal.

On a fait allusion tantôt à l'objet esthétique, parce qu'il me semble présenter une espèce d'objet se distinguant de façon privilégiée de l'objet scientifique. Observons tout d'abord que l'une et l'autre espèce, lorsqu'on les considère du point de vue du savant et de l'artiste, sont des *créations.* Le mot de connaissance se référant à l'une ou à l'autre de ces espèces, et qui est assurément plus particulièrement pertinent du point de vue du "récepteur" de science, l'est tout aussi bien, quoique plus secrètement, appliqué au "récepteur" d'art. La différence des deux espèces de connaissance ne doit pas laisser croire qu'il existerait alors un obstacle dirimant à l'emploi dans les deux cas du mot "objet". La différence radicale est ailleurs. Elle consiste, croyons-nous, essentiellement en ceci, que l'objet de la connaissance scientifique est, comme on l'a dit, une construction abstraite, un *modèle* des phénomènes ; l'objet esthétique est au contraire restauration d'une expérience concrète, soit qu'elle vise à reproduire *avec style* une expérience déjà éprouvée, soit qu'elle tente la production d'une expérience toute nouvelle, mais au moyen de matériaux sensibles existants, directement ou indirectement évoqués.

4. Dans la science comme dans l'art, le rapport opérations-objet est déterminant, et c'est en quoi, à ce qu'il me semble, objets scientifiques et objets esthétiques relèvent bien du même métaconcept. Mais il s'agit assurément alors d'opérations de genres différents. Caractériser les modes et les styles opératoires de la pensée scientifique, ainsi que les formes d'objets corrélatifs, c'est l'affaire de l'épistémologue, reconnaître les modes opératoires à l'œuvre dans la constitution des objets esthétiques, c'est la tâche de l'esthéticien, du critique, du philosophe des Beaux-Arts. Je me bornerai ici à noter un trait tout à fait visible et grossier qui contribue à faire la différence : c'est le statut des *contenus* dans l'une et l'autre espèce d'objets. Pour l'objet scientifique, les contenus se présentent d'abord comme un obstacle à réduire, le travail de la science consistant, d'une certaine manière, à produire des formes à des niveaux de plus en plus "profonds". Cette notion même de contenu, originairement empirique, réapparaît pourtant en mathématiques sous les espèces de "contenus formels", révélant alors de façon assez éclatante la signification et les limitations essentielles de cette réduction des contenus. Dans l'objet esthétique au contraire, bien loin que le travail de l'artiste — ressenti et retrouvé par l'interprète et l'amateur d'art — soit orienté vers une réduction des contenus, il vise à leur instauration. L'application du mot "objet" à de tels résultats opératoires risque peut-être de paraître impropre ; le produit des opérations est alors un nœud de *significations*. Acceptons néanmoins ici, afin de faire ressortir la présence du métaconcept, de parler des opérations esthétiques comme visant à susciter des "objets" ; mais c'est, répétons-le, par instauration de *contenus*. Non pas, bien entendu, que le pôle formel de l'objet s'affaiblisse : la corrélation des contenus aux formes y est tout au contraire particulièrement prégnante ; mais le rôle de la forme, qui était premièrement dans l'objet scientifique d'*exténuation* des contenus, surtout des contenus sensibles, est alors de production et d'*exaltation* de ceux-ci.

5. Quant aux objets des mathématiques ce sont de purs corrélats d'opérations effectuables dans un système de symboles, imaginées comme détachées du sensible. Le support sensible y est psychologiquement sans doute indispensable mais non pertinent. Mais la libre construction des systèmes opératoires est *limitée par les contraintes logiques*, et par l'incapacité de ces systèmes à *dominer complètement les productions d'objets* (théorèmes de limitation). Ces limitations sont les contreparties, dans le domaine des objets mathématiques, des résistances des phénomènes empiriques ; elles constituent l'une des révélations des contenus formels mentionnés plus haut. De ce point de vue, les mathématiques, sont une théorie des *formes possibles d'"objectité"*.

6. La philosophie, en ce sens, n'a pas d'objets : elle ne vise pas à construire des modèles de l'expérience, ni à produire des formes possibles d'"objectité". Au sens le plus strict, la philosophie n'est donc pas une science, pas même au sens non empirique des mathématiques. Car contrairement à celles-ci elle a des *contenus* au niveau de l'expérience. Mais à la différence des sciences de l'empirie elle ne les représente pas au moyen de structures abstraites. Elle les organise en totalités virtuelles en leur donnant des *significations*.

Opérations

1. La notion d'opération, en tant que métaconcept, est indissolublement liée à celles d'objet et de forme. Dans la perspective adoptée, les opérations sont *créatrices d'objets*, qu'elles ne dominent pourtant que partiellement, même en mathématiques. Seuls des sytèmes d'opérations logiques *stricto sensu*, c'est à dire le calcul des énoncés, dominent totalement les objets-corrélats, mais ce sont alors les formes vides de l'objet sans qualité.

Afin de parvenir à entrevoir un métaconcept d'opération, nous partirons encore une fois de la forme la plus abstraite qu'elle revêt comme *concept* dans la pensée scientifique, en logique et en mathématique. Se donnant la notion d'*ensemble* et la notion d'*application*, on peut aisément fomuler de l'opération une définition abstraite, qui est celle de l'application d'un produit cartésien d'ensembles d'opérandes dans un ensemble de résultats. Mais la question se pose alors de décider si la notion d'"application" elle-même ne relève pas du concept d'opération. Il semble donc que la mise en forme offerte par la théorie des Catégories analyse plus profondément le concept, et nous rapproche en quelque sorte du métaconcept que nous voudrions mettre au jour.

On part dans cette théorie d'une notion d'"objets" indéterminés *a parte ante*, simples supports ou points d'application des "flèches" ou "morphismes" qui introduisent justement, me semble-t-il, les opérations comme concept mathématique sous leur forme la plus abstraite et la plus générale. La classe des flèches d'une Catégorie n'est déterminée en effet que par des propriétés formelles minimales. Ce sont alors les spécifications attribuées à un système de flèches qui déterminent la nature des objets correspondants. Il semble donc que le point de vue de la théorie des Catégories privilégie l'aspect opératoire de l'univers mathématique (alors que le point de vue ensembliste privilégie plutôt l'aspect objectal). Mais le fait épistémologique essentiel c'est bien la codétermination des opérations et des objets.

2. Pourtant, à ce niveau d'abstraction, qui est celui des mathématiques, c'est principalement le système opératoire qui est formulé par les axiomatiques ; les objets n'y apparaissent seulement que, pour ainsi dire, comme des concrétions possibles aux nœuds du réseau dessiné par les opérations. Mais la dualité opérations-objets est alors, comme on l'a rappelé plus haut, imparfaite, apparaissent des *contenus* formels que ne peut en principe dominer complètement l'aspect opératoire des démonstrations.

D'autre part, même en mathématiques, les opérations sont aussi source de styles, dans la mesure où plusieurs systèmes opératoires conduisent aux mêmes objets-corrélats avec des variantes non pertinentes sur un premier plan structural.

3. En philosophie, puisque l'on vise seulement à constituer et organiser les significations de notre expérience, les opérations ne sauraient être corrélatives d'objets, pas même d'objets non empiriques comme ceux du mathématicien. Les opérations *logiques*, qui ne sont corrélatives que de formes d'objets vides, n'y peuvent cependant être ignorées ; elles demeurent nécessairement appliquées à la *tactique* de construction du discours, mais l'organisation des significations elles-mêmes (des contenus philosophiques) n'y est pas nécessairement soumise. Il est un autre niveau opératoire concernant la manière d'aborder, analyser ou décrire ces contenus, déceler des analogies, etc. Les opérations spécifiques ainsi mises en œuvre constituent le "système" d'un philosophe, irréductible à une formulation axiomatique parce que lié aux contenus. La pensée formelle en philosophie, aussi importante sans doute qu'en mathématiques, n'est ici architectonique que pour des significations, non pour des objets. L'historien de la philosophie tente de décrire et d'interpréter ce système, lui-même agissant alors comme philosophe.

La mathématique, à l'extrême frontière de l'empire des sciences, *crée* des formes, librement mais non arbitrairement, car ces formes doivent être formes d'*objets*. Elle est la théorie des systèmes opératoires déterminant les formes d'objet en général. La philosophie crée également des formes, mais non pas des formes d'objets. La libre imagination du philosophe est pourtant elle aussi contenue, car ces formes doivent pouvoir donner consistance et unité à la multiplicité toujours inachevée de nos expériences, non pour les constituer en objets, mais pour les proposer comme *totalité virtuelle*, lui conférant ainsi un sens.

Le platonisme transcendantal et le problème de l'objectivité

par Jean Petitot
EHESS, Paris (France)

"To be a mathematician
is to be an out-and-out Platonist".
*David Mumford**

1. Faire droit à la connaissance mathématique

Le travail et la technique mathématiques n'ont plus besoin de philosophie. Le lien rationnel privilégié, miraculeux, qui s'était noué dès l'origine grecque entre mathématiques et philosophie et qui avait traversé les siècles s'est brutalement dénoué il y a un demi-siècle. Depuis ce temps, l'opérativité mathématique se

* Préface à l'ouvrage de C. Parikh, *The Unreal Life of Oscar Zariski*, Academic Press, 1991.

déploye de façon autonome dans l'oubli complet du souci de la philosophie. Est-ce un bien, est-ce un mal? Je ne sais pas. Mais c'est un fait historique.

En revanche, le lien des mathématiques à la *rationalité* a, quant à lui, le plus grand besoin d'une auto-réflexion philosophique. Il en a besoin pour traiter les points évoqués par Hourya Sinaceur dans la présentation de cette Table Ronde, et en particulier pour clarifier le statut très particulier et très problématique de *l'objectivité* des idéalités mathématiques dans leur rapport au sens et à l'intelligible, pour expliciter la relation entre démonstrations et objets, pour penser la dialectique entre vérité objective et évolution historique, pour justifier le rôle constitutif des mathématiques dans la physique théorique, bref pour participer à l'élaboration d'une théorie de la connaissance.

En général, on prend pour philosophie des mathématiques une philosophie des diverses *conditions d'accès* aux objets mathématiques (accès symbolique-syntaxique, ou psychologique-cognitif, ou social-institutionnel, etc.). On réduit les objets à ces modes d'accessibilité épistémiques ou sociologiques. Puis l'on explique alors, à partir de thèses nominalistes et finitistes, que les mathématiques sont sans objets propres et que la croyance réaliste en l'existence d'idéalités est archaïque et mystique. Mais l'on omet en général soigneusement de rappeler que tout nominalisme est inséparable depuis Ockham d'une ontologie d'étants singuliers substantiels existant à titre d'entités individuées, indépendantes et séparées. Or, il est évidemment impossible d'attribuer une telle ontologie aux idéalités mathématiques. Le nominalisme ne saurait donc être en l'occurrence une position philosophique consistante.

Dans la philosophie des mathématiques telle que je la conçois et la pratique, il s'agit, au contraire et avant tout, *de faire droit à la réalité mathématique*. Les mathématiques constituent le *factum rationis* par excellence. C'est de cela qu'il faut partir pour répondre non pas à la question : "que doit être la réalité mathématique pour que telle philosophie positiviste, nominaliste, matérialiste, etc., ait raison ?", mais bien plutôt à la question : "de quelle philosophie avons-nous besoin pour faire droit à la réalité mathématique ?"

Mais pour faire droit philosophiquement à une telle réalité, encore faut-il évidemment poser la *question de droit* : "comment les mathématiques peuvent-elles justifier leur prétention à être une connaissance objective et vraie portant sur des objets idéaux abstraits ?". Prétention que pratiquement toutes les philosophies de ce siècle leur dénient. Cette question de droit est, on l'aura reconnue, la question critique, la question transcendantale.

Philosophiquement, je cherche à développer, dans une perspective critique transcendantale, une *apologie* de la réalité mathématique. J'aimerais à ce propos aborder trois points.

2. L'Entre-expression et l'Inter-traductibilité des Mathématiques

Le premier point concerne ce que j'appelle *l'entre-expression* et *l'inter-traduction* des théories mathématiques.

Dans son débat avec Jean-Pierre Changeux, *Matière à Pensée*[1], Alain Connes a souligné certains critères d'objectivité pour les mathématiques. Par exemple :

1. La possibilité de *classifier* exhaustivement les objets définis par une axiomatique (les corps finis, les corps localement compacts, les groupes finis simples, etc.).
2. Le fait que les théories mathématiques intéressantes ont *un contenu informationnel infini* comme le montre le théorème d'incomplétude.
3. La *cohérence et l'harmonie* inter-théoriques *globales* des théories mathématiques. Bien qu'"inexpliquées" et constituant un problème central, elles sont incontestables et objectives.

Ce dernier point est particulièrement important. Or force est d'admettre qu'il constitue le continent inconnu de la philosophie des sciences. Il relie démonstration, vérité et objectivité à une dimension *herméneutique* intra-mathématique[2]. Sur le plan philosophique, il n'y a guère qu'Albert Lautman qui l'ait bien compris : c'est ce qu'il appelait *l'unité* des mathématiques. Lautman a thématisé le fait qu'une partie considérable des mathématiques modernes consiste à traduire certaines propriétés de certaines structures par des propriétés ou par l'existence d'autres structures (cf. la topologie algébrique, la théorie analytique des nombres, etc.).

De théories en théories, les mathématiques s'*auto-interprètent* indéfiniment et c'est pourquoi la déduction n'y est que *locale*. C'est la *compréhension* de cette entre-expression qui régule, et même souvent domine, la mécanique démonstrative. Il existe bien comme une "herméneutique intrinsèque" des mathématiques pures qu'il est essentiel de thématiser philosophiquement[3]. Elle ouvre à une philosophie des *contenus*, à une auto-réflexion philosophique *interne* aux mathématiques.

3. Les symétries en physique, l'esthétique transcendantale et la philosophie de la géométrie

Un autre point philosophique que j'aimerais évoquer très brièvement concerne le rôle des mathématiques dans les théories physiques et, plus précisément, la façon

---

[1] Changeux-Connes [1989].

[2] Pour la redéfinition du concept d'herméneutique en sciences, cf. Petitot [1990a], [1991a], [1991b], [1992b] et Salanskis [1991].

[3] Pour quelques exemples concernant les théories respectives des surfaces de Riemann, des groupes et des algèbres de Lie, des nombres, des singularités, de la cohomologie des faisceaux, etc., cf. Petitot [1979–1982] et [1990a].

dont des principes de relativité, des principes de symétrie, y sont transformés en principes *dynamiques* constitutifs de l'objectivité physique.

Ce *factum rationis* impose une conception philosophique de l'*objectivité physique* qui n'a plus guère à voir avec l'ontologie ou le sens commun. Comme j'ai tenté de le montrer ailleurs[4], une telle doctrine est de nature transcendantale. Elle repose sur une *philosophie de la géométrie* [5].

Si je privilégie pour ma part la philosophie transcendantale c'est parce qu'elle est la première philosophie (et l'une des très rares jusqu'ici) à avoir compris cela, et en particulier trois caractéristiques fondamentales des théories physiques.

1. Le fait que l'existence de symétries est constitutive du concept d'objectivité physique et disjoint celui-ci de toute ontologie substantialiste d'étants singuliers individués existant à titre d'entités séparées indépendantes. D'où, comme en mathématiques, la non pertinence d'une doctrine nominaliste.
2. Le fait que les principes de relativité imposent des contraintes de covariance aux descriptions physiques. La covariance garantit la possibilité de l'objectivité. Elle la définit et lui confère un contenu non ontologique. Tel est le sens moderne d'une *Esthétique transcendantale.*
3. Le fait qu'il existe une *corrélation* entre une telle Esthétique transcendantale et les *objets* de la théorie physique, autrement dit que, comme le dira beaucoup plus tard Hermann Weyl en introduisant le concept d'invariance de jauge, des principes de relativité, de covariance et d'invariance peuvent être transformés en principes *dynamiques*. Chez Kant, cette corrélation correspond à la notion de *construction des catégories dynamiques* (substance, causalité, interaction). Il s'agit d'interpréter ces catégories *géométriquement*, c'est-à-dire à partir d'un enrichissement des principes de symétrie, soit, en dernière instance, à partir d'une Esthétique transcendantale enrichie.

Dans mon étude *Transcendental Æsthetics for Modern Physics*[6], j'analyse trois exemples spectaculaires.

(a) Le théorème de Nœther et la construction de la catégorie de substance.
(b) La relativité générale et la construction de la catégorie de cause (du concept de force).
(c) Les théories de jauge et la construction de la catégorie d'interaction.

Voilà quelques thèmes philosophiques que nous sommes quelques uns à nous efforcer de redéployer. Pourquoi la réflexion à leur sujet, encore très riche au début du siècle s'est-elle tarie ? Je pense que cela est dû à une obstruction positiviste.

---

[4] Cf. Petitot [1990a], [1991b], [1992a].

[5] Pour la philosophie de la géométrie, cf. *1830–1930 : A Century of Geometry* (Springer, *Lecture Notes in Physics* #402). Sur la pertinence d'une épistémologie *géométrique*, cf. le bel exposé de G. Longo à la Table Ronde.

[6] Petitot [1992a].

4. L'obstruction positiviste à la philosophie des mathématiques

La plupart des débats de philosophie *générale* en mathématiques concernent les apories du réalisme platonicien. Dans sa forme *non technique*, la question traditionnelle du réalisme platonicien est celle de l'acceptabilité d'une ontologie d'entités abstraites. Il est actuellement débattu en général à partir de cinq préjugés qui, au nom d'un *a priori* naturaliste, veulent conjurer l'attribution à la cognition mathématique de facultés magiques supposées permettre d'accéder à des entités suprasensibles et qui, au nom d'un *a priori* logiciste, pensent le réalisme en termes exclusifs de sémantique vériconditionnelle[7]. Ces préjugés sont les suivants :

(i) "objectivité" et "réalité" signifient "référence à des objets" existant à titre d'êtres individués, séparés et indépendants;
(ii) conformément à la théorie causale de la référence, la référence est la relation converse de la façon dont un sujet est causalement affecté par des objets physiques externes;
(iii) l'ontologie des objets physiques est une ontologie de choses matérielles;
(iv) il est sensé et légitime d'utiliser les méta-concepts de réalité extérieure, de matière, de chose, d'objet, de causalité, etc. indépendamment de toute constitution préalable de domaines d'objectivité spécifiques;
(v) la logique est universellement et indifféremment applicable aux objets d'un domaine quelconque.

Ces préjugés obligent évidemment à dénier toute réalité aux objets et aux structures mathématiques pour la raison triviale que si "exister objectivement" signifie "exister physiquement dans le monde externe en tant que chose matérielle", alors il est impossible que nous puissions posséder un accès épistémique (un apprentissage, des croyances, des croyances vraies, des croyances vraies rationnellement justifiées, c'est-à-dire des connaissances) à des entités externes qui sont *abstraites*, et ne peuvent par conséquent posséder *aucun efficace causal*.

C'est cette aporie qui monopolise le débat. On cherche en général à la dissoudre de façon éliminativiste. Un bel exemple en est celui de l'antiplatonisme matérialiste de J.P. Changeux[8]. Un autre exemple en est celui de l'antiplatonisme instrumental d'Hartry Field.

Dans *Science without numbers*[9], Hartry Field abandonne complètement le problème de la vérité en mathématiques, c'est-à-dire celle des "truth-makers". Reprenant l'ancienne problématique positiviste de l'éliminabilité des termes théoriques depuis Mach (Hempel, etc.) et s'inspirant des nombreux théorèmes

---

[7] Sur ce point, cf. l'exposé de J. Dubucs à la Table Ronde.

[8] Cf. Changeux-Connes [1989]. Nous avons analysé en détail ce débat dans notre étude *Idéalités Mathématiques et Réalité Objective*, (Petitot [1991a]).

[9] Field [1980].

de logique mathématique affirmant qu'une théorie formelle "forte" est en fait conservative sur une théorie formelle "faible", il essaie de montrer que les théories scientifiques utilisant des mathématiques sont *conservatives* sur des théories nominalistes *sans* mathématiques[10].

De par son aspect technique, la thèse d'Hartry Field est évidemment très séduisante. Mais elle n'est pas véritablement convaincante. En effet, pour éliminer par exemple les nombres entiers, Field est obligé d'introduire des *quantificateurs numériques* : $\exists_x^k$ etc. Les énoncés numériques du genre $\# \operatorname{Ext}(A) = k$ deviennent alors trivialement des *contreparties abstraites* (éliminables) d'énoncés logiques $\exists_x^k A(x)$. De même, en ce qui concerne l'espace, Field est obligé d'admettre que les *points* et les *régions* de l'espace-temps sont des objets *physiques* concrets (sic!) sur lesquels on peut quantifier (au premier ordre pour les points, au deuxième ordre pour les régions). Il introduit alors des relations de comparaison, d'incidence, de congruence entre points, segments, etc., et montre trivialement que la métrique (la distance) en est la contrepartie abstraite (éliminable).

Faire droit à la réalité mathématique, c'est avant tout faire droit au platonisme. Mais pour ce faire, il faut remettre en cause les 5 préjugés dogmatiques évoqués plus haut. Selon nous, la sémantique logique ne possède *aucun contenu cognitif ni ontologique*. Sa validité ne concerne exclusivement que les idéalités mathématiques (relation syntaxe-sémantique en théorie logique des modèles)[11]. Elle ne peut donc prétendre expliquer les rapports référentiels et intentionnels entre un système cognitif et une réalité externe. De même, il est scientifiquement illégitime de parler de réalité, d'existence objective ou de causalité en dehors d'une théorie particulière où ces concepts se trouvent schématisés et construits (au sens transcendantal).

La conséquence de cette critique transcendantale du dogmatisme logiciste est que, en ce qui concerne le statut des idéalités mathématiques, le problème princeps n'est pas celui de leur existence et de leur réalité *mais celui de leur objectivité*. La confusion "objectivité $\cong$ ontologie" est une catastrophe épistémologique.

J'aborde donc, pour conclure ces remarques, un exemple où la question de l'objectivité me paraît pertinente.

5. Approche transcendantale du platonisme de Gödel[12]

Partons du programme de recherche anti-gödelien de la "reverse mathematics" développé par Takeuti et Feferman. Il s'agit, on le sait, d'un programme

---

[10] Cf. Chihara [1990] pour un résumé et une discussion des travaux techniques de Field.

[11] Sur la philosophie de la théorie des modèles, cf. Sinaceur [1991], Salanskis [1991] et Petitot [1979–1982], [1989].

[12] Pour des détails, cf. mon étude *Continu et Objectivité* (Petitot [1990c]).

réductionniste qui cherche à réaliser partiellement le programme de Hilbert en réduisant des parties significatives de l'arithmétique du deuxième ordre à des systèmes plus faibles faisant intervenir les axiomes d'existence — i.e., les axiomes de compréhension, les axiomes d'induction et les axiomes de choix — les plus faibles possibles, les moins engagés ontologiquement.

Enormément de théories classiques sont ainsi réductibles à des systèmes faibles constructivement justifiés.

Mais il existe pourtant des arguments solides en faveur d'un platonisme non réductionniste, à la Gödel, c'est-à-dire en faveur d'une ontologie ensembliste riche. Ils concernent en particulier les résultats *d'indépendance* pour des énoncés *combinatoires strictement finitistes* qui sont des versions finitistes de théorèmes classiques et qui ne sont pourtant pas arithmétiquement prouvables, et cela *à cause du phénomène d'incomplétude.*

Un exemple particulièrement spectaculaire d'un tel phénomène est fourni par la version finitiste du théorème de Kruskal. Le théorème de Kruskal (1960) démontre que si T_n est une suite infinie dénombrable d'arbres finis (avec une racine), il existe nécessairement $i < j$ et un homéomorphisme injectif $h : T_i \rightarrow T_j$ (c'est à-dire une injection préservant les *inf* de l'ordre partiel défini par la structure d'arbre).

La version *finitiste* de ce théorème est due à Harvey Friedman :

Théorème F: $\forall k > 1 \, \exists n \; A(k, n)$ *où* $A(k, n)$ *est l'énoncé : si* $T_1, \ldots, T_n$ *est une suite d'arbres de cardinal* $\#T_i \leq k + i$ *alors* $\exists i < j \leq n$ *et un homéomorphisme injectif* $h : Ti \rightarrow T_j$.

Une preuve *non* constructive dans ZFC s'obtient trivialement au moyen d'un argument de compacité. Mais l'énoncé F est démonstrativement *extrêmement puissant.* Il est en fait *indépendant* de l'arithmétique formelle. Il fournit un excellent exemple d'énoncé $\Pi^0_2 \forall k \; \exists n \; A(k, n)$ qui n'est pas arithmétiquement démontrable alors que $\exists n \; A(k, n)$ l'est pourtant pour tout k. Ce phénomène *d'immanence de l'incomplétude à des énoncés finitistes* est dû à la croissance vertigineuse du $n = n(k)$ dont l'existence est affirmée et de la longueur des preuves successives de $\exists n \; A(k, n)$. *Une telle croissance échappe complètement aux limites de l'arithmétique prédicative.*

L'énoncé F implique la consistance de l'analyse et transcende donc les principes prédicatifs. Bien qu'il soit un énoncé "concret" de combinatoire finie, il n'admet que des preuves très imprédicatives. De tels résultats plaident incontestablement en faveur d'une ontologie ensembliste "riche". En effet, comme l'admet Feferman, "the most favorable interpretation to be placed on these results is that they tend to support Gödel's doctrine"[13]. Harvey Friedman, pour sa part, pense que de tels résultats *ouvrent une nouvelle crise des fondements*, dans

---

[13] Feferman [1989], p. 34.

la mesure où ils montrent que "strong abstract set theory will prove to play an essential role in a variety of more standard finite mathematical contexts"[14].

La réponse de Feferman à cet argument solide en faveur de la doctrine de Gödel est *la réduction* possible à des systèmes "non platoniciens" *constructivement justifiés*. Et du fait que, grâce aux réductions de la "reverse mathematics", "higher set theory is dispensable in scientifically applicable mathematics", il conclut à une position philosophique radicale : "I am convinced that the platonism which underlies Cantorian set theory is utterly unsatisfactory as a philosophy of our subject, despite the apparent coherence of current set-theoretical conceptions and method. To echo Weyl, *platonism is the medieval metaphysics of mathematics*; surely we can do better"[15].

Je pense que cette conclusion de Feferman n'est philosophiquement acceptable que pour autant qu'elle concerne une forme vulgaire et naïve (transcendante) de platonisme *ontologique* selon laquelle ZFC serait une théorie qui, dans le cadre d'une théorie réaliste de la dénotation, porterait sur "a fixed and definite world"[16]. Mais tout change dès que l'on tient compte du fait que ce qui est en cause est la structure du *continu* et que celle-ci relève philosophiquement d'une *Esthétique transcendantale* et pas du tout d'une ontologie.

L'argument de Feferman n'est pas *philosophiquement* convaincant pour la raison suivante. Soit l'on restreint l'existence à l'existence logique, soit on lui attribue un contenu transcendant les mathématiques. Dans le premier cas, on ne saurait lui attribuer aucun contenu ontologique. Ce n'est que le préjugé selon lequel la même logique vaut pour les mathématiques et le monde réel qui peut induire et entretenir l'illusion d'un contenu ontologique intra-mathématique de la quantification. Dans le second cas, le moins est alors de s'interroger un peu sérieusement sur le statut de réalité externe du continu. Or, celui-ci n'est pas un objet. C'est *une intuition pure*, une *forme* de donation et de manifestation de la réalité objective externe. Il constitue le fondement d'une Esthétique transcendantale objectivante et, à ce titre, son réalisme empirique ne saurait se doubler d'un réalisme transcendant. Le statut de "réalité" du continu externe est celui de *l'idéalité transcendantale*. Cela signifie qu'il doit être *déterminé* par des idéalités mathématiques et qu'il ne confère à celles-ci *aucun contenu ontologique*.

L'idéalité transcendantale implique que la détermination mathématique ne peut pas être une détermination *complète*. Cette dernière ne constitue qu'un *horizon* de détermination. Le caractère logique de l'idéalité transcendantale est que les thèses *problématiques* d'existence appartiennent elles-mêmes au processus (incomplet) de détermination. Elles ne peuvent donc pas porter sur des étants singuliers individués fixes et définis mais seulement sur des idéalités dont la valeur

[14] Friedman [1986], pp. 92-93.
[15] Feferman [1989], p. 42.
[16] Ibid. p. 43.

et la réalité objective sont en proportion de la portée opératoire de leur fonction de détermination.

J'appelle *platonisme transcendantal* ce platonisme objectif qui *comprend l'existence mathématique dans un processus transcendantal de constitution d'objectivité.* Le platonisme transcendantal est "négatif" et non pas "positif" comme le platonisme ontologique vulgaire. Il permet *d'inverser* les affirmations philosophiques de l'anti-platonisme *et de transformer les phénomènes d'incomplétude et d'indécidabilité en arguments en faveur du platonisme.*

C'est pourquoi, pour conclure, j'aimerais dire un mot sur la justification en quelque sorte *physique* du platonisme gödelien et, en particulier, de l'introduction d'axiomes d'existence de grands cardinaux pour élaborer une bonne théorie du continu.

Dans les travaux sur les grands cardinaux il ne s'agit pas d'élaborer un modèle du continu qui soit compatible aux contraintes constructivistes, mais bien au contraire de tenter de *reconstruire* au mieux, à l'intérieur même des mathématiques, la transcendance externe de cette réalité *donnée intuitivement* comme forme de l'extériorité physique.

L'Axiome de constructibilité est trop contraignant. Il implique en effet qu'il existe un bon ordre Δ^1_2 sur $\mathbb{R}$ et donc qu'il existe un ensemble Δ^1_2 non Lebesgue-mesurable. Or, un bon ordre sur $\mathbb{R}$ devrait être hautement non constructible et indéfinissable. De façon générale, l'axiome du choix AC (qui reste vrai dans un univers constructif) implique l'existence d'ensembles très compliqués et très irréguliers qui sont néanmoins projectifs. Ces ensembles devraient être hautement non constructifs. Or l'axiome de constructibilité les force à exister dans la hiérarchie des projectifs. D'où une inversion complète de stratégie.

On cherche à démontrer de bonnes propriétés de régularité des projectifs $A \subseteq \mathbb{R}$: Lebesgue-mesurabilité, propriété de Baire, propriété de l'ensemble parfait ou, plus généralement, propriété de détermination au sens de Gale et Stewart[17]. On veut généraliser les résultats de Luzin et Suslin : tout Σ^1_1 et tout Π^1_1 sont Lebesgue-mesurables ; tout Σ^1_1 a la propriété de l'ensemble parfait. Gödel et Cohen ont montré que dès le niveau des Σ^1_2 et des Π^1_2 les difficultés de démonstration sont *méta-mathématiques* et sont en fait des conséquences de *l'incomplétude.* D'où l'idée gödelienne d'enrichir les axiomes de la théorie des ensembles *en précisant la grandeur de l'univers.* Pour cela, il faut introduire des axiomes platoniciens d'existence de grands cardinaux (inaccessibles, mesurables, etc.). Un fait fondamental est alors que la "bonne" structure du continu dans un univers de ZFC dépend très fortement de ces axiomes d'existence de grands cardinaux. Elle en est *la contrepartie.* On peut alors, comme l'ont proposé Gödel et D.A. Martin, justifier ces axiomes comme des sortes d'*hypothèses physiques*, portant

[17] $A \subseteq \mathbb{R}$ est déterminé si dans le jeu où deux joueurs choisissent à tour de rôle un entier (ce qui, à la limite, donne un $r \in \mathbb{R}$), l'un des joueurs a une stratégie gagnante pour $r \in A$.

non pas sur un monde fixe et complètement prédéterminé ontologiquement mais sur *les formes de l'objectivité* (les intuitions pures).

Bibliographie

[1] Changeux J.-P., Connes A., *Matière à penser*, Editions Odile Jacob, Paris, 1989.

[2] Chihara Ch. S., *Constructibility and Mathematical Existence*, Clarendon Press, Oxford, 1990.

[3] Feferman S., "Infinity in Mathematics : Is Cantor Necessary ?", *Philosophical Topics* **XVII**: 2 (1989), 23–45.

[4] Field H., *Science without Numbers*, Princeton, 1980.

[5] Friedman H., "Necessary Uses of Abstract Set-theory in Finite Mathematics", *Advances in Mathematics* **60** (1986) 92–122.

[6] Petitot J., "Infinitesimale", "Locale/Globale", "Unità delle matematiche", *Enciclopedia Einaudi*, Einaudi, Turin, VII, 1979–1982, 443–521; VIII, 429–490; XV, 341–352; XV, 1034–1085,

[7] Petitot J., "Refaire le 'Timée'. Introduction à la philosophie mathématique d'Albert Lautman", *Revue d'Histoire des Sciences*, XL, 1, 1987(a), 79–115.

[8] Petitot J., "Rappels sur l'Analyse non standard", *La Mathématique non standard*, Editions du CNRS, Paris, 1989, 187–209.

[9] Petitot J., "Logique transcendantale, Synthétique a priori et Herméneutique mathématique des Objectivités", *Fundamenta Scientiæ*, (numéro en l'honneur de L. Geymonat), 10, 1, 1990(a), 57–84.

[10] Petitot J., "Continu et Objectivité", *Le Labyrinthe du Continu*, J.M. Salanskis, H. Sinaceur (éds.), Colloque de Cerisy, 1990(c), Springer-Verlag, Heidelberg, Berlin, New York, 1992.

[11] Petitot J., "Idéalités mathématiques et Réalité objective. Approche transcendantale", *Hommage à Jean-Toussaint Desanti*, G. Granel éd, (éd.), Editions TER, Mauvezin, 1991(a), 213–282.

[12] Petitot J., *La philosophie transcendantale et le problème de l'objectivité*, Entretiens du Centre Sèvres, Père F. Marty (éd.), Editions Osiris, Paris, 1991(b).

[13] Petitot, J., "Actuality of Transcendental Aesthetics for Modern Physics", *1830–1930, A Century of Geometry*, L. Boi, D. Flament, J.M. Salanskis, (eds.), Springer, Berlin, 1992(a).

[14] Petitot J., *Physique du Sens*, Editions du CNRS, Paris, 1992(b).

[15] Salanskis J.-M., *L'Herméneutique formelle : L'Infini — Le Continu — L'Espace*, Editions du CNRS, Paris, 1991.

[16] Sinaceur H., *Corps et Modèles, Essai sur l'histoire de l'algèbre réelle*, Vrin, Paris, 1991.

Reflexions on Formalism and Reductionism in Logic and Computer Science

by Giuseppe Longo
LIENS (CNRS) - DMI, Ecole Normale Supérieure, Paris (France)

Many logicians have now turned into applied mathematicians, playing a role in Computer Science that is increasingly acknowledged even in industrial environments. This fact is gradually changing our understanding of Mathematical Logic as well as its perspectives. In this note, I will try to sketch some philosophical consequences of this cultural and "sociological" change, largely influenced by Computer Science, by a critique of the role of formalism and reductionism in Logic and Computing.

In mathematics, we are mostly used to a dry and schematic style of presentation: numbered definitions and theorems scan the argument. This may add effectiveness and clarity, though nuances may be lost: the little space allowed here requires me to stress effectiveness.

Themes

Three main aspects relating Computer Science and the logical foundation of mathematics will be mentioned below, namely,

1. the growth of a pragmatic attitude in Logic
2. revitalization and limits of formalism and constructivism
3. the role of space and images.

1. Pragmatism in Logic

1.1 Tool vs. foundation. In Computer Science, Mathematical Logic is no longer viewed as a foundation, but as a tool. There is little interest in setting on firm grounds Cobol or Fortran in a similar way that Logic aimed at founding Number Theory or Analysis by, possibly complete, axiomatic systems. The actual work is the invention of new programming languages and styles, or algorithms and architectures, often by using tools borrowed from Mathematical Logic. This also may originate in attempts to have a solid foundation based on clear grounds known as constructs, but the ultimate result is usually a novel proposal for computing. Functional and Logic Programming are typical examples of this.

This "engineer's approach" in applied Logic is helping to change the philosophical perspective in pure Logic as well.

1.2 An analogy: the completeness of mechanical systems. The interests in the foundation of mathematics, in this century, have been a complex blend of technical insights and philosophical views. We must acknowledge first that philosophical *a priori*, in Logic, may have had a stimulating role, in some cases. Yet ideologies and the blindness of many attitudes have been a major limitation for knowledge, an unusual phenomenon in scientific research. They have been the source of

— wrong conjectures (completeness, decidability, provability of consistency),

— false proofs, by topmost mathematicians (just to mention the main cases: an inductive proof of the consistency of Arithmetic, by Hilbert, refuted by Poincaré in 1904–05; a second attempt, based on a distinction of various levels of induction, debated by Hermann Weyl in the twenties).

The wrong directions taken by the prevailing formalist school may be understood as a continuation of a long lasting attitude in science and philosophy. On the shoulders of last centuries' giants, Newton and Laplace, typically, the positivist perspective believed in perfect and complete descriptions of the world by classical mechanics, namely by sufficiently expressive systems of partial differential equations. Similarly, in Logic, adequate axiomatic systems were supposed to describe completely Analysis or the whole of mathematics. These were two levels of descriptions, both exhaustive, or complete in the sense of Logic: one of the world by mathematical equations, the other of mathematics by (finitely many) axioms.

Still, this positivistic vision in Logic was not compelled by the times. Hermann Weyl conjectured the incompleteness of number theory and the independence of the axiom of choice in *Das Kontinuum* in 1918. Poincaré rejected purely logical or linguistic descriptions as the only source for mathematics and stressed the role of geometric insight. As I will try to hint below, Poincaré's distinction between analytic work and the intuition of space as well as his approach to the foundation of mathematical knowledge may be today at the base of a renewed foundational work, similarly as his work on the three body problem is at the origin of contemporary mechanics (Lighthill[1986]).

1.3 Foundation of mathematical knowledge. Another mathematician should be quoted among those who did not accept the reductionist and formalist attitudes of the first part of this century: Federico Enriquez. Also in Enriquez's philosophical writings, the interest in the interconnections of knowledge and in its historical dynamics suggested more open philosophical perspectives. It may be fair to say, however, that Poincaré, Weyl and Enriquez were interested in the foundation of mathematical knowledge more than in the technical foundation of mathematics.

The difference should be clear: by the first I mean the epistemology of mathematics and the understanding of it "as an integral part of the human endeavor toward knowledge" (to put it in Weyl's words). Not a separated transcendence, isolated in a vacuum, but an abstraction emerging from our concrete experience of the real world of relations, symmetries, space and time (Poincaré [1902, 1905], Enriquez [1909], Weyl [1918, 1952]).

In contrast with this broad attitude, the purely technical, internal foundation, as pursued by formalistic and reductive programs in their various forms, somewhat reminds one of a drawing of a book on a table and of the belief that the table really supports the book, while they are designed by us by the same technique and the same tools. This cognitive circularity is at the source of negative results in Logic.

1.4 The unity of formal systems. The first step towards a more open attitude comes with the need for a variety of systems and the understanding of their interconnections. The layman's attitude inspired by applications and the suggestions coming from geometric intuitions are at the origin of recent inventions of new systems of Logic, where unity is given not by a global, metaphysical, system, but by the possibility of moving from one system to another, by changes in the basic rules and by translations or connecting results. Indeed, Girard's focus on the structural rules in Girard [1987] and his seminal work in Girard [1992] are largely indebted to a pragmatic attitude that views Logic as part of (applied) mathematics, with no special "meta-status". In this perspective, geometric structures and applications suggest formal systems, as they guide toward relevant changes, and propose comparisons in the common mathematical style in which connections and bridges preclude ideological closures within one specific frame. In this sense, its unity is a deep mathematical fact, as much as Klein's unified understanding of Geometry.

2. Formalism and Constructivism in Computer Science (and their limitations)

2.1 Linguistic notations. The volume in Combinatorial Logic by Curry and Feys contains many pages on the renaming of bound variables and related matters (in set-theoretic terms, $\{x \mid P(x)\}$ is the same as $\{y \mid P(y)\}$). I believe that the foundational relevance of these pages, if any, may be summarized in about three lines. The formalist treatment is a typical example of purely symbolic manipulation, where meaning and structures are lost (see Curry's book on a formalist foundation of mathematics for a firm view in that direction). This opinion is shared by all working mathematicians who simply ignored that discussion, and explain the problem to students in twenty seconds on the blackboard. Still, it happened that variable binding is a crucial issue in computer programming. Thus, the discussion in Curry et al. has been largely and duly developed in functional programming and it is at the core of detailed treatments in implementations.

This example is just a small, but typical one, of the revitalization of formalism and constructivism due to Computer Science. It happens that computers proceed as our founding fathers of Logic described the parody of mathematics: linguistic definitions and formal deductions, with no meaning as a guidance. Meaningless, but effective constructions of programs, more than unifying insights and concepts. Mathematical invariants are lost, but denotations are very precise. This requires technically difficult insights into pure calculi of symbols and, sometimes, brand new mathematics. However, the branching of methods and results, due to translations and meaning, which are at the core of knowledge, may be lost within extremely hard, but closed, games of symbols. This is part of the practionner's daily experience even in the theory of computing.

2.2 Denotational semantics. Fortunately, though, even programming has been affected by meaning. In the last twenty or so years, various approaches to the semantics of programming languages embedded programming into the broader universe of mathematics. Here is the main merit of the Scott–Strachey approach, as well as of the algebraic or other proposals, most of which are unifiable in the elegant frame of Category Theory (this is partly summarized in Asperti & Longo [1991]). In some cases, the meaning of formal systems for computing, over geometric or algebraic structures, suggested variants or extensions of existing languages. More often, obscure syntactic constructs, evident at most to the authors, have been clarified and, possibly, modified. As a matter of fact, in the last decade, computer manuals have slowly begun to be readable, as they are moving towards a more mathematical style, that is towards rigor, generality and meaning simultaneously. We are not there yet, as most hackers think in terms of pure symbol pushing and are supported in this attitude by the formalist tradition in Logic. Many still do not appreciate, as one learns from mathematics that the understanding and, thus, the design of a strictly constructive, but complicated system may also derive from highly non constructive, but conceptually simple, intellectual experiences.

2.3 Resources and memory. Brouwer, the founding father of intuitionism, explicitly considers human memory as "perfect and unlimited" for the purposes of his foundational proposal (Troelstra [1990]). This is implicitly at the base of the formalist approach as well. Indeed, computer memories are perfect and, by a faithful abstraction, unlimited. This has nothing to do with human memory, and mathematics is done by humans. One component of mathematical abstraction, as emerging from our "endeavour towards knowledge", may precisely derive from the need to organize language and space in least forms, for memory saving purposes. A "principle of minima" may partly guide our high level organization of concepts, Sambin [1990]. Moreover, imperfection of storage is an essential part of approximate recognition and analogy building.

For the aim of founding mathematical knowledge, we need exactly to understand the emergence of abstraction, the formation of conceptual bridges, and of methodological contaminations among different areas of mathematical thinking. There may be a need for psychology and neurophysiology in this approach. Good: the three mathematicians quoted in 1.3 in their opening remarks have been often accused of "psychologism", of "wavering between different approaches". The proposed "one-way" alternatives lead to deadlock where formalism and reductionism brought us in understanding mathematics (and the world). Moreover, so much happened in this century in other areas of knowledge that we should start to take them into account.

2.4 Top down vs. bottom-up. There is no doubt that formalism and reductionism have been at the base of Computer Science as it is today and of its amazing progress. In particular, top-down deductions and constructive procedures set the basis for the Turing and Von Neuman machines as well as for all currently designed languages, algorithms and architectures. Yet, there is a growing need to go beyond top-down descriptions of the world, even in Computer Science. The recent failures of strong Artificial Intelligence are the analogue of incompleteness and independence results in Logic: most phenomena in perception and reasoning escape the stepwise-deductive approach. Partly as a consequence of these failures, there is an increasing interest in bottom-up approaches. Relevant mathematics is being developed in the study of the way images, for example, organize space by singularities, or how the continuum becomes discrete and reassembles itself, in vision or general perception, in a way which leads from quantitative perception to qualitative understanding (Petitot [1992]).

3. Space and Images

3.1 Denotations and geometry. In the practice of mathematics, formal notations and meaning are hardly ever distinguished. Indeed, one may even have symbolic representations where, besides the geometric meaning, there exists a further connection to Geometry at the notational level. Relevant examples of this are given by Feyman's and Penrose's calculi or by Girard's proof nets.

In Feyman's calculus, planar combinations of geometric figures allows computations representing subatomic phenomena. Penrose extends familiar tensorial calculi over multidimensional vector spaces in a very powerful way: bidimensional connections between indices explicitly use properties of the plane to develop computations. A more recent example may be found in Girard's Linear Logic. In this system, formal deductions are developed by drawing planar links between formulae in a proof tree. Proofs (and cuts) are carried on by an explicit use of the geometric representation, by modifying the links. As the proof-theoretic calculus is essentially complex, according to recent complexity results, the use of the geometric representations comes in as an essential tool for the formal computation.

In a sense, all these calculi derive from (physical) space or Geometry and, after an algebraic or syntactic description, end up in geometric representations, possibly unrelated to the original one. As a matter of fact, even Linear Logic originated in Geometry, as it was suggested by the distributive or linear maps over coherent spaces (Girard [1987]), and ends up in a Geometry of proofs.

3.2 Geometric insight. I would like to mention here the possible relations of the recent mathematical approaches to vision mentioned in 2.4, and similar ones in other forms of perception, to the wise blend of linguistic, or analytic, and geometric experiences required by the practice and the foundation of mathematics.

It should be clear that in mathematics, synthetic explanations may provide an understanding and a foundation as relevant as stepwise reductionist descriptions. The drawing on a blackboard may give as much certainty as the search for least axioms for predicative Analysis. The point now is to understand what is behind the drawing, which intellectual experiences give to it so much expressiveness and certainty. The point is to turn this practice of human communication, by vision and geometric insight, into a fully or better understood part of knowledge. This is where our endeavor towards general knowledge cannot be separated from the foundation of mathematics. Mathematics is just an uppermost human experience in language and space perception, unique in generality and objectivity, but part of our relation to the world.

A few examples may be borrowed from neurophysiology (see Ninio [1991], Maffei [1992]). It seems that only human beings perform interpolations (vertices of a triangle or of a square seen as complete figures, sets of stars as a constellation...). Apparently this is done by minimal lines which complete incomplete images. We seem to interpolate by splines when required. More: there are neurons which recognise (send an impulse) only in the presence of certain angles, or others which react only to horizontal or vertical lines. This is recomposed in intellectual constructions which are at the base of our everyday vision and of the so-called optical illusions (which are just attempted reconstructions of images). So, why not, at the base of our geometric generalizations!

But how? How can we make this "composition of basic mental images" part of a new foundation of mathematical knowledge in the same way as formal, linguistic axioms, have been describing part of the analytic developments in mathematics?

For now, I can only mention the problem, and stress what is really missing, the possible source of incompleteness, namely, the lack of Geometry and images in basic studies. A modern rediscovery of these aspects may be at the core of an understanding of image recognition which goes together with an appreciation of geometric abstraction in mathematics.

In a sense we should enrich the insufficient attempts to deduce all of mathematics by linguistic axioms, by adding, at least, the knowledge we have today of space perception and of the process of image formation. This may help to focus the way in which mathematics develops, admittedly by compositions of elementary components (lines and triangles mentioned before), but also by "synthesis" and reorganization of space, as mentioned in 2.4. In this, a renewed Artificial Intelligence, far from the prevailing formalist one, may be a novel contribution of Computer Science to the foundation of mathematical knowledge, and conversely. The difficult point is to be able to move, in foundational studies and everyday work, from local, quantitative and analytic approaches to global, qualitative and geometric perspectives and still preserve the crucial (informal) rigor of mathematics.

3.3 The continuum and minima: more about reductionism. In *Das Kontinuum* (Weyl [1918]), Hermann Weyl raises the issue of the continuum of Analysis vs. the continuum of time. The understanding of the latter is based on the simultaneous perception of past, present and future. In this irreducible phenomenological intuition of time, it is not possible to isolate the temporal point, in contrast to the analytic description where this can be done by reduction to linguistic abstractions, that is, to symbols and (derivable) properties. Weyl, a mathematician working also in relativity theory, expresses his dissatisfaction and raises a major point for mathematical knowledge: the convenient analytic unity of space and time does not correspond to our fundamental experience (see also Petitot [1992]). This problem has not been sufficiently studied since at that time, we were mostly concerned with formalist reductions and the search for complete and (self-)consistent Set Theories, as a basis for Analysis. These formal theories have not been able to tell us anything even about the cardinality of an arbitrary set of reals (independence of the Continuum Hypothesis), let alone the profound mismatch between time and the analytic description of space, as given by the real line.

This need of ours to "fill up the gaps", possibly by continuity, may probably go together with the principles of minima, mentioned in 2.3 and 3.2 as a possible description of some aspects of abstraction (memory optimization and the formation of images, respectively). These principles are usually very complex in mathematics and, when referring to them, we depart from reductionism. Yet another relevant mathematical experience, then, has to be added to the continuum of time, which seems to escape reductionism. Reductions are surely a relevant part of scientific explanations; however, they are far from proposing complete methodologies or providing the only possible foundation of knowledge.

In conclusion, we need to focus on alternative approaches to formalism and reductionism both in applied as well as in theoretical approaches to cognition. In 2.4 and 3.2, the role of current inverse paradigms is mentioned with respect to the prevailing top-down, deductive formalizations: bottom-up descriptions, for example, which may give a complementary account of perception and conceptual abstraction. What really matters now is to extend, not to keep reducing our tools. Our rational paradigms must be made to comprehend the mathematical, indeed human, intuition of space and time. In other words, we need to lower the amount of magic and mystery in these forms of intuition, and bring them into the light of an expanded rationality.

References

[1] Asperti A., Longo G., *Categories, Types and Structures*, M.I.T. Press, Cambridge, MA, 1991.

[2] Van Dalen D., "Brouwer's dogma of languageless mathematics and its role in his writings", *Significs, Mathematics and Semiotics*, Heijerman (ed.), Amsterdam, 1990.

[3] Enriquez F., *Problemi della Scienza*, 1909.

[4] Girard J.-Y., "Linear Logic", *Theoretical Comp. Sci.* **50** (1987), 1–102.

[5] Girard J.-Y., "The unity of logic", *Journal of Symbolic Logic*, 1992.

[6] Goldfarb W., "Poincaré Against the Logicists", *Essays in the History and Philosophy of Mathematics*, W. Aspray and P. Kitcher (eds.), Minn. Studies in the Phil. of Science, Minnesota, 1986.

[7] Lighthill J., "The recent recognized failure of predictability in Newtonian dynamics", *Proc. R. Soc. Lond.* A 407, 1986, 35–50.

[8] Longo G., "Notes on the foundation od mathematics and of computer science", *Nuovi problemi della Logica e della Filosofia della Scienza*, Corsi, Sambin (eds.), Viareggio, 1990.

[9] Longo G., "Some aspects of impredicativity: notes on Weyl's philosophy of Mathematics and on todays Type Theory", *Logic Colloquium* 87, Studies in Logic, Ebbinghaus et al. (eds.), North- Holland, 1989.

[10] Maffei L., "Lectures on Vision at ENS", in preparation, Pisa, 1992.

[11] Ninio J., *L'empreinte des sens*, Seuil, Paris, 1991.

[12] Petitot J., "L'objectivité du continu et le platonisme transcendantal", Document du CREA, Paris, Ecole Polytechnique, 1992.

[13] Poincaré H., *La Science et l'Hypothèse*, Flammarion, Paris, 1902.

[14] Poincaré H., *La valeur de la Science*, Flammarion, Paris, 1905.

[15] Sambin G., "Per una dinamica dei fondamenti", *Nuovi problemi della Logica e della Filosofia della Scienza*, Corsi, Sambin (eds.), Viareggio, 1990.

[16] Troelstra A.S., "Remarks on Intuitionism and the Philosophy of Mathematics", *Nuovi problemi della Logica e della Filosofia della Scienza*, Corsi, Sambin (eds), Viareggio, 1990.

[17] Weyl H., *Das Kontinuum*, Leipzig, 1918.

[18] Weyl H., *Symmetry*, Princeton University Press, Princeton, 1952.

Pourquoi les mathématiques sont-elles difficiles ?

par Jacques Dubucs
CNRS, Paris (France)

L'objectif de la philosophie des mathématiques est de fournir une conception cohérente des relations qui unissent le sujet mathématicien, les démonstrations qui sont le produit de son activité, et l'univers des objets auxquels ces démonstrations réfèrent. Une telle philosophie devrait, me semble-t-il, satisfaire à deux conditions :

(i) s'en tenir à un cadre *naturaliste*, c'est-à-dire ne pas imputer aux mathématiciens la capacité d'entretenir des relations "directes" avec des entités supra-physiques :

(ii) rendre compte de la totalité de l'expérience mathématique, à la fois dans son étendue (on se contentera pas d'une méditation assidue sur des vérités arithmétiques du type "$7 + 5 = 12$") et dans ses qualités distinctives, qui sont pour l'essentiel l'objectivité et l'applicabilité au monde physique.

Dans ce bref exposé, je me propose

1) de dresser une vue cavalière de la situation actuelle en philosophie des mathématiques :
2) de montrer en quoi, dans la situation présente, aucun système de philosophie des mathématiques ne remplit pleinement les deux conditions que je viens d'énoncer, et notamment qu'aucun ne rend compte adéquatement d'une dimension trop négligée de l'expérience mathématique, à savoir sa *difficulté*.

1.

On constate que dans la période récente l'essentiel des discussions en philosophie des mathématiques s'est concentré sur le problème fondamental de l'objectivité des mathématiques, et plus particulièrement sur la question de savoir si l'on peut ou non assurer cette objectivité sans faire référence à un univers d'objets ou de faits spécifiquement mathématiques.

1.1. Sommairement décrite, la controverse est la suivante. Pour le réaliste (ou le "factualiste"), les théorèmes mathématiques sont vrais, le mot "vrai" signifiant ici ce qu'il signifie partout ailleurs : "conforme aux faits". Les énoncés mathématiques cherchent à décrire une réalité que nous n'avons pas construite, et ils sont vrais lorsque les objets auxquels ils se réfèrent ont bien les propriétés qu'ils leur attribuent. Le réalisme introduit donc une agréable unité dans les affaires de la science, puisque les mathématiques sont objectives au même sens que la physique ou la cartographie. De plus, le réalisme semble massivement attesté par l'expérience du travail mathématique, qui est (en grande partie) de l'ordre de la contrainte et de l'hétéronomie, du sentiment d'être affecté, d'avoir à s'acclimater à un univers d'entités déjà là. C'est l'exemple de la répartition des nombres premiers, aussi heurtée que les côtes du Morbihan. Ou celui des définitions canoniques : plonger une structure dans une autre, ce qui paraît un libre décret, et s'apercevoir ensuite que toute autre définition aurait donné une structure isomorphe, comme si cette possibilité était déjà réalisée en un objet idéal. Le réalisme éclaire bien cette propension à réifier. Deux objections majeures expliquent cependant qu'une conception aussi séduisante ne soit pas partout reçue.

(i) Etant donné le genre d'indétermination dont la référence des termes mathématiques est insurmontablement affligée (cette référence est définie à un isomorphisme près au mieux), les objets invoqués par le réalisme sont proliférants ("2" peut référer aussi bien à $\{\{\emptyset\}\}$, construction de Zermelo, qu'à $\{\emptyset, \{\emptyset\}\}$, construction de von Neumann). Et surtout cette luxuriance d'objets qui restent *causalement inertes* n'a aucun titre à figurer dans une explication naturaliste de l'activité mathématique. C'est ce que l'on appelle le "dilemme de Benacerraf" : si nous voulons maintenir pour les mathématiques une *sémantique* de même nature que celle qui convient pour les autres sciences, c'est-à-dire une sémantique véríconditionnelle, nous sommes apparemment contraints d'introduire un nombre indéfini d'objets qui transcendent les capacités cognitives les plus larges qu'une *théorie naturaliste de la connaissance* est disposée à nous reconnaître[1].

(ii) S'agissant des faits mathématiques allégués par le réalisme, (a) ou bien ils sont, d'une certaine manière, aussi *contingents* que les faits physiques eux-mêmes, et dans ce cas l'*applicabilité* des mathématiques au monde réel semble relever du miracle : comment expliquer, sinon par une déraisonnable *chance*, que le monde physique fasse justement partie de ceux des mondes possibles dans lesquels les faits mathématiques en question sont avérés ? (b) Ou bien ils sont *nécessaires*. Mais alors il peut difficilement s'agir d'une nécessité *logique*, qui impliquerait non seulement que les énoncés déduits des axiomes mathématiques en soient des conséquences logiques (ce que l'on est, après tout, en droit d'espérer...), mais que les *axiomes* eux-mêmes soient logiquement valides, ce qui est (paix à Russell) une thèse éminemment douteuse : l'axiome de choix, l'axiome de l'infini, qui sont des axiomes existentiels, ne sont pas des vérités logiques *stricto sensu*. Et quant à la notion d'une nécessité spécifiquement *mathématique*, intermédiaire si l'on ose dire entre la nécessité logique et la contingence physique, il faut reconnaître qu'elle reste à l'heure actuelle un concept tout à fait obscur.

1.2. A ces deux difficultés essaient de répondre un certain nombre de perspectives *rivales* du factualisme réaliste, qui ont en commun d'explorer l'*autre* ver-

[1] Cf. Benacerraf 1983.

sant du dilemne de Benacerraf, et refusent de fonder l'objectivité mathématique sur l'existence de faits qui transcenderaient nos capacités cognitives les plus libéralement conçues :

(i) L'intuitionnisme est une forme de factualisme non réaliste, qui maintient, en un sens, l'idée de vérité mathématique comme conformité à des faits, mais à des faits "immanents", *en principe* cognitivement accessibles par leur nature même, à savoir les constructions mentales du mathématicien (A est vrai si et seulement si il existe une construction mentale — ou une preuve — de A).

(ii) Un ensemble hétérogène de doctrines partagent le rejet de toute forme de factualité mathématique :

— La conception "syntaxique" des mathématiques issue du *Tractacus* de Wittgenstein (ou en tout cas de l'interprétation qu'en ont donnée les membres du Cercle de Vienne), et selon laquelle les mathématiques se réduisent à des tautologies sur la base de conventions linguistiques.

— Le fictionnalisme de H. Field, qui fonde une conception purement instrumentale des mathématiques sur un résultat de conservativité selon lequel si N est une théorie "nominaliste" (dont le language ne comporte aucune référence à un objet mathématique), alors tout énoncé de N prouvable dans $ZF + N$ est déjà prouvable dans N seul. En d'autres termes : les mathématiques sont un instrument *en principe superflu* (quoique pratiquement utile) dans la preuve des énoncés nominalistes, et l'on peut rendre compte de l'applicabilité des mathématiques sans prendre leur ontologie au sérieux[2].

— L'anti-réalisme radical issu des écrits du dernier Wittgenstein, qui refuse toute distinction tranchée entre les raisons pour lesquelles un énoncé mathématique *est* vrai et les raisons pour lesquelles nous *croyons* qu'il l'est. Cette approche radicalise la doctrine intuitionniste en limitant le principe de bivalence aux seuls énoncés pour lesquels une procédure de décision peut être *pratiquement* mise en œuvre (et non pas seulement aux énoncés décidables *en principe*).

2.

Je ne discuterai pas ici *directement* la question controversée de savoir si les différentes perspectives anti-réalistes (ou anti-factualistes) que je viens d'évoquer sont ou non capables de rendre compte raisonnablement de l'objectivité des mathématiques, mais j'aborderai cette question de façon oblique, en demandant si elles sont ou non aptes à expliquer un trait des mathématiques qui paraît être curieusement considéré par bien des philosophes comme tout à la fois secondaire et marginal, à savoir leur *difficulté*. Il me semble que la plupart des philosophes dont je viens de parler ont tellement insisté pour arracher le voile de mystère dont

---

[2] Cf. Field 1989.

ils croient, à tort ou à raison, que le mathématicien aime à entourer son activité, et pour convaincre leurs lecteurs que les énoncés mathématiques doués de sens sont précisément ceux dont les ressources cognitives les plus ordinaires ne sauraient manquer de venir à bout, qu'ils ont rendu presque incompréhensible la situation pourtant tout à fait commune dans laquelle un individu se trouve embarrassé, ignorant, ou incompétent en cette matière (naturellement le réalisme factualiste, quant à lui, est tout à fait indemne de ce reproche : lui a au moins le mérite de rendre compte impeccablement de la *dureté* des mathématiques, au double sens de la solidité et de l'opacité).

2.1. Je ne prendrai qu'un exemple des conséquences pernicieuses d'une doctrine de ce genre, en examinant le cas de la philosophie post- tractarienne des mathématiques, la tradition du Cercle de Vienne. Il est clair, jusqu'à un certain point, que cette tradition représente un infléchissement des thèses défendues dans le *Tractacus* : la signification d'un énoncé n'est plus ici identifiée à la classe des mondes possibles où il est vrai, mais à sa méthode de vérification empirique. Toutefois cet infléchissement laisse intacte la détermination du contenu de signification des théorèmes mathématiques : ce contenu est vide dans les deux cas, soit parce que les théorèmes, vrais dans tous les mondes possibles, n'en déterminent aucune sous-classe propre, soit parce qu'aucune expérience sensorielle n'est requise pour les vérifier. Quant au postulat proprement positiviste d'une nette démarcation entre la composante factuelle (empirique) et la composante linguistique dans les facteurs qui concourent à la détermination de la valeur de vérité des énoncés, il suggère bien une solution à la question de la *difficulté* des (inférences) mathématiques, question que Wittgenstein pour sa part abandonnait purement et simplement à la "psychologie" : un individu qui n'aperçoit pas que *B* découle logiquement de *A* ne manque pas d'une information *factuelle* (sinon *A* n'entraînerait *B* qu'en présence du fait additionnel rapporté dans cette information, et n'en serait pas une conséquence logique), mais d'une information de type *linguistique* (il ne comprend pas comme il le devrait les termes qui figurent dans *A*). Or une telle explication est tout à fait inacceptable. D'une part parce qu'une caractéristique saillante des mathématiques est justement l'abondance des énoncés qui sont *difficiles* à décider tout en étant assurément à la portée d'une maîtrise, même modeste, des conventions linguistiques (exemple typique : le "dernier théorème" de Fermat), et qu'il est donc nécessaire de préserver partout la possibilité de distinguer entre l'incompétence bien compréhensible de celui qui n'a pas vu la preuve, et celle, plus énigmatique, de celui qui n'a pas compris l'énoncé. D'autre part parce qu'en supposant que l'aptitude à reconnaître les théorèmes mathématiques comme tels est une compétence de nature essentiellement linguistique, on s'engage inévitablement à faire de la nécessité même (de l'"inexorabilité") des mathématiques une propriété qui découle de leur nature conventionnelle, donc à fonder cette nécessité sur une contingence. Enfin parce que cette thèse du caractère "linguistique" des théorèmes mathématiques est *réfutable* lorsqu'elle est présentée sous sa forme la plus précise — donc la plus vulnérable —, selon laquelle les théorèmes sont "analytiques" au sens où ils pourraient se ramener à des variantes du principe d'identité moyennant des transformations définitionnelles : le fait que toutes les identités arithmétiques du type "$7 + 5 = 12''$ puissent être ainsi *effectivement* réduites moyennant les définitions usuelles de "successeur" et de "plus" ne peut bien évidemment pas être extrapolé à l'ensemble des théorèmes arithmétiques, qui est indécidable.

2.2. L'échappatoire suggérée par cet échec consiste naturellement à admettre une notion d'information qui ne soit ni empirique ni linguistique, mais de type proprement mathématique. Etant acquis qu'un individu (ne) donne généralement son assentiment à un énoncé (qu') après en avoir vu une preuve — ou encore, que c'est la mise au point de la preuve qui est difficile, non la vérification du fait que la preuve est bien une preuve —, un candidat naturel pour cette notion d'information est évidemment représenté par la composante de la preuve qui ne se réduit ni à la réitération des prémisses ni aux transformations définitionnelles routinièrement applicables aux termes qui y figurent. Il n'est pas aisé de donner de cette composante informative une définition unitaire, et je me contenterai ici de deux remarques.

a) Bien qu'il existe évidemment des cas où la preuve d'un théorème résulte d'une simple *élucidation* de la situation à laquelle l'énoncé fait référence (il en va ainsi, typiquement, de la preuve que deux droites concourantes déterminent deux angles congruents), il en va différemment dans la plupart des preuves non triviales, qui requièrent la construction d'objets "auxiliaires" absents de la situation mentionnée dans les prémisses (exemple canonique[3] : la preuve euclidienne que la somme des angles d'un triangle est égale à deux droits, immédiate *une fois* une certaine droite auxiliaire tracée). Notre incapacité de "voir" la conclusion d'un argument à partir de la situation exposée dans les prémisses n'a souvent pas d'autre origine qu'un défaut objectif de constructions auxiliaires, et l'expérience accumulée depuis un demi-siècle dans le domaine des "machines à démontrer" permet sur ce point de préciser le type d'"ingéniosité" requis par le raisonnement mathématique[4] : si l'on peut concevoir une règle mécanique permettant d'engendrer de nouveaux objets, il ne saurait, en revanche, en exister une pour déterminer combien sont nécessaires à la preuve. Cette impossibilité de prédire récursivement le nombre d'objets à introduire pour prouver un théorème (et, conséquemment, de fixer une borne supérieure à la longueur de la preuve

[3] Cf. Aristote, *Métaphysique*, Θ 9, 1051 a 21–30.

[4] Cf. A. Robinson 1983, qui est le premier à avoir fait le rapprochement entre les constructions auxiliaires de la géométrie et les constantes neuves mentionnées dans le théorème de Herbrand.

cherchée), qui est, par exemple, la contrepartie de l'*indécidabilité* de la géométrie usuelle, semble fournir une explication correcte de la difficulté *computationnelle* des mathématiques.

b) L'idée que la mise au point d'une preuve pourrait résulter d'une simple analyse "linguistique" des termes contenus dans l'énoncé apparaît rétrospectivement comme une aberration philosophique peu disséminante, et la complexité computationnelle des processus démonstratifs a été très tôt et très constamment reconnue. Une discussion mieux engagée concerne le point de savoir si cette complexité est la seule en jeu. On rencontre ici un précepte méthodologique vénérable, celui de la *pureté*, en vertu duquel :

— il existe une stratification naturelle dans laquelle les objets mathématiques viennent s'ordonner par rang de complexité croissante :

— la justification d'un énoncé mettant en jeu des objets de rang donné ne devrait pas faire référence à des objets de rang plus élevé.

Le principe de pureté entraîne notamment l'existence d'un segment initial, "élémentaire", de la hiérarchie des objets et des concepts mathématiques, jouissant d'une propriété remarquable de *stabilité épistémique* : chaque problème *énonçable* en termes élémentaires devrait également être *résoluble* en termes élémentaires. Or il est significatif que les domaines atteints par les tentatives répétées pour cerner l'*élémentarité mathématique* (ensemble des objets constructibles à la règle et au compas pour les Grecs, ensemble des courbes algébriques pour Descartes, classe des énoncés finitistes pour Hilbert) *n'*aient *jamais* vérifié cette propriété de stabilité : même dans les domaines pouvant être raisonnablement considérés comme "élémentaires", certaines preuves sont nécessairement impures, c'est-à-dire qu'elles introduisent non seulement des objets non mentionnés dans l'énoncé du théorème (complexité computationnelle), mais encore des objets d'un *type* différent de ceux auxquels l'énoncé fait référence (complexité conceptuelle).

Bien que l'on soit aujourd'hui fort loin de savoir articuler systématiquement ces deux types de complexité, certains résultats de théorie de la démonstration (théorème d'élimination des coupures de Gentzen, résultats de "speed-up" de Gödel) permettent de comprendre qu'elles varient, d'une certaine manière, en sens inverse l'une de l'autre : grossièrement parlant, on fait croître de façon exponentielle la longueur d'une preuve lorsqu'on la "purifie" (quand ceci est possible). Ceci suggère naturellement une mesure *intrinsèque* de la complexité d'un théorème : le *produit* de la complexité computationnelle par la complexité conceptuelle de l'une quelconque de ses preuves[5].

D'une analyse plus précise de ces deux composantes dans la complexité d'une preuve dépend l'établissement d'une notion correcte de *l'état d'information* associé à un théorème (la donnée des types d'objets, et, dans chaque type, la donnée

[5] Cf. G. Kreisel,1971, p. 259.

des objets requis par une preuve du théorème en question), et donc la construction d'un concept de *synonymie* entre énoncés mathématiques (deux énoncés étant considérés comme synonymes si les états d'information associés coïncident). Privée d'un tel concept, et par là-même réduite à une taxinomie grossière qui la contraint à identifier les membres de toute paire d'énoncés mathématiques possédant même valeur de vérité dans tout monde possible, c'est-à-dire prouvablement équivalents, aucune philosophie ne saurait rendre compte de la difficulté de l'expérience mathématique.

Bibliographie

[1] Benacerraf P., "Mathematical truth". In *Philosophy of Mathematics, Selected Readings*, Benacerraf and Putnam (eds.), 2nd ed., 1983, p 403–420.

[2] Dubucs J., "Die Sogenannte Analytizität de Mathematik: für eine Radikalinierung der Theorie Hintikkas," *Grazer Philosophische Studien*, XXXII, 1988, pp 83–112.

[3] Field H., *Realism, mathematics and modality*, T.J. Press Ltd., Padstow, Cornwall, 1989.

[4] Kreisel G., Compte rendu des *Collected Papers* de G. Gentzen, *Journal of Philosophy*, LXVIII, 1971, p 259.

[5] Robinson A., "Proving a Theorem (as Done by Man, Logician or Machine)". In *Automation of Reasoning I. Classical Papers on Computational Logic*, 1957–1966, J. Siekmann & G. Wrightson (eds.), Springer, 1983, p 74–76.

Le geste bien plutôt que la règle

par Gilles Chatelet
Université Paris VIII, Paris (France)

"Heureusement pour les chercheurs,
à mesure que les brouillards se dissipent
sur un point, c'est pour se reformer sur un autre".
A. Weil

A. Weil n'a pas manqué la cible : "les troubles reflets", les "futures caresses", les "brouillards" qui excitent le désir confus de renouer charnellement avec ce qui est perçu comme une totalité mutilée par la dispersion technique, une véritable nostalgie de la puissance magique exaspérée par l'incapacité du rationalisme classique à saisir tous ces tours de main, toutes ces "recettes", toutes ces expériences de pensée, ces figures et ces diagrammes, toutes ces dynasties de problèmes capables, semble-t-il, du "miracle" de la réactivation et qui s'exercent précisément aux points sensibles mais aveugles de l'entendement, aux charnières-horizons où se déploient les systématicités *naissantes*, en tous ces lieux où l'orientation ne s'obtient pas à titre gracieux et où le vrai ne se laisse pas saisir comme le vérifiable.

Gestes et problèmes guident l'œil et la main à l'insu des géomètres ou des philosophes, mais ils *font époque*, et leur exemplarité dynastique ne se laisse jamais épuiser par le constat ou l'hypothèse "d'influences" : quelque chose retentit avec Galois, Faraday, Maxwell ou Heisenberg. Mais les substances, les catégories constituées, les structures, les objets, les formes ne *retentissent* pas...

Une philosophie offensive ne saurait se contenter de ratiociner indéfiniment sur le "statut" des objets scientifiques et doit se situer résolument aux avant-postes de l'obscur, en ne considérant pas l'irrationnel comme "diabolique" et réfractaire à l'articulation : une science qui resterait fascinée par la prédiction serait bien incapable de relever le fameux défi de Heidegger, et c'est pourquoi une philosophie qui prétend ignorer la *contemplation opérative* dont se réclament les plus grands créateurs est bien vite réduite à s'agenouiller devant l'opérationnel et la "faisabilité".

Comment penser cet effet de catalyse des "brouillards" de Weil sans les émasculer par des paraphrases ou des clichés de dépannage ? Un mathématicien d'une telle trempe sait bien que la science s'appuie sur deux béquilles :

— la béquille officielle du texte littéral qui donne le procès verbal de l'effectuation des opérations et qui assure la transmission du savoir,

— la béquille réservée aux initiés qui savent lire entre les lignes, où se joue la réactivation des gestes qui multiplient le savoir et où se fomente le passage du pressentiment à la certitude.

La science dans son fonctionnement ordinaire et pour sa transmission "fait semblant" d'user surtout de la première béquille. Ainsi, on peut très bien anticiper, prédire, stocker des données et surtout "effectuer les opérations", c'est à dire tirer le meilleur parti de la disjonction artificielle du produit et de la productivité, en oubliant l'âpreté de la conquête de l'homogénéité réclamée par l'opération pour assurer l'extériorité de la relation à ses termes.

Il y a une espèce de pudibonderie ingrate de la "clarté" opératoire qui s'efforce toujours de voiler, d'aplatir ou d'oublier que le geste qui découpe une forme d'articulation — une liaison ou une fonction — se déploie en faisant surgir à son insu une *autre* articulation : l'artifice qui dispense une forme possède toujours un point aveugle : "si un brouillard se dissipe sur un point, c'est pour se reformer sur un autre".

La peur de l'irrationnel se nourrit certes de notre défaut de savoir, mais surtout de ces prurits et de ces points d'oubli, d'une certaine impatience qui nous empêche de pénétrer au cœur de l'opérativité en jugeant de l'apprendre par la virtuosité à consommer au plus vite un "contenu informatif". Mais apprendre, c'est se préparer à apprendre ce que l'on connaît déjà en quelque manière, et donc se mettre dans l'état où la connexion des choses retentit sur la connexion de l'esprit.

L'opération n'est pas d'abord donnée comme une flèche qui relie une source à une cible, mais émerge aux lieux où les variables se confondent et s'enchevêtrent sans être policées par des parenthèses. C'est ici que bat le cœur de l'opérativité, que peut être atteinte la relation donnée implicitement comme "courbe" $Q(x, y) = 0$, où la détermination s'enveloppe elle-même et libère ensuite en ses points singuliers des relations externes (relations "paramétriques" ou "fonction" du type $y = P(x)$), où la séparation entre les deux membres — donc entre donnée et résultat — est déjà consommée.

D'autre part, il arrive qu'une présentation nouvelle ébranle tout un continent mathématique. Un exemple flagrant est celui de la "redécouverte" des nombres "imaginaires" par l'introduction du plan complexe. Cette nouvelle rencontre du géométrique et de l'algébrique et donc du *visible et du calculable*, imposera les travaux de Hamilton et Riemann (et de bien d'autres), tout en retentissant aussi sur la notion même d'application des mathématiques : la physique, la géométrie, l'algèbre vont coopérer pour constituer un véritable espace électrogéométrique dont s'inspirent les recherches les plus contemporaines.

Il y a bien urgence pour la philosophie à pénétrer plus avant dans ce goulet où s'articulent les figures qui innervent la contemplation, et les formules qui actualisent les opérations de détermination. Mais il s'agit d'un endroit stratégique déjà hanté par deux autres combattants au front de l'obscur : l'Art et la Science. Ici, par le diagramme, la métaphore amorce la mue qui la métamorphosera en opération et c'est pourquoi ce recoin grouille de clichés qui s'efforcent de convier au spectacle d'une opérativité retrouvée.

Le cliché dissout dans un rapport de ressemblance le geste qui avait su déceler patiemment une identité de rapports : il prétend ainsi attendrir par une chaleureuse confusion, la technicité de l'opération et son extériorité "froide".

Mais de tels clichés viennent toujours trop tard ! Ils seront toujours dépréciés et dénoncés comme ce qui usurpe la place prestigieuse de l'opération. Les superstitions n'ont rien à craindre de cette opérativité du pauvre qui laisse intacte l'extériorité de l'opération. Le cliché n'entame rien de l'arrogance du fonctionnel mais entend bien disqualifier toute pensée un peu conceptuelle comme le montre la panoplie puérile charriée par le Big-Bang (qui ne cesse de commencer), le Chaos (qui neutralise tout), la Catastrophe (qui n'effraie plus grand monde), le Fractal (qui fascine surtout les esprits un peu simples), le Hasard (dont il est bien connu que Dieu l'a en horreur), etc.

Dans le passage consacré à l'expérience physique de sa *Méthode axiomatique et formalisme*, J. Cavaillès exige que soit retrouvée "l'intuition centrale saisissable dans l'action d'une théorie" et que la comprendre c'est "en rattraper le geste et pouvoir continuer" [2].

Ce concept de *geste* nous semble crucial pour approcher le mouvement d'abstraction amplifiante des mathématiques qui échappe aux paraphrases ratiocinantes — toujours trop lentes —, aux clichés et à leurs fascinations confuses.

Ce geste ne se laisse pas "attraper" devant nous — il ne lance pas des ponts entre nous et les choses, ni "derrière" nous —; aucun algorithme ne commande sa mise en scène. Il faudrait plutôt parler *d'une propulsion qui se referme en une impulsion*, qui décape une structure en réveillant en nous d'autres gestes.

Le geste n'est pas substantiel : *il gagne de l'amplitude en se déterminant.* Sa souveraineté est à la mesure de sa pénétration et c'est pourquoi, on évoque la "sûreté" du geste : le retentissement de son adresse est gagé par la précision de sa frappe. Le geste inaugure une lignée de gestes, alors que la règle n'énonce que des "instructions", qu'un protocole de décomposition de l'action en actes répétables indéfiniment.

Le geste possède une exemplarité historiale : si on peut parler *d'accumulation* du savoir au cours de la succession des *générations*, on doit parler de gestes *inaugurant des dynasties* de problèmes. Le geste n'est pas un simple déplacement spatial : il décide, libère et propose une nouvelle modalité du "se mouvoir". Hugues de SaintVictor définissait le geste comme : "le mouvement et la figuration des membres du corps selon la mesure et les modalités de toute action et attitude". Le geste renvoie à une discipline de distribution de la avant tout transfert : *on se pénètre du geste avant de le savoir* [3].

Le geste est élastique, il peut se ramasser sur lui-même, sauter au delà de lui-même et retentir, alors que la fonction ne donne que la forme d'un transit d'un terme extérieur à un autre terme extérieur et que l'acte s'épuise dans son résultat : le geste a donc à voir avec le pôle implicite de la relation.

Le geste enveloppe avant de saisir et esquisse son déploiement bien avant de dénoter ou d'exemplifier : ce sont les gestes déjà domestiqués qui font référence.

Un geste sait mettre en réserve toutes les virtualités provocatrices de l'allusion sans la dégrader en abréviation.

On ne sera donc pas surpris d'être surtout attentif aux lignées de gestes de découpe, aux diagrammes qui les saisissent au vol, aux expériences de pensée qui débusquent la charnière où les principes portent à faux, et surtout aux dispositifs de symétrisation construits pour solliciter une dissymétrie moins attendue et qui oriente donc un projet plus radical.

Un diagramme peut immobiliser un geste, le mettre au repos, bien avant qu'il ne se blotisse dans un signe et c'est pourquoi les géomètres ou les cosmologistes contemporains aiment les diagrammes et leur pouvoir d'évocation péremptoire : ce sont les sourires de l'Etre.

Les diagrammes sont un peu les complices de la métaphore poétique. Mais ils sont un peu moins impertinents — il est toujours possible de trouver refuge dans le tracé ordinaire en une opération qui les sauvent de l'usure. Comme la métaphore, ils bondissent pour créer des places et réduire des écarts : ils bourgeonnent de pointillés pour déborder les images déjà figurées en traits gras.

Mais le diagramme ne s'épuise pas comme la métaphore : s'il immobilise un geste pour déposer une opération, c'est en esquissant un geste qui en découpera un autre. Le pointillé ne renvoie ni au point et à sa désignation discrète, ni à la ligne et à son tracé continu, mais à la pression de la virtualité qui inquiète l'image déjà disponible pour faire place à une dimension nouvelle : le mode d'existence du diagramme est tel que sa genèse fait partie de son être. On pourrait parler à son propos d'une *technique d'allusions*.

Nous savons comment G. Simondon [4] a su dégager des procès de convergence et d'adaptation du geste technique qui autorisent à parler de *lignée technique*. Il y a donc des lignées de diagrammes aux allusions de plus en plus ajustées et de plus en plus ambitieuses, comme il y a des lignées de plus en plus autonomes et de plus en plus concrètes et, pas plus que l'objet technique ne vient après un savoir, le diagramme n'illustre ou ne traduit simplement un contenu déjà disponible.

Les diagrammes ont donc à voir avec l'expérience et se révèlent capables de s'approprier et de véhiculer "tout ce parler avec les mains" (qu'il faudrait peut-être mieux appeler ce parler *dans* les mains), dont les physiciens sont si fiers et qu'ils réservent, bien sûr, aux initiés. Une philosophie du physico- mathématique ne saurait ignorer cette pratique symbolique en amont du formalisme, pratique de condensation et d'amplification d'intuition.

Le diagramme est un projet qui vise à ne s'appuyer que sur ce qu'il esquisse : cette exigence d'autonomie en fait le complice naturel des expériences de pensée. Celles-ci ne prétendent pas prédire ou vérifier une loi, ni heurter de front la causalité des choses : elles recherchent plutôt à s'y soustraire, à faire le vide des lourdes évidences et des intuitions. L'expérience de pensée n'est pas "idéale" ou "théorique" parce qu'impossible à effectuer avec des instruments "réels", mais parce qu'elle prétend questionner ou découvrir des pratiques d'idéalisation. Elle est à sa manière très pratique : *c'est l'épreuve par laquelle le physicien-philosophe prend sur lui de se désorienter*, de connaître la perplexité inhérente à toute situation où le discernement ne va nullement de soi. Il s'agit pour lui d'orchestrer une subversion des habitudes associées à des clichés sensibles (la "réalité" décidant alors pour lui), et de se transporter par la pensée dans les enceintes hors causalité, à l'abri des forces pour se laisser flotter entre Mathématique et Physique.

L'expérience de pensée menée à son terme est une expérience diagrammatique, où il devient manifeste que le diagramme est à lui-même sa propre expérience.

Il existe certains dispositifs, délibérément producteurs d'ambiguïté, qui suscitent des expériences de pensée pénétrant plus avant au cœur de la relation et de l'opérabilité : ce sont ce que nous pourrions appeler les *stratagèmes allusifs*. Un stratagème allusif compresse et déploie mais ne se confond ni avec une abréviation, ni avec une explication. Les stratagèmes appréhendent l'opération dans sa naissance même, comme c'est l'ambition de toute théorie des singularités. Cette théorie se propose de maîtriser les points où le choix canonique d'un paramètre — ce qui permet de conduire un parcours et d'écrire $y = f(t)$ — ne "va plus de soi". Ces points imposent donc l'invention d'un continuum susceptible d'accueillir *toutes* les solutions d'une équation algébrique ou fonctionnelle.

Les stratagèmes allusifs ne se limitent pas à la relation adulte, en train de "fonctionnner", c'est à dire de proliférer comme séquences de termes à enrober dans des parenthèses suivant une règle déjà établie, mais aménagent des expériences de pensée — géométriques ou physiques — qui *dispensent la relation comme courbe implicite* et transportent au cœur de l'opérativité *comme projet et comme préjugé innocent* de toute causalité.

Avec ces dispositifs, la Science détient les pièces à conviction qui attestent, mieux que les infinis quantitatifs ou les multiplicités les plus exotiques de la Théorie des Ensembles, qu'elle maîtrise l'infini. Ici, veillent les stigmates des gestes farouches et amples qui ne sont jamais "produits", mais découpent les formes, dressent et écartent les axes, s'installent aux horizons, immobilisent ce qui, "par définition" serait censé être mobile, posent le signe = qui sépare et lie les deux "membres" d'une équation, imposent un *rythme de déroulement* — un "paramètre" — aux indivisions les plus têtues.

Dans sa marche ordinaire, la Science semble se limiter aux gestes qui assurent la maintenance du savoir, et laisser sommeiller le patrimoine de ceux qui l'embrassent et le multiplient. Ceux-ci pourtant la sauvent de l'accumulation et de la stratification indéfinies, de la niaiserie des positivités établies, du confort des transits de "l'opérationnel" et enfin de la tentation de se laisser boucler par une grammaire.

Ils montrent l'urgence d'une pensée authentique de l'information qui ne serait pas seulement commise à la communication, mais viserait à une saisie rationnelle de l'allusion et de l'apprendre sur l'apprendre. Celle-ci, bien sûr, serait assez éloignée d'une certaine barbarie neuronale qui s'épuise à débusquer le récipient de la pensée et à confondre l'apprendre avec une razzia sur un butin informatif. Schelling voyait peut-être plus clair : il savait que la pensée n'était pas encapsulée dans une cervelle, qu'elle pouvait être partout, ... dehors,... dans la rosée du matin.

Bibliographie

[1] A. Weil, "De la Métaphysique aux mathématiques", *Œuvres Scientifiques*, Springer-Verlag, II, p 408–412.

[2] J. Cavailles, *Méthode axiomatique et formalisme*, Hermann, 1981, p. 178.

[3] J.-C. Schmitt, *La raison des gestes*, Gallimard 1990, p 177.

[4] G. Simondon, *Du mode d'existence des objets techniques*, Aubier 1969, p 19–49; p 158–177.

Discussion

Fabrice Pataut (Chargé de cours à l'Université Paris I)

Trois questions à Gilles-Gaston Granger, qui intéressent aussi bien Jacques Dubucs et Jean Petitot, puisqu'elles se rapportent à la question générale de l'objectivité mathématique et concernent les relations entre sujet connaissant (idéalisé), prouvabilité (en principe) et "objets" mathématiques.

Pouvez-vous apporter des éclaircissements sur le concept (ou, comme vous dites, le "métaconcept") d'objet auquel vous avez recours pour défendre l'idée que le (méta)concept d'opération fonctionne comme médiateur entre le (méta)concept de forme et celui d'objet ?

Vous dites, en particulier, que les opérations mathématiques sont créatrices d'objets et que les objets mathématiques sont des corrélats d'opérations pures. J'ai deux questions à ce propos. Pensez-vous qu'il faille, de manière générale, défendre une conception constructiviste — en un sens large — des mathématiques ? En particulier si les opérations sont créatrices d'objets, une opération logique crée un objet. Lequel ?

Gaston-Gilles Granger

1) Le métaconcept d'objet, qui se réalise aussi bien dans les entités mathématiques que dans les objets des sciences empiriques, n'est défini selon moi qu'en corrélation — en dualité — avec le métaconcept d'opération, qui se réalise dans les systèmes opératoires mathématiques aussi bien que dans ceux des sciences empiriques.

Les formes, dans les deux cas, apparaissent comme des formes d'objets, mais ne le peuvent justement que dans la mesure oú des systèmes opératoires déterminent les propriétés des objets. D'une certaine manière le métaconcept de forme n'est rien que la détermination de l'objet. En logique, à la limite, l'objet n'est qu'opération et que forme. En mathématiques, l'objet comporte une opacité qui est son contenu formel.

2) De sorte qu'en logique *stricto sensu* (calcul des propositions), on a le degré zéro du contenu formel. C'est ainsi qu'on pourrait interpréter l'aphorisme du *Tractatus* de Wittgenstein que "les constantes logiques ne sont les représentants de rien" (4.0312).

3) Le constructivisme en mathématiques, oui, étant donné la corrélation opération/objet; mais pas au sens oú la réalité de l'objet mathématique serait complètement épuisée par la procédure de construction (par exemple formulée dans un système d'axiomes et de règles). Je pense qu'il y a toujours en mathématiques un débordement de l'opératoire par l'objectal, qui est le contenu formel. C'est en somme la vérité partielle du "platonisme".

Michel L. Lapidus (mathématicien, University of California)

Question à Jean Petitot : connaissez-vous un critère objectif qui permette de distinguer entre les "idées" mathématiques appartenant au paradis platonicien et celles qui n'y appartiennent pas ?

Jean Petitot

J'ai précisément essayé de montrer que le platonisme ne concerne pas, comme on le dit en général, une ontologie d'idéalités qui seraient comme des entités individuées, séparées et indépendantes appartenant à un troisième monde (ni physique, ni psychologique). J'adhère donc à la critique philosophique traditionnelle du réalisme des universaux. Selon moi, le platonisme concerne uniquement le statut de l'objectivité propre aux idéalités mathématiques. Et j'ai opposé objectivité à ontologie (ou a existence au sens ontologique). Parler de "paradis platonicien" c'est admettre un mode ontologique de l'existence et de la réalité mathématiques. Parler de "critère objectif d'appartenance" à un tel paradis c'est admettre qu'il existe une caractérisation de ce mode d'existence qui précéderait les mathématiques et serait en quelque sorte absolu. Or ces deux hypothèses n'ont aucun sens dans une doctrine transcendantale comme celle que j'ai présentée sous le nom de "platonisme négatif". Mais il existe bien, en revanche, des critères d'objectivité des idéalités mathématiques, même si "objectivité" ne signifie pas "appartenance" à un "paradis". J'en ai signalé un certain nombre dans mon exposé.

Fabrice Pataut

Question à Jacques Dubucs : Le dilemme de Benacerraf fait valoir que si nous disposons d'une sémantique vériconditionnelle unifiée, alors, nous héritons, dans le cas mathématique, du problème épistémologique insoluble de l'inefficacité causale des objets, autrement dit du genre de problèmes qui motive en général les entreprises nominalistes.

Supposons que nous expliquions la signification des phrases en termes de leurs conditions de vérité. dans le cas des phrases mathématiques, ces conditions seront satisfaites par des objets : nombres, fonctions, ensembles, etc., qui n'ont aucune propriété spatio-temporelle et avec lesquels nous ne pouvons donc avoir aucune interaction causale. Le résultat est que si nous voulons une sémantique vériconditionnelle homogène, nous avons une épistémologie mathématique irrecevable. Supposons, maintenant, que nous disposions d'un argument convaincant contre la possibilité d'une sémantique vériconditionnelle, et que nous pensions que les raisons pour lesquelles une proposition mathématique est vraie sont très exactement les raisons pour lesquelles nous *sommes justifiés à croire* qu'elle est vraie. Nous n'en héritons *pas moins* du problème épistémologique de l'inefficacité causale.

Dans la mesure où il semble ainsi que la véricondititionnalité joue un rôle neutre dans le dilemme de Benacerraf, et dans, la mesure, par ailleurs, où ni l'intuitionnisme — que vous avez justement qualifié comme un "factualisme", une sorte de réalisme des preuves — ni, plus généralement, l'anti-réalisme ne permettraient de résoudre l'énigme posée par le dilemme de Benacerraf, on peut se demander si le fauteur de troubles ne se trouve pas ailleurs. N'est-ce pas la conception de la vérité-correspondance qui est en cause ici ? Pensez-vous, par ailleurs, qu'un fictionnalisme du type envisagé par Hartry Field, qui distinguerait entre (i) croire qu'une proposition mathématique est *littéralement* vraie et (ii) croire qu'une proposition est vraie *conformément aux mathématiques standard*, dispose d'une solution à ce problème ?

Jacques Dubucs

Les solutions proposées par le fictionnalisme et par l'anti-réalisme au dilemme de Benacerraf sont en effet très différentes. Les deux s'accordent pour considérer que le renoncement à une sémantique véricondititionnelle unifiée est une chose à peu près inévitable. Mais le fictionnalisme renonce *également* à l'idée d'une sémantique unifiée pour le langage de la science, alors que l'anti-réalisme tente de la préserver en abandonnant *partout* l'idée de véricondititionnalité.

S'agissant du fictionnalisme, la situation à laquelle on aboutit est au fond assez semblable à celle à laquelle conduisait le programme de Hilbert : les énoncés de la classe jugée non problématique (énoncés nominalistes chez Field, finitistes chez Hilbert) sont les seuls à se voir conférer ce que vous appelez une signification *littérale*, et les autres sont tenus pour de simples "façons de parler" n'acquérant de signification que *dérivativement*, via leur rôle dans l'extraction des vérités de la classe non problématique. Mais dans une perspective de ce genre, la difficulté est toujours le tracé de la frontière entre les deux classes d'énoncés promises à des sorts si différents : l'efficacité de la construction est suspendue à ce choix. Or la démarcation proposée par Field semble aberrante : on ne peut certainement pas exprimer la physique contemporaine dans un langage purement nominaliste, d'où seraient éliminée toute mention de nombres, de fonctions, d'espaces, etc. Les développements récents de la logique mathématique (Feferman, Sieg, Simpson), qui conduisent à des réalisations "partielles" du programme de Hilbert, suggèrent un choix beaucoup plus raisonnable : les mathématiques "applicables" (topologie élémentaire, analyse réelle et complexe, etc.) peuvent être développées dans des extensions conservatives de l'arithmétique de Peano. Mais dans ces conditions, les théorèmes arithmétiques eux-mêmes devraient être considérés comme des vérités "littérales", et le problème de Benacerraf continue à se poser : on l'a circonscrit, mais non pas résolu.

La situation de l'anti-réalisme est fort différente, et je ne vois pas que l'on puisse abruptement nier, comme vous le faites, qu'il dispose d'une solution à ce problème. On pourrait même dire que la principale ambition de l'anti-réalisme est justement de faire face à des difficultés de ce genre en admettant que des contraintes de nature épistémologique s'exercent sur la sémantique. Car ce qui est ici rejeté, c'est l'idée *générale* de conditions de vérité qui transcenderaient nos capacités de recognition. Et moyennant une conception suffisamment étroite ("réaliste", au sens vulgaire de "lucide") de ce que sont ces capacités, cette décision a des conséquences sémantiques générales, aux termes desquelles les mathématiques cessent d'apparaître comme un secteur *exceptionnel* du savoir humain : de ce point de vue, certains énoncés contingents (par exemple les généralisations de la physique, ou les énoncés relatifs aux états mentaux d'autrui) soulèvent des difficultés tout à fait comparable à celles que l'on rencontre en mathématiques. En d'autres termes, l'anti-réalisme radical (le "finitisme strict"), qui se propose de remplacer les conditions de vérité par des conditions d'assertabilité définies en termes de procédures effectivement implémentables *hic et nunc*, me paraît bien, contrairement à ce que vous dites, dissiper l'apparence "insolite" des mathématiques. Naturellement, le prix à payer pour cela est très élevé. A mes yeux, *trop* élevé : on sacrifie l'objectivité de la science en réduisant le vrai au justifiable *hic et nunc*. Mais c'est évidemment une autre affaire...

Ali Benmakhlouf (Université de Technologie de Compiègne)

Question à Jacques Dubucs: Serait-il possible de tenter, sur la base d'une théorie de l'information, une unification des trois points que vous avez avancés pour une "théorie de l'ignorance mathématique" : contenu informationnel des preuves (qui renvoie à la complexité computationnelle et à la complexité conceptuelle), synonymie des énoncés, survie des concepts mathématiques ?

Jacques Dubucs

Ce que j'ai avancé sur ce point est de nature évidemment très programmatique. Mais la suggestion est de considérer comme synonymes les seuls énoncés dont l'équivalence peut être établie par une preuve de complexité modeste (l'expérience montre que la définition carnapienne de la synonymie par l'isomorphisme intensionnel, c'est-à-dire par l'équivalence prouvable *plus* l'identité de structure syntaxique, quant à elle, n'est guère viable). Ou, en d'autres mots, d'utiliser ce que nous savons de la structure des démonstrations pour construire un critère fin ("hyperintensionnel") d'identité entre propositions.

Dominique Lambert (F.U.N.D.P., Namur)

Question à Jacques Dubucs : S'ils existent, les faits mathématiques à la fois "durs" et contingents pourraient-ils être considérés comme ceux qui possèdent la plus forte probabilité d'être appliqués en physique ? Dans ce cas pourrait-on penser qu'il existe un critère de démarcation entre les énoncés applicables à la physique et les autres ?

Jacques Dubucs

Je me garderai bien de parler de probabilités dans un tel contexte. S'agissant de la démarcation (un mot vraisemblablement trop fort) entre les mathématiques applicables à la physique et les autres, le problème est le suivant. Il paraît difficile de considérer que les théorèmes reflètent la manière dont les objets mathématiques sont agencés "en fait" sans admettre également que ces objets pourraient être agencés autrement qu'ils ne le sont (le propre d'un fait, c'est qu'il pourrait ne pas se produire). Dans ces conditions, les théorèmes mathématiques eux-mêmes pourraient différer de ce qu'ils sont. Dès lors, l'explication factualiste de *l'objectivité* des mathématiques tend à rendre inexplicable leur *applicabilité*. Car si les théorèmes mathématiques sont applicables à la physique, c'est, si l'on ose dire, parce qu'ils sont ce qu'ils sont : des théorèmes différents seraient inapplicables. C'est pourquoi le factualiste ne peut guère compter que sur un miracle pour expliquer le succès des mathématiques : l'idée que les mathématiques sont vraies (en un sens non minimal, c'est-à-dire dans une acception convoyant l'idée d'une correspondance aux faits), bien loin d'expliquer, comme on le dit souvent, l'applicabilité des mathématiques, la rend totalement inintelligible. Là est l'essentiel. Et c'est une question tout à fait distincte de celle de savoir s'il est possible de tracer, *à l'intérieur des mathématiques "réelles"*, une ligne de partage entre les énoncés applicables et les autres. En particulier on ne saurait tenir les exemples familiers d'applications surprenantes (résultats subtils de théorie des nombres obtenus "pour l'honneur de l'esprit humain", et utilisés ensuite en cryptographie, etc) pour des objections à l'argumentation qui précède.

Raphaël Douady

Remarque : Si j'ai bien compris il y a deux démarche en philosophie des mathématiques, l'une qui s'intéresse à *l'état* de la pensée mathématique d'un point de vue statique, l'autre qui s'intéresse au *mouvement* de cette pensée.

Gilles Châtelet

Il n'y a pas d'opposition entre le statique et le dynamique. Les vérités mathématiques sont bien "éternelles", mais leur conquête, en vue de leur stratification dans un savoir, présuppose des gestes qui les *font ressortir pour les arracher* au tissu de l'Etre (les mathématiques sont l'histoire humaine de l'Eternité dirait Alain Badiou). Ces gestes se déploient dialectiquement dans des lignées — ce sont les "brouillards" de Weil qui surgissent fatalement lors de la découpe d'une forme nouvelle. Ces lignées s'associent à des dispositifs allusifs, cruciaux pour saisir le physico-mathématique, qui réactivent des appels d'intention : le théorème de

Pythagore “retentit” vingt cinq siècles plus tard avec Hilbert et avec la mécanique quantique.

Marcel Guillaume (mathématicien)

Observation à propose de la “pureté des méthodes” : André Lichnérowicz fait remarquer que la communauté mathématique agit toujours de façon à repousser les limites fixées, soit de l’extérieur soit par les mathématiciens eux-mêmes, à la légitimité de son activité. Par exemple, en ce qui concerne la définissabilité — dont on remarquera en passant qu’elle emprunte beaucoup aux mathématiques constructivistes — : P. Cegielski a eu besoin d’utiliser le théorème de Dirichlet sur une infinité de nombres premiers dans une progression arithmétique, qu’on ne sait démontrer que par des méthodes transcendantes. Mais Cegielski a pu l’obtenir dans l’arithmétique de Peano, par le biais d’extensions définissables de celles-ci. Je veux dire que la définissabilité permet ici un détour par des types supérieurs, et il se pourrait que ce soit le cas dans beaucoup de preuves de ce genre.

Bernard Tessier (DMI, ENS)

Observation : Comme le dit René Thom, la limite de la vérité n’est pas l’erreur mais l’insignifiance. Les philosophes ont tendance à considérer les mathématiques comme une machine à produire de la vérité. Ce n’est pas cela. C’est un univers dans lequel les intuitions de base sont soumises à l’action d’un formalisme extrêmement puissant, mais cet univers semble comparable s à celui de l’art ou de la poésie : on y cherche non pas la vérité mais le sens du monde.

Table Ronde G : 8 juillet 1992

MATHÉMATIQUES ET SCIENCES SOCIALES

organisée par Pierre-André Chiappori [1] et Roger Guesnerie [2]

DELTA (CNRS, EHESS, ENS) Paris, (France)[1,2]

Introduction au débat

par Roger Guesnerie

Si le rôle des méthodes statistiques dans l'analyse de la vie sociale est plutôt convenablement compris par les spécialistes d'autres domaines, et en particulier ceux qui viennent des sciences dites dures, la contribution de la modélisation mathématique à la compréhension des phénomènes sociaux prête souvent à confusion. C'est cette confusion que, j'espère, les propos des orateurs dissiperont. Ces commentaires introductifs visent à leur préparer le terrain.

Mon premier commentaire porte sur le rôle des méthodes statistiques dans l'analyse de la vie sociale. L'utilité de telles méthodes devrait être évidente à chacun. Les Sciences Sociales, comme les autres Sciences, utilisent des données chiffrées — prix et quantités pour l'économie, populations, fréquences de divers attributs en démographie et en sociologie — et il n'est pas surprenant que l'analyse de ces données requiert le recours à des outils statistiques. Il arrive même que ces outils soient très perfectionnés d'un point de vue mathématique ; par exemple, l'analyse des séries temporelles est un sujet classique de l'économétrie qui utilise des résultats avancés en théorie des probabilités ; la recherche en ce domaine est étroitement imbriquée à la recherche en mathématiques appliquées.

Après le rôle des mathématiques dans l'analyse empirique de la vie sociale, il faut aborder le sujet plus ambigu de leur rôle dans l'analyse théorique, c'est-à-dire le rôle de la modélisation mathématique dans les sciences sociales. La discussion qui suit s'essaie à lever les malentendus sur les trois sujets suivants, d'abord le volume de la contribution de la formalisation mathématique aux sciences sociales ensuite la nature de cette contribution et enfin sa relation avec l'hypothèse de rationalité.

Discutons tout d'abord l'*importance* de la contribution mathématique en mettant l'accent sur la modélisation mathématique plutôt que sur les problèmes de vérification empirique déjà évoqués dans mon premier commentaire. De ce point de vue, le rôle des mathématiques est extrêmement variable au travers des champs thématiques : il est modeste pour ne pas dire plus en anthropologie, timide en sociologie, plus important en sciences politiques et tout à fait central en économie. Le cas de l'économie, qui est aussi celui que je connais le mieux, mérite qu'on s'y arrête. Le rôle de la modélisation mathématique a crû de manière régulière depuis la fin de la deuxième guerre mondiale. Si l'on prenait au hasard des membres des départements d'économie d'universités telles que Berkeley, Harvard, Princeton mais aussi Bruxelles, Londres, Bonn, Barcelone,... l'on constaterait que dans presque tous les tirages de l'échantillon une vaste majorité de ces économistes universitaires acquiescerait à l'idée que le débat théorique dans la discipline doit être organisé autour de modèles mathématiques qui fournissent une description stylisée des phénomènes discutés. A vrai dire, même si l'on incluait Paris dans la liste, la majorité à laquelle je faisais allusion resterait écrasante ! Cette situation, que confirme immédiatement la lecture des revues les plus influentes, est évidement très nouvelle dans l'histoire de la discipline où l'introduction de la formalisation s'est heurtée à de vives réticences. Elle est pourtant à mon sens irréversible, même si la forme que prendra la modélisation et les modalités de son insertion dans le débat intellectuel sont susceptibles de variations. Ces faits qui sont bien connus dans le milieu des économistes, un peu moins connus dans le milieu des sciences sociales mais très souvent ignorés au dehors, apparaîtront en filigrane des interventions des économistes de cette table ronde — P.A. Chiappori, le coorganisateur de cette table ronde et F. Hahn. La situation est différente en sociologie : nous avons la chance d'avoir ici un sociologue respecté, P.M. Menger qui nous en dira plus sur le sujet.

La *nature* de la contribution de la modélisation mathématique au débat dans les sciences sociales, même dans le cas favorable où son importance est perçue, est souvent mal comprise. Elle est pourtant simple dans son principe : les sciences sociales, comme les autres sciences, soulèvent des questions logiques délicates ; leur analyse requiert le recours à l'outil conçu pour pallier les piètres performances spontanées de notre esprit dans la maîtrise du raisonnement. Pour illustrer cette affirmation, de façon trop rapide mais qui sera reprise plus en détail par P.A. Chiappori, je prendrais un exemple tiré de l'histoire des théories économiques. Au 19ème siècle, les économistes ont beaucoup débattu de la relation entre ce qu'ils appelaient les valeurs des biens et leurs prix de marchés ; ce faisant ils sont devenus de plus en plus conscients des difficultés logiques du problème ; à vrai dire, quelques diagrammes permettent d'éclairer sa solution quand le nombre de biens économiques à l'examen est au plus de deux, mais cette hypothèse est bien évidemment inadaptée à décrire la complexité réelle de la question. Il est pourtant difficile d'aller plus loin sans recours à une analyse matricielle qui bien que n'étant pas d'une difficulté mathématique particulière fait appel à des propriétés dont l'énoncé est postérieur au commencement du débat économique qu'ils permettent d'éclairer. La prise de conscience des difficultés logiques des théories économiques n'est d'ailleurs pas récente ; à la fin de sa vie, K. Marx, dont certains disciples ont donné une image fort hostile à l'usage des mathématiques, ne s'était-il pas mis à l'étude de l'algèbre ?

Il doit être souligné que le degré de sophistication des instruments mathématiques susceptibles d'être utiles aux sciences sociales en général et à la science économique en particulier est très variable. Le diagramme dit IS-LM autour duquel les manuels de macro-économie ont pendant longtemps présenté une synthèse des idées keynésiennes est un instrument à la fois extrêmement simple — deux courbes dans le plan qui résument l'ajustement de l'offre et de la demande globale au taux d'intérêt sous les axiomes de la théorie — et extrêmement efficace sur le plan pédagogique. En revanche, l'effort de vérification du champ de validité et du degré de généralité de certaines théories, telle la théorie walrasienne de la concurrence, passe par le recours à des outils assez, voire très sophistiqués : théorèmes de point fixes, énoncés de topologie différentielle, etc... Dans certains cas, les outils nécessaires à la solution des problèmes rencontrés n'existent simplement pas, et dans d'autres cas — rares à ce jour mais existants — des principes purement économiques ont contribué à la résolution de problèmes mathématiques ouverts[1]... Il y a donc un large spectre d'usages possibles des mathématiques en économie : langage commode ou outil d'investigation indispensable, substitut d'une imagination défaillante ou outil irremplaçable du test de la généralité et de la robustesse des théories. F. Hahn dont les contributions à l'économie mathématique ont une grande notoriété, connaît bien ce sujet et nous l'écouterons avec attention.

Le dernier point que je voudrais évoquer est le rôle de *l'hypothèse de rationalité* et son rapport avec le développement de la formalisation mathématique. En économie, modélisation mathématique et hypothèse de rationalité, telle qu'elle apparait dans la fiction de l'homo economicus, ont été des compagnons de route. Il n'y a en principe rien de nécessaire dans compagnonnage et nombre d'applications des mathématiques à l'économie ne font pas référence à la rationalité. Il importe pourtant de s'interroger sur le caractère plus ou moins profond de la relation historique entre rationalité et modélisation. Personnellement, je ne suis souvent pas satisfait de la réponse des épistémologues à la question évoquée ni d'ailleurs de leurs analyses des forces et des faiblesses de l'hypothèse de rationalité. Pour nous éclairer sur ce sujet nous avons la chance d'avoir parmi nous J.Elster, un des analystes contemporains les plus lucides et les plus rigoureux de la rationalité et de ses applications aux sciences humaines.

En exergue de cette introduction, trois questions seront posées auxquelles le débat apportera, j'en suis sûr, des éléments de réponses :

— Jusqu'à quel point le succès de la modélisation mathématique en économie s'étendra-t-il à d'autres sciences sociales ?

---

[1] Voir la thèse de J.C Rochet (Université de Paris-Dauphine, 1984) pour une importation du "principe de taxation" à la solution d'équations aux dérivées partielles.

— Qu'est la nature réelle de la modélisation mathématique en économie et en sciences sociales et comment évoluera-t-elle ?

— Jusqu'à quel point et pourquoi l'hypothèse de rationalité a-t-elle été et restera-t-elle, d'une part, utile à la modélisation et d'autre part féconde ?

On the use of Mathematics in Social Sciences

Some Remarks

by Pierre-André Chiappori

Two roles of mathematics can be emphasized in the field of social sciences. On the one hand, mathematics is used as a *conceptual tool*; it can help to clarify a subject, illuminate unforeseen logical consequences, and protect against errors of logic. On the other hand, mathematics can also be seen (and utilized) as an *empirical tool*, not only because empirical tests in social sciences currently require sophisticated statistical techniques, but also because the derivation of empirically testable consequences from a general and abstract theoretical core usually rely on mathematical formalization.

1. Mathematics as a Conceptual Tool

When he was asked why he did "mathematical" rather than "literary" economics, Paul Samuelson used to answer he wished he would be able to do literary economics — but this was just too difficult for him. This feeling is, I believe, shared by all economists nowadays: most, if not all problems in economic theory are simply too intricate to be thought of without the help of at least some formalization.

Elementary as this claim may seem, it still suggests two remarks. First, its relevance is not limited to economics. Rather, I would submit, as a general principle, that the analysis of social behavior will systematically face problems of considerable complexity (much more complex, as a matter of fact, than those encountered in natural sciences); and that mathematics, or more precisely some well-chosen subfield of mathematics, potentially constitutes a very useful tool for dealing with this complexity. Of course, which area of mathematics will be needed remains to be seen; it will depend on the problem, and may in some cases require entirely new developments.

Secondly, even within the field of economics, the interest of mathematics is not restricted to some particular model or approach. In particular, formalization need not refer to specific assumptions about individualism or rational behavior. This point is to be emphasized, since mathematical economics is sometimes viewed as intrinsically associated with the so-called "homo oeconomicus" framework. In fact, history shows that *all* economic theories have, sooner or latter, faced difficulties that could only be solved with the help of mathematical reasoning.

The best example is probably given by the foundations of the theory of value, a problem that was the focus of attention of most 19th century economists. The basic intuition that has been put forth by classical economists (and especially D. Ricardo) is that the value of a commodity should depend on the quantity of labor that was needed to produce it. Of course, it was clear, even at that time, that one should consider not only "direct" labor (i.e., the time spent by a worker to directly produce the commodity), but also "indirect labor" — i.e., the time spent to produce the machines which were used during the production process. For instance, farming requires ploughs, which in turn have to be produced using labor; this indirect labor, embodied within the tool, has to be taken into account when computing the labor content of one ton of wheat.

Here is where things became intricate. For most production processes, this argument will lead to a "cyclical" relationship. For instance, iron is needed to produce hammers, but hammers are needed to produce iron. At this point, classical economists were in trouble, because apparently one had to know the labor content of iron to determine the labor content of hammers, and conversely. Did that mean that it was simply impossible to define value on the basis of labor content in a consistent way?

For any mathematician, the answer is obvious. Let V be the vector defining the values of the different commodities. From the knowledge of production processes, the labor content argument allows us to derive one constraint on V for each produced commodity. Moreover, V is defined up to a multiplicative constant, hence can be normalized. Thus for N commodities, among which n are produced and $N - n$ are given initially (labor and natural resources), we have n constraints for $N - 1$ unknowns; and we know that $N - 1$ is greater than or equal to n.

The only problem is to show that the system can be solved. For linear production functions, such as those considered by classical economists. This requires some kind of full rank condition. In a more general framework, involving non linear techniques and more sophisticated demand behavior, convex analysis and fixed point theorems will have to be used. General equilibrium theory, for instance, defines the "excess demand" vector field by associating, to each price vector within some compact set, the vector of differences between demand for and supply of each commodity at these prices. An equilibrium is then defined as a zero of this vector field; several techniques (Brouwer theorem, Kakutani's extension to correspondences, differential topology,...) can then be used to derive the conditions under which an equilibrium does exist[1].

---

[1] For a more precise assessment of the mathematical techniques used in economics, see the round table on this topic.

Simple as this argument may seem for a mathematician, it must be remembered that basically all great economists of that time simply failed to answer the question, not because they were not clever enough, but rather because they did not master such elementary tools (by our present standards) as matrix theory or convex analysis. In fact, they were not even able to conceptualize the problem in an adequate way. One had to wait till Walras, at the end of the century, for a consistent formalization to be given — from which the solution did follow. Here, we have an example of a crucial problem that could not be solved until it was formalized; when analyzed in "literary" terms, it was simply too complex to be worked out.

The intuition provided by this example is quite general. Of course, technical problems may remain here and there; some mathematically difficult issues can be raised. The point is, however, that once an adequate formalization has been adopted, *conceptual* difficulties essentially disappear. The problem is well-defined, and, under "suitable" assumptions, can be proved to have a solution. Conversely, a violation of the predictions of the model has to be attributed to the failure of one of the assumptions. Then one has to investigate which hypothesis should be removed, and with what consequences. A new research program may thus emerge. For instance, an entirely new field appeared when an economist tried to remove the assumption that information was symmetric across agents.

Besides being useful for the resolution of given problems, mathematics is also indispensable for checking the consistency of new insights with existing theories. Take the notion of *endogenous macroeconomic fluctuations*. The idea that can be traced back to Keynes (or even before) is that economic fluctuations need not result from *exogenous* shocks regarding preferences, technologies, natural resources, or whatever. Rather, they may stem from the agents' beliefs themselves; i.e., booms and slumps may appear simply because people anticipate that they will, and behave accordingly.

Interesting as this vision might seem, it was first rejected on the grounds that it was not compatible with individual rationality. Rational agents, it was sometimes argued, will never behave in such a counterproductive way as to provoke a depression simply by their expectations. Though all economists were not convinced by this objection, most of them agreed that, in any case, intuition had to be reconciled with the standard framework of rational decision makers before it could be accepted as a consistent theory.

At first sight, the issue was quite complex, if only because the notion of rational behavior was not easy to extend to expectations. One of the main achievements of economic theory during the last twenty years has been to formalize and carefully investigate the concept of rational expectations. Broadly speaking, rational expectations can be seen as a consistency hypothesis: agents' expectations cannot be self-defeating. Specifically, let x_t denote a vector of relevant economic variables; assume that x_t follows some random process, and let I_t be available information at date t, including past values of x_t. Agents' expectations about x_{t+1} can be represented by a (conditional) probability distribution $\mu_t = \mu(\, . \, / I_t)$. At each date, equilibrium requires that some mapping $Z(x_t, \mu_t)$ be zero; the precise form of Z depends on the model, and reflects the characteristics of the economy at stake. A rational expectation equilibrium (from now REE) will be defined as a process (xt) such that:

(i) μ_t is the *actual* distribution of x_{t+1}, conditional on past and present values of the process

(ii) for each realization x_t° of the process at date t, the equilibrium condition $Z = 0$ is fulfilled.

In a word, (**i**) says that agents do not make systematic mistakes when predicting the future (technically, they use the true probability distribution), while **(ii)** states that this behavior is always compatible with equilibrium. With the help of this definition, one may investigate whether, in a given model, REE exists, and is unique. If, in particular, a "truly random" REE exists while the fundamentals of the economy are non-random, we have demonstrated the possibility of endogenous fluctuations under rational expectations.

Several techniques can be used here, from fixed point theorems in functional spaces to bifurcation theory. Using these (sometimes rather sophisticated) tools, we have learned that even with a very strong and demanding definition of rational expectations, and in an economy where all agents are utility — or profit-maximizers, different expectations may indeed be independently self-fulfilling. This establishes that the idea of endogenous fluctuation is in fact *fully consistent with rational behavior*. In addition, the formal models did provide a list of necessary conditions that helped us to better understand in which situations such phenomena are likely to appear. Again, this would have been out of reach if one had had to rely on "literary" reasoning alone.

2. Mathematics as an Empirical Tool

Apart from the conceptual role just discussed, mathematics is essential in the process of empirical testing, which, according to Popper, lies at the core of scientific approaches. In this respect, the role of mathematics is twofold. For one thing, the notion of empirical testing in economics, and more generally in social sciences, has to be understood in a statistical sense. Social and psychological interactions are by far too complex to allow for exact predictions at the individual level. Or, to be more precise, theoretical predictions have to rely (sometimes implicitly) on some kind of "other things equal" assumption that is extremely difficult to check at the individual level. As a result, an economist will never predict that an increase in non labor income will lead Mr. So-and-so to work less. He will simply announce that increases in non-labor income rather tend, other things being equal, to reduce the propensity to participate or to accept overtime work. Obviously, a single counter example — say, the observation that Mr. Smith did actually enter the labor market after receiving a bequest — will not be sufficient to falsify the prediction. What we need to test is whether, on a sufficiently large sample of households — so that individual peculiarities average out — increases in non-labor income do turn out to have a statistically significant negative effect on participation or working time.

Even under this apparently less demanding form, such a prediction can be quite tricky to test, especially since severe constraints exist on available data. For instance, we may have data on average weekly working time for a sample of individuals. From this set, we may try to check whether individuals with higher non-labor income do actually work less than others. But a naive investigation could be quite misleading. People differ not only by non-labor income, but also by age, family status, education, place of residence, and others. For instance, it may be the case that old people have accumulated more wealth, hence receive more asset income, while they work less because of their age; this effect does obviously not confirm the theory, though it goes in the same direction and may be quite difficult to distinguish from it. Even unobservable differences may play a role. Take two apparently identical individuals, X and Y, and assume that their preferences are different — say, X's disutility of labor is lower than Y's. Then X is likely to work more now, but also to have worked more in the past, hence to have accumulated more wealth; in this case we may observe that larger non-labor income is associated, in the case of X, with *more* labor supply (this is a typical example of what we call an *endogeneity bias*).

These examples clearly suggest that many biases may appear during the estimation process. The only way to avoid erroneous conclusions is to correct them in an adequate way, which may (and usually does) require sophisticated techniques. Recent progress in econometric theory is precisely motivated by this research of more accurate testing procedures. And probability theory is indispensable for modelling this kind of problem.

A second role of mathematics lies at a deeper level. Empirical tests are possible only insofar as empirically testable consequences of the theory have been put forth and precisely characterized. This is by no means an easy task. Most models are constructed to account for a few "stylized facts" that were left unexplained by the previous literature. Testing the model, in this case, implies deriving what Popper calls "unexpected consequences" of the framework — namely, predictions which did not belong to the initial stylized facts but are still (ideally) empirically testable from available data. Formalized models are needed at this stage because they facilitate a more systematic exploration of the various implications of a given theoretical argument.

An interesting example is provided by the analysis of careers in an hierarchical organization. To explain the diversity among careers of individuals with equivalent education, several mechanisms have been alternatively emphasized. Some models stress the incentive effect; in this view, promotions are essentially needed for inducing people to work more, and will favor those individuals who are more eager to perform well (given that some degree of randomness cannot be avoided in the measurement of effort level through performances). Other theories concentrate upon those differences among individual productivities that cannot be observed at the outset, but gradually become revealed as the production process takes place. The question, at this point, is whether we can decide among the various approaches at stake from the observation of actual behavior, as recorded in the data sets at our disposal.

For instance, different theories predict different relationships between performances and promotions. If the incitation motive plays a key role, then it can be shown under adequate assumptions (and again with the help of mathematical developments) that efficiency requires old performances to be forgotten; promotion should essentially depend on recent results. Conversely, should promotion essentially be a selection process, then the whole history of performances should be taken into account. While these conclusions may seem intuitively plausible, models are still needed, not only to check formally their consistency, but also to get a precise statement of the assumptions they require to hold true — a key step, since scientific discussion can then focus on the relevance of these assumptions.

Finally, the data sets can be revealed as rather incomplete, thus making the process even more difficult. Typically, data on careers will not include records of performances, if only because these are based upon the partly subjective appreciation of some supervisor. This fact, however, does not preclude empirical tests to be performed. The trick, for instance, can be to derive from the model some predictions about the time autocorrelation of the promotion process, which can then be detected with econometric techniques. But, in any case, formalization intervenes at each stage of the process.

In conclusion, it should be emphasized that formalization is no more than a tool at our disposal. Indispensable as formal models may be, they do not in themselves constitute all of theory. What social scientists are talking about are theories about individual and social behaviors, not mathematics. In particular, the temptation is always present to concentrate exclusively upon those models which turn out to be easier to formalize. This is something we should definitely resist; rather, we should systematically propose alternative approaches, even if their formalization is difficult at the first stage. But introducing a new approach will unavoidably imply the development of formal models, for all the reasons given above. This, in fact, is the essence of scientific progress.

La sociologie face à la mathématisation

par Pierre-Michel Menger
EHESS, CNRS, Paris (France)

La position de la sociologie à l'égard des mathématiques peut être analysée en trois temps. D'une part, un bilan peut être dressé des formes prises par la mathématisation de la sociologie, à partir des développements qu'a connus depuis une trentaine d'années, et essentiellement aux Etats-Unis, le courant de la sociologie mathématique. En évoquant ce domaine de recherche comme un courant, on peut comprendre déjà que les tentatives les plus poussées de mathématisation de la sociologie ne sont qu'une des formes d'activité dans la recherche sociologique : le recours à la formalisation mathématique et aux propriétés heuristiques de la modélisation mathématique est en effet très loin d'occuper en sociologie une position comparable à celle de la mathématique dans la science économique. Tel sera mon deuxième point : évoquer la mathématisation des sciences sociales revient pour une bonne part à circonscrire l'analyse à la comparaison avec la discipline pivot, sous ce rapport, l'économie.

Nous concentrerons le débat sur les conditions auxquelles doivent satisfaire la construction et la mesure des faits en économie pour que puissent être inventés modèles et formalismes mathématiques, et sur les différences qui séparent alors le travail et les outils conceptuels et techniques de l'économiste de ceux du sociologue. Cette comparaison conduirait tout droit à un constat alarmant pour la sociologie, décidément très loin d'approcher le niveau de formalisation aujourd'hui présent dans les articles d'économie, si la posture épistémologique des deux sciences, notamment au regard de la "mathématisabilité" des faits sociaux et des comportements humains, n'était pas profondément différente. D'où notre troisième point : les recours moins étendus à la formalisation dans la recherche sociologique peuvent être compris autrement qu'en termes négatifs d'immaturité scientifique, par qui défend l'originalité épistémologique de la sociologie.

1. La sociologie et ses mathématiques

En sociologie, certains des pionniers de la mathématisation comme J. Coleman ou H. White ont présenté leur agenda de recherche comme la construction d'une théorie mais la plupart des sociologues situés hors du champ de la sociologie mathématique voient plutôt en eux des méthodologues. La théorie sociologique, telle que la pratiquent ceux qui s'intitulent théoriciens en sociologie, est, à de rares exceptions, très éloignée de la modélisation mathématique.

L'essentiel de ce qui s'appelle théorie sociologique n'est pas traduit en termes mathématiques. Ce qui est surtout pratiqué en théorie sociologique, c'est l'histoire de la sociologie, la métathéorie, l'élaboration conceptuelle de ce qu'on appelle norme, capital, rôle, position, espace social, etc, sans nécessité de formulation mathématique originale. C'est la raison pour laquelle, selon certains, la théorie sociologique n'aurait pas beaucoup progressé ces dernières années.

Il importe pourtant de décrire ce qu'a été l'effort pour développer les outils mathématiques en sociologie et pour construire une branche particulière de la sociologie, la sociologie mathématique, même si l'importance de cette dernière est très loin d'atteindre celle qu'obtient en économie l'économie mathématique[1].

La distinction principale qui orientera cette description sépare les modèles de processus sociaux des modèles de structure sociale. Schématiquement, les premiers s'attachent à modéliser le changement des valeurs que prend une variable ou un ensemble de variables, comme fonction du temps ; les seconds visent à déterminer la structure des relations entre les valeurs des variables dépendantes et indépendantes en un point donné du temps.

1.1 Les modèles déterministes de structure sociale

L'analyse des structures causales par la mathématique des équations algébriques constitue la forme la plus connue et la plus usuelle de formalisation en sociologie. L'intensité de la liaison entre les variables, mesurée à partir des coefficients de régression, renseigne sur la relation de causalité.

L'analyse causale des données statistiques et les multiples techniques de calcul statistique ont été développées initialement en biologie et en sociologie, puis surtout en économétrie et ont conduit notamment à traiter les structures causales complexes où les variables explicatives sont elles-mêmes liées, à déterminer les réseaux de relations causales sous-jacents aux variables considérées, en étendant les méthodes de calcul des variables quantitatives aux variables qualitatives. L'analyse multivariée et l'analyse de dépendance abondamment employées par les sociologues depuis 20 ans permettent de modéliser les relations de causalité par étape, par le choix approprié des variables tests introduites. Mais dans tous les cas, qu'il s'agisse de corrélations simples ou de réseaux complexes de relations entre variables, la variété des techniques élaborées ne débouche pas sur une modélisation adéquate si la formulation d'hypothèses sur les relations d'interaction et de dépendance causale n'est pas assez précise et discriminante pour engendrer des conclusions interprétatives fécondes. La complexité des relations entre variables que cherche à restituer un système d'équations algébriques risque toujours en effet toujours d'être réduite à un raisonnement causal simplificateur, quand les limites de traductibilité en langage naturel de la formalisation mathématique cumulent leurs effets avec le caractère déterministe de l'explication structurale causale.

---

[1] Nous nous référerons notamment aux travaux de Sørensen (1977, 1978) et de Boudon (1967) pour la présentation de la sociologie mathématique qui suit.

Bien qu'utilisée généralement dans un sens probabiliste, la notion de causalité soutient en effet des modèles déterministes où, à la différence des modèles stochastiques, les valeurs prises par les variables dépendantes ne sont pas exprimées par des distributions de probabilités. L'aléa est introduit dans les modèles d'analyse causale à travers la présence d'un terme d'erreur rendant compte d'erreurs de mesure ou de l'existence de variables non mesurées. Le comportement de ce terme d'erreur est crucial dans les régressions linéaires pour le choix des procédures d'estimation : la recherche sociologique a adopté et développé des méthodes et des techniques d'estimation familières à l'économétrie.

A bien des égards, rien ne distingue fondamentalement les méthodes sociologiques d'analyse multivariée et d'analyse par simulation des techniques de calcul propres aux économistes. Plutôt que de modélisation mathématique caractéristique de la formalisation en sociologie déterministe, il s'agit ici de méthodes statistiques communes aux sciences sociales confrontées aux données empiriques. C'est en effet ce domaine de la sociologie mathématisée qui s'apparente le plus au contenu de la méthodologie quantitative et de la statistique, le développement de modèles se confondant souvent avec le choix de techniques appropriées d'analyse des données.

1.2 L'analyse sociologique des processus sociaux

L'étude des processus est centrée sur le changement : des variables caractérisant des individus, des groupes sociaux ou des structures sociales sont soumises à un changement dont on veut identifier la cause. Ce qui implique de spécifier le mécanisme de changement, par la modélisation du processus de changement et, d'évaluer les influences causales exercées par le mécanisme de changement. La représentation du changement dépend de trois caractéristiques : la forme temporelle du changement (temps continu, conduisant à l'emploi d'équations différentielles, ou intervalles temporels discrets, supposant le recours à des équations aux différences finies), le caractère continu ou discret de la variable étudiée, la conception stochastique ou déterministe du processus de changement, qui rend compte de l'évolution temporelle d'un système soit en décrivant les probabilités pour un système de se trouver dans tel ou tel état à un moment donné (modèle stochastique) soit en décrivant ses états successifs, une fois précisées les conditions initiales et les lois du changement (modèle déterministe).

Les modèles stochastiques de mobilité

Les modèles stochastiques ont depuis longtemps fourni la majorité des contributions à la sociologie mathématique. Leur attrait s'explique par l'utilisation qu'ils permettent de modèles probabilistes, isomorphes au caractère mouvant des comportements individuels, et par les ressources de modélisation qu'ils fournissent pour l'analyse des variables discrètes, très fréquentes en sociologie, quand l'utilisation d'équations différentielles est impossible.

Les valeurs d'une variable sont considérées comme un espace d'états : les changements dans un tel espace d'états sont l'objet des processus stochastiques, que ces mouvements soient ceux d'un individu, d'un groupe, ou d'une autre entité. Le résultat du mouvement est donné comme une distribution de probabilités sur l'espace d'états. Les deux classes principales de modèles stochastiques se distinguent par la nature de l'espace d'états considérés : les modèles markoviens s'appliquent à des valeurs ou catégories de variables qualitativement différentes (e.g., préférences politiques, position professionnelle, attitude de consommation), les modèles de processus de Poisson concernent la probabilité d'occurrence d'événements (e.g., survenue d'accidents du travail).

Les modèles simples de chaîne de Markov ont été beaucoup utilisés en sociologie pour analyser les phénomènes de mobilité sociale, professionnelle ou géographique (cf. Stewman, 1976). Il s'agit de saisir les données à l'aide d'une matrice de rotation montrant combien de personnes passent d'une position sociale, professionnelle ou géographique i à une position j dans un intervalle de temps donné : le modèle markovien simple conduit à calculer les taux de transition ou les probabilités de transition d'un état à l'autre. Dans leur forme simple, qui s'applique sous les conditions d'homogénéité temporelle (la loi d'évolution du système est invariante dans le temps) ou d'homogénéité de la population d'individus sur les variables indépendantes, les modèles markoviens permettent de modéliser des phénomènes sociaux relativement élémentaires (e.g., les probabilités de transition entre types d'intention de vote politique) et s'utilisent surtout dans des analyses théoriques (e.g., les changements temporels dans la structure des réseaux sociométriques, le pouvoir social) ou dans des études de situations expérimentales très simplifiées (e.g., les situations d'apprentissage). Pour tous les faits sociaux plus complexes, i.e., survenant dans des systèmes sociaux qui ne sont plus considérés comme suffisamment dégagés des influences extérieures pour que la matrice de transition markovienne puisse être tenue pour constante, il importe de relâcher l'hypothèse d'homogénéité sous sa double espèce pour dépasser les limites empiriques d'adéquation et les limites théoriques de pouvoir explicatif du modèle simple.

Différentes stratégies ont été élaborées pour inclure séparement ou simultanément hétérogénéité interindividuelle et non-stationnarité : segmentation d'une population hétérogène en sous-populations homogènes, introduction de variables à dépendance temporelle en position de variables indépendantes, modèles semi-markoviens plus flexibles, mais mathématiquement plus complexes puisqu'ils dotent en quelque sorte d'une épaisseur temporelle (temps passé dans un état) le système de paramètres décrivant l'état occupé et sa relation avec l'état de destination, modèles à états latents destinés à séparer termes d'erreur et vrais changements dans les observations de processus markoviens, modèles à équations différentielles simultanées pour les cas où deux variables ou plus d'un processus considéré s'influencent mutuellement.

Les modèles markoviens ont été massivement utilisés en sociologie pour analyser les phénomènes de mobilité sociale et professionnelle, tant intra- qu'intergénérationnelle. Nous retrouvons là les deux problèmes posés par l'emploi de chaînes markoviennes simples : l'identité sociale des acteurs n'est pas prise en compte, et le postulat de base demeure que c'est la mobilité entre les emplois ou entre les statuts qui engendre la structure et non les caractéristiques structurales d'un système qui orientent la mobilité. En sociologie du travail et des professions, l'une des tentatives les plus originales pour dépasser ces limites fut celle des chaînes de vacance de poste de H. White (1970). [2]

Les modèles d'influence, de contagion et de diffusion sociale

Ce sont des modèles fondés sur le décompte des états dans un espace constitué par des événements affectant une population d'individus, par des actes individuels ou par le nombre de sujets accomplissant une certaine action ou placés dans une situation donnée. Le point de départ mathématique est le processus

---

[2] Les mouvements de mobilité peuvent, dans les conceptions sociologiques les plus répandues, être engendrés par deux ensembles de forces :

— un premier ensemble formé des attributs individuels tels qu'âge, éducation, aptitude, qui influence la probabilité pour la personne d'aller dans telle ou telle direction;

— un second ensemble de forces est constitué des caractéristiques structurales qui déterminent la disponibilité des emplois dans les professions potentielles de destination, la pression à quitter la profession d'origine, la distance ou les affinités entre les groupes professionnels de départ et d'arrivée. Il s'agit ici de modéliser la manière dont la structure engendre le taux et la direction de la mobilité. C'est ce que fait pour sa part le modèle de chaîne de postes vacants de White. Celui-ci part d'un problème fondamental du modèle markovien de mobilité individuelle, qui suppose que les mouvements sont indépendants et que les changements professionnels d'une personne n'affectent pas ceux des autres. Mais cette hypothèse est incompatible avec l'idée que la mobilité vers un état donné i dépend du nombre d'emplois disponibles en i. I Dans un système où la plupart des emplois sont occupés et seuls un petit nombre d'emplois sont disponibles, la mobilité d'une personne affectera celle des autres. Le modèle de White est une chaîne de Markov en temps discret. Une chaîne de postes vacants est engendrée lorsqu'une personne quitte le système ou qu'un nouvel emploi entre dans le système, et cette chaîne est terminée lorsqu'une nouvelle recrue entre ou qu'un emploi disparaît. La longueur de la chaîne des mouvements est importante : c'est le nombre de mouvements qu'une vacance va provoquer avant que la chaîne soit terminée.

Le modèle de chaînes de postes vacants montre comment la structure engendre la mobilité dans des systèmes où les vacances de postes ont l'initiative pour produire la mobilité et où les individus sont homogènes. L'étape suivante serait de pouvoir appréhender les différences entre individus dans leur capacité à utiliser les postes vacants.

de Poisson, modifié pour engendrer soit des modèles de contagion interindividuelle (e.g., comportements d'achat ou de vote se diffusant dans une population donnée par l'influence d'un individu ou d'un groupe sur la probabilité pour que les autres agissent de même) soit des modèles de diffusion sous l'influence d'une source constante. Ces modélisations sociologiques sont dérivées d'outils employés par les biologistes et les démographes. Le processus de contagion aura une allure exponentielle si la population est nombreuse et non limitée a priori, et une allure logistique si la population est limitée, le phénomène de diffusion s'épuisant alors dans le temps. Ces modèles supposent généralement une homogénéité des populations étudiées : ils ont été mises en concurrence avec des analyses multivariées partant de l'hétérogénéité des acteurs pour expliquer, par exemple, des phénomènes de violence dans les villes, ou la probabilité du divorce dans une population de limites données. D'autres modèles de diffusion, s'inspirant de ceux de l'épidémiologie, ont été conçus pour analyser la propagation de maladies, d'informations, de rumeurs, d'attitudes ou de pratiques (e.g., l'adoption d'innovations) : les hypothèses de non-récursivité des états et surtout d'interaction équiprobable de chacun avec tous dans un milieu isotropique doivent être modifiées pour rendre compte des phénomènes d'érosion de l'influence de la source ou d'interruption du processus de contamination avant que l'ensemble d'une population ait été touché. [3]

---

[3] Ainsi, dans une recherche célèbre de Coleman, Katz et Menzel (1966) sur la diffusion d'un nouveau médicament dans la communauté médicale et sa prescription par les médecins, la probabilité d'interaction et de communication entre les membres de la population étudiée est affectée par des facteurs structurels liés au contexte et à l'organisation de la pratique professionnelle. La stratégie consiste, ici aussi, à segmenter la population en sous-populations homogènes sous les critères choisis comme dominants, et, en l'espèce, à distinguer entre praticiens exerçant en cabinet privé, informés essentiellement sur les nouveautés pharmaceutiques par la presse spécialisée, et auprès de qui l'innovation se diffuse selon un processus d'équation $dn/dt = k(N-n)$ (n étant le nombre de personnes informées, N le nombre total de praticiens, k une constante et t le temps écoulé) et médecins hospitaliers, s'informant sur la nouveauté à travers leur réseau dense de liens interpersonnels et adoptant celle-ci selon un processus de contagion d'équation $dn/dt = kn\,(N-n)$.

Hägerstrand (1965, cité d'après Boudon, *op. cit.*) a apporté, dans son étude de la diffusion d'une innovation agricole, une double modification au modèle logistique de diffusion : d'une part, la probabilité de rencontres interpersonnelles ou d'échanges d'informations est fonction de la distance spatiale entre les individus ; d'autre part l'efficacité de ces échanges pour l'adoption de l'innovation est variable selon les individus, l'inégale résistance au changement pouvant être caractérisée selon une distribution simple du nombre de communications interpersonnelles nécessaires à la persuasion. Ces deux hypothèses de non-isotropie (la distance pouvant être spatiale, mais aussi sociale dans d'autres applications) et distribution non-dégénérée des résistances offrent des gains substantiels en réalisme sociologique, avec l'introduction d'un grand nombre de paramètres captant l'hétérogénéité du "milieu" étudié, mais au prix d'une complication de la formalisation et du traitement mathématiques.

1.3 Les modèles de structure

A la différence des analyses stochastiques ou déterministes décrites plus haut, les modèles structuraux cherchent à modéliser non pas les processus de changements (exprimés par les distributions de probabilités sur les variables ou par les valeurs prises par celles-ci), mais des systèmes de relations entre des entités élémentaires (individus, groupes, organisations, clans, sociétés, nations, rôles) constitutives d'un ensemble le plus souvent conçu dans un état d'équilibre. La caractérisation des structures qui en résulte fournit une description statique relativement abstraite de la stabilité des relations entre variables.

Le défi que constitue la spécification rigoureuse du vocabulaire des structures, aussi omniprésent qu'imprécis dans les sciences sociales, suppose que soit récusé l'ancrage de l'analyse de l'action dans la simple caractérisation et agrégation des comportements individuels, que celles-ci soient le produit de l'individualisme méthodologique, qui tient les acteurs pour analytiquement indépendants et maximisateurs d'utilité, ou le produit des postulats holistes, qui voient dans les "structures sociales" des principes de détermination causale des comportements collectifs et individuels. L'analyse formalisée des structures sociales implique 1) que tout acteur soit saisi comme participant d'une système social dont les autres membres agissent sur ses perceptions, croyances, décisions et actions, et 2) que soient mis en évidence les différents niveaux de structuration d'un système social, une structure se définissant par les régularités dans la configuration des relations entre entités concrètes. L'hypothèse fondatrice est que la mesure des relations interindividuelles capte des propriétés émergentes des systèmes sociaux inaccessibles à une approche par agrégation des attributs individuels et que ces propriétés affectent et le comportement du système et le comportement des entités élémentaires. Par exemple, la structure des liens informels d'amitié et d'antagonisme dans des groupes organisés de travail peut engendrer des variations dans les taux de productivité individuels qu'une analyse par les caractéristiques individuelles des acteurs ne permet pas de cerner.

Deux formes principales d'application ont dominé la recherche : la construction de concepts structuraux permettant la description de réseaux de relations — e.g., les concepts de clique, de clan, d'équivalence structurale, de hiérarchie — et la modélisation des systèmes de parenté à des fins de classification et de comparaison, d'emploi limité et que nous ne discuterons pas ici. Les outils mathématiques employés sont la théorie des graphes, l'algèbre matricielle (surtout dans le premier domaine), l'algèbre abstraite de la théorie des groupes de permutations (dans l'étude des systèmes de parenté). Les contenus substantiels des relations le plus communément étudiées sont : les transactions, les communications et échanges d'informations, les formes d'interpénétration entre groupes, par le jeu des cumuls de position de certains membres (e.g., entre équipes dirigeantes ou entre conseils d'administration d'entreprises), les liens affectifs, les relations d'autorité et de pouvoir, les relations de parenté.

Les études sociométriques jouent pour les modèles structuraux le rôle qu'ont les études de mobilité dans la modélisation stochastique des processus sociaux : elles sont pourvoyeuses de données empiriques relativement aisées à obtenir, mais les propriétés structurales sont malaisées à identifier sans un cadre mathématique rigoureux.

On peut distinguer trois grandes stratégies de représentation des structures sociales (Marsden, Laumann, 1984) :

— les modèles topologiques s'inspirent des concepts sociologiques de position sociale et de rôle, et sont fondés sur l'idée mathématique d'équivalence structurale : deux objets (e.g., acteurs) dans un ensemble sont équivalents si et seulement si leurs relations (de quelque type qu'elles soient) à tous les autres éléments de l'ensemble sont identiques. La définition de l'équivalence structurale fournit un critère théorique pour la construction d'une réduction homomorphique des acteurs dans des positions sociales, sur la base de réseaux multiples de relations ou de types de lien (e.g., liens entre organisations, transferts d'information, liaisons entre entreprises par les multipositions des dirigeants, transactions d'achat et de vente).

— les modèles de théories des graphes formalisent la notion sociologique d'un groupe en recourant à l'analyse mathématique d'un sous-graphe entièrement connecté. Au centre de la modélisation figure le concept de cohésion. La structure sociale est conçue comme un ensemble de groupes. Un groupe correspond à un ensemble d'acteurs partageant des relations effectives les uns avec les autres. La théorie des graphes fournit les moyens d'identifier des ensembles solidaires d'acteurs à travers la mesure de la densité des relations. Les données de base sont contenues dans un digraphe (ensemble de relations non réflexives, directes et binaires définies sur un ensemble de noeuds). L'identification de groupes plus fortement connectés que le reste du réseau se fait à travers la recherche de sousgraphes à connexion maximale, appelés cliques : chaque acteur d'une clique est impliqué dans des relations de forte intensité avec tous les autres membres de la clique.

— les modèles spatiaux de stratification sociale recourent à l'idée mathématique d'une métrique pour situer les acteurs dans un espace multidimensionnel de positions et pour construire des représentations rigoureuses des distances sociales entre les acteurs, engendrées par leurs positions sur les différents axes de différenciation sociale. Comme les instruments utilisés pour mesurer la distance sociale n'ont pas toujours la propriété d'une métrique, la technique des échelles multidimensionnelles se fonde sur l'hypothèse d'une métrique spatiale sous-jacente et situe les acteurs dans un espace multidimensionnel où les distances entre les points figurent la distance sociale.

Ce domaine de recherche, qui demeure généralement circonscrit à l'étude de groupes sociaux de faible taille et d'un nombre limité de problèmes sociologiques, apparaît relativement isolé : ses mathématiques sont plus ésotériques dans les analyses les plus avancées, les formalisations structurales ne permettent pas d'établir des points de contact et d'échanges avec les autres branches de la sociologie et limitent les perspectives d'intégration méthodologique et théorique.

1.4 Les modèles sociologiques de l'action intentionnelle et rationnelle

Les modèles stochastiques et déterministes de processus sociaux concevaient le changement dans le temps comme le produit de forces agissant sur les individus et les groupes et représentées par des variables causales. C'est la conception la plus commune dans la recherche sociologique empirique. Et c'est la conception qui prévaut dans la plupart des modèles de sociologie mathématique.

Mais la théorie sociologique ne saurait se réduire à ce paradigme. On trouve dès l'origine de la sociologie moderne, chez Weber (1971), une conception plurielle de l'action et de la rationalité des acteurs, et notamment une théorie des processus sociaux conçus comme réalisation des actions par des acteurs intentionnels, rationnels en finalité ("zweckrational") : des acteurs cherchent à obtenir un état futur désiré en choisissant les actions qui leur permettent de maximiser leurs chances de réussite.

On fait l'hypothèse que l'acteur a des préférences ordonnées qui sont transitives et qu'il choisit l'action qui lui procurera le résultat optimal : il agit rationnellement. Si l'on affecte aux différentes solutions des valeurs numériques représentant l'utilité de l'acteur, agir rationnellement consiste à maximiser son utilité. Un tel comportement est soit postulé, comme en théorie économique positive, soit prescrit, comme en théorie normative. La famille des théories fondées sur la rationalité maximisatrice du comportement individuel fournit une axiomatique puissante et parfaitement accordée à une formalisation mathématique, comme le montre le développement très homogène de la théorie économique à partir de ce socle.

Dans les modèles qui intéressent le plus les sociologues, le principal objectif de l'analyse est en fait d'établir ce qui constitue le comportement rationnel. Si la définition est fixée, c'est-à-dire robuste dans différents contextes de décision et d'action, des tests de la théorie peuvent être effectués pour découvrir si les actions correspondent ou non au principe de comportement postulé. Ce qui implique ordinairement des hypothèses lourdes quant à la connaissance par les acteurs des situations où ils se meuvent et à leur connaissance des résultats de leurs actions, et quant à la définition des utilités maximisées. Surtout, les écarts à la rationalité ne sont généralement pas conçus comme des termes d'erreur exprimant l'influence aléatoire de facteurs sociaux et psychologiques trop mal connus pour être mesurés. L'hypothèse de maximisation est, dans les variantes élaborées de l'individualisme méthodologique, considérée comme l'un des vecteurs du comportement individuel, lorsque la situation de l'action ou l'environnement de l'acteur sollicitent une formation rationnelle des choix et des croyances. L'individualisme méthodologique, qui est le paradigme dominant en économie, connaît aujourd'hui des développements en sociologie qui font coexister des éléments de comportement ancrés dans le postulat de rationalité et des explications de l'action par les normes sociales : si les normes combinent, comme le suggère Elster (1989) dans une typologie d'allure wébérienne, des éléments de tradition, des éléments de rationalité en valeur et des dimensions affectives et émotionnelles, l'explication par l'intérêt individuel ne couvre plus le champ entier du comportement.

En théorie classique du comportement individuel rationnel, la prise de décision est, par hypothèse, située dans un contexte de certitude, les acteurs contrôlant entièrement les résultats des événements qui résultent du cours de leurs actions. Dans ce cas, seule une spécification des préférences est exigée, et le principe de maximisation de l'utilité n'est pas nécessaire au développement de la théorie. Quand les résultats des événements ne sont pas entièrement sous le contrôle des acteurs, mais influencés par un environnement, les choses se compliquent. Cet environnement peut être soit hors d'influence de ses actes, soit être constitué d'acteurs réagissant à ses actions. Dans le premier cas, il s'agit de décision en incertitude. Dans le second cas, on distinguera les situations selon le degré de conflit entre les préférences des acteurs. S'ils sont totalement ou partiellement en conflit, nous aurons une situation décrite par la théorie des jeux et ses extensions, et qui intéresse directement la sociologie formalisée.

La théorie des jeux fournit en effet un cadre adéquat de formalisation de l'interaction sociale en termes individualistes, sous ses trois espèces : interaction entre les actions (par l'analyse des interdépendances stratégiques entre des acteurs analytiquement indépendants), interaction entre actions et résultats (par le pré-engagement de soi pour se fermer certaines options accessibles dans une négociation, afin d'agir sur les anticipations d'autrui et en tirer parti), interaction entre résultats eux-mêmes (à travers des mécanismes de liaison interpersonnelle comme l'envie ou l'altruisme) (Elster, in Swedberg, 1990).

Les divers types de jeux peuvent être caractérisés selon le nombre d'acteurs impliqués (deux ou plus) et selon le degré de conflit d'intérêt, conflit pur ou conflit partiel. Parmi les jeux à deux joueurs, ceux auxquels les sociologues s'intéressent plus particulièrement sont les jeux à conflit partiel d'intérêts, comme dans le dilemme du prisonnier, alors que les nombreux développements théoriques consacrés aux jeux purement conflictuels à somme nulle n'ont pour la sociologie qu'un intérêt secondaire. Quant à l'analyse des jeux à acteurs multiples, elle impose, outre une complexité mathématique plus grande, des hypothèses plus arbitraires sur le réalisme des comportements et offre des possibilités plus fragiles d'application empirique.

Dans les limites d'acceptabilité de l'hypothèse de rationalité telle qu'elle a cours dans l'analyse sociologique, y compris dans le paradigme individualiste le plus accueillant à une formalisation poussée du comportement des acteurs, la théorie des jeux constitue davantage un langage qu'une ressource de théorisation. Par ailleurs, quand des comportements relativement élaborés sont envisagés dans des contextes d'interaction stratégique, les solutions sont généralement multiples et le critère de rationalité devient plus malaisé à définir, le raisonnement ne débouchant pas forcément sur des modèles d'anticipation stable par l'acteur du comportement d'autrui (Elster, in Swedberg, 1990). D'où l'importance des paradoxes, des solutions contre-intuitives et non-optimales. La sensibilité des résultats au cadre de règles retenu comme à la structure d'information prise en compte, la part d'arbitraire que suppose le traitement de problèmes à solutions multiples, sont des caractéristiques qui expliquent la fonction surtout heuristique des applications sociologiques.

Pour l'essentiel, au regard de sa carrière remarquable en mathématiques, en économie, en psychologie sociale et en recherche opérationnelle (cf. Shubik, 1982), la théorie des jeux n'intéresse qu'à son niveau d'élaboration élémentaire les sociologues, qui n'en ont jamais fait un outil de modélisation de puissance comparable à celle de la mathématique des processus stochastiques ou des réseaux.

2. La sociologie et l'économie face à la mathématisation

Les limites du développement de la formalisation en sociologie tiennent tout à la fois à l'hétérogénéité théorique forte de la discipline, à la relative indistinction entre modélisation et traitement statistique pour tout ce qui touche à l'analyse des données, à la part importante qu'occupent en sociologie d'autres aspects de méthodologie tels que les techniques de recueil et de traitement des données qualitatives et le problème de la construction des matériaux quantitatifs. Ensemble, ces facteurs peuvent expliquer aussi pourquoi la culture mathématique des sociologues est beaucoup plus inégalement distribuée que dans d'autres sciences sociales, comme l'économie ou la psychologie expérimentale. Arrêtons-nous au point de la diversité théorique, afin d'établir, par comparaison, les conditions de possibilité d'un haut degré de formalisation tel qu'il prévaut en économie.

Les principaux théoriciens de la sociologie ont, à peu d'exceptions près, ignoré les mathématiques dans la construction des systèmes qui ont le plus durablement influencé la discipline depuis les origines jusqu'à l'époque contemporaine. La publication récente par Coleman, qui compte parmi les rares théoriciens pionniers et tenants de la mathématisation de la sociologie, d'un livre conçu comme fondateur d'une nouvelle théorie sociologique largement formalisée (Coleman, 1990), constitue plutôt une exception, sans qu'on puisse préjuger de l'évolution future de telles tentatives, malgré le pessimisme, évoqué à l'instant, de certains des tenants de la formalisation mathématique en sociologie.

La position de l'économie est ici radicalement opposée. Les conflits les plus aigus autour de l'économie mathématique remontent à la fin du 19ème siècle et à la première moitié du 20 ème siècle, quand, en France notamment, a été contestée la fécondité de la théorie walrasienne (Zylberberg, 1990). Depuis quelque quarante ans, la mathématisation de la science économique apparaît comme une tendance forte et irrésistible, qui retentit à tous les niveaux du travail scientifique : les progrès de la théorie économique ont partie liée avec l'invention de nouveaux outils mathématiques, notamment depuis l'invention de la théorie des jeux et la formalisation mathématique par Debreu de la théorie néoclassique de l'équilibre général. Il est aujourd'hui hasardeux pour un économiste de concevoir une quelconque proposition théorique qui ne reçoive pas de formulation mathématique. Le travail empirique ordinaire de la majorité des économistes, et d'abord de ceux qui situent leurs contributions dans la famille des paradigmes néoclassiques, largement dominants dans la discipline, est fondé sur l'enchaînement des séquences suivantes : position d'un problème, construction d'un modèle, linéarisation, analyse statistique formelle sur des données existantes, estimation des paramètres du modèle ou test des hypothèses, conclusion, le raffinement des méthodes l'emportant sur le besoin d'une distance critique et de corrections éventuelles de l'axiomatique fondamentale (Solow, in Swedberg, 1990).

De toutes les disciplines des sciences sociales vouées à l'analyse du monde contemporain, l'économie est la plus proche de la sociologie, mais aussi la plus concurrente depuis qu'elle étend le champ des faits sociaux, des comportements individuels et collectifs et des réalités institutionnelles et organisationnelles auxquels s'appliquent ses théories et modèles. C'est donc l'exemple du haut degré de formalisation de la science économique qui entretient en sociologie une attitude ambivalente de fascination et de réticence (due aux risques d'épuisement conceptuel) à l'égard de ce mouvement apparemment irrésistible vers la formalisation.

A priori, la formalisation ne peut guère apparaître comme une option scientifique parmi d'autres, si l'on songe aux gains en précision, en rigueur et en calculabilité qu'offre la construction d'une architecture logique des propositions à partir d'un ensemble de concepts et d'une syntaxe aussi univoques que possible. Son emploi est systématique en économie, et distingue très fortement économie et sociologie dans leur teneur respective en langage naturel. L'éloge qu'en font Walliser et Prou (1988), après avoir observé qu'elle conditionnait l'application de tous les outils mathématiques nouveaux — systèmes d'équations linéaires, puis systèmes d'équations différentielles, jadis, outils topologiques (théorèmes de point fixe, théorèmes de séparation) et systèmes dynamiques (théorie des catastrophes, théorie des bifurcations) aujourd'hui — est sans ambiguïté :

> "Toute théorie économique se présente sous forme d'un système hypothético-déductif, qui associe à des hypothèses de départ des conclusions dérivées selon des règles d'inférence logique : calcul logicomathématique de base (symboles primitifs, règles de formation des propositions, axiomes formels, règles de déduction) et calcul propre au domaine économique (symboles exprimant des concepts, axiomes exprimant des propositions spécifiques). (. . .)

> Le travail des économistes consiste, dès lors, en un jeu formel visant à exhiber des résultats significatifs, c'est-à-dire à trouver des conditions généralement suffisantes (mais non forcément nécessaires) pour justifier une conclusion : "une entreprise à rendements décroissants ayant un comportement concurrentiel vend au coût marginal". (...)
>
> Nombre de résultats argumentés littérairement peuvent recevoir une démonstration analytique rigoureuse avant d'être retraduits à nouveau en théorèmes qualitatifs, le détour ayant pour avantage de baliser leur domaine précis de validité. (...)
>
> Le raisonnement formalisé permet d'enchaîner (...) logiquement les démonstrations les unes aux autres et d'enclencher ainsi une combinatoire particulièrement féconde : à partir d'une axiomatique implicite des choix individuels (réflexivité, transitivité, continuité), on peut d'abord exprimer le comportement sous forme explicite et intensive (maximisation d'une fonction d'utilité sous contraintes), puis sous une forme explicite et extensive (fonction de réaction liant l'action à l'environnement).
>
> Le raisonnement formalisé permet de rechercher des résultats de plus en plus généraux, en particulier de découvrir des hypothèses de moins en moins restrictives justifiant une conclusion donnée : l'existence d'un équilibre concurrentiel a d'abord reposé sur la continuité des fonctions d'offre et de demande des agents, elle-même obtenue pour des préférences et des technologies convexes, mais elle a pu ensuite être démontrée en présence de non-convexités. Il permet enfin de changer de niveau de spécification et de construire des théories générales dont les théories spécifiques ne sont que des cas particuliers ; l'exemple de la théorie de l'équilibre est ainsi immergé dans la théorie des jeux, dont les résultats sont aussi bien utilisés dans les modèles micro-économiques de concurrence parfaite ou imparfaite que dans les modèles macro-économiques de relations Etats-agents." (Walliser, Prou, 1988, p. 78 sqq).

Les bénéfices scientifiques de la formalisation ont été obtenus en économie sous des conditions axiomatiques qu'il importe d'examiner pour comprendre pourquoi la formalisation a connu un sort si différent en économie et en sociologie. Une série de principes théoriques ou de concepts recteurs organise en effet les affinités entre conceptualisation, modélisation et mathématisation. Ils forment l'armature de la théorie économique néo-classique, i.e., de celui des paradigmes de la science économique qui est le plus influent et le plus puissant mathématiquement : nous ne pouvons pas ignorer, en évoquant "la" théorie néo-classique, ce qu'une telle généralisation a d'indigent, eu égard à la multiplicité des courants théoriques qui, sur tous les points évoqués, peuvent discuter la doctrine ainsi évoquée. Mais l'une des évidences majeures de la comparaison entre sociologie et économie est bien le degré beaucoup plus élevé d'homogénéité paradigmatique de l'économie, ce qui nous autorise à rappeler schématiquement les points d'ancrage de la famille des théories néo-classiques.

Nous avons déjà évoqué le postulat de la rationalité du comportement de l'acteur, dont les économistes débattent notamment depuis Herbert Simon pour l'amender ou l'enrichir, tout en conservant cette pierre angulaire de l'axiomatique du comportement individuel. Ce postulat n'a pas la même position centrale en sociologie. C'est même l'une des cibles favorites de la critique de l'économie par ceux des sociologues qui combattent le plus vigoureusement l'individualisme méthodologique, avec lequel ils l'assimilent.

La théorie néo-classique est une théorie de l'équilibre, ce qui permet d'assurer la compatibilité entre les diverses actions intentionnelles d'un même agent, sous des contraintes données, et, plus radicalement, entre les actions des divers agents, la première hypothèse pouvant se réclamer d'un certain réalisme dans la représentation du comportement individuel alors que la seconde est une nécessité mathématique, qui permet de définir un système de relations entre les valeurs prises par les grandeurs considérées (Malinvaud, 1991). L'un des intérêts majeurs du concept d'équilibre est de permettre la construction de modèles candidats à une capacité prédictive. L'outillage conceptuel des sociologues ne contient aucun postulat aussi radicalement simple et aussi indispensable à la mathématisation, puisqu'en sociologie, l'individualisme méthodologique et le postulat de la rationalité de l'acteur, essentiels à l'économiste pour rendre opératoire le concept d'équilibre, ne sont pas aussi largement acceptés comme des pièces cardinales de l'axiomatique.

Solidairement, la théorie néo-classique recherche des solutions mathématiques au problème de l'agrégation des comportements d'agents élémentaires pour passer sans biais de la diversité à la synthèse intégratrice. La solution est d'autant meilleure que les conditions quant à l'hétérogénéité des comportements et des environnements des acteurs sont plus restrictives. La sociologie construit des concepts intermédiaires pour capter l'hétérogénéité interindividuelle et accorde à la logique des situations d'action un poids plus grand dans l'analyse des hétérogénéités composées qui distinguent des acteurs ou ensembles d'acteurs diversement selon le contexte de l'action.

De même, la modélisation économique opère généralement par une décontextualisation beaucoup plus poussée que la modélisation sociologique. L'environnement des phénomènes étudiés est schématisé, par réduction des coordonnées spatio-temporelles et des dimensions qualitatives, et traité comme un élément exogène. D'où la fréquence des modèles raisonnant sur deux biens, deux agents, deux périodes ou, au contraire, sur une infinité d'agents et de périodes. Si le coût de la réduction schématisante apparaît trop élevé, la construction de modèles à validité locale rencontre, comme en sociologie, des problèmes de synthèse ou des limites de complexité dans le traitement mathématique des problèmes si l'on superpose les facteurs d'hétérogénéité interindividuelle (e.g., asymétries d'informations) et intertemporelle (e.g., cadre dynamique et aléatoire).

Plus généralement, le partage entre éléments endogènes et éléments exogènes aux systèmes étudiés distingue fortement l'économie de la sociologie. Si l'on s'en tient aux seules dimensions psychologiques, sociales et culturelles du comportement, les préférences et les représentations des agents sont traitées comme exogènes par l'économiste et comme endogènes par le sociologue. L'endogénéisation, qui accompagne une extension de l'approche économique à des domaines nouveaux (e.g., famille, travail, organisation) pose des problèmes de traitement dynamique, de modélisation de l'hétérogénéité des comportements (e.g., optimisation ou non) et d'écart par rapport au schéma théorique d'équilibre concurrentiel.

Enfin, la théorie économique néo-classique s'appuie sur une distinction entre microéconomie et macroéconomie qui n'a pas d'équivalent théorique aussi ferme en sociologie. Il s'agit d'une différence d'échelle dans l'étude des phénomènes économiques (e.g., niveau des prix/structure des prix), mais aussi d'une différence d'objet et d'une différence d'objectifs, la macroéconomie ayant une vocation plus directement empirique et appliquée mise en action dans l'orientation des politiques économiques, et la microéconomie représentant plutôt le pôle de l'analyse fondamentale. Non qu'il s'agisse d'une division du travail harmonieuse, puisque les divergences théoriques existent dans l'orientation dominante des deux branches de la discipline, mais la dualité de la discipline exprime bien les deux termes de l'activité de modélisation : modélisation abstraite, universalisante et anhistorique en microéconomie et modélisation numérique en macroéconomie, caractérisant une économie située et datée par des relations entre des variables mesurables et spécifiables à l'aide de paramètres numériques. La partition entre microsociologie et macrosociologie engage fréquemment, sans s'y laisser réduire, une distinction plus radicale entre des paradigmes concurrents, l'individualisme et le holisme.

3. La sociologie, science historique ?

Dans les regards croisés que portent sociologues et économistes sur la discipline de l'autre, tels qu'ils nous sont restitués par exemple, et selon l'arbitraire du choix des représentants des deux sciences, dans le livre composé par Swedberg (*op. cit.*), l'économie apparaît comme une science largement unifiée autour d'un corps de postulats et d'hypothèses de base, ceux de la théorie néo-classique ; aussi radicales que soient les critiques internes contre les limites de celle-ci, la conviction demeure dominante dans la communauté des économistes qu'il n'y a pas d'alternative crédible. La sociologie apparaît à l'inverse plus fragmentée sous l'influence d'une concurrence entre les paradigmes candidats à la prééminence, l'individualisme méthodologique ne jouant pas en sociologie, et de loin, le rôle pivot qu'il a en économie. La capacité de modélisation et de mathématisation et la prétention à la cumulativité varient avec le degré d'homogénéité paradigmatique, comme on l'a esquissé plus haut.

Le défaut d'unification de la sociologie, et la relative indifférence de nombre de ses principaux théoriciens à la traduction mathématique de leurs concepts et systèmes, s'ils étaient mesurés à l'aune de l'économie, se réduiraient à des déficits et des manifestations obstinées d'immaturité scientifique. Ce qui supposerait qu'économie et sociologie relèvent d'un même ensemble scientifique, dont l'une incarne mieux les progrès que l'autre. Or, on peut soutenir, concurremment, que la diversité des paradigmes dans la théorie sociologique exprime une situation épistémologique différente, comme l'a fait récemment Passeron (1991), dont nous présentons brièvement les arguments ci-dessous.

Dans l'hypothèse généreuse d'altérité "non-déficitaire" de la sociologie, il convient de distinguer des sciences sociales à ambition synthétique (histoire, sociologie, anthropologie) qui, dit Passeron, "se donnent pour objet des moments ou des aires du cours du monde historique considérées dans l'infinité potentielle de leurs traits pertinents" et des sciences sociales comme l'économie, la linguistique ou la démographie, qui "autonomisent un sous-ensemble de faits sociaux présentant une forte systématicité de fonctionnement. (...) A condition de ne pas trop s'interroger sur le fondement 'quasi expérimental' de leurs assertions générales, on peut considérer que le rôle des paradigmes dominants dans l'organisation de la recherche y ressemble davantage à celui qu'ils jouent dans les sciences de la matière ou de la vie. Mais c'est au prix d'un affaiblissement de la teneur historique de leurs assertions, puisque ces assertions doivent s'entendre implicitement à contexte historique constant et doivent par conséquent être reformulées en changeant de contexte." (Passeron, 1991, pp. 363–364).

La conception, d'inspiration nettement webérienne, de la sociologie comme une science pleinement historique a une double signification : la sociologie se donne des objets dont la descriptibilité n'est pas finie, toutes ses observations sont indexées sur un contexte historique chaque fois singulier et qui ne se répète jamais intégralement dans le temps ou l'espace. Et une double implication épistémologique :

— la sociologie relève d'une épistémologie non-poppérienne :

> "Les analyses permettant de généraliser les constats empiriques d'une enquête au-delà de son contexte singulier relèvent d'un raisonnement naturel : les constats sont opérés dans des contextes qui ne peuvent être dits équivalents que selon une typologie les apparentant par hypothèse et non par nécessité. (...)
>
> L'analyse des variations qui recourt au langage des variables permet de tenir, dans un contexte supposé constant, des raisonnements formellement expérimentaux mais elle reste toujours tributaire , dans les sciences sociales, de l'interprétation du sens des variations en fonction du contexte. (...)
> Les concepts qui constituent la sémantique du langage de description du monde en sociologie ont un statut logique irréductible à celui qui rend opératoires les définitions des sciences axiomatico-formelles ou des sciences expérimentales. (...)
>
> La mise à l'épreuve empirique d'une proposition théorique ne peut jamais revêtir en sociologie la forme logique de la réfutation au sens poppérien." (Passeron, *op. cit.*, p. 368 sqq)

— aucune théorie sociologique ne peut espérer fonctionner comme un paradigme scientifique dominant, parce qu'"il ne peut exister de langage protocolaire unifié de la description empirique du monde historique" (*ibidem*, p. 363). La stabilité paradigmatique n'est pas un idéal régulateur, mais une impossibilité. Et l'évolution de la sociologie démontre amplement la non-convergence des théories.

Une telle épistémologie a l'intérêt de rendre compte de la diversité des théories et des pratiques de la sociologie autrement qu'en termes d'impuissance à dégager un paradigme dominant, ancré dans une conception physicienne de la scientificité et de l'expérimentation. Mais elle ne permet pas de discriminer de manière univoque les mérites des théories concurrentes dans l'explication des "mêmes" catégories de faits et de comportements sociaux.

Faute de pouvoir discuter ici toutes les implications de cette position épistémologique, et de pouvoir établir qu'elle n'a rien d'une célébration anarchiste et relativiste, retenons un point essentiel. En sociologie, les abstractions et les généralisations sont toujours incomplètes, puisqu'elles se réfèrent immanquablement à un contexte spatio-temporel. Les concepts ne peuvent être que des idéal-types, le raisonnement sociologique opère par va-et-vient entre analyse statistique et contextualisation historique. D'où la moindre prise du calcul et de la formalisation sur le raisonnement sociologique, sur ses procédures typiquement comparatives, sur son expression dans le langage naturel. A sa manière, la sociologie illustre avec le plus grand relief la tension entre la substance historique de ses observations et l'exigence de règles de description, de comparaison, de généralisation par conceptualisation typologique.

Si l'économie se situe au pôle opposé dans le champ des sciences sociales, en procédant par abstraction, simulation, modélisation, raisonnement hypothético-déductif, au plus loin du langage naturel, elle n'échappe pas aux difficultés du traitement de l'hétérogénéité contextuelle. L'un des indices en est la complication des relations entre macro- et microéconomie.

Sans exagérer la portée heuristique du rapprochement, on peut se demander en effet si l'écart ou même la concurrence observés entre macroéconomie et microéconomie contemporaines ne signalent pas, dans une discipline pourtant beaucoup plus homogène, une tension interne due pour partie à l'ambition d'embrasser une plus grande diversité de phénomènes pourvus de leurs caractéristiques contextuelles. On assiste en effet à un enrichissement continu des hypothèses et des domaines d'analyse en microéconomie, notamment quant au comportement des éléments premiers du système (endogénéisation des préférences et représentations des agents, complication de l'hypothèse de rationalité, prise en compte de régulations normatives et institutionnelles, traitement de l'incertitude et des asymétries d'information, affaiblissement de la fonction pivot de l'hypothèse de la concurrence pure, analyse du comportement des entreprises, ...). Cet enrichissement ne ruine-t-il pas l'idéal d'une construction unitaire de la science économique, la microéconomie posant les fondements et la macroéconomie démontrant empiriquement la validité et la puissance des outils théoriques sur ce qui est mesurable et calculable ? C'est l'hypothèse de Malinvaud (1991) :

> "La démarche consistant à partir systématiquement de l'agent individuel pour aboutir à une compréhension des lois régissant le système économique entier, sans passer par des représentations agrégées simplificatrices (. . .) a la préférence de la plupart des économistes mathématiciens rigoureux. Cette préférence les a conduits à envisager le processus d'acquisition de la connaissance en économie comme devant s'effectuer par phases successives, chacune ajoutant à la précédente un élément de complexité du monde réel. (. . .)
>
> Ce programme s'avère irréalisable. D'une part, on s'aperçoit vite que la théorie devient trop peu spécifique pour fournir la base exclusive qu'on espérait y trouver. Beaucoup de ses résultats sont négatifs, montrant tantôt que l'intuition n'est validée que pour des cas particuliers, tantôt qu'un résultat positif ne résiste pas à l'introduction d'une nouvelle complication. D'autre part, la démarche s'empêtre. Loin d'être définitivement acquis, ses pas successifs doivent souvent être repris, pour étudier des propriétés dont l'importance n'avait été perçue que par la suite. Chaque pas nouveau oblige à explorer plusieurs directions, dont certaines exigent de véritables programmes de recherche. En somme, le moment où l'on croit que les phénomènes macroéconomiques seront vraiment appréhendés s'éloigne à mesure qu'on avance. Le travail donne à la théorie des prix plus de réalisme et de sophistication, mais il ne suffit pas à construire une véritable théorie de la croissance, de l'emploi ou de l'inflation.
>
> C'est donc par une approche plus directe des phénomènes macroéconomiques que l'essentiel des progrès se fera." (Malinvaud, *op. cit.*, p. 146–147)

A mesure que se complique l'analyse microéconomique des agents (individus, entreprises,. . .), il devient plus difficile, pour passer au niveau agrégé, de maintenir, par commodité analytique, des conditions restrictives de moins en moins réalistes sur le comportement identique des agents (résumé dans celui du célèbre agent représentatif) ou sur les variations homothétiques des environnements. La qualité de l'agrégation opérée devient plus discutable et la conscience des risques d'erreur plus aiguë.

Par un paradoxe qui n'est qu'apparent, la complication de l'analyse microéconomique est mue par le même souci de formalisation et de modélisation que le maintien des exigences de calculabilité en macroéconomie, mais les mathématiques dominantes ne sont pas les mêmes : la théorie des jeux fournit à la microéconomie l'épure de l'analyse contextuelle, si l'on appelle contexte l'environnement d'un agent peuplé d'autres agents liés à lui par des relations d'interdépendance, l'économie de l'information introduit l'asymétrie et les différences interindividuelles de connaissance et de capacité d'action, le traitement de l'incertitude confère au temps son irréductible densité.

Bibliographie

[1] Boudon, Raymond, *L'analyse mathématique des faits sociaux*, Plon, Paris, 1967.

[2] Coleman, James, *Introduction to Mathematical Sociology*, The Free Press, New York, 1964.

[3] Coleman, James, *Foundations of Social Theory*, Belknap Press, Cambridge, 1990.

[4] Coleman, James, Katz, Elihu, and Menzel, Herbert, *Medical Innovation*, Bobbs–Merrill, Indianapolis, 1966.

[5] Elster, Jon, "Social Norms and Economic Theory", *Journal of Economic Perspectives* **3**:4 (automne 1989).

[6] Hägerstrand, Torsten, "A Monte-Carlo Approach to Diffusion", *Archives Européennes de Sociologie*, vol. 6, 1965.

[7] Malinvaud, Edmond, *Voies de la recherche macroéconomique*, Editions O. Jacob, Paris, 1991.

[8] Marsden, Peter and Laumann (Edmond O.), "Mathematical ideas in social structural analysis", *Journal of Mathematical Sociology*, vol. 10, 1984.

[9] Passeron, Jean-Claude, *Le raisonnement sociologique*, Editions Nathan, Paris, 1991.

[10] Shubik, Martin, *Game Theory in the Social Sciences*, MIT Press, Cambridge, 1982.

[11] Sørensen, Aage B. and Sorensen, Annemette, "Mathematical sociology", *Current Sociology*, n° 2, 1978.

[12] Sørensen, Aage B., "Mathematical Models in Sociology", *Annual Review of Sociology*, n° 4, 1978.

[13] Stewman, Shelby, "Markov models of occupational mobility: theoretical development and empirical support", *Journal of Mathematical Sociology*, vol. 4, 1976.

[14] Swedberg, Richard, *Economics and Sociology. Redefining their Boundaries : Conversations with Economists and Sociologists*, Princeton University Press, Princeton, 1990.

[15] Walliser, Bernard and Prou, Charles, *La science économique*, Edition du Seuil, Paris, 1988.

[16] Weber, Max, *Economie et société*, trad. fr., Plon, Paris, 1971.

[17] White, Harrison C., *Chains of Opportunity*, Harvard University Press, Cambridge, 1970.

[18] Zylberberg, André, *L'économie mathématique en France, 1870–1914*, Economica, Paris, 1990.

Round Table H: July 9, 1992

MATHEMATICS AND INDUSTRY

by Julian C.R. Hunt[1] and Helmut Neunzert[2]
UK Meteorological Office and University of Cambridge[1], Universität Kaiserslautern, [2]
Cambridge (United Kingdom)[1], Kaiserslautern (Germany)[2]

1. Introduction

It was recognised at the founding meeting of the European Mathematical Society that, for the future well-being of mathematics, it is essential to examine and improve the relationship between Mathematics and Industry. It was therefore an excellent proposal by the organisers (especially M. Karoubi) of the European Congress of Mathematics to hold a round table at the congress on this subject. The two co-chairmen who were appointed have had broad experience with applying mathematics to industry in Europe and North America, and with bringing together academics, industrialists, and relevant government agencies for conferences, short courses, etc., for example, those organised by the European and national organisations such as ECMI (European Committee for Mathematics and Industry), ERCOFTAC (European Research Community On Fluid Turbulence and Combustion), GAMM, IMA (Institute of Mathematics and its Applications), and SMAI (Societé Mathématiques Appliqués et Industriels).

At the round table, speakers chosen from industry, universities and government agencies mainly focussed on the broad organisational issues, and only briefly touched on the details of the mathematics. In this account of the proceedings the contributions are reported approximately as they were given. The concluding contributions by the chairmen summarised the main points that emerged, especially those that seemed to be of common concern.

2. Views from Industry

T.R. Auton from ICI Central Toxicological Laboratory, Cheshire, spoke on "Hard numbers from soft data". His essential interest in mathematics in industry is to improve risk assessment and risk management which involves using biological data, modelling and statistics to help make decisions.

There are many sources of data; toxicity in animals, epidemiology, human exposure, results from basic science, and from in vitro methods (on lower cells).

The latter data are useful because human risk assessment is often based on the analysis of subsystems in order to predict the effects on a whole human.

The way that mathematics is used together with these large quantities of data to help decisions involves

(a) Statistics,
(b) Modelling,
(c) Design of new experiments, or linking experiments.

An important aspect of modelling is that it is useful for predicting toxicological effects from very small quantities that are difficult to measure. There are many public demands to improve toxicology; the particular challenge for mathematicians is to utilise all the available data. It is also essential that the results of the mathematical studies are communicated well both in the industry and to the public beyond.

To illustrate one of the greatest difficulties confronting those applying mathematics in industry, namely the problem of explaining the results to non-mathematicians, Auton ended with a nice quote of Goethe (1820):

> "Mathematicians are like the French; when you tell them something they repeat it in their own tongue, and immediately it sounds quite different".

Dr. Stoufflet from Dassault Aviation, Saint-Cloud, focussed his discussion on the mathematics required for the design of advanced aeroplanes and spacecraft. The aerodynamics, thermodynamics and chemistry controlling the airflow lead to PDEs. Before detailed solutions are attempted it is desirable to study the functional behaviour, uniqueness, bifurcation, asymptotic behaviour and other general mathematics of the equation together with the initial and appropriate level of approximation and discretisation for numerical solution of the problems. Applied mathematical analysis plays a vital role in the choice of the method of discretisation, for example in how the nodes of the pressure and velocity files in the fluid flow should be related to each other. Such analysis is also needed in the development of appropriate numerical methods with low numerical diffusion. Numerical methods are currently developing rapidly, for example by increasing accuracy in particular regions, and, by the use of higher order approximation as well as by mesh deformation, and refinement. Numerical methods may make use of wavelets rather than the standard spectral or finite different methods.

Other topics needing theoretical development for industrial purposes are the analysis of nonlinear hyperbolic systems in several space dimensions and the analysis of inverse problems and coupling of systems. Stoufflet emphasized that the latest large aerospace projects depend on the mathematical input and the collaboration of large numbers of teams, including mathematical groups. The final product from the modelling is a set of large computer codes. These have to be developed by industry because of their experience in maintaining and assuring the quality of the codes; it is no longer possible for software to be written in universities for major industrial projects. However the research, including mathematical research, by university groups is found by many organisations in the aerospace and other industries to be essential for the successful development of the codes. The Hermes aerospace project involves more than 50 research teams.

Following his presentation, a question was addressed to Dr. Stoufflet from the audience asking him why so much industrial research was oriented towards military applications and was so masculine in its gender orientation. He and others commented that only some aeronautical research was military. Industry is attempting to bring more women into mathematical research; however there are usually relatively more women in **industrial** research than in academic research as is shown by the large numbers in Dassault Aviation's team. (Some women have stated that they prefer the practical work of industry to more open-ended (and less well-paid) academic research.)

Dr. Gilg of Siemens began his presentation by asking "Do we need more Mathematics in Industry?" His answer was "Yes", because first, the kernels of most modern software systems now used in industry are mathematical, for example, the solution of complex electronic circuit equations of the form $F(y(t), \dot{y}(t), t) = 0$ that form a set of implicit, nonlinear, differential algebraic systems are "stiff", very sparse and unstructured. Secondly, software is increasingly more important than hardware, and thirdly, because the complexity of systems is exploding exponentially.

To emphasize his second point about the increasing importance of mathematics in industry, Gilg described the change in the distribution of business activity of Siemens. In 1970, 70% came from conventional electrical engineers and consumer electronics with 30% from industrial electronics and software, whereas it is projected that in 1995 these figures will be 33% and 67% respectively.

Given the **importance** of mathematics in industry, it might be asked whether it is necessary to recruit mathematicians. In fact it is found in the electronics industry that the percentage of mathematicians in the research and development staff is decreasing! To answer this, consider the required profile of an industrial mathematician. He/she must be able to adjust his/her work to meet quickly moving targets, and work with an interdisciplinary team. Gilg raised the question of whether mathematicians are able and willing to move, change or expand their fields according to the requirements of industry. In other words, are they interested in developing their career in the interests of being more useful to industry, rather than in developing themselves purely as mathematicians? His words were "should an industrial mathematician be a multi-legged being?"

He then explored whether universities are educating mathematicians with skills and attitudes that are most appropriate from an industrial point of view; he phrased his question as: "Is academia miseducating its students?" It is clear that the study of pure mathematics is inevitable and necessary for the continuing progress of the subject, but he questioned whether there is any need for increasing effort in this branch of mathematics in terms of manpower. He questioned whether Applied Mathematics is only involved in filtering industrial problems by simplifying or even over-simplifying them in order to apply well-understood theories rather than solve the actual problem, which is the subject of Industrial Mathematics. In his words, "Applied Mathematics is **not** Industrial Mathematics!" He thought that industrial mathematics is not emphasized enough by universities, and that this will only be overcome in the long term.

Gilg reminded us of the useful report on mathematics published in the USA and edited by Edward E. David (President of Exxon Research and Engineering). Two important conclusions were that "For mathematics to be renewed, mathematics must be perceived as it historically has been: as an essential contributor to science and technology", and "Apparently, too few people recognize that the 'high technology' that is celebrated today is essentially mathematical technology". These thoughts may help guide universities in the future development of mathematics.

3. Academic Views

Prof. Cercignani of Politecnico of Milan spoke on "The different uses of mathematical models in industry". His presentation is given here verbatim:

Even a superficial examination of a modern society shows the importance of the role played by mathematics. If one remains at the surface, this role may appear to be limited by the ubiquitous computer, which conditions our life in vital areas, such as new technologies, medical tests, banking, business, transport, and the scientific progress itself. The use of this tool (as well as its invention), however, depends on mathematical ideas which were or are invented by mathematicians. This use also needs, in order to be creative and lead to further developments, a wide knowledge both of the mathematics to be exploited and of the problems which can be mathematised in the various areas of application.

The industries (and more generally the users of mathematics) have long since realised that the use of powerful computational techniques to solve complicated equations or optimization problems has dramatically changed both the organization and the simulation of the industrial processes and the design of new products. In more recent years they have begun to realize, more and more frequently, that new abilities are required in order to exploit these powerful tools. These new abilities are related to a higher vision of what should be called applied mathematics. Too frequently, in fact, by this term one understands the use of formulas and methods, which are more or less recent and more or less appreciable, to problems of interest for applications. While this type of applications deserves respect and is of notable importance, the creative moment of mathematics is not absent from them. To appreciate this aspect means to realise what is the mathematics done by a professional mathematician, as opposed to the mathematics, also appreciable, usually applied in the industrial world.

One of our chairmen, Helmut Neunzert, in a talk given nine years ago in Oberwolfach, summarized his findings on Mathematics in German Industry in three theorems:

1) In industry the role of mathematics is perceived neither as developing theories deductively nor as supplying convergence proofs to numerical algorithms, but as formulating and evaluating mathematical models of given practical problems.
2) If mathematics is understood as in Theorem 1, the increase of its importance for industrial research and development during the last twenty years must be called revolutionary.
3) The biggest part of mathematical work in the research and development departments of German industries is done by engineers.

I don't know what the actual situation is nine years later in Germany, though I guess it has changed because of the very energetic engagement of H. Neunzert himself, who more than anyone else tried to modify the situation prevailing at that time. Certainly the three theorems illustrate rather well the situation prevailing in Italy at least up to some time ago. I am glad to report, in fact, that a few months ago, the top manager of the research branch of an important automotive industry told me that they find mathematicians more useful than engineers in their advanced research programmes.

What is the situation in other European countries? I guess that on the continent the picture outlined by the third of Neunzert's theorems was prevailing until some time ago, though the situation may be changing now. On the other hand, information supplied by the other chairman, Julian Hunt, seems to indicate that in the UK mathematicians not only do mathematics but also engineering and physics! This certainly has much to do with the difference in the university curricula and may be worth further exploration.

What is the mathematics which has received more attention by industry? Modelling is certainly very important; in this respect one understands the dichotomy between the continent and the UK, because, according to the rather harebrained curricula of most universities on the continent, a mathematician is supposed (to use the terminology of Neunzert's first theorem) to develop theories deductively or to supply convergence proofs to numerical algorithms and not to formulate and evaluate mathematical models of given practical problems. The teaching of probability is also lagging behind, at least in Italy, with respect to the actual needs, [and also the UK: ed].

It is my opinion, however, that curricula are not the entire story. Doing applied mathematics (I mean applied "applied maths" and not pure "applied maths") frequently requires a team of a certain number of persons who collaborate on the same project. This may be difficult in a university unless there is a certain degree of organization and a certain reward (either in terms of career or of money) for people who deal with the tedious details of a big project. Clearly certain countries have an advantage here with respect to others.

The chairmen suggested that I should highlight both mathematical successes in my field and key issues of general interest to a mathematical audience. To do this in a short time is very hard but I will try. My field has to do with the Boltzmann equation (BE) for the distribution function of the molecules in a gas, a field which has certainly become more known in the last few years because of several (perhaps not unrelated) circumstances. The general reason for the increased interest in this type of equation is that the particle nature of many media that used to be treated as continua is coming to the fore not only when we deal with ordinary gases; one might mention, in passing, semiconductor devices.

The BE is a rather complicated integro-differential equation: the unknown depends upon seven independent variables, since to the usual four space-time coordinates one has to add the three velocity components of a molecule. As such, its numerical solution on a computer is rather hard; it is however needed in applications which may range from environmental problems, when one wants to understand and control the formation, motion, reactions and evolutions of particles of varying composition and shapes, ranging from a diameter of the order of $0.001\mu m$ to $50\mu m$, as well as their space-time distribution under gradients of concentration, pressure, temperature and the action of radiation, to the flow of air past a space shuttle which, during the re-entry, travels at an altitude where, because of the rarefaction degree of the atmosphere, the Navier–Stokes equations are no longer valid.

The theory of the Boltzmann equation is concerned with such topics as existence and uniqueness of solutions. This was first investigated in the thirties by the famous Swedish mathematician T. Carleman. Since then many people have studied the very difficult initial and boundary value problems for the BE. Recently, R. DiPerna and J.- P Lions provided an existence theorem (without uniqueness) for the general case of the Cauchy problem; their proof is quite clever and makes use of a compactness lemma by Perthame, Golse, Sentis and Lions to overcome the difficulties met by other authors.

Approximate solutions of analytic nature have been very useful in helping us form qualitative ideas on the solutions of practical problems and may be of great help for the problems of rarefied gas dynamics which arise in the study of the environment, but in general they do not provide us with the detailed and precise answers to the sort of question that is posed by the space engineer. Various numerical procedures exist which either attempt to solve for the distribution function by conventional techniques of numerical analysis or efficiently bypass the formalism of the integro-differential equation and simulate the physical situation that the equation describes (Monte Carlo methods). Only recently have proofs been given that these partly deterministic, partly stochastic games provide solutions that converge (in a suitable sense) to solutions of the Boltzmann equation. In these methods the intermolecular collisions are considered on a probabilistic rather than a deterministic basis. Furthermore, the real gas is modelled by some thousands of simulated molecules on a computer. For each of them the space coordinates and velocity components (as well as the variables describing the internal state, if we deal with a polyatomic molecule) are stored in the memory and are modified with time as the molecules are simultaneously followed through representative collisions and boundary interactions in the simulation region of space. These methods have been vigorously pursued in Europe by Helmut Neunzert‘s group in Kaiserslautern in connection with the Hermes shuttle project.

The question raised following Cercignani's contribution, was how does industry ensure that mathematical problems are actually solved. A number of different answers were given by the panel of speakers. Auton (ICI) said that it is essential to have good mathematicians employed in the company, and it is not sufficient to rely solely on external consultants, although they can certainly contribute. Hunt, (UK Met Office) described one way of helping to ensure that the mathematics used is up-to-date, by having a "technical audit". These might involve a panel of outside experts coming in to an organisation to review and advise on how mathematics is being used. The Met Office recently benefitted from such a visit of a panel of expert statisticians.

L.M.B.C. Campos of Max-Planck Institut and Lisbon University spoke on the Acoustics of Transportation Systems in the general context of mathematics and industry. Industry is interested in mathematical and physical modelling as a means of optimizing the design of a vehicle (car, train, plane, ship) or a process (manufacture, repair, operation, monitoring). Such modelling requires both a good knowledge of mathematical physics (physical description of phenomena and mathematical methods of solution) and familiarity with the practical problem (including technical experience and "jargon"). This raises the question of "who can do the modelling"? Usually on the continent this would be an engineer who had been trained in a 5-year course which included 2 years of maths and physics, and 3 years of technology. This contrasts with the British approach where an applied mathematician or physicist learns about solving practical and engineering problems at the same time as learning about a 3-year degree course.

Since modelling is an interface between exact sciences (such as mathematics and physics) and engineering or other quantative technologies, the model must speak both languages to some extent. After listing the usual stages of modelling from problem identification to validation of the solution and its application, Campos discussed the problem of the use of computing in modelling.

He concluded that for industrial problems computing is a powerful and indispensable tool but it must be accompanied by theory, experiment and interpretation. Regrettably the latter element is often missing and then "brute force" computations often end in failures because the domain of validity is unclear, results cannot be explained, and one does not know if it is true.

He commented that in the fields of acoustics and aerodynamics, while computational methods had developed rapidly (often with the use of mathematical theory), many calculations used for industrial designs still rely more on data derived from well-designed experiments. However their planning and interpretation always depends on a certain level of theoretical understanding; the classical applied mathematician is still of great value in raising that level.

One of the reasons why a straightforward computational approach is inappropriate in acoustics is because of the rapidly oscillating field. (Crighton, 1992). Mathematical theory is needed to develop a suitable form for computation. In some cases also such theory can even be used for the very practical purpose of designing methods for noise suppression (e.g., "anti-noise" e.g., Ffowcs-Williams, 1988).

Other acoustic problems involve such a wide range of phenomena (e.g., turbulence, materials, complex geometry) that it is essential to use mathematical physical theory to indicate the orders of magnitude of different effects and their interaction before any detailed calculations are performed, for example on the optimum shape of the source of sound.

H.W. Engl of Johannes-Kepler University, Linz, Austria, spoke on "Inverse problems in industry". Here is the verbatim account of his talk.

Such a problem may be defined as the determination of a cause of a desired or an observed effect. From this general definition, it is already quite clear that inverse problems arise naturally in many industrial contexts: Nondestructive testing is the means of seeking for the cause (e.g., a crack) of an **observed** effect (e.g., perturbations in a magnetic or electrical field); if one wants to control the solidification of steel or polymers by appropriate cooling, one looks for the cause (namely the cooling strategy) of a **desired** effect (namely a given solidification front).

Inverse problems almost always lead to ill-posed problems in the sense of Hadarmard, i.e., problems do not have solutions in the strict sense and/or their solution may not be unique and/or may not depend continuously on the data. This lack of continuity of course leads to serious numerical difficulties when trying to solve an inverse problem in a "naive" way. Many aspects of inverse problems are covered in Hoffmann (1986), Baumeister (1987), Engl & Groetsch (1987), Banks & Kunisch (1989), Louis (1989). Recent reviews are Engl (1991) and Engl (1993). While engineers in companies are nowadays very skilled in using numerical methods (e.g., finite element methods) for direct problems (at least for more or less standard situations), the special mathematical features involved in solving inverse problems are not yet so widely-known in these circles. It still happens that inverse problems are "solved" in industry with methods that do not address the question of ill-posedness, which results in computer codes that yield nearly arbitrary results. However, this is quickly changing: engineering societies hold conferences on inverse problems in engineering (e.g., a IUTAM Symposium on Inverse Problems in Engineering Mechanics took place in Tokyo in May 1992 and was co-sponsored by many other, especially Japanese, engineering societies). It is interesting to note that one of the first conferences organised mainly by mathematicians devoted to inverse problems in engineering science was also held in Japan, namely in Osaka in August 1990 (see Yamaguti et al. (1991)). It is clearly desirable to bring engineers and mathematicians working in inverse problems together also in Europe and in the USA. A conference where this will be attempted will be co-organised by Avner Friedman and the author (sponsored by SIAM and ECMI) in 1993.

In spite of the growing interest of engineers in inverse problems, it is safe to say that, currently, the field of inverse problems is still one where mathematicians can successfully cooperate directly with industry. It is my opinion that such direct cooperation with industry can and should be undertaken by mathematicians whenever possible, since it increases the understanding in industry that mathematics plays a direct role in solving industrial problems; this might not become clear if mathematicians work with industry indirectly via engineering departments.

After these general remarks, I would like to describe briefly some inverse problems that arose during our cooperation with industry.

Linz is the headquarters of VOEST Alpine, the largest Austrian company, whose core activities are in iron and steel making and in engineering (e.g., building whole steel mills including continuous casting machines and rolling mills). The first inverse problem we treated together with them was the determination of a strategy for secondary cooling in the continuous casting of slabs with the property that a desired solidification front should be approximated as closely as possible. Problems of this type are widely studied in the mathematical literature. However, in many of the mathematical papers on this topic, the problem is formulated as a Stefan problem (mostly for a linear heat equation). The first thing one notices when discussing with engineers and reading the relevant engineering literature about continuous casting is that the enthalpy in steel as produced in a continuous casting machine does not change abruptly, but increases gradually over a "solidification interval". Therefore, in spite of its mathematically more interesting features, a Stefan problem (which assumes a discontinuity) is not necessarily the right model for a real continuous casting process. As discussed in Engl & Langthaler (1988), the problem can be treated as a sideways (nonlinear) heat equation, which is of course ill-posed. Thus, regularization techniques have to be used; Engl et al. (1987), Engl et al. (1989), and Binder et al. (1990).

Another problem we treated (in cooperation with Hilti AG in Liechtenstein) was to design a method for the identification of steel reinforcement bars in concrete from measurements of a distorted magnetostatic field. This problem is closely related to what is called "impedance tomography" in the mathematical literature. However, while in most mathematical papers, the problem of identifying a spatially varying conductivity from the Dirichlet-to-Neumann-map is discussed (e.g., Kohn & Vogelius 1985, Sylvester & Uhlmann 1988), one does not have (an approximation of) the Dirichlet-to-Neumann-map available in this practical problem, but one might have only one measurement. Mathematical results for the inverse problem of identifying a spatially varying conductivity from one measurement have been obtained by Friedman & Isakov (1989).

The mathematical results just quoted can be used to prove uniqueness for the practical problem outlined (Engl et al. (1989)); however, numerical simulation revealed that this result is not practically useful, since an actual determination of the location of the reinforcement bars would require measuring the magnetic field at the construction site at a far higher accuracy than possible in practice. These numerical simulations by our partner company reduced the time taken by experiments and led to the design of a new measurement technique (Engl & Isakov (1992)), which is the basis of a machine which had been successfully tested in the field and will be built by Hilti.

A third project which has been undertaken together with an Austrian company whose name should not be revealed for confidentiality reasons was the inverse problem of designing a three-dimensional reflector in lighting which produces a prescribed illumination distribution on some plane. Again, a successful treatment of this problem involved a careful analysis of (in-)stability issues (Engl et al. 1991).

Of course, it is legitimate to ask why a university institute should undertake cooperation with industry in the way we do. In my case, the trivial answer would be that I hold a "Chair for Industrial Mathematics". In Linz, cooperation with industry has a long tradition. It was started by the late Hansjörg Wacker, who was a pioneer in this field and one of the founding members of ECMI. His original motivation was to convince local industry that mathematics could really contribute to the solution of their problems, so that it would be rewarding for them to create interesting jobs for mathematicians. At that time, everybody with a mathematics diploma got a job (mostly as a programmer). As a result of the success of Wacker's efforts, many jobs have been created in our local industry where graduates can do worthwhile mathematical work. As a result Austrian industry has accepted the contribution that mathematics makes to their success. I do not want to hide the fact that since we work on a contract basis with industry, industrial cooperation brings in money which can be used for buying equipment, hiring temporary assistants, bringing in visiting scientists and travelling to conferences and to other universities. Considering the usual budgetary constraints, this is of course very welcome. But the most valuable results from cooperation with industry have been the many interesting mathematical problems that arose and which subsequently led to mathematical research (which often extended beyond the specific problem which motivated it). These studies have been published in diploma theses, Ph.D theses, and in the scientific literature. I have been pleased that we could use our mathematical knowledge for contributing to the solution of the problems of "the real world"!

John Norbury of the University of Oxford and Chairman of the Educational Committee of ECMI, spoke about the training of those working or intending to work on mathematical problems in industry.

For over 20 years Oxford Centre for Industrial and Applied Maths at Oxford University under the leadership of Alan Tayler has been developing new training methods.

1) The first of these are the industrial study groups, which typically last for 1 week, and are devoted to examining 8-12 problems. As many as 60 to 100 people participate; with 20 from industry, 20-40 graduate students, 30 faculty (from a number of different universities with different specialist skills). It is usual for the problems to be described on Monday and for progress to be presented on Friday. Sometimes the problems are solved, but more usually the result of the study is that the problem is better formulated, is shared with others, and perhaps is undertaken as a longer term research problem. The most difficult part is **mathematical modelling** of the aspects of the problem that **industry wants**, not the aspects of the problem that mathematicians want!

2) The second development is the training of graduates on M.Sc and Ph.D programmes (1-3 years) to work in **mathematical modelling** of industrial problems that may arise from study groups, e.g., rolling steel, pulling nylon threads (fibres), stock market behaviour/option pricing/finance, flooding rivers, grinding lenses, nuclear safety, stainless steel corrosion, soil pollution/seepage, flow in harbours, weather prediction, freezing fish/meat, drying extruded food/plastics, electricity production/distribution, chemical reactions, etc.

   Some of the mathematical aspects of these industrial problems have involved the standard equations of elasticity, heat, fluid flows, coupled reaction and diffusion processes, partial differential equation of the usual form, but also mixed (parabolic-elliptic etc.), numerical methods, Stefan problems (see Tayler 1986).

3) The third development is to encourage academics to be interested in mathematical and practical problems that arise/are distilled from the industrial study groups and other contacts with industry. It has been found that **new** mathematics arises from difficulties encountered in industrial applications, perhaps because easier problems can already be solved by engineers using modern computer packages.

One of the questions is whether any given centre can provide expertise in all areas of industrial mathematics. The answer is "No". Oxford has focussed on problems in continuum mechanics, but has also worked on problems involving mathematical biology, numerical analysis, statistics, etc.

4) The fourth development is to provide courses for people in industry lasting from 3 days to 1 week.

During the past 5 years, ECMI was established to develop cooperation first of all among centres in Oxford, Linz, Kaiserslautern, Eindhoven, Paris, Florence. Recently Helsinki, Goteborg, Trondheim, Lyngby, Glasgow, Grenoble, Dresden have joined, and there is now strong interest from Eastern Europe (Warsaw, Budapest, Riga) and the Iberian Peninsula. ECMI has been particularly concerned to develop graduate training, industry–university cooperation in mathematical modelling and the **retraining of** industry and university staff to improve cooperation. Currently, ECMI receives some EC money and the goodwill of individuals. However it needs industrial help to grow effectively (Tayler 1990).

The conclusions from all these developments is that mathematical modelling for industry is a **new** skill and needs special training. It is a combination of using conventional mathematical analysis together with mathematical language and ideas. As other speakers have explained, mathematical analysis is needed before computation to identify singularities, etc.

It is also clear that mathematical analysis helps develop the ideas that prevent problems being drowned by numbers of data, and also help identify laws **and** parameters.

4. An Historical Overview, Conclusions and Recommendations

Many of the points made at the round table concerning mathematical modelling and of the contributions that mathematicians can make to physical and industrial problems had been made nearly 100 years ago. Of H. Poincaré's many insights into the application of mathematics two seem most appropriate:

> "The mathematician must not be simply a provider of formulae for the physicist; on the contrary there must be an intimate collaboration between them. Mathematical physics and pure analysis are not merely adjacent 'powers', with good diplomatic relations; rather they are inextricably interwoven and their underlying spirit is the same. — physics receives from mathematics — and mathematics, in return, borrows from physics". (Poincaré 1897);

> "There we meet the physicist or the engineer, who says, 'Will you integrate this differential equation for me; I shall need it within a week for a piece of construction work that has to be completed by a certain date?' 'This equation,' we answer, 'is not included in one of the types that can be integrated, of which you know there are not very many.' 'Yes, I know; but, then, what good are you?'

More often than not a mutual understanding is sufficient." (Poincaré, Science and Method, 1914).

L. F. Richardson, pioneer of numerical methods and numerical weather forecasting, explained why classical analysis is not sufficient and that another approach is needed:

> "Further than this, the method of solution must be easier to become skilled in than the usual methods (i.e., analytical solutions). Few have time to spend in learning their mysteries. And the results must be easy to verify – much easier than is the case with a complicated piece of algebra. Moreover, the time required to arrive at the desired result by analytical methods cannot be foreseen with any certainty. It may come out in a morning, it may be unfinished at the end of the month. It is no wonder that the practical engineer is shy of anything so risky." (Richardson, 1906.)

But he also warned mathematicians that some things are beyond them,

> "Mathematical expressions have, however, their special tendencies to pervert thought: the definiteness may be spurious, existing in the equations but not in the phenomena to be described; and the brevity may be due to the omission of the more important things, simply because they cannot be mathematised ..." (Richardson, Mathematical Psychology of War, 1919).

Just as mathematics has not yet begun to be seriously applied by politicians and diplomats, nor has it yet been used by industry for the psychological and sociological aspects of management in the way that Richardson indicated – perhaps because the mathematical treatment of these topics does not yet yield practical and unambiguous solutions (e.g., Bennett, 1985).

There were common themes in the contribution to this round table, but all the contributions were interesting because they were from slightly different view points.

At the lowest level it was agreed that a majority of industrial mathematicians merely tend to be users or advisors of standard software packages. It is worth remarking that this is a remarkable transformation from the late 1960s when in many industries the mathematicians were mainly concerned with the search for mathematical problems (as Richardson pointed out) while the engineers were experimenting with computers. (Of course there were many exceptions, notably in aeronautics) (Hunt 1989).

At the higher level it is now generally acknowledged that the mathematician working with the engineers and physicists (à la Poincaré and the contributions to this discussion) are making essential contributions to industry, for example in the rigorous posing of problems, the interpretation and design of physical or statistical experiments, the analysis of proposed designs, or, by inverse methods, the development of designs to meet specified criteria. Also their handling of large quantities of data, and (following Poincaré's second quotation) interpreting the results of computation and analysis so as to produce new concepts that have wider application than to the original problem, for example, the results of the mathematical theory of chaotic dynamical systems have provided valuable concepts of wide application in industry and science.

A particular contribution that mathematical analysis can make to industry is to develop new techniques for quantitative and qualitative analysis. Probably the most important of these developments in this century were the numerical solutions of differential equations (Runge 1904, Richardson 1910), the analysis of stochastic processes and signal analysis (e.g., by Wiener (1930), Khinchin (1934), Shannon (1948)) and control theory and signal analysis (Nyquist (1932)). (Most of these developments were made by scientists and engineers rather than mathematicians!)

Most of the mathematical techniques used in modelling physical systems have stemmed from 19th century developments. Contrasting the contributions by Gilg and Campos, it is clear that the 19th century mathematical and physical methods may have more use for some industrial problem (e.g., acoustics) than for other (e.g., electrical engineering)!

It is invidious to pick out which current developments in mathematics are or will have the greatest impact in industry. However of all the new mathematical methods and concepts that are being considered seriously, it does appear that those most likely to be used are the new methods.

This might seem strange to those who think industrial mathematics is only concerned with predicting how particular designs will function; but perhaps it is not so surprising because for complex functions the characterisation and representation analysis is a necessary first step before the design of a new system. Recent mathematical methods for describing the behaviour of dynamical systems play an essential part in the control of a large industrial plant and distribution systems.

Although mathematically trivial, it is now commonplace in aeronautics to use topological terminology to characterise the complex pattern of flow over aerofoils (e.g., Tobak & Peake (1982)) that may be calculated or measured experimentally. Methods for the optimum presentation of packets of information, which may be signals, or eddies, or bubbles, etc., are now based on the use of the "wavelet transform" (Grossman & Morlet 1985) whose mathematical properties and practical possibilities have begun to be defined in more detail. For the many kinds of complex random signals that are self similar on a wide range of length or time scales, the wavelet technique is related to another modern method for the analysis and representation of functions in terms of their "fractal dimension" (Mandelbrot 1982, Farge et al. 1993).

The most important mathematical question that concerns industry and even policy making on global issues is the estimation of the errors in using current mathematical and computational techniques for quantitative analysis of large systems. Do new mathematical ideas exist to help reduce this uncertainty, for example the analysis of idealised dynamical systems with small errors (Lanford 1981)? The modelling of large systems involve considering many subcomponents and the data for them; they might be a computer controlled aircraft or the atmosphere–ocean system. In the former case the mathematical description of the subsytem and the data are quite well-defined, but in the latter case both are quite approximate. Despite these uncertainties, the mathematical, scientific, industrial and government communities have to continue designing and calculating the behaviour of complex systems and making decisions about them; the usual policy seems to be to modelling each system in as much detail as is practically possible, adapt the modelling as performance data and measurements become available and be cautious about making long-range predictions.

The essential points to guide future action on mathematics and industry may be summarized rather provocatively:

i) Most mathematics performed **in** industry consists of applying, improving, and developing software packages.

ii) If modern mathematics is to be used in industry, it is necessary for the research community to advertise and explain it (for example by continuing education and by advice to industry) and, where necessary, to transform the modern developments into new software. (Most industrial mathematicians are hard pressed and do not have the time to delve deeply into new techniques in order to write the software themselves).

iii) If universities are serious about wanting to meet the challenge of introducing modern mathematical techniques into industry they need to encourage and promote those staff who are knowledgeable in both mathematics and industrial problems. Also these academics must be interested in continuing education and the software needs of industry.

iv) Most of the innovative mathematics in industry is performed by engineers and applied physicists. Regrettably mathematicians are often used merely as computer scientists. Very little is done by mathematicians (except in the UK where applied mathematicians are similar in many respects to continental engineers and physicists).

In order for mathematicians to contribute most effectively to industry, universities have to educate the students in such a way that they are able to communicate with engineers, economists, etc., as well as to work in an interdisciplinary way.

It is also of critical importance that a new attitude be encouraged within the mathematical community so that mathematical studies should consider mathematical modelling and scientific computing as worthy tasks (as emphasized in the US "David" report).

It is important to realise that there are different national mathematical cultures in Europe (as we noted in (iv)). Therefore, to improve their contributions to European industry, mathematicians in education and research must recognise their diversity and work together in a multicultural community. To help achieve all these goals, the European Consortium of Maths for Industry was formed. As Dr. Norbury pointed out, it is already making substantial progress.

Other valuable contributions to mathematics and industry in Europe are made by the national societies interested in the industrial applications of mathematics. These societies now hold regular bilateral or multilateral meetings on subjects ranging from wavelets to computational modelling in aeronautics. (e.g., Farge et al., 1993, Vos et al., 1992; Fezoui et al. 1993). There have been other European associations for promoting different aspects of technology and applied science, some of which have a strong mathematical element. In September there was the first meeting of ECCOMAS (the European Collaboration for Computational Methods in Applied Science) which was cosponsored by ECMI, ERCOFTAC, national applied mathematical societies, and other organisations. This was a most successful meeting ground of academic and industrial researchers involved in the application of mathematics. (See Hirsch et al., 1992).

The first European mathematical congress has also played its part in bringing mathematics and industry together. However we hope that at the next congress this aspect of mathematics will have greater emphasis both because of its intrinsic interest and contribution to mathematical ideas, and because of its importance for the future well-being of both industry and mathematics.

References

[1] Banks, H. and Kunisch, K., *Parameter Estimation Techniques for Distributed Systems*, Birkhäuser, Boston, 1989.

[2] Baumeister, J., *Stable Solution of Inverse Problems* Vieweg, Braunschweig, 1987.

[3] Bennett, S. (ed.), "Mathematical modelling of conflict and its resolution". Proc. IMA conference, Clarendon Press, 1985.

[4] Binder, A., About Two Problems Arising in Continuous Casting of Steel, Doctoral Thesis, Johannes Kepler Universität, Linz, 1991.

[5] Binder, A., Engl, H.W. and Vessella, S., Some inverse problems for a nonlinear parabolic equation connected with continuous casting of steel: stability estimates and regularization. *Numer. Funct. Anal. and Opt.* **11**(1990), 643–671.

[6] Crighton, D.G., Computational aeroacoustics for low Mach number flows. Proc. ICASE Workshop on Computational Aeronautics (eds. J. Hardin and Y. Husaini), (1992). To appear.

[7] Engl, H.W., Inverse und inkorrekt gestellte Probleme, in Chatterji, S.D., Kulisch, U., Laugwitz, D., Liedl, R. and Purkert, W. (eds.), *Jahrbuch Überblicke Mathematik*, Vieweg, Wiesbaden 1991, pp 77–92.

[8] Engl, H.W., Regularization Methods for the Stable Solution of Inverse Problems, *Surveys on Mathematics for Industry*, Springer-Verlag, Heidelberg, Berlin, New York, 1993.

[9] Engl, H.W. and Groetsch, C.W. (eds.), *Inverse and Ill-Posed Problems*, Academic Press, Orlando, FL, 1987.

[10] Engl, H.W. and Isakov, V., On the identifiability of steel reinforcement bars in concrete from magnetostatic measurements. *European Journal of Applied Mathematics* **3** (1992), 255–262.

[11] Engl, H.W. and Langthaler, T., Control of the solidification front by secondary cooling in continuous casting of steel. In: *Case Studies in Industrial Mathematics,* Engl, H.W., Wacker, H. and Zulehner, W. (eds.), Teubner and Kluwer, Stuttgart, Dordrecht, 1988.

[12] Engl, H.W., Langthaler, T. & Manselli, P. (1987) On an inverse problem for a nonlinear heat equation connected with continuous casting of steel. In: *Optimal Control with Partial Differential Equations II*, Hoffmann, K.H. and Krabs, W. (eds.), ISNM 78, Birkhäuser, 1987, pp 67–89.

[13] Engl, H.W. and Manselli, P., Stability estimates and regularization for an inverse heat conduction problem in semi-infinite and finite time intervals. *Numer. Funct. Anal. and Opt.* **10**(1989), 523–540.

[14] Engl, H.W. and Neubauer, A., On an inverse problem from magnetostatics. In: *Computational Techniques and Applications: CTAC-89*, Hogarth, W.L. and Noye, B.H.J. (eds.), Hemisphere Publishing Corporation, New York, 1990, pp 3–15.

[15] Engl, H.W. and Neubauer, A., Reflector design as an inverse problem, in: Heiliö (ed), Proc. of the Fifth European Conference on Mathematics in Industry, Teubner and Kluwer, Stuttgart, Dordrecht, 1991, pp 13–24.

[16] Farge, M., Hunt, J.C.R., & Vassilicos, J.C., Wavelets, Fractals and Fourier Transforms. *Proc. IMA. Conf. of Dec. 1990*, Clarendon Press, Oxford, 1993.

[17] Ffowcs-Williams, J.E., Anti-sound. *Proc Roy Institution*, **60** (1988), 187–193.

[18] Fezoui, L., Periaux J., and Hunt, J.C., Proc IMA Conference on Computational Aerodynamics, OUP, 1993, to appear.

[19] Friedman, A. & Isakov, V., On the uniqueness in the inverse conductivity problem with one measurement. *Indiana University Mathematics Journal*

38 (1989), 563–579.

[20] Grossman, A., Morlet, J., Decomposition of functions into wavelets of constant shape, and related transforms. *Math & Phys Lectures* L. Streit (ed.), World Scientific, **1** (1985), 135–165.

[21] Hirsch, C., Périaux, J. and Kordulla, W. (eds), Computational Fluid Dynamics '92 Volume 1. Elsevier, 1992.

[22] Hofmann, B., *Regularization for Applied Inverse and Ill-Posed Problems*, Teubner, Leipzig, 1986.

[23] Hunt, J.C.R., Some changes in the doing and the application of applied mathematics 1964-1989. *IMA Bulletin* **25** (1989), 97–103.

[24] Isakov, V. and Powell, J., On the inverse conductivity problem with one measurement., *Inverse Problems* **6** (1990), 311–318.

[25] Khinchin, A.I., Korrelatsiontheorie der stationären stochastischen Prozesse. *Math. Ann.* **109**(1934), 604.

[26] Kohn, R. & Vogelius, M., Determining conductivity by boundary measurements II: Interior results. *Comm. Pure and Appl. Math.* **38**(1985), 643–667.

[27] Lanford, O.E., Introduction to the mathematical theory of dynamical systems. Les Houches, *Rev. Fluid Mech.*, **14** (1981), 347–364.

[28] Louis, A., *Inverse und schlecht gestellte Probleme*, Teubner, Stuttgart, 1989.

[29] Mandelbrot, B., *The Fractal Geometry of Nature*, W.H. Freeman, New York, 1982.

[30] Nyquist, H., Regeneration theory. *Bell System Tech. Journal*, **11** (1932), 126–147.

[31] Poincaré, H., Les rapports de l'analyse et de la physique mathematique. *Revue Generale des Sciences Pure et Appliqués* **8** (1897), 857–861.

[32] Poincaré, H., *Science and Method*, Thomas Nelson, London, 1914.

[33] Richardson, L.F., A freehand graphic way of determining streamlines and equipotentials. *Proc. Phys. Soc.* **21** (1908), 88–124.

[34] Richardson, L.F., *Mathematical Psychology of War*, W. Hunt, Oxford, 1919.

[35] Runge, C.D.T., Uber die numerische Auflösung von differential-gleichungen. *Math. Ann.* **46** (1895) .

[36] Shannon, C.E., A Mathematical Theory of Information. *Bell System Tech. Journal* **27** (1948), 379–423.

[37] Sylvester, J. and Uhlmann, G. , Inverse boundary value problems at the boundary – continuous dependence. *Comm. Pure and Appl. Math.* **41**(1988), 197–219.

[38] Tayler, A. *Mathematical Models in Applied Mechanics*, Clarendon Press, 1986.

[39] Tayler, A. *IMA Bulletin on ECMI*, 1990.

[40] Tikhonov, A. and Arsenin, V., *Solutions of Ill-Posed Problems*, Wiley, New York, 1977.

[41] Tobak, M. & Peake, D.J. , Topology of three-dimensional separated flows. *Ann. Rev. Fluid Mech.* **14** (1982), 61–86.

[42] Vos, J.B., Rizzi, A. & Ryhming, IL (eds.), Proceedings of the Ninth GAMM-Conference on Numerical Methods in Fluid Mechanics, Vieweg, 1992.

[43] Yamaguti, M., Hayakawa, K., Iso, Y., Mori, Nishida, T., Tomoeda, K. and Yamamoto, M. (eds.), Inverse Problems in Engineering Sciences, *ICM-90 Satellite Conf. Proc.*, Springer-Verlag, Tokyo, 1991.

[44] Wiener, N., Generalised harmonic analysis. *Acta Math.* **55**: 117 (1930).

Round Table I: July 9, 1992

DEGREE HARMONIZATION AND STUDENT EXCHANGE PROGRAMMES

by Christian Berg,[1] Hans J. Munkholm,[2] Ivan Netuka,[3]
David Salinger,[4] and Vladimir Souček[5]

Copenhagen (Denmark)[1], *Odense (Denmark)*[2], *Prague (Czech Republic)*[3]
Leeds (United Kingdom)[4], *Prague (Czech Republic)*[5]

1. Preface

The Round Table *Degree Harmonization and Student Exchange Programmes* was organized by three of the present authors (HJM, IN and VS) who decided at the outset that IN and VS would be concerned mainly with the question of harmonization, while HJM would concentrate on the exchange programmes. The need for a moderator and a co-organizer with the personal experience of student exchange programmes became apparent during our preparations, so the two remaining authors (CB and DS) became involved.

This final report should be viewed as a joint effort by all five authors. However, a word about its origin is in order. At the Round Table itself, a preliminary report, prepared by the three original organizers, was distributed to the participants. This preliminary report has now been augmented by DS's account of the ERASMUS scheme which he coordinates (Section 4.2), by Professor Cioranescu's account of her TEMPUS scheme, and verbatim accounts of part of the ECM Round Table discussions. It has also been subjected to minor revisions (linguistic and otherwise). It now forms most of Sections 2–4.

The Preliminary Report contained suggested recommendations, which were also summarized in the Abstract of the Round Table, distributed to all participants at the Congress. It is a testimony to the lively discussion which took place at the Round Table that the recommendations presented in Section 6 of the present report differ significantly from the earlier version, especially in the matter of a European Standard/European Degree in Mathematics.

The discussions at the Round Table have been interpolated in Sections 3.3, 3.8, 4.6 and 5, corresponding to the points at which they occurred during the presentation of the preliminary report. Every effort has been made to check that the account given here is a true picture of what each contributor wanted to say. But the five authors accept responsibility for the final wording.

2. Objectives

Mathematics has been an international endeavour for several hundred years, and mathematicians have been involved with student mobility long before official exchange programmes were created. Nevertheless, because of these programmes and the events leading to them (including the recent political changes in Eastern Europe and the advent of the EEC Internal Market) it is expected that the number of students doing part of their studies abroad will increase dramatically over the next decade and that "mobility" will become part of everyday life at universities.

With this in mind, the Organizing Committee of the ECM decided to include a Round Table entitled *Degree Harmonization and Student Exchange Programmes.* It should be emphasized that this was not an attempt to organize the unification of programmes on a European level. The intention of the Round Table was rather to analyze the present state of mathematics teaching at European universities and to judge to what extent, if at all, it would be useful, suitable or possible to harmonize programmes of study in order to encourage the easier mobility of students.

3. Harmonization of Degrees

3.1 Cooperation with EMS member societies

The member societies of the EMS were invited to contribute information about:

- General characteristics of programmes of study;
- Vertical division of programmes of study, that is, degrees or titles awarded, and internal divisions (e.g. by examinations);
- A description of the systems of examinations and/or other means of verification of knowledge gained from lectures and/or exercises;
- Opinions about a period of study abroad including conditions under which such a period abroad could be officially acknowledged by the home university;
- Detailed information on plans of study for basic courses during the first parts of programmes of study.

Thanks to the activity of colleagues from most European countries, much relevant material was collected. In particular, we received reports (in many cases detailed ones) from colleagues in the following countries: *Austria, Belgium, Czechoslovakia, Denmark, Estonia, Iceland, Ireland, Lithuania, Luxembourg, Portugal, Sweden, Switzerland* as well as from the *Edinburgh Mathematical Society.* From other sources we had (at least partial) access to information on the systems of study in *France, Germany, Great Britain, and Italy.* The information obtained is summarized and some conclusions are drawn in Sections 3.3-3.5 below.

For purely practical reasons, we decided to restrict ourselves to the study of mathematics in the narrow sense: we did not discuss informatics or the training of future mathematics teachers.

3.2 The European student's record: a brief history

Attempts to make the exchange of students easier are by no means new. As we have learned from G. Choquet and F. Hirzebruch, around 1960 a European committee (whose members included E. Artin, H. Cartan, G. Choquet and F. Hirzebruch) prepared a booklet for students called the Student's European Academic Record Booklet. The booklet was multilingual (English, French, German and Italian) and its aim was expressed as follows:

> "The purpose of the present booklet is to enable the student who moves from one University or College to another during the course of his studies to let those concerned know exactly what knowledge he has already acquired.
>
> The attestations given in this booklet will be supplied by the professors or teachers directly in charge of the student. They are only intended for the information of other teachers, and in no way engage the responsibility of any Universities or Governments.
>
> Each teacher should indicate the number of those parts of the Mathematics syllabus in which the student is proficient (please, refer to the numbers of the syllabus printed at the end of the present booklet). If teachers think it advisable, they will also have the opportunity to mention other subjects in which the student has rather good knowledge though they do not appear in this syllabus. Teachers are entirely free to express their appreciation as they think best.
>
> The information supplied may relate to a successful examination; if there is no examination, information should be given by the teacher who knows the student.
>
> Note on the syllabus. The syllabus printed in this booklet is obviously not intended to be compulsory or limiting in any way. It is mainly designed to be used as reference".

Syllabi for the first level include *fundamentals of algebra, analytic geometry and classical differential geometry in two and three dimensions, first level linear algebra, elementary differential calculus, integral calculus, series, differential equations, numerical analysis, kinematics and dynamics,* and *introduction to the calculus of probability.*

Second level syllabi are divided into two parts: *pure mathematics and applied mathematics.* The former includes *algebra of sets, second level algebra* as well as *linear algebra, general topology, function spaces, second level integration, differential calculus, functions of a complex variable, differential geometry of curves and surfaces,* while the latter contains *advanced algebra, functions of a complex variable, advanced integral calculus, function spaces, integral equations, ODE's, PDE's, calculus of variations, distributions* (including *Laplace* and *Fourier transforms*) and *special functions.*

Here are some samples of the syllabi.

First Level

1. Differential equations. Fundamental concepts of differential equations; trajectories of a vector field, initial value problems, boundary value problems. Illustration of these problems by equations which can be solved by integration; linear equations with constant coefficients.
   Linear superposition of the solutions of homogeneous systems of linear equations with variable coefficients. Non homogeneous equations.
2. Numerical Analysis. Systems of linear equations, elimination methods and the method of successive approximations. Linear optimization; approximation in the sense of Tchebycheff and Gauss. Simple algorithms for the calculation of proper values.
   Polynomials and division algorithms, as simple examples of algorithms. Majoration and calculation of roots, Newton approximation, interpolation by polynomials, continued fractions.
   Numerical integration processes. Ordinary differential equations, iteration method, the Runge–Kutta method for initial value problems. Difference methods for boundary value problems.
   Practical exercises on computers.

Second Level (pure mathematics).

3. General topology. Topology of R, of R^n, the Borel–Lebesgue theorem. General definition of a topological space (by open sets or by closed sets); example of metric spaces. Continuous functions. Products of topological spaces.
   Compact spaces; classical theorems. Locally compact spaces.
   Connected spaces; image of a connected space by a continuous mapping. Metric spaces (several examples). Criterion of compactness for metric spaces. Uniform continuity; the case of a continuous mapping of a compact metric space into a metric space.
   Complete metric spaces (without treatment of completion). Method of successive approximations.
   Summable families in a normed complete space; normal convergence.
4. Function spaces. Distance of uniform convergence on the space of mappings into a metric space. The case where the latter space is complete, case of continuous transformations. Normed vector spaces, Banach spaces. Examples: norm of uniform convergence on a vector space of numerical functions, various norms on function spaces, defined by means of integrals.
   The Stone–Weierstrass theorem, or at least the Weierstrass theorem (approximation by polynomials).
   Pre-Hilbert spaces; examples: The space L^2 is complete (with Lebesgue integrals). Inequalities. Projection on a complete subspace, and more generally on a complete convex subset. The projection is a contraction. Pre-Hilbert

spaces with enumerable basis. The Schmidt orthogonalization. Applications: sequences of special polynomials, Fourier series.

Second Level (Applied Mathematics)

5. Partial differential equations. A first order quasi-linear partial differential equation. The Cauchy problem, characteristics.
   Definition of the characteristics of a quasi-linear system of two equations with two variables.
   Pfaff equation in the case of complete integrability.
   Equations of second order: separation of variables.
   Equations of the second order with constant coefficients.
   Elliptic equations: The Laplace equation $\Delta\varphi = 0$.
   Mean value theorem, elementary solution, uniqueness for the Neumann and Dirichlet problem, Green's function, the Poisson formula for the sphere, the energy integral (Dirichlet).
   The Helmholtz equation $\Delta\varphi + k^2\varphi = 0$, the radiation condition, reduction to integral equation.
   Hyperbolic equations: the wave equation in 1, 2, and 3 space variables; elementary solution, boundary value problems, the Poisson–Kirchhoff formula. The method of descent (Hadamard). The energy integral.
   Parabolic equations: the heat equation with one space variable. Elementary solution, boundary value problems.

3.3 Discussion of the student's record at the Round Table

M. Karoubi, Paris:

I would like to understand why, as Henri Cartan said, this booklet was not so successful. Maybe Professor Hirzebruch can answer this?

F. Hirzebruch, Bonn:

Well, I can hardly give an answer to that. It was distributed to many European societies. I remember that I myself introduced it to the German mathematical society, but then the professors in the various universities maybe did not cooperate well enough. Maybe one should have sent it directly to the universities, and then whenever students left the university to go somewhere else, they should have used it. Maybe students were afraid that some records in the booklet would have given them bad grades; but the booklet was not connected with grades, so I think it was missing information.

I sometimes used it myself in German committees, when there was discussion about the minimum requirement for certain courses. This was also done in other countries when one was participating in official reform committees for the syllabus. And in that sense it was certainly helpful.

J. Winkler, Berlin:

Your work ignores a very serious problem, not so very essential for us at the university or for students working in a scientific direction, but for all other students. Starting next year in the EEC, a degree will have a new value for those students. In fact, a degree which is a professional qualification will be valid as such in every EEC country: for example, for becoming a teacher or for working in industry. For a student who intends to work in science at a university this is not so important, but if he wants to work in industry or outside the university, this is a very essential question.

R. Mlitz, Vienna:

I just want to point out that we are discussing the problem as representatives or members of mathematical societies, but, in the various countries, the content of university studies is usually not prescribed by the members of the societies. Such a booklet would be good — but we cannot impose it. I think this is one of the main problems.

M. Karoubi, Paris:

I think that this booklet could not—and should not—be imposed in any way. The general idea was to suggest a programme which could be harmonized throughout all Europe so that the students could travel easily. That's what I understood. Now, I think the problem is more practical. First of all, is it necessary to come back to this idea? And secondly, if there is some general agreement that this idea is still good, even now, how can it be made practical? That is the question.

C. Berg, Copenhagen:

I think we will stop the discussion on this booklet now. We can come back to it in the general discussion afterwards, when we have more time. Of course there are many problems with such a book, but it could be easier for all of us, instead of just sitting and writing the whole syllabus every time a student comes or goes abroad, just to mark that this or that has been covered.

3.4 Vertical division of mathematics programmes and degrees

Within European countries, there is a great variety of systems of vertical division of study. Historically, in countries of the former eastern block, university study usually took five years without any subdivision. But now there is a clear tendency to adapt to conditions common in western European countries.

In general, the mathematics programme is divided into three parts, though the division may not be explicit. It is hard to choose a self-explanatory terminology, since similar expressions in various languages may refer to different situations.

For the purpose of this report we use

- First cycle.

Students enroll in the university at the age of eighteen or nineteen. Depending on the country, the first cycle takes 2–3 years. In addition to mathematics, there is usually another subject (frequently physics, informatics, economics. Sometimes other choices are possible).

- Second cycle.

  This takes 1 or 2 years. There is a great variety of programmes. Students specialize more and are led to individual work.

- Ph.D study.

  Students choose their supervisor and a subject for their own scientific work from a wide spectrum of rather specialized fields. They usually have to pass some examinations and one of the main goals is to submit a thesis containing new mathematical results.

As illustrated below, the greatest variety of titles appears in the first cycle. In some countries a title corresponding to B.Sc. has a long tradition and is recognized on the job market. In other countries, on the contrary, no title of this sort is awarded. A continuous spectrum of possibilites can be traced in between.

The situation is more uniform after the second cycle. The majority of universities award a title corresponding more or less to M.Sc.

The doctoral degree is awarded to students having finished their Ph.D. This is also beginning to be adopted in former socialist countries, where the title "Candidate of sciences" has been in use for several decades.

In order to emphasize how the systems in European countries differ, we give a couple of examples.

In *France*, university study is divided into *premier cycle* ending with *Diplôme d'Etudes Universitaires Générales.* The *second cycle*, which also takes two years, is accompanied by two titles: *licence* at the end of the first year and *maîtrise* at the end of the second year.

In *Belgium*, the first diploma (*candidature*) is awarded after two years, but has no real significance. It is only the diploma of *licence* after four years that is recognized by the job market. Unlike the French system, third cycle programmes have a specialist orientation: the most common ones are in mathematics applied to insurance (e.g., a 3-year programme in actuarial science, a 1-year programme in statistics or business).

The first degree normally awarded in *Scotland* is either a B.Sc. or an M.A. About twenty-five years ago the majority of mathematics students would take an M.A. degree, whereas now the overwhelming majority take a B.Sc. degree. The above degrees are four-year honours degrees. It is also possible to take a three-year ordinary B.Sc. or ordinary M.A.

In *Lithuania*, an M.Sc. degree is awarded to students having average results greater than 7.5 out of 10.

In *Germany* there is no title corresponding to B.Sc. On the other hand, students have to choose a minor subject (*Nebenfach*).

Similarly, in *Denmark*, mathematics as a major subject most often appears in combination with physics, computer science, statistics or business administration/economics. Other combinations are possible, though. The vertical division is: B.S. (3 years of study after high school), Cand. Scient. (corresponding to M.Sc. with a thesis, further 2 years) and Ph.D . (3 years as a rule). The ratio of major subject to minor subject for 5 years of study is about 70:30.

In *Switzerland* as well, the study of mathematics is usually accompanied by one or two non-mathematical minor subjects. Titles vary from university to university, for example, Dr.phil., Dr.phil.nat., r.sc.math., Dr.ès sciences, Dr.ès sciences techniques.

In *Italy* there is (or shortly will be) a degree called *Diploma* (roughly equivalent to B.Sc.); two other degrees are *Laurea* and *Dottorato di richerca.*

The general study programme in mathematics and natural science leads in *Sweden* to an M.Sc. degree, although single subject courses may be combined, under certain conditions and after at least three years of study, into a programme leading to a B.Sc. (Sw. Filosofie Kandidatexamen).

In the *Portuguese* universities the B.A. in mathematics called *licenciatura* now takes 4 years (5 between 1964 and 1986). A few years ago courses for the masters degree (two years more) were introduced.

Almost all university institutions in *England and Wales* offer three-year mathematics degrees. In the past, two types of degree predominated: *honours mathematics* in which other disciplines were not studied or were given very little attention, and *Ordinary* or *General* degrees which treated mathematics less deeply. Now honours degrees predominate, but a variety of "joint" honours degrees are offered, particularly at newer universities. Oxford and Cambridge are in many respects exceptional within the English university system.

3.5 Examinations

For the purpose of exchanging students, a survey of examination systems seems to be of great importance. Once again, there is a great variety of schemes and approaches.

In some academic systems, examination is more or less a continuous process of evaluation. Frequently, students take examinations at the end of each separate course; work in examples classes may also be taken into account. There are cases where mid-semester exams or the like are organized. These patterns of evaluation may or may not be terminated by a final exam. The situation also varies as far as the contribution of a written thesis to the final assessment is concerned.

On the other hand, the traditional examination system in some countries is based on a few examinations covering quite a large number of subjects taught during one or two (sometimes even more) years of study.

In most European countries the examination system lies between these two extreme cases.

Let us give some examples.

In *Denmark*, each course is evaluated by an examination at the end of the semester or the year. Each student typically has to take 2 to 4 exams at the end of each semester. There is a B.Sc. thesis which is usually a report based on the individual study of some mathematical literature. During the second cycle, courses are usually evaluated by an oral examination at the end. The thesis is evaluated by a committee appointed in each individual case. The M.Sc. thesis is typically a report on some recent development in the adviser's area of interest. It need not (and in most cases does not) contain original research. An interesting feature is that typically the examination in Denmark is given by the teacher of the course in question plus an *external examiner*.

Scottish universities are responsible for granting their own degrees with quality assessed by a system of external examiners. Apart from the annual degree examinations, there are often one or two class examinations. A few courses may have some form of continuous assesment.

In *Belgium*, the general idea is that the basic unit is 1 year and each year the student must pass the examination on each course followed. In the first two years there are written tests in January, which sometimes count as final examinations (at the end of the course) or sometimes contribute to the final mark.

Oxford examines students at the end of their first and third years. Division of degrees into "classes" is based on performance in eight three-hour conventional written examinations taken within the space of a week. These examinations test all material taught in the compulsory parts of the three-year course as well as the options offered in later years. Students, then, have at one and the same time to show a grasp of a very wide range of "basic" mathematics plus more specialized knowledge.

Other English and Welsh universities, however, differ considerably from Oxford. Assessment is usually done at the end of each year and is principally by written examination. In some cases, however, examinations are held at the end of every semester (and this is likely to be the pattern for more universities in the future). Moreover, a variety of techniques may be used in addition to conventional examinations and can all contribute to the final mark. These include open-book examinations (where books or lecture notes may be consulted), coursework, essays, take-home papers, and investigations (but oral examinations are rare).

In *Portugal*, in most basic disciplines, the usual rule is to have one written test every semester and, for those students who do not get a satisfactory mark, a final written examination (followed by an oral, or not, according to the marks obtained).

The *German* system includes two main examinations (the so called Vordiplom-Prüfung and Diplom-Prüfung) at the end of the first cycle and second cycle, respectively. In addititon, there is a more or less continuous system of evaluation of the student's work in individual subjects.

The reports we have received indicate, however, that the diversity of examination systems is not a bar to the exchange of students.

There seems to be no difficulty in incorporating courses from abroad into degree requirements, provided that an adequate description of the courses in question is available.

Experience so far indicates that a student should spend a whole "unit" of study at another university, which, in practice, means a semester or a whole year. The efficient exchange of students depends, of course, on the efficient exchange of information about the programmes of study at the universities involved. A student intending to study abroad for part of his/her university career will look for a list of prerequisites needed for a smooth continuation of study in another country. The procedure would be greatly simplified if a list of basic requirements, at least for the first cycle, were created on a European level. Such a list would surely help students prepare for a stay abroad and would provide a point of reference for universities when giving information to applicants.

3.6 Programmes of study

Programmes vary greatly from university to university. In particular, it is practically impossible to trace common features in the second cycle. One can even find programmes that are almost mutually disjoint. The situation is, however, different in the first two years of study, where quite a large portion of basic mathematical knowledge appears in all curricula.

The information from individual member societies was rather unhomogeneous. In some cases, detailed syllabi were provided, while in others only a few key words were given. A few basic subjects (frequently under different names) appear in all plans of study: *mathematical analysis, algebra and linear algebra, geometry, numerical analysis, informatics, probability and statistics.*

The first two subjects are studied more extensively than the others. The last three are sometimes integrated into a minor subject or into a part of the programme devoted to applied mathematics; often they are just introductory.

We tried to isolate the main topics usually covered by analysis, algebra and geometry. The distribution of individual topics among the courses is by no means uniform. Elementary set theory is taught in the framework of algebra, sometimes in analysis, or may even be included as a separate course. Similarly, analytic geometry may form a part of linear algebra, or may be included in a course on geometry. Surface integrals and Stokes' theorem are usually treated in analysis, but may sometimes form a part of the differential geometry of curves and surfaces.

Nevertheless, several topics are present in most programmes and are grouped into three parts which we present as follows.

Mathematical analysis. Real and complex numbers. Sequences. Fundamental theorems on limits. Functions of a real variable, elementary functions, limits, continuity, fundamental theorems on continuous functions on an interval, uniform continuity. Derivatives, calculation of derivatives, mean value theorems, maxima and minima of functions of a real variable, Taylor's formula, monotone functions, convex and concave functions. Series of complex numbers, convergence tests, power series, series of functions, uniform convergence. Fundamental concepts of Fourier series. Elementary theory of ordinary differential equations. Indefinite integrals, the Riemann integral, fundamental theorem of integration, various methods to calculate indefinite integrals, integrability conditions. Elementary topology in Euclidean spaces, compactness, connectedness. Functions of several variables, continuity, differentiablity, partial derivatives, chain rule, Taylor's formula, inverse and implicit function theorems, maxima and minima of functions of several variables. Multiple integral, Fubini's theorem, change of variables.

Algebra and linear algebra. Vector spaces, linear dependence, bases, dimension. Linear maps, matrices, determinants, systems of linear equations, Gaussian elimination. Eigenvalues and eigenvectors, diagonalization, Jordan normal form. Linear forms, dual spaces, bilinear and quadratic forms, inner product spaces. Groups, rings, fields, quotient structures, the fundamental homomorphism theorems, group action, permutation groups. Polynomials, the division algorithm, factorization, unique factorization domains, ideals.

Geometry. Euclidean geometry, angles and distances, volumes. Analytic geometry, examples of quadrics. Curves and surfaces, length of a curve, curvilinear and surface integrals, Green's, Gauss' and Stokes' theorems.

It would be desirable to continue tracing common features of European university mathematics teaching by collecting more information, particularly on subjects not covered above.

3.7 Reflections on a European standard in mathematics

The preceding sections indicate a rich diversity of systems of study, plans of study and degree titles, as well as of examinations and methods of teaching. Nevertheless, analysis of the information we received and exchange of opinion with numerous colleagues have led us to the conclusion that, during the first cycle, the common features of the courses at the various European universities are more important than their differences.

Further, as another part of this report shows, students typically spend their time abroad during the third and fourth years of study, not earlier.

These two observations, combined with initiatives from various European mathematicians, raise the question of whether it would be helpful to create a European standard for the amount of mathematical knowledge to be acquired during the first cycle. The Round Table could serve as a place for the exchange of ideas on this question. The existence of a European standard could provide an inspiration for European universities and might have a positive influence on the teaching of mathematics at the university level.

It is not without interest to compare these new initiatives with the European Student's Record mentioned in Section 3.2. There is a general consensus that the underlying idea could help in tracing the main features common to and typical of existing systems of study. There is not the slightest intention of imposing some kind of administrative unification of teaching programmes. Even if a European standard could be agreed upon, each university would have to have complete freedom to decide whether or not to conform to it. No doubt, there will be a broad spectrum of opinions about a European standard. Some mathematicians have reacted enthusiastically, others have had reservations, while yet others have been vehemently opposed.

It is hard to predict what the reaction of the mathematical community will be. The organizers hope that many colleagues will make their voices heard at the Round Table concerning the general principle as well as practical details.

It is our opinion that, among others, the following ideas should be discussed at the Round Table.

1. A large increase in student exchanges forces Mathematical Departments to produce descriptions of courses (in English?) for foreign students.

In this situation it is very natural to reconsider the idea of an updated version of the European Student's Record described earlier.

As follows from discussions among mathematicians, such a booklet should concentrate mainly on the first three years of study. For each subject, a "core curriculum" should be given (containing topics appearing in the curricula of practically all universities). To this part, topics which may or may not be included should be added.

If such a booklet were available, the teacher could simply mark the topics mastered by a foreign student without producing a lengthy description of the course.

2. The use of the booklet, in practice, could contribute to the formulation of a certain "European standard".

This standard should consist of the topics which most frequently appear in the curricula of European universities. On a local basis, such attempts have already existed, as we shall hear in the next report: some ERASMUS as well as TEMPUS projects include curriculum development. It goes without saying that the development of a European standard should profit from the experience gained on these projects.

3. Suppose that the above mentioned European standard were to be created; the next step would be to consider establishing a European Degree in Mathematics based on the European standard. So far, however, discussions of this idea

suggest that such a project might involve the danger of a complex bureaucracy.

Further analysis and discussion will show to what extent such an idea should be developed.

3.8 Discussion on harmonization at the Round Table

A. Van der Sluis, Utrecht:
Would you say that on the whole a reasonable degree of harmonization already exists?

V. Souček, Prague:
I would not say so. It seems that the systems are really substantially different in the various countries, but that, at the moment, this does not prevent the exchange of students. Certainly, some sort of harmonization would be very good in the future, if it could be attained.

L. Lemaire, Brussels:
I am not sure that a real harmonization of a vertical division is possible. I am not even sure it is desirable. Because the major problem for teaching is really within each country. Universities have to be financed, diplomas have to have a legal value, and also a diploma of mathematics cannot be really different from the diploma in other sciences, in law or in any other branch, or it would probably become illegal in the country. So I do not think it would make any sense for the EMS to try to impose a substantial modification of the vertical division of the study of mathematics in the various countries. Thinking of Belgium, I would even object to such a proposal, because any modification would probably have an adverse effect on the financing of the universities and therefore on the support of mathematical research by teaching. So I think it is much more practical to see how a curriculum could be disseminated like the booklet and could be applied year after year. Certainly we cannot impose a substantial modification of the law in the various countries.

C. Berg, Copenhagen:
I agree completely with you in that respect. It is difficult enough within each country to get an agreement between the different universities on how things should be done.

A. Hajnal, Budapest:
Firstly, I apologize that we did not supply you with a curriculum. Ours is quite different from the one you read in the booklet, because in the first level we teach a lot of combinatorics, and even set theory is included. But I question whether harmonization is really good. I understand that it makes it possible for a student to travel, but if they learn the same thing everywhere why should they travel? I doubt that harmonization should be an aim.

M. Karoubi, Paris:

I do not think that we shall impose anything on every country: it is too complicated. Maybe it would be good to have another diploma which would be different: a European Diploma which would be given by universities in France or in other countries, but which would not be congruent to the diploma of any one country. The students want to travel. But what do I mean by travel? I mean that when we receive a student from another country we know what he has done. And harmonization does not mean that every country has to teach the same thing; it just means that you have to know about the teaching in each country. For example, when we receive a student from Hungary, we ought to know what his background is and what recommendations we can give him for improving his studies.

C. Berg, Copenhagen:

Concerning travel—of course it is nice to travel, and there is an important byproduct: we learn some languages, and we all need to know several languages when we study mathematics.

P. Michor, Vienna:

I think that an updated version of the red booklet would be just a least common denominator of existing curricula. Maybe it would be a good idea to have a model curriculum which is published somewhere and which is not aimed at being adopted by European universities. It would be just a model, and could be updated every 5 years and published so that universities could have a model to look at and explain why they deviate from it. They could see what the mathematics curriculum in an ideal university would look like. Of course no ideal university really exists.

C. Berg, Copenhagen:

I think that is a very good point, and I am also sure that such a booklet will have an influence. As mentioned in Cartan's article in the Gazette[1], the Danish mathematicians Fenchel and Jessen participated in the discussion with Cartan and others about curricula in Europe, and I am sure that it had a lot of influence on the new programmes in Denmark around 1960.

A. Van der Sluis, Utrecht:

You mentioned the description of the courses in English, thereby bypassing the difficulty of explaining what the influence of languages really is. Actually as the situation is now, each country has its own university courses given in its own language. In my country, the Ministry of Education has tried to promote the idea of giving university lectures in English. We got into a hell of a row, because so many people feel that in the language you find the soul of the nation. Languages are a very touchy subject within the EEC, so I think it is not to be expected that lectures will be given in languages different from the language spoken in the country where the lectures are given. So it might well be wiser to circulate a description of courses in the language in which the courses are given, since if the student is not able to

[1] H. Cartan: *Quelques souvenirs.* Gazette des mathématiciens, **53**, Juin 1992.

read that, he will not be able to attend the lectures either.

R. Mlitz, Vienna:

I do not agree with my colleague from Vienna that something could be an ideal university, because if you asked 10 people what an ideal university was, you would get 10 different answers, and if you take the intersection, the programme would be quite small. And another thing that might be of interest is that in our new laws in Austria for studies at technical universities, we have now imposed the amount of eight semester hours of mathematics lectures in English for mathematics students. This is a first step. You cannot require the students to do everything in English, because in the beginning they have problems understanding the mathematics, and these problems should not be increased by having to learn in a foreign language.

H.J. Munkholm, Odense:

One of the questions we asked the European mathematical societies was precisely whether they would be willing to teach in a foreign language, if the need arose because of exchange of students, and the answers were pretty uniform, but I should qualify the statement by the fact that actually some large countries with Roman languages did not answer—not all, but some. But the answers were pretty much uniform:—God forbid—for the first 2 years—no way. Up to the Masters degree level, practically everybody says that if there is one student who wants tuition in English we will teach in English—and English is preferred, even by those Roman language countries that did answer. And at the Ph.D level it is completely uniform. So this little questionnaire seems to indicate that at the beginning level – no way can you teach in a foreign language systematically. But as soon as you come to the second cycle, or higher, then English seems to be a pretty good guess. But I would have liked to have those answers from a couple more of the Roman language countries.

I. Netuka, Prague:

I want to comment on the language question. We are a small country (practically no foreigner speaks Czech or Slovak), so the language problem is important for our students. For this reason some more advanced seminars are now held in English, just in order to train both teachers and students to be good at English mathematical communication. So I do not see any serious obstacle, at least in small countries.

A. Pelczar, Krakow:

I think it is very important that we are talking about this subject since the Conference of Rectors of European universities (CRE), that is, the conference of rectors, presidents and vice chancellors of European universities, is discussing this problem in a very broad framework, looking at curricula and the evaluation of curricula with respect to general aspects of a European education. It is part of this problem that we are presently discussing. We each have opinions about our own fields, and I think that there should be some connection between the permanent committee of the CRE, referred to above, and the Executive Committee of the EMS.

4. Exchange of Students

4.1 ERASMUS

As is probably well known, ERASMUS (**E**u**R**opean community **A**ction **S**cheme for the **M**obility of **U**niversity **S**tudents), is an EEC programme to stimulate cooperation across borders in university education. It works mainly by funding so-called ICPs (**I**nter-university **C**ooperation **P**rogrammes). Such an ICP involves a number of universities (from at least two member states). As far as mathematics is concerned there are four types of activity:

- Student exchange (practically all ICPs have this component),
- Teacher exchange (only one in five ICPs has this component),
- Curriculum development (about one in fifteen ICPs has this component),
- Intensive courses (fewer than one in fifteen ICPs have this component).

A gold mine of (statistical) information about ERASMUS is the *ERASMUS AND LINGUA ACTION II DIRECTORY*[2] [3] It gives a short description of each ICP (in 1991/92 the total number of ICPs is 1794). The style can be seen in Figure 1.

It also contains a large number of statistical and other tables as can be seen from the partial list of contents shown in Figure 2.

The table reproduced in Figure 3 shows that the programme is divided into roughly 20 main subject areas, one of which is called Mathematics/Informatics. It is seen that *3.8% of the total number of 1794 ICPs fall into the Mathematics/Informatics category.*

Mathematics/Informatics is further divided into 7 subsections, as shown in Figure 4.

In the following we only treat the first two of the above subsections. Together these two areas (mathematics and mathematics/informatics) account for

- 2.1% of the approved ICPs; but
- only 1.3% of the total number of students.
- around 1 million ECU (assuming an even distribution of money per student, roughly 100 ECU/month/student + other budget items).

4.1.1 Mathematics and mathematics/informatics in ERASMUS

---

[2] Published yearly by the Commission of the EC, and available from the Office for Official Publications of the European Communities, L2985 Luxembourg.

[3] Another valuable source for information is the *ERASMUS NEWSLETTER*, published by the ERASMUS Bureau, 70, rue Montoyer, 1040 Brussels, Belgium.

ICP-91-B-1176/11 (Student Mobility, Teaching Staff Mobility)

Subject Area(s): Mathematics, Informatics, Computer Science

B	Rijksuniversiteit Gent	Prof. dr. R. J. Delanghe
D	Rheinisch-Westfälische Technische Hochschule Aachen	Herr Prof. Dr. G. Jank
F	Université des Sciences et Techniques du Languedoc (Montpellier II)	Prof. S. Ciulli
NL	Technische Universiteit Eindhoven	Prof. dr. J. De Graaf
P	Universidade de Aveiro	Prof. Dr. J. D. Vieira
P	Universidade de Coimbra	Prof. Dr. J. A. F. De Carvalho
UK	University of Kent at Canterbury	Prof. J. S. R. Chisholm

Description of the Programme:

This renewal of ICP-90-B-0039 will enable up to 34 students to study for periods of between 3 and 9 months in another Community country. The estimated pattern of student exchange between the countries involved will be as follows:

Home country to:	B	D	F	NL	P	UK
B	0	4	0	2	0	0
D	1	0	0	0	1	4
F	1	0	0	0	0	5
NL	0	0	0	0	0	2
P	4	2	0	0	0	2
UK	2	0	4	0	0	0

The structure of the study abroad programme in this ICP includes language preparation in the host country, regular academic coursework (lectures, seminars, etc) and thesis preparation or project work.

Staff mobility: arrangements made as part of this programme will involve an estimated 20 members of staff from the participating universities teaching courses in partner universities for a total of 100 weeks. The pattern of staff exchanges is as follows:

Home country to:	B	D	F	P	UK
B	0	0	1	3	0
D	1	0	0	1	0
F	0	0	0	0	3
NL	0	0	0	1	0
P	2	1	0	0	1
UK	1	0	3	2	0

Figure 1.

STATISTICAL TABLES 915 *1. Approved ICPs by primary subject area/type ofcooperation* 917
2. ICPs: General overview by Member State 918
3. Approved ICPs by Member State participation/subject area 919
4. Student months for home country/host country 920
5. Student numbers for home country/host country 920
6. Expected Student numbers in approved ICPs by subject area/home country 921
7. Student flows per Member State 922
8. ICP Applications from 1987/88 to 1991/92 by country of coordinating institution 923

INDEXES
1. ICPs by subject/Member State/type of cooperation 925
2. ICPs by Member State/institution 973
3. Alphabetical list of ICP programme coordinators and directors 1041

Figure 2.

Table 1. Approved ICPs by primary subject area/type of cooperation

Distribution des PIC acceptés par domaine d'étude/type de coopération

Subject Area	Number of ICPs / Nombre de PIC: nb	%	Type of cooperation / Type de coopération: Student mobility — Mobilité des étudiants	Teaching staff — Mobilité des enseignants	Curriculum development — Développment d'enseignements en commun	Interactive programmes — Programmes interactifs	Domaine d'étude
Agriculture	47	2.6	45	9	3	4	Agriculture
Architecture	57	3.2	56	14	8	6	Architecture
Fine Arts	63	3.5	57	11	5	4	Beaux-Arts
Business Management Science	153	8.5	148	25	8	3	Gestion
Education/ Teacher Training	49	2.7	38	10	8	6	Education/Formation des enseignants
Engineering/Technology	243	13.5	228	33	20	8	Ingénierie
Geography/Geology	53	3.0	47	15	5	4	Géographie/Géologie
Humanities	104	5.8	90	21	5	7	Sciences/humaines
Languages/Literature/ Linguistics	231	12.9	222	47	9	9	Langues/Littérature/ Linguistique
Law	105	5.9	99	15	4	7	Droit
Mathematics/ Informatics	69	3.8	64	17	9	6	Mathématiques/ Informatique
Medical Sciences/ Psychology	107	6.0	93	17	8	7	Médecine/ Psychologie
Natural Sciences	140	7.8	130	19	6	10	Sciences naturelles
Social Sciences	167	9.3	150	38	15	12	Sciences sociales
Communication/Information	15	0.8	13	5	2	2	Communication/Information
Other areas of study	16	0.8	12	6	3	2	Autres
Framework Agreements	26	1.5	25	7	3	3	Accords-cadre
Lingua	149	8.3	144	32	1	1	Lingua
Total	1 794	100.0	1661	341	122	101	Total

The category 'Framework Agreements' involves inter-university Cooperation programmes in several different subject areas, the main link between the different subjects being that a formal agreement exists between all the participating institutions at inter-institutional level.

La catégorie "Accords-cadre" regroupe des programmes interuniversitaires de coopération dans différents domaines d'étude, le point commun étant l'existence d'un accord interinstitutionnel formel entre tous les participants.

Note: One ICP may contain more than one type of cooperation, which explains why the total number of ICPs does not correspond to the sum of the totals of the different types of cooperation.

Note: Un PIC peut contenir plus d'un type de coopération; c'est pourquoi le nombre total des PIC ne correspond pas à l'addition des totaux des différents types de coopération.

Figure 3.

MATHEMATICS, INFORMATICS

11.0 Mathematics, Informatics
11.1 Mathematics
11.2 Statistics
11.3 Informatics, Computer Science
11.4 Artificial Intelligence
11.5 Actuarial Science
11.9 Other Mathematics, Informatics

Figure 4.

In the 1991/92 ERASMUS directory there are 38[4] ICPs under this heading.

We have assembled the information given in the directory into the following three tables[5]. In Figure 5, the "actual" number of student exchanges is based upon the generally accepted estimate that 60% of all expected exchanges are realized.

	Number of universities, by country														
Year	B	DK	D	G	E	F	IR	I	L	NL	P	UK	total	expect no. of stud.	"actual " no. no. of stud.
90/91	6	3	20	3	10	16	3	18	0	5	5	22	111	557	ca. 350??
91/92	6	4	27	5	11	24	4	13	0	5	6	27	132	866	ca. 500??

Figure 5. ERASMUS Networks. Mathematics and Mathematics/Informatics, Total Figures 1990/91 and 1991/92.

Figures 6 and 7 are based on the information given in the 1991/92 ERASMUS directory for the programmes listed under the headings "Mathematics"[6] and "Mathematics/Informatics", respectively. Each ICP is identified by a country code and a number, e.g. B1176. For each ICP, the table lists the number of universities involved (total, as well as by country); the expected number of students moved; and the projected period of study abroad for each student. In addition, if the ICP in question involves other topics than mathematics and/or mathematics/informatics, such other topics are mentioned. Also, for those ICPs which involve an exchange of teachers, the number of teachers exchanged, and the *rough average* duration is indicated. Similarly, the number of students or staff involved in intensive programmes [e.g. summer schools] can be seen.

In the 1990/91 ERASMUS directory each ICP also listed during which study year the students were supposed to go abroad. *The list shows a clear preference for the third and fourth years.* Such information is no longer in the 1991/92 edition, so it is not given below in the tables.

4.1.2 A questionnaire

In the 1990/91 edition of the ERASMUS directory there were 33 ICPs listed under mathematics or mathematics/informatics. A questionnaire shown in the appendix was sent to the main coordinator of each of these ICPs. We received answers from 16 of the 33. The results can be summarized as follows:

---

4 There were, in fact, at least 39 such ICP's, see below.

5 In Figure 5, the numbers have been revised so as to take account of universities involved in more than one ICP.

6 ICP UK1539 did not appear in the directory, but has been included in this table.

ICP	Number of UNIVERSITIES by COUNTRY													Expect. no. of stud.	Months per stud.	Other topics besides MATH and INF	Teacher exchange	Intensive Program stud. & staff
	B	DK	D	G	E	F	IR	I	L	NL	P	UK	total					
B1033	1		1		1	2		2				5	12	14	4-6	CS ENG TECH	7 a 6 weeks	
B1142	1		3		1	1	1			1		1	9	40	3-12	CS	8 a 5 weeks	
B1170	1							1			1		3	18	6	CS ENG TECH	6 a 2 weeks	
B1185	1		1	1				2		1		1	7	21	4	CS	9 a 5 weeks	
E1068			1		1								2	12	6	PHIL HIST		
E1229			1		2			1					4	14	9		5 a 2 weeks	
F0044			1			1	1					2	5	13	6-9	AI STAT		
F1209	1					1					3		5	17	4-12	AI STAT		
F1280			1		2	1					1	1	6	22	4	PHYS CHEM		
F1336		1	1			1		1				2	6	24	4-12			
G1010	2		1	3	2	8				2		3	21	84	12	PHYS	25 a 4 weeks	40 & 16
I1279			3			3		3					9	14	6-12	CS		
I1302	1		2			2		1		1			7	29	3-12	CS AI		
IR1020			1	1			1					1	4	37	5-12	CS		
IR1022			2			1	1						4	35	3-11	ENG TECH NS		
NL1148			2			1		1		1		2	7	29	4	STAT	13 a 2 weeks	51 & 9
P1032						1					1		2	6	10-11			
UK1013	1	1	3			1		1				1	8	32	5-11	STAT		
UK1192			1		1	1						1	4	16	10	CS		
UK1231		4	1				1	1		1		2	10	Curriculum development in MATH INF COMM EDU TEACH				
TOTAL	9	6	26	5	10	25	5	14	0	7	6	22	135	477				

AI = Artificial intelligence; COMM = Communication science; CS= Computer Science; EDU = Education; ENG = Engineering; HIST= History; NS = Natural Science; PHIL = Philosophy; PHYS = Physics STAT = Statistics; TEACH = Teacher training; TECH = Technology.

Figure 6. ERASMUS Networks in Mathematics and Mathematics/Informatics 1991/92.

ICP	Number of UNIVERSITIES by COUNTRY													expect. no. of stud.	Months per stud.	Other topics besides MATH	Teacher exchange	Intensive Program stud. & staff
	B	DK	D	G	E	F	IR	I	L	NL	P	UK	total					
B1176	1		1			1				1	2	1	7	34	3-9	INF CS	20 a 5 weeks	
D1080			1			1							2	19	4-11	INF CS PHYS		
D1260			3	1				1				2	7	38	5-6	HIST	13 a 4 weeks	
DK1013		1								1			2	10	3-4	INF CS	4 a 6 weeks	
E1023					2	1							3	6	9			
E1184					1	1		3		1	2		8	19	3-10	INF PHYS		
F0459		1	2		1	2		1		1		1	9					52 & 2
F1132	1					1		1				1	4	13	6-10			
F1392			2			1							3	14	10	INF CS STAT		
G0027			1	2		1						2	6	18	12	INF CS TEACH		
G0046				3								2	5	10	4			
G1031				2	4	3		1					10	49	4-9	INF		35 & 6
I1010			3			1		2				3	9	39	4-10	INF CS PHYS		
I1150			2			2		1					5	16	3		6 a 4 weeks	
UK0079						1						2	3	8	5			
UK0257			1			1						1	3	14	4-10	INF CS STAT		
UK0398		1										1	2	6	9-11	ENG TECH		
UK1487			1									1	2	8	11-12	INF CS STAT		
UK1539		1	1			2		1		1		1	7	68	5-11	CS STAT		
TOTAL	2	4	18	8	8	19	0	11	0	5	4	18	97	389				

INF = Informatics; CS = Computer Science; ENG = Engineering; HIST = History;
PHYS = Physics; STAT = Statistics; TEACH = Teacher training; TECH = Technology.

Figure 7. ERASMUS Networks in Mathematics 1991/92.

Question 1. Does the programme run roughly according to plan or did essential changes have to be made?

Several ICPs report that the number of students actually exchanged is lower than the expected number; in some cases much lower. The findings do not really support any numerical estimate. More or less official sources in the EC bureaucracy report that overall around 60% of the planned exchanges are realized. We see no reason to believe that mathematics deviates significantly from this figure. Some quotes:

- "We requested 19 grants for student mobility, and obtained approximately 15, of which *we only used 7*".
  "Lately it became hard to find students to go abroad".
  "... runs essentially according to plan".
- "... runs roughly according to plan".
- "... runs roughly according to plan, but fewer than 27 postgraduate students were exchanged (actually 12)".

Question 2. As a coordinator do you feel that the results are worth the effort you (and others) put into it?

Almost everyone answers YES, but also indicates that the effort has been a considerable one. Here are some quotes to show the diversity of opinions:

- "Although it takes a lot of administration and there are only a few of our students interested, I still feel that it is worth the effort".
- "... the Mathematics Departments of different EC countries have come to know each other better with respect to teaching ... at this moment the harmonization and homogenization of degrees in mathematics is a goal we could set ourselves, at least within certain university networks".
- "I still think it is worth the effort but, taking into account all the work that has to be done, an accurate cost/benefit analysis would perhaps show otherwise".
- "The effort that coordinators put into the programme should not be underestimated. ... It is worth reiterating that departmental support is vital ..."
- "In general the results are worthwhile. The students on the exchange certainly significantly improve their foreign language".

Question 3. If financial support from the EC disappears, do you think the programme would be continued (by using funds from the universities involved or by students paying their own way)?

The answer here is almost unanimously NO, even though several organizers point out that the stipends are actually quite meagre, and that extra support is needed.

Again, here are some quotes

- "Financial support from the EC (even considering that our students receive around 300 ECUs per month) has in many cases a kind of 'symbolic' attraction to the students (and their parents). As in 'bargain' sales, you think it is a big deal and therefore you are willing to spend more".

- "It takes a lot of overhead costs to organize and coordinate the programme, which cannot be offered by our university".
- Even now the parents of our students complain that staying abroad is still expensive even with ERASMUS support".
- "The amount of the ERASMUS grants is obviously ridiculous".
- "The European *label* is even more important than financial support, but the two are connected".
- "If financial support from the EC stops, ... [there] will be fewer students to apply ..., they will be older and much more specialized, which for some professors might be a positive development".

Questions 4a and 4b. How many mathematics students from your university go abroad as part of the programme? And how many go abroad to study as part of different programmes or unrelated to any official programme?

The purpose of these questions was to estimate the relative importance of ERASMUS for the student mobility of the Mathematics Department in question. Quite a few coordinators report difficulties in obtaining good estimates, so the answers are hard to summarize in a precise numerical form, but there is a clear indication that

- ERASMUS related mobility is much larger than all other forms put together.

Of course, it might be that universities that are not part of any ERASMUS scheme find other ways to create mobility. An attempt to answer that question was made in a questionnaire sent to (most of) the national mathematics societies in Europe (see below).

4.1.3 Conclusions

- Involving 232 mathematics professors and 132 university administrators (counted with a certain multiplicity, but still!) in moving roughly 500 students seems a very inefficient way of administering a total sum of at most 1 million ECU.
- Nevertheless, the psychological impact of ERASMUS can hardly be questioned.
- In particular, a lot of important personal contacts have undoubtedly been created.
- Any attempt to describe/compare European mathematics degrees should take advantage of the expertise found in the ERASMUS networks.

4.2 ERASMUS: a case study

The ERASMUS ICP which now involves the universities of Copenhagen, Leeds, Paris 7, Paris-Sud, Rome "La Sapienza", Saarland and Utrecht, had its origin before the ERASMUS project started. From 1983 to 1987, Myriam Dechamps of Paris-Sud organized a programme under a previous EEC initiative. After some preliminary visits, this took the form of short exchanges of small groups of students and slightly longer staff teaching visits. Only the universities of Paris-Sud, Leeds and Rome took part at this stage.

These first exchanges were centered on the theme of computers in mathematics teaching. The students worked on projects using computers. The staff members helped with initial programming classes. The activities culminated in a conference on Educational Computing in Mathematics held in Rome in 1987[7].

When ERASMUS started, Leeds agreed to coordinate a continuation of the programme. The number of universities in the scheme increased, arriving by stages at the present position. The whole of mathematics (including applied mathematics, statistics, and computer science) was to be the field for both student and staff exchanges.

From 1988 to 1991, the programme included staff teaching visits of four to six weeks. These visits were a necessary condition for developing experience of and confidence in each others' teaching methods before embarking on student exchanges.

The first students, three of them, arrived at the host universities in February 1990. In 1990–91, 28 students travelled abroad and in 1991–2, the figure was 49. Most of the students were away for a full academic year, the rest for half. They had all spent at least two years at their home university.

ERASMUS has led to several innovations at the universities concerned. Leeds has introduced a Mathematics (European) BSc degree which involves language training in year 2, the year abroad is year 3 and students return for their final year to Leeds. Several of the other universities have appointed official tutors for the ERASMUS students; several have given priority to ERASMUS students for accommodation. In general, the host university has responsibility for seeing that students have somewhere to live. Several of the universities have started clubs to help with social integration. Even with such help, finding friends in a foreign country takes time. A year at the host university (the maximum allowed) is just long enough.

The provision of language training has received a great stimulus from ERASMUS in general. Most of the support money in this programme goes towards the

[7] M. Dechamps, *A European Co-operation on the Use of Computers in Mathematics* in *ECM/87 Educational Computing in Mathematics*, T.F. Banchoff et al. eds., 197–209, North Holland, Amsterdam, 1988

provision of language courses. So far, students in the programme have always attended lectures and taken examinations in the language of the host country. This is very hard for them at first. The structure of the Leeds degree allows students to learn a new language from scratch, if necessary (as is usually the case for Danish, Italian or Dutch.) Students from other universities generally have less than a year to learn or revise their language skills, which makes it more difficult, but not impossible, to recruit students to go to Copenhagen and Utrecht.

One of the conditions of ERASMUS is that the period abroad must count towards the student's final degree. As may be imagined, it is a non-trivial task to adapt the examination results in one country to another university's system. That it can be done at all is thanks to the annual meeting of staff, where each student's performance is discussed, and to an official table of equivalences drawn up by the home university. Most students do as expected, but there are a few disasters and some outstanding successes.

One of the things the staff involved has learned is that there are considerable disparities between different countries' grading schemes, which are not arbitrary, but reflect disparities between the various educational systems. For example, in some countries, students are not allowed to retake examinations; in others they are; in yet others, they need not take examinations until they feel they are ready to take them. Nearly all students have difficulty in adjusting to the different types of examination systems.

The students' evaluation of their time abroad is almost always favourable, even in cases of relative academic failure. Typical comments are: "extremely worthwhile, both from an academic point of view and socially";

"Lectures in Italian were a nightmare at first... (but) I have certainly, during my exams, only encountered problems of a mathematical nature" (this refers to oral as well as written exams.)

"I find there is more actual teaching here than I am used to... but... the pressure on the students... (is) very high."

Turning to finance, the largest sum in this programme goes to the students as travel and subsistence grants. A grant is of the order of 1000 ecu per student, but varies widely, depending on which country the student comes from. Student grants are not administered by the programme coordinator, who, in this case, has a much smaller sum to support the organization of the programme. More than two thirds of the latter sum will be spent on language support. Most of the rest will go for the annual meeting of organizers and a very small sum will be spent on minor expenses such as photocopying and telephone bills. The European Community gets good value for its money, because universities bear much of the administrative cost of the scheme and usually subsidise the language training.

In conclusion, the ERASMUS scheme enables a significant number of students to live abroad, to learn mathematics in a different tongue and in a new cultural setting. Those students who do not go abroad meet those coming from elsewhere. In administering the scheme, staff learn to appreciate different teaching and examining methods. The cross-fertilisation has already been productive. Most of the staff who work with the programme would agree with the students that ERASMUS is extremely worthwhile.

4.3 TEMPUS

A pamphlet, called *TEMPUS — List of accepted projects academic year 1991–92*, lists roughly 350 projects. The descriptions given are very brief, as can be seen in Figure 8 which reproduces the information given for those 7 projects that appear to have an essential mathematical content. A questionnaire sent to the main coordinators of these 7 is reproduced in the appendix. Only 3 answers were received, and no conclusions can be drawn. In comparison with ERASMUS projects, it seems clear, though, that the typical TEMPUS project involves a much larger grant, in part because TEMPUS also covers expenses for equipment, e.g. computers.

D. Cioranescu, Paris:

There are 7 TEMPUS programmes in Mathematics. I am coordinator of one of them: the programme JEP 2797 "Development of Applied Mathematics in Romania". TEMPUS programmes have to involve at least two western countries and one eastern country. The programme I am talking about actually involves 32 universities or institutions from Romania and 8 countries from the European Community. Five Romanian universities participate in the project. The western countries are France, Germany, the United Kingdom, Spain, Portugal, Ireland, Italy and Belgium. Three institutions also participate: the Société Mathématique de France, the Société des Mathématiques Appliquées et Industrielles (France) and the International Centre of Pure and Applied Mathematics which depends on UNESCO. As we can see, JEP 2797 is quite large as regards the number of participants. It is also large from the financial point of view: we had a grant for the first year of the order of 320 000 ECU. The activities in the programme are the following:

- 12 fellowships of one year
- 12 fellowships of 6 months
- specialized courses in the field of Applied Mathematics in each Romanian university given by western mathematicians
- organization of 3 summer schools in Romania in Applied Mathematics
- publication of a mathematical review.

Another part of our activity concerned the furnishing of computers, photocopying equipment and books to the Romanian universities.

Our scientific committee (formed by 14 mathematicians representing the 9 countries) selected the students who were to spend a fifth year in a western country

JEP-1980-91 **Unification and Development of Experimental Environment of the University Mathematical Education in Europe**
Contact: Mr. S. Negrepontis, National University of Athens, Mathematics
Panepistemiopolis, GR-15784 ATHINA
Tel.: 30/1/7243241; FAX: 30/1/7243502

Eligible countries	E. C. Countries	G-24 countries
BC CS H	E F GR I	A

JEP-2092-91 **Applications of discrete mathematics**
Contact: Mr. J. Siemons, University of East Anglia, School of Mathematics
University Plain, UK-NORWICH
Tel.: 44/603/56161; FAX: 44/603/58553, Telex: 975197

Eligible countries	E. C. Countries	G-24 countries
BG H	D I UK	——

JEP-2306-91 **H/01/91 Development of new curriculum strategies and materials for the training of mathematics teachers**
Contact: Mr. R. Gwyn, Manchester Polytechnic, Office of European Development
All Saints, UK-MANCHESTER M 15 6BH
Tel.: 44/61/2471033; FAX: 44/61/2367383; Telex: 667915 POLLIB G

Eligible Countries	E. C. Countries	G-24 countries
H	D UK	——

JEP-2692-91 **Algebraic and categorical methods in computer science**
Contact: Mr. J. Veiga, Universidade de Coimbra,
Paço das Escolas, P-3000 COIMBRA
Tel.: 351/351/3935448; FAX: 251/351/39325841

Eligible Countries	E. C. Countries	G-24 countries
Cs H	B D I P UK	USA

JEP-2775-91 **Statistics and Operations research for Applications**
Contact: Mr. Prof. K. Ritter, Technische Universität München,
Institüt für Angewandte Mathematik und Statistik
Arcisstrasse 21, D-8000 MÜNCHEN 2
FAX: 49/89/2105828

Eligible Countries	E. C. Countries	G-24 countries
R	D F GR	——

JEP-2797-91 **Développement des Mathématiques Appliquées en Roumanie**
Contact: Ms D. Cioranescu, Université Pierre et Marie Curie (Paris VII)
Dépt. d'Analyse Numérique
Tour 55-65, ét.5, 4 Place Jussieu, F-75252 PARIS CEDEX 05
Tel.: 33/1/44274298, FAX: 33/1/44274001

Eligible Countries	E. C. Countries	G-24 countries
R	D E F I IRL P UK	——

JEP-2814-91 **Développement de l'enseignement dans la Faculté des Sciences à l'Université de Craiova**
Contact: Mr. Y. Bouveret, Université Jean Monnet,
34, rue Francis Baulier, F-42023 SAINT-ÉTIENNE CEDEX 2
Tel.: 33/77421700, FAX: 33/77421799

Eligible Countries	E. C. Countries	G-24 countries
R	B F	——

Figure 8

and at this point we encountered some difficulties concerning the harmonization of European degrees. The students in France all took the DEA (diplôme d'études approfondies). It was the same in Italy, not exactly the same in Germany, but totally different in the United Kingdom. This remains a problem for the future. We also had bureaucratic problems, for instance the level of fees which varied with the different countries. The majority of French universities did not require any fees while the British ones asked for quite substantial sums. This is another problem which needs harmonization.

The students with 6 months fellowships are all now preparing a Ph.D thesis. In the second year of the project we will select 12 new students for a one year fellowship and the 24 students selected this first year will all have 6 months fellowships to begin or continue a Ph.D thesis. All the theses have two directors: one from a Romanian university and another from a western university, the Ph.D degree being the one supposedly recognized by both universities.

As planned, in each Romanian university we provided a course delivered by western professors as part of the curriculum.

Finally let me speak about the three summer schools organized in Romania, with the help of CIMI-UNESCO. One was on statistics (with France as coordinator), one on fluid mechanics (coordinators: the United Kingdom and Ireland), and the third on numerical analysis (coordinator: France). The support of CIMI-UNESCO allowed students from Poland, Lithuania and Russia to attend these schools.

Our aim is that these schools should continue regularly each summer with a large participation from the other eastern countries.

We hope by the end of this project to have a good basis for a Romanian school of Applied Mathematics.

4.4 Other programmes

NORDPLUS is a special ERASMUS-like programme involving the Scandinavian countries. Mathematics is almost totally absent in this programme according to the 1991 report.

The EC programme SCIENCE has involved a lot of student exchange, apparently mainly at the advanced Ph.D student and the postdoctoral level. The same will be true for the new programme HUMAN CAPITAL AND MOBILITY. These programmes are among the topics covered by the Round Table on *European Science Policy for Mathematics*, so they have been excluded from the present Round Table.

The main purpose of the COMETT programme is to sponsor the placement of students in an industrial "apprenticeship". It was, therefore, felt to fit better into the Round Table *Mathematics and Industry* even if it does involve some activities of a mathematical nature. One activity supported jointly by COMETT and ERAS-

MUS is ECMI (the European Consortium for Mathematics in Industry) which has created a 2 year postgraduate programme. Attempts to get a representative of ECMI to present this programme unfortunately failed (in part because of electronic communication errors). At the time of writing, ECMI can be contacted at the following address: ECMI secretariat, Prof. R. Mattheij, Den Dolch 2, Postbus 513, 5600 MB Eindhoven, The Netherlands.

4.5 Exchange outside programmes

4.5.1 A questionnaire

A questionnaire [reproduced in the appendix] was sent to mathematical societies in the following countries. Those listed in boldface answered.

Austria	**Belgium**	Bulgaria	**Czech Republic**	**Denmark**	**England**
Ireland	**Estonia**	Finland	France	Georgia	Germany
Greece	Holland	Hungary	**Iceland**	**Italy**	**Lithuania**
Luxembourg	**Norway**	**Portugal**	Roumania	**Scotland**	Spain
Sweden	**Switzerland**	Yugoslavia			

The main purpose was to find out how many mathematics students go abroad for a significant part of their studies unrelated to programmes like ERASMUS, TEMPUS and NORDPLUS.

In the end, the response rate was quite reasonable. However, many answers came in rather late and were incomplete. The results leave us with two clear conclusions:

- The questionnaire was not really well thought out; but also
- Most mathematical societies in Europe are ill-prepared for answering such questions.

Nevertheless, some features of the answers are worth recording:

General comments

- "To be frank, it is difficult to answer the questions you raise even within a [local university] level and I should not know where to turn for national figures. ... It may well be that this [national, mathematics education-related] committee will wish to carry out a national survey to discover more about current practice".
- "... moves towards more modular course structures. If this happens, exchange of students with European universities might become easier and increase".

Questions 1 and 1′. Additional, international exchange programmes? Eastern European networks?

- Estonia: "... there existed such programmes between ministries of higher education of USSR and Eastern European countries [and] academies of sciences. We could use this network quite efficiently for Post. Doct. studies and sometimes also for graduate studies. *Now nothing replaces this system.*

- Switzerland mentions ETAP (European Thesis Abroad Programme).
- Portugal mentions ESEP (European Science Exchange Programme).

Question 2. How many mathematics students studied abroad (at different levels)?

No solid figures appear, but it is the impression that the numbers are (generally) quite small, and that ERASMUS has indeed increased the amount of exchange drastically at all levels below Ph.D . Here are some quotes:

- "[exchange of students may sometimes] mean that a Ph.D student went away because he found no hope of a career in XXXX and never came back. This is not mobility in a good sense".
- "... ERASMUS has a strong psychological effect. Even if the fellowships are insufficient, they induce students to move".
- "I think that things are going to be better with new programmes like ERASMUS".

Question 3. Desire to study abroad (at various levels).

- "Because of the present chaos in our economics, everybody wants to stay abroad (in the West) at any possible level. I believe that about a hundred of our students (all well qualified) would go immediately abroad if they could. The same is true for members of the faculty. ... the needs of our university ... 5 students at the M.Sc level, 2 at the Ph.D level and 5 at the Post Doct. level each year".
- "... increasing interest among students concerning exchanges with North America rather than Europe".
- "... recently the overwhelming majority of these [Ph.D students] prefer (in order of preference) U.S., British, French and German Universities".

Question 4. Willingness to teach in a foreign language (at different levels).

At the M.Sc. and the Ph.D level, almost everybody indicates ability and willingness to teach in English. Very few indicate other possible foreign languages. One should note, though, that the other major European languages are not strongly represented in the sample.

At the undergraduate level, the situation is equally clear: Almost nobody sees any need or desire to teach in a foreign language.

An interesting comment from Austria: "... the new Austrian law for studies at technical universities forces every mathematics-student to follow at least 8 semester-hours of mathematical lectures held in English".

Conclusions

- Even though mathematics has been a very international endeavour for a long time, there appears to have been rather little "spontaneous" exchange of students below the graduate level.

- ERASMUS is seen to have a large impact, even if the administrative work relative to the number of students moved appears large.
- Eastern Europe is likely to be left behind unless we improve the possibilities for their participation in ERASMUS and other programmes, e.g. by creating the possibility of combining ERASMUS and TEMPUS.

4.6 Discussion of student exchange at the Round Table

A. Van der Sluis, Utrecht:

The number of exchange students is disappointingly low. Have any investigations been carried out as to what kind of students applied for this programme? Actually, I could well imagine that the weaker students would be rather doubtful about going and spending some time in other universities, where they study subjects that do not fit in very nicely with what they have already learned. Also in some countries the time which a student may spend at his university is limited. In my country, for instance, you may not study longer than six years or else there is a big stick standing behind the door. And you could well imagine that the weaker student who already has problems in meeting deadlines will not apply for this project. So I was just wondering what kind of students apply for this programme?

H.J. Munkholm, Odense:

I think Salinger has more to say about what kind of students participate in the exchange, but I would like to add that mathematics builds so much on the preceding years at any given stage. You always build on what you were told last year and the previous year, and that is different in the humanities and probably even in physics, although it should not be, so maybe that is one reason why we are scoring low figures, as I think we are.

D. Salinger, Leeds:

We started by excluding weaker students, for obvious reasons, but a number of them have, in fact, profited from the scheme. Such students are quite willing to apply.

H. Martens, Trondheim:

I wanted to comment on the ECMI programme since you brought it up, and to put a slightly different, and not quite as pessimistic, perspective on the situation as Munkholm has done. The ECMI programme is run by a consortium, a group of institutions which co-operate in a programme of training mathematicians for industry. Part of that programme involves the exchange of students between the institutions. In fact, the idea is that ECMI will certify the training and that the cooperating institutions will carry out a programme that roughly conforms to ECMI's specification. Now, in this case it is a question of perhaps seven or eight institutions. The COMETT and ERASMUS programmes are handled by a common administration. I do not have the data on the number of students that are moved yearly—I am not an official representative of ECMI—but I think the order of ten students are moved every year between the institutions under the COMETT programme, which gives a slightly different perspective of the efficiency.

H.J. Munkholm, Odense:

INPUT	OUTPUT	CONCLUSIONS
1 million ECU		Efficient?
	500-600	
225 universities		Bureaucratic !
	students	
225 administrators		BUT ALSO
	moved	
225 mathematicians		IMPORTANT

This table[8] is meant to be provocative. I do know that there is multiplicity in the 225, but still I think we should think about it.

A. Figà-Talamanca, Rome:

I planned to talk only at the end in the general discussion, but I cannot resist making a comment on the problem of efficiency. I would like to point out that, even though mathematicians are not allowed to decide where to spend the money, with 1 million ECU one could support about a hundred Ph.D students. A hundred Ph.D students are probably one third or one fourth of the Ph.D students in mathematics, which is reasonable for a country of about 50 million, so it is not much money, but we know how many countries have difficulties in financing graduate studies, so that is not a trivial amount of money. Perhaps too much to spend just for moral reasons and because it is important to talk about community and to let students travel. I also feel, because of the structure of the fellowships, because they are very small, and for many other reasons, that this programme is very discriminating in the sense that it is a very nice complement to the education of students coming from an upper middle class background.

R. Mlitz, Vienna:

I just wanted to remark that the ERASMUS is a programme, mainly based on the European Community, so for other states — for Austria — it is much more difficult. There are some agreements, but then when Austrian universities need to be involved with at least 2 or 3 European Community universities, participating involves quite a lot of bureaucracy.

J.F. Rodrigues, Lisbon:

I believe that my university was maybe one of those 225[9] universities, and

[8] Based on the unadjusted totals, see footnotes 5 and 6.

[9] The figure should be 132, see previous footnotes

I support professor Figà-Talamanca in demanding the question mark in the efficiency. And I would like to ask about the percentage of the one million ECU used for fellowships, for bureaucracy and for professors. Are there any numbers about this?

H.J. Munkholm, Odense:

There are figures in the ERASMUS Newsletter of the total level. As far as I remember, of the total budget of 65 million ECU for ERASMUS in 1991/92, something between 37 and 40 million ECU, abount 60 to 65% went directly in student grants. So it may not be top heavy in that respect[10].

D. Salinger, Leeds:

Of the money that is allotted to universities for the support of the programmes, not more than 10% is allowed to be spent on photocopying, telephone calls, or minor administration. In practice one spends less. Certainly, in my scheme, the majority of the money goes on supporting language training.

5. General Discussion

The discussion was intended to concentrate on the following questions:

1. Updated version of student's record book.
2. Formulation of a European Standard.
3. European Degree in Mathematics.
4. EMS–committee linking EEC programmes to the mathematics community. Lobbying. Guiding.
5. Is student exchange worth the trouble and bureaucracy?
6. Should the exchange of students be restricted to the research level (Ph.D students)?
7. Should the EEC create a Ph.D supporting programme?

A. Figà-Talamanca, Rome:

There is a bit of information I want to give. Italy has a two year diploma, a first cycle. This is not completely true: we are going to have it soon, but it will take some time for the universities to adopt the new regulations.

I just found out that the Italian Mathematical Union did not answer the Questionnaire simply because it was not sent to the right address.

I want next to talk about the problems with the programmes and the booklet.

I think it is a very difficult problem, and one of the difficulties which has not been mentioned is that when mathematicians sit down at their desk and think about what should be taught at a certain level, they jump into a fantastic world, because they do not have the students in front of them.

---

[10] Added after the Round Table: In 1990/91 the total budget was 61.2 MECU; direct student grants amounted to 32 MECU, i.e. 52%. The incorrect figures given at the Round Table were based on a mix up between 1990/91- and 1991/92 figures in ERASMUS Newsletter, No. 11, 1991.

It very often occurs that programmes which are stated, printed and circulated do not correspond to what students actually learn. This is quite common in many countries. In French high schools, this is true to an incredible degree, and perhaps at the university level as well, and it is certainly true in Italy. So we should be aware that it is very difficult to write a booklet containing minimal requirements for a degree in mathematics which corresponds to what actually happens.

On many occasions, the people who make these recommendations are mostly concerned with teaching at a relatively high level or in the best universities in their own country. Therefore they do not actually meet students with weak backgrounds from minor universities or in the first years of study.

So a harmonization which will be only formal is probably going to damage the system and will not be an improvement. Another difficulty about harmonization is that sometimes programmes are complementary to each other in various countries. There is a certain amount of mathematics which is really necessary in order to go on, both in science and in mathematics, say, calculus or linear algebra. In many cases, what is required to go on in mathematics, is mathematical maturity, which can be achieved by studying different topics.

It may very well be that someone achieves maturity by studying probability and another student will acquire it by studying algebraic geometry. And they will both be mature enough to open a book and understand things at a higher level.

Consequently we should concentrate on the very elementary and basic things in mathematics at the university level. So your recommendation to look at what happens in the first three years of study should be limited to the first two years. This corresponds to the first cycle in French universities.

If we are trying to make a correspondence between different universities, it is very difficult to do this course by course. People may have studied different things and yet be at the same level and have the same degree of maturity. My recommendation is that we attempt to have a uniform degree after the first two years which signifies a certain maturity, and a certain level of abstraction. After all this is perhaps not so important due to the small number of student exchanges.

But there is another more important question which perhaps should be taken up by the EMS. It is the question already mentioned about a European Community law concerning professional activities and access to the various professions.

First of all, in mathematics, this concerns only the profession of teaching or perhaps actuarial mathematics, or in some cases also business mathematics, because mathematics is not a profession but a science, a discipline. But it is very important that we do understand that—for the EEC countries now and most probably and hopefully for the other countries which join the European Community or become affiliated—if somebody has a three year degree which gives access to teaching mathematics at the secondary school level, this same person can teach mathematics, even as a state employee, in any other European country. The practical problem now is to make certain that programmes for secondary school teachers give them a sufficiently sound basis.

There may not be much migration of mathematics teachers from country to country, but this could be an opportunity, given by European Community law, to intervene in support of mathematicians who are trying to improve teacher education in mathematics. The EMS could arrange meetings in which these problems are discussed, and it could influence policy.

A. Bak, Bielefeld:

Who outside the mathematics community will be interested in harmonization of degrees? and what is their interest?

I. Netuka, Prague:

It is hard to react immediately, but I think we should discuss the notion of harmonization more extensively. To me the basic goal is to make student exchange easier, and this could be a consequence of a harmonization of programmes and degrees. There is hardly anybody outside the mathematics community interested in these questions.

J. Musielak, Poznan:

One should not have the exchange in only one direction from East to West. It is also a possibility in the other direction, and this point of view should be worked out.

D. Salinger, Leeds:

I am happy to report that we have sent an undergraduate student to Poland this year.

A. O'Farrell, Dublin:

I think that the concept of a European standard for a degree in mathematics is a terrible idea. Basically it has the prospect of creating a straitjacket. The simple adoption of such a standard by a body like the EMS, would put terrific pressure on people to conform to such a standard. I can see that down the road we will end up like the Americans: having to teach the same calculus course. We have at the moment a terrific diversity of programmes which we should treasure. It is fine to communicate with each other about what the students have done if they go from one place to another, but a European Standard Degree is awful.

C. Berg, Copenhagen:

I think that many are against such a European degree, but will anybody speak in favour of it?

R. Rentschler, Paris:

I think it would be a good idea to have at least some hints for a European standard, not as a straitjacket but to know an average level, so my opinion is slightly different from O'Farrell's.

M. Karoubi, Paris:

I am unhappy that some of my students are not here, because when I hear of the figure for rate of success in exams in England, only 5% fail[11]. This would create a revolution here. My students would say that I am a poor teacher with 40% failing. Harmonization for my students would be: try to be a better teacher and try to have a system like in England. Let me put this as a question to Salinger: How do you obtain these 5% ?

D. Salinger, Leeds:

This is not a simple matter. My personal reply is certainly not backed up by hard evidence, but my feeling is that many of the students getting third class results in Britain would not get their degree in a foreign country in the minimum number of years. However, there is almost no possibility of taking extra years in the UK. So we do not have students going on for 5 or 6 years to get their degree. This simply cannot be done. A very few take one year extra, but with no financial support. They are not allowed to take the final examination more than once. So there is a tremendous pressure to let the students pass, even if we feel that they have not understood very much of the programme. You are not under that pressure in France, and I sometimes envy you that position.

C. Berg, Copenhagen:

I should like to draw your attention to point 4 of our discussion: A committee which could influence the science policy of the EEC. Have you any comments on this?

F. Hirzebruch, Bonn:

The EMS already has such a committee. The chairman is Figà-Talamanca and I am very pleased to convey this recommendation to our executive committee, and I think we should develop our existing committee to look at all the different programmes which exist. In our newsletter we have already discussed the programmes on Human Capital and Mobility and cooperation with the eastern European countries. Professor Luc Lemaire of Brussels is also a member of our committee, and he is, as it were, our man in Brussels. So we are happy to take this recommendation as a good guideline.

J. Stegeman, Utrecht:

Concerning question 6, I think that everybody agrees that the answer is no: we should not restrict student exchanges to the Ph.D level. The experience of ERASMUS is very positive. And I also want to mention another positive effect. The students staying at home meet foreign students visiting. Concerning question 5 above, I would like to point also to what we, the staff, learn from all this: we see how varied the teaching programmes can be, and they all lead, hopefully, to good mathematicians, so there is no need to make one uniform programme. On

[11] This relates to a somewhat speculative table, not reproduced here, comparing third year examination results in England and France.

the contrary, it is very good that it is the way it is. It is one of the fruits, already, of the ERASMUS programme that we have become, I think, on a much larger scale, aware of the variety, and this can have a very good influence.

H. Ibisch, Nantes:

I am afraid that there is a lot of misunderstanding about the ERASMUS programme.

First some facts. The ERASMUS programme was created by the European Community to develop the exchange of students between European universities. The objective was that about 30% of European students should visit a foreign university during the course of their studies. This programme started 5 or 6 years ago and, what is much less well known is that it will stop 4 or 5 years from now.[12] It is not an eternal programme. It is an opportunity given to the universities to develop student exchanges.

With respect to the initial object you can say it has failed, because only about 10% of the students participate in the exchange programmes. I think one of the reasons why it has failed is that university professors have not become sufficiently involved.

I am a coordinator of one of the programmes. Why did I take up this responsibility, which is a lot of work? I did it because our students asked me to. They wanted to go abroad. Right now 15 of our 120 mathematics and informatics students go abroad during their third year. Another reason why I accepted this responsibility is that the money accompanying these exchange programmes allows us to reinforce scientific relations with our European partners.

As I said before, to my knowledge there will be no phase 4 of the ERASMUS programme, so if you have not yet developed an ERASMUS exchange programme, you'll have to get a move on. Our university is looking for alternative financial resources to maintain the exchange programmes after phase 3. We have started to get additional support from regional funds.

One last remark. We were asked which students go abroad. In our university it is usually the best students.

D. Cioranescu, Paris:

I would like to say a few things about your question 6: Should the exchange of students be restricted to the research level? Surely not: from my own experience with the TEMPUS programme. It turns out that it is much more worthwhile for students to do their last year of study in a western country and start the Ph.D thesis afterwards. Ph.D studies are harder for people and, in applied mathematics, we even had the difficulty of knowing what is meant by applied mathematics in Romania compared to France, for example. In Romania we had only people in informatics who wanted to study applied mathematics in France.

---

[12] (added in proof) Prof. Ibisch was too pessimistic. ERASMUS will continue, as part of the new SOCRATES programme.

In conclusion it is very important not to restrict the exchange of students to the research level.

C. Berg, Copenhagen:

I shall comment on question 7. It seems to me that we send many of our best students to the United States for Ph.D studies. Of course there are often very good scientific reasons for this, but another point is that, in the United States, students can be financed to a certain extent by being teaching assistants. A similar system in Europe would be difficult to achieve because of language problems, for example, in small countries like Denmark. This makes it difficult to attract foreign Ph.D students, and they are certainly an important factor in a vital research group.

The EEC has recently launched the Human Capital and Mobility programme which is at the postdoctoral level. I think we need a European programme to support Ph.D studies in Europe.

L. Lemaire, Brussels:

I think it is possible to argue that in some cases the Human Capital programme can be used to finance Ph.D students, for example, in branches with a certain shortage of people.

6. Recommendations

Taking account of the views expressed at the ECM Round Table, the organizers make the following recommendations.

1. The European Mathematical Society should take steps to create an updated version of the European Student's Record and ensure that it is distributed widely to all European university mathematics departments.
2. On the one hand, a considerable majority of the Round Table participants was opposed to the idea of harmonization of degrees or of the imposition of a European degree in mathematics, or of any kind of European standardization.
3. However, where the European Community has taken steps, through directives, or by other means, to create certification procedures for university courses involving mathematics, the European Mathematical Society should ensure that mathematicians' voices are heard. This is potentially particularly important with regard to the certification of mathematics teachers.
4. The European Mathematical Society should monitor the participation of European mathematicians in European Community related programmes such as ERASMUS and TEMPUS for the purpose of increasing the visibility of such programmes for European mathematicians. Prominent in the domain of interest should also be the new European Community programme HUMAN CAPITAL AND MOBILITY.
5. The EMS should take active steps to ensure that Eastern European mathematics students can participate in all existing exchange programmes on an equal footing with citizens of the European Community.

7. Appendix

This appendix contains copies of the questionnaires mentioned in the main text.

7.1 Questionnaire concerning harmonization

Dear Professor . . . ,

Thank you very much for your willingness to collaborate with us on behalf of your society. As you have been informed, several Tables Rondes will be organized during the European Congress of Mathematics, Paris 1992. We were appointed, together with Prof. Munkholm (Odense, Denmark), to be the coordinators of Table Ronde: Exchange programmes. Harmonization of European degrees in mathematics.

The main motivation for this is to contribute to an increase of mobility of mathematics students among individual European countries. We wish to emphasize that there is no attempt to organize any unification of programmes of studies on European level. We would like to collect relevant information, to make a comparison and to find common features. In our opinion, it would be desirable to clarify conditions under which some well-defined parts of the study in one European country would be acknowledged in the other one. We are fully aware that even within one country the plans of studies vary university from university. Therefore, it seems to be necessary to consider an "idealized" situation, it means to think in terms of a "typical" student of mathematics.

We would appreciate very much if you could elaborate on a report concerning the system of the study of mathematics at universities in your country. We feel that at this moment we should concentrate on the study of mathematics in the narrow sense, it means to leave aside the study of informatics as well as the training of future mathematics teachers. By the way, there is another Table Ronde "The role of mathematics in educational policies in Europe" (organized by Prof. M.-F. Roy). We hope that the overlapping should not be too large.

The report (which should not exceed 10 pages) should contain the following main points:

1. The general characteristics of the orientation of the study.
2. The vertical division of the study, i.e. degrees or titles awarded, including the internal division (e.g. by means of a special examination, final examination or so).
3. A description of the system of partial examinations and/or other means of the control of the knowledge obtained in lectures and/or exercises.
4. Specify, please, which requirement, in your opinion, could be ful-

filled by partial study abroad and under what conditions. Do you think that there is a prevailing opinion within your universities that such partial study at another place could be officially acknowledged in future (if it is not the case at present).

5. It is most probable that plans of study at various universities are overlapping very much in the first, say, three years. Could you provide us with a detailed plan of study (syllabus) for basic courses for this period? On the basis of this information, a "standardized" list of items (such as concepts, main theorems or methods...) could be compiled. It should be mentioned here that such an effort had already been undertaken about 30 years ago. As a result, a booklet "European student record" was produced. The booklet did not find a practical application, it seems that it was too premature. Fortunately, the situation in Europe has essentially changed since. So why not try to reconsider the old idea nowadays.

Besides your report, we would be happy to get any further material concerning the subject (a booklet for students, the list of lectures, regulations for the study as well as for examinations, etc.). If e-mail is available at your university please let us know your e-mail address.

We would like to express our conviction that the project could significantly contribute to a comparison and a better understanding of various programmes of study and could make student mobility easier.

Let us thank you very much for your assistance in this matter.

With best regards,
Ivan Netuka and Vladimir Souček

7.2 Questionnaires concerning exchange of students

7.2.1 To ERASMUS coordinators

Dear Professor...,

As you probably know, at the European Congress of Mathematics in Paris, July 6–10, 1992, there will be a number of Round Table presentations including one on "Exchange Programmes and Harmonization of Degrees" which will be co-chaired by myself and Ivan Netuka and Vladimir Souček from Praha. I am in charge of the exchange part of this Round Table, and in that capacity I am currently trying to ascertain how the ERASMUS programme works for mathematics. Since you are listed as the co-ordinator of one of the ICP's involving mathematics I hope that I can get you to answer a few questions.

First of all, I should tell you that my information about the ICP you are co-ordinating comes from the 1990/91 ERASMUS handbook

(ERASMUS and LINGUA action II Directory). Thus I have a 4–10 line description including an expected number of students involved, the expected duration for each student, and the approximate study level targeted. For your information I enclose a photocopy of the relevant page of the ERASMUS handbook. I am particularly interested in the following questions:

1. Does the programme run roughly according to plan or did essential changes have to be made?
2. As a co-ordinator, do you feel that the results are worth the effort you (and others) put into it?
3. If financial support from the EEC disappears, do you think the programme will be continued (by using funds from the universities involved or by students paying their own way)?
4. Can you give an estimate for the following figures?

   a) How many mathematics students from your university go abroad as part of the programme?

   b) And how many go abroad to study (for a similar period or longer) as a part of different programmes (which?) or unrelated to any official programme ?

I should be very grateful to get your answer before February 15. Your answers will of course be treated confidentially, and only statistical information will be given at the Round Table.

Yours sincerely,
Hans J. Munkholm

PS. If in connection with the establishing of the ERASMUS network you had occasion to make a comparison of the mathematics programmes in the various universities I should be very interested in receiving also a copy of such a description.

PSPS. If you prefer, the answers can also be sent by email.

7.2.2 To TEMPUS co-ordinators

Dear Professor . . . ,

As you may know, at the European Congress of Mathematics (July 6-10, 1992 in Paris) there will be a number of Round Table presentations, including one on Exchange programmes and harmonization of degrees where I am in charge of the exchange part. In that context I am trying to collect information on mathematics in relation to the TEMPUS programme.

You are listed as contact person for one of the 1991/92 TEMPUS projects which includes mathematics (at least it looks like that to me). I hope you will be willing to answer the following questions (or you may have available a description of the project from which I could dig out the information myself).

The information you give will, of course, be treated confidentially (unless we agree otherwise later on). I hope that you will be able to send your answer no later than Feb. 15.

Question 0: Please give a (short) description af the overall purpose of the project, including which parts of mathematics is involved (and whether non-mathematical subjects are also involved).

Question 1: (Optional if you do not want to supply money information). The contract extends over ...... years, starting 19......; it involves a total of ... ECU's.

Question 2: If the project involves exchange of students, please tell

- how many students
- from where to where
- for how long a period
- at which level of study
- do the students work towards a degree in their home country or the host country?

Question 3: If the projects involves exchange of teachers, please tell

- how many teachers
- from where to where
- for how lang a period
- at which level do they teach?

Question 4: Which other activities are involved?

Question 5: Has the risk of "brain drain" appeared in your consideration, and if so what have you done to avoid it?

Question 6: Are you aware of other TEMPUS projects (or other cooperative projects with Eastern Europe having a similar aim) related to mathematics excluding JEP-1980-91, JEP-2092-91, JEP-2306-91, JEP-2692-91, JEP-2775-91, JEP-2297-91, JEP-3814-91?

Yours sincerely,
Hans J. Munkholm

7.2.3 To European national/regional Mathematical Societies

Some time ago your society was contacted by professors Netuka and Souček who asked you to appoint a person to help them (and me) with some information concerning the Round Table on Exchange Programmes and Harmonization of Degrees at the European Congress next summer. If you already found such a person, please pass this letter on to her/him.

[This letter is being sent to

1. those persons who promised to collaborate with professors Netuka and Souček concerning harmonization of mathematics degrees in Europe
2. Those national/regional mathematical societies for which no such contact person has been appointed.]

As you know, a series of Round Tables will be organized during the European Congress of Mathematics, Paris, July 6–10, 1992. One of these Round Tables, called "Exchange of Students and Harmonization of Degrees" is chaired by professors Netuka and Souček from Praha, and myself. We have divided the responsibility so that I take care of the exchange part.

I have contacted the EEC offices in Brussels to obtain information about the participation of Mathematics in the exchange programmes ERASMUS and TEMPUS as well as other EEC research and teaching programmes. Similarly I will get information from the nordic countries concerning NORDPLUS.

Since these EEC schemes are relatively recent inventions and mathematics has been an international subject for at least a few hundred years, it is clear that there is a lot of exchange that takes place through other channels.

I hope that you can help me get an impression of the relative importance of such other channels. In particular, I would like answers to the following questions:

1) Are you aware of any additional (relative to those mentioned above), formalized, international exchange programmes involving mathematics?

1′) In particular some of you must know whether there exists/existed an Eastern European network of some kind.

2) Is it possible to estimate how many mathematics students from your country studied abroad

   a) for at least half a year during the undergraduate studies
   b) for at least half a year during their graduate studies (i.e. studies for a Master's Degree or a Ph.D.)

c) for their full graduate degree (M.Sc. or Ph.D.)
d) for at least one year as a Post.Doc.

[The time period covered by the information should also be given, perferably for the period 1988/89–1990/91]

I am aware that it may not be possible to find such information at a national level. If so, can you give me some figures from your own university or from a group of universities that you know of.

3) Would more students be interested in a stay abroad at the various levels? Can you estimate the demand?

   a) at the undergraduate level
   b) at the M./Sc. student level
   c) at the Ph.D. student level
   d) at the Post. Doc. level.

4) Would your university be willing to teach in a foreign language in order to accommodate foreign mathematics students?

   a) at the undergraduate level?
   b) at the M.Sc. level?
   c) at the Ph.D. student level?

   If so, which foreign language?

5) If you have any further information concerning exchange possibilities that you might want to share with me, I should be most interested.

Yours sincerely,
Hans J. Munkholm

Round Table J: July 7, 1992

THE PYTHAGORAS PROGRAMME

A European Policy for Mathematics

by Daniel Gabay

CNRS, Université Paris IX, Paris (France)

Between 1988 and 1992, the European Community allocated 10 million ECU to nearly forty different research projects with a strong mathematical content that involved several research teams working cooperatively. This represented 6% of the SCIENCE programme budget, whose goal was to stimulate the exchange of researchers between various European research institutions.

The HUMAN CAPITAL & MOBILITY programme, which takes over in the new Community R&D Framework Programme with reinforced financial means should allow even more mobility for young researchers through post-doctoral grants and result in the creation of more research networks. Prof. Hirzebruch, President of the European Mathematical Society, emphasized in his opening address before the Congress, how important it is that mathematicians participate massively in this programme and why it would be desirable that the programme be extended as quickly as possible to include Eastern and ex-Soviet Union countries.

This is just a first step, however, in the construction of a real European policy for the development of the potential for mathematical research and education in our various countries in order to maintain its place of excellence worldwide. Such a policy still has to be conceived and organized.

The round table on European policy organized during the European Congress of Mathematics focused on future prospects. The round table provided the opportunity for a comparison among several national experiments and an exchange of ideas among various elements of the scientific community (academics, industries, institutions). The round table proposed sketches for strategic choices called — even before its conception — the *Pythagoras Programme* in reference to the origin of mathematics as a key to understanding nature.

Mathematics today is extremely dynamic. New themes emerge, areas of applications expand rapidly. This was well demonstrated in the lectures during the Congress. However, the prospects for the development of our discipline, which has become a "strategic resource" for our technological society, seem to be threatened by a foreseeable lack of mathematicians in the near future if no appropriate strategy is undertaken.

This danger was first recognized in the United States at the beginning of the 1980s. The mathematical community mobilized itself through several initiatives, the most famous of which was the *David Report*[1], proposing an ambitious ten year plan to ensure the renewal of American mathematics. The recommendations from the *David Report* have been at least partially applied and seem to have resulted in the inversion of a negative trend.

The American example had echoes in France where the *Société Mathématique de France* (SMF) and the *Société de Mathématiques Appliquées et Industrielles* (SMAI) jointly organized the *Mathématiques à venir* colloquium[2] (Mathematics to Come), a prospective and consciousness-raising meeting that was a complete success. In effect, authorities have since then made catching up in favor of mathematics research teams, and more specifically of young researchers, a priority.

The deterioration of resources of the educational and research system in Great Britain prompted British mathematicians to evaluate the current situation and future perspectives in the framework of the Science and Engineering Research Council (SERC). The results of this evaluation were published in the *Kingman Report.*[3] It seems that unfortunately the minimal recommendations set forth in 1991 have not yet been implemented.

This new awareness of the dangers that threaten the future development of mathematics has also spread elsewhere in Europe. In Holland, a report was issued in 1992 at the request of the Minister of Education and Science[4]. In 1989, a "German David report" was envisioned in West Germany. The fall of the Berlin wall and reunification seem to have suspended this report in order to first face the situation in the new *Länder* of the ex-GDR.

In effect, the concept of solidarity with Eastern colleagues has considerably evolved since the time of support committees for the few dozen mathematicians who were persecuted or interned due to political or racist motives. The opening of borders has resulted in a drain of scientists attracted by the working and living conditions in the West. There is a lofty concern that such a massive brain-drain will further worsen the economic crisis of these countries, which are thus losing part of their potential and investments in highly qualified human resources. There is also a growing competition between the scientific departments and even companies of industrialized countries to grab the best teams at the lowest price. Links of solidarity are turned into databases that are used for a detailed analysis of the

---

[1] *Renewing U.S. Mathematics: Critical Resource for the Future*, National Academy Press, Washington D.C. , 1984.

[2] *Mathématiques à venir : quels mathématiciens pour l'an 2000*, Colloque SMF-SMAI, contributions gathered by K. Chemla and I. Ekeland. Special issue of the *Bulletin de la SMF*, t. 115, 1987.

[3] *Mathematics, strategy for the future*, Report of the Mathematics Strategy Review Panel chaired by Prof. J. Kingman, SERC, 1991.

[4] *Mathematics: ordering a complex world*, Report of the Advisory Committee on Mathematics, Ministry of Education and Science, 1992, in Dutch.

scientific elite of the military-industrial complex of the former Soviet Union in order to extract its strategic elements. Such a logic is in accordance with the primary recruitment concern of western universities: where to find the educators that are needed for the next decade since the number of students in science — and notably in mathematics — has decreased in the 1980s?

This brief introduction tried to point out a few essential problems that hinder the future developments of mathematics and concern the European scientific community at large, in spite of the diversity of specific national circumstances. These problems arise at a time when the scientific community must accomplish a profound change in its behavior to meet a rapidly growing demand for new tools and concepts and for experienced mathematicians, a demand caused by the explosion of mathematical modeling in all areas and linked to that of computing possibilities.

In a context of vague unease among mathematicians, European institutions can cease to appear as simple complements of national funding agencies to come to play a major role in triggering a development policy for mathematics. Such a role would consolidate the cultural unity of Europe in a spirit of solidarity and knowledge.

The present report was prepared with the help of the documents quoted above and of the results of a questionnaire sent to all national mathematical societies. The report also makes use of the discussions during the round table "European Science Policy for Mathematics" that mainly dealt with research, while other important aspects such as education, the general public, student exchanges, developing countries, libraries and so on were the subject of specific round tables.

The report follows the usual stages in the conception of a scientific policy. In the first part, we try to define a few indicators of the situation of mathematics in Europe. The second part is concerned with the formulation of objectives for the proposed policy and touches upon some aspects of its implementation. As a conclusion, we propose a series of recommendations in the form of a programme proposal, the *Pythagoras Programme*.

I. Some Data on the Situation of Mathematics in Europe

We give some quantitative data here on the current state of the educational/research system in mathematics taken from the above mentioned national reports and from a rather cursory questionnaire sent to all European mathematical societies. Unfortunately, the result is only fragmentary since there were few useable answers: France, Great Britain, Italy, Netherlands, Sweden as well as some data from Poland and the former Czechoslovakia. We are far from a panoramic comparative picture of the situation and we must be content with this sample and try and extract a topology from it. Besides, this experience leads us to wonder about the dynamism and representativeness of certain learned societies and about the desired evolution of the European Mathematical Society of which they are a member.

These lacunae are also signs of a frequent lack of information on human and financial resources for mathematics at the national level. Such a knowledge is indispensable however in raising awareness about the problems we are facing. Our first recommendation is thus to encourage national scientific communities to undertake inquiries and analyses of their situation. To give an example, the committee in charge of the *David Report* defined the quantitative objectives of its plan on the basis of a questionnaire sent to all heads of departments of mathematics in American universities.

The situation of mathematics in Europe presents a great deal of variety in terms of actors and institutions. We will nonetheless try to detect the common features and significant trends of the dynamics while presenting a few specific national examples for concrete case analysis.

We will thus establish a diagnosis that is rather similar to that of the *David Report*, in spite of the differences between the various management modes in European educational/research systems and the market-like decentralized mechanisms that prevail in the United States. We will find the problems of renewal and recruitment posed by an aging population of researchers and professors and by the lack of advanced students attracted by mathematics. We will also find a degradation of material conditions inherited from an historical tradition of mathematics as a "no laboratories" science, whereas today, it requires increasingly sophisticated computer equipment for research as well as teaching. Also the teaching of mathematics now includes a dimension of services towards other disciplines such as computer science, engineering, economics and commerce, biology, and so on.

1. The demography of mathematicians

Table 1 sums up the data from our sample concerning the population of mathematicians who are employed full-time in higher education and research institutions, that is essentially in the mathematics departments in universities. It was not possible to obtain reliable data on researchers in industry and services who work in mathematics. Differences in national ways of classifying disciplines further reinforce the ambiguities and difficulties found in studying this population. Most figures refer to the academic year 1990–1991. They are often only estimates of the population of university mathematicians that would need to be further increased by 20 to 30%[5] in order to include the mathematical activities of theoreticians from other disciplines.

This partial data shows that the European mathematical population is larger in numbers than the American population, *i.e.*, 10,000 according to the *David Report*. Let us recall that the same report considers that half of this population is productive and a quarter is highly productive in terms of research results.

[5] The 30% estimate is mentioned in the *David Report* for the United States.

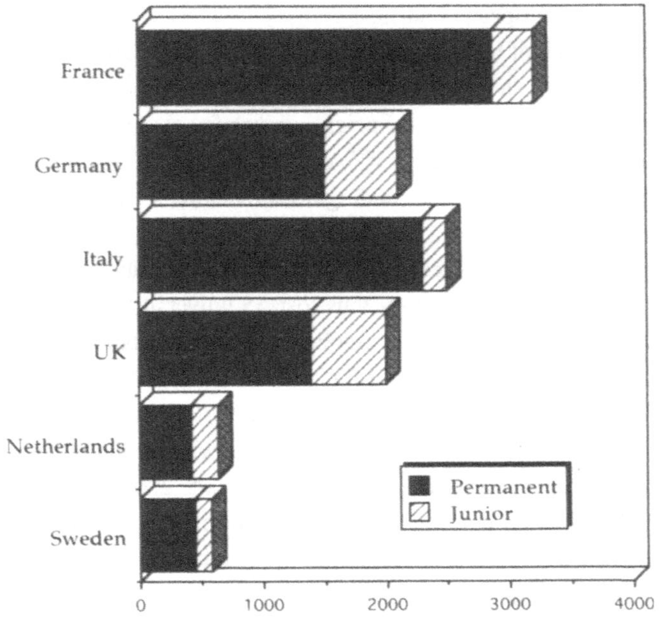

Table 1. Number of mathematicians in six European countries.

Table 1 shows a dichotomy between the four largest countries of the European Community. On the one hand, France and Italy have 3,150 and 2,500 mathematicians respectively, while Germany and Great Britain only have 2,150 and 1,800 respectively. The contrast is even sharper if differences in status and careers are taken into account. In France and Italy, positions are most often permanent, whereas only 1,500 positions are permanent in Germany and Great Britain. Moreover, "young" mathematicians from the latter countries can only obtain temporary teaching positions or post-doctoral grants for three to five years in the best of cases[6]. The result of this situation is a tension that is not favorable for the renewing of the academic body and provokes a non-negligible emigration of young talent towards the United States where they can obtain assistant professorships while waiting for a possible chance for a position in their home country. This danger is likely to become worse due to the recruitment problem in American universities, which is caused by a diminishing domestic market and the insufficiency of the Asian communities basin. American universities tend to target their hunt for young mathematicians explicitly towards Europe, and especially toward

[6] In Germany there are still 113 mathematicians who obtained their *habilitation* before 1986 and who still do not hold a permanent position. About a hundred obtained their *habilitation* before 1990 and are still in precarious positions today.

Great Britain and Germany which do not present the same cultural and linguistic obstacles as Eastern European countries.

Moreover, the lack of human resources in Germany and Great Britain affects the functioning of the system of training of future mathematicians. The national survey of the London Mathematical Society[7] shows for example that British departments of mathematics only devote 7% of their teaching activities to their own graduate students, but 60% to their undergraduate students and 30% to students from other departments.

The difference in situations seems to be increasing if we examine the dynamics. The British mathematical community thus lost 55 university positions between 1988 and 1991, whereas the efforts accomplished in France from 1988 on resulted in an increase in permanent positions from 2,500 in 1988 to 2,830 in 1991.

These effects influence the demographic structure of the mathematical population. The age pyramid demonstrates the preocupation with the aging phenomenon in most cases and points to an acute crisis for the decade 2001–2010. More than 51% of the mathematicians currently holding a position today in France and Great Britain will reach retirement age in this period. The situation is even worse in Germany where the average age for a professor is 52. Italy stands out as an exception due to three massive waves of national recruitment that followed each other in the 1980s. These waves reduced waiting lists and significantly lowered the age of a first appointment for a professor. Among the 1,825 professors of mathematics in the universities, only 346 will retire between 2001 and 2010.

Given the terms of this problem, statistics on doctorates are useful in evaluating the possibilities for renewal. Once again, there is a broad diversity of situations in Europe, starting with the degrees themselves, in spite of the reforms undertaken in several countries to harmonize them on the model of the American Ph.D. We have already alluded to the preservation of the tradition of *habilitation* in Germany. We can also mention the specific case of the Italian reform according to which access to doctoral studies is limited by national competitive exams. The organization of these exams took several years. This had the perverse effect of increasing the average age of "young" researchers during the first stage. It also created a paradox in that half of them hold a permanent academic position before having completed their Ph.D. This last factor explains the abnormally low number of degrees awarded under the new system (35).

France and Great Britain have a more reasonable productivity with more than 200 Ph.Ds in 1990, a significant fraction of which goes to foreigners originating mainly from the former colonial empires. According to detailed demographical analyses in both countries, the flux of national dissertations is scarcely sufficient to ensure a stationary level of renewal of the academic mathematical population (which would require 80 recruitments of researchers aged under 30 per year).

[7] *Mathematics Departments Survey 1991*, LMS Report, 1992.

This flux will most likely not be sufficient to meet the additional demand for mathematicians from other disciplines as well as explicit or latent, from industry and services. The situation is even more serious in Germany where the number of *habilitations* fell to 40 in 1990 and is likely to reach even lower levels — only 22 foreseen in 1995 — to a point where the annual ratio of professorship positions to *habilitations* is getting dangerously close to one[8].

Table 1 also shows the statistics concerning the Netherlands and Sweden, which are representative of smaller European countries (one is a member of the EEC and the other a future member). The mathematical communities are smaller than in the previous countries, which does not prevent them from obtaining excellent results in certain areas related to a solid historical tradition or to advanced technology from large industrial groups. These communities do not have the same problem of aging as their larger neighbors, but they have less concerned themselves with attracting new talents, hence the low number of Ph.Ds and recruitments until recently. As we mentioned in the introduction, a reform of doctoral studies and their funding is currently being implemented in the Netherlands as well as in Sweden.

Finally, we include some data from Poland. Poland is typical of the critical phase of political and economical transition that is drastically changing all of Eastern Europe. This particularity affects the validity and relevance of the statistics. These statistics nonetheless underline the presence of a significant mathematical community with 750 mathematicians in 1991 in a sample of six Polish universities (Warsaw, Wroclaw, Katowice, Gdansk, Poznan). Poland maintains a flux of about 20 Ph.Ds in mathematics per year.

The answers received from Eastern countries all emphasize the precariousness of the material situation of mathematicians, whose net income fell to 140–200 ECU a month (estimates for Hungary, ex-Czechoslovakia and Poland) and even less than 40 ECU a month in Russia now due to the crash of the ruble. Hence the fear of a massive brain-drain toward the West, in particular among youth who are likely to be forever lost to mathematics as they most often are employed in different activities, such as software development.

2. The funding of mathematical research

Historical and cultural differences, which are characteristic of Europe, have created a broad variety in the organization of research. This variety is increased by the variable degree of decentralization of national administrations. Concerning mathematics, we can nonetheless distinguish two main agents in every country:

---

8 "Survey on the age pyramid of professors in mathematics and their replacement when they retire," University of Bielefeld, 1992.

- the Ministry of Education[9] which funds universities, possibly in connection with regional agencies,
- a national establishment which funds research activities of essentially university teams: CNRS (National Center for Scientific Research) in France, DFG (German Science Foundation) in Germany, SERC (Science and Engineering Research Council) in Great Britain, CNR (National Research Council) in Italy, NWO (Netherlands Organization for Scientific Research) in the Netherlands, and so on.

For simplicity, we do not take into account various research contracts submitted by other administrations (defense, transportations, informatics, telecommunications, health) that are limited in Europe and difficult to track down.

Generally speaking, the lack of management structures specific to mathematics does not facilitate the study of national budgets for mathematics. They have to be reconstructed by breaking down budgetary lines and sometimes by careful analysis of the decisions of the various committees in charge of mathematics. The administrative dichotomy makes the task even more difficult insofar as research teams receive funds from both sources even for the same projects. There is no clearcut separation between teaching expenses and research expenses, and this is not necessarily a bad thing when you think, for example, of doctoral studies and computer equipment. The trouble is more in the lack of coordination between the two sources, the lack of harmony and transparency in decisions and the useless double efforts in the evaluation of proposals and teams.

There is another difficulty due to the differences in the organization and status of mathematicians. For example, the Ministries of Education in France or Italy separate research funds, whereas the public financing of British universities by UFC (University Funding Council) includes a component for research that is determined by complex formulas depending on the university, which makes it almost impossible to establish a global budgetary line. The differences in status mentioned above imply that junior permanent positions (*chargés de recherche*, assistant professors) are accounted for as personnel expenses while temporary positions such as postdoctoral grants are included in research budgets. Finally, there are ambiguities in fiscal regimes, infrastructure expenses, and so on, all of which have a significant impact on the volume of funds.

The estimates presented in Table 2 must thus be regarded with caution. We have tried to homogenize the data, but there is too much uncertainty to allow for in-depth comparisons.

We have tried when possible to distinguish between research funds and funds

---

[9] In France, and less so in Germany, we must also take into account the Ministry of Research which funds notably the post-doctoral grants programme as well as research institutions that fall under its tutelage.

for doctoral studies and postdocs. We also included some information about infrastructure expenses corresponding essentially to computer equipment. Let us mention that it was not possible to establish the public research budget for Great Britain and that the sums indicated correspond to the total amount of research funds and contracts from all sources, public, private or international obtained by British universities.

Table 2 demonstrates the importance of budgets that are devoted to making mathematical research attractive again for students. See Table 3 for more details on doctoral scholarships and postdoctoral grants each year in some countries (this corresponds to expenses for three to four years).

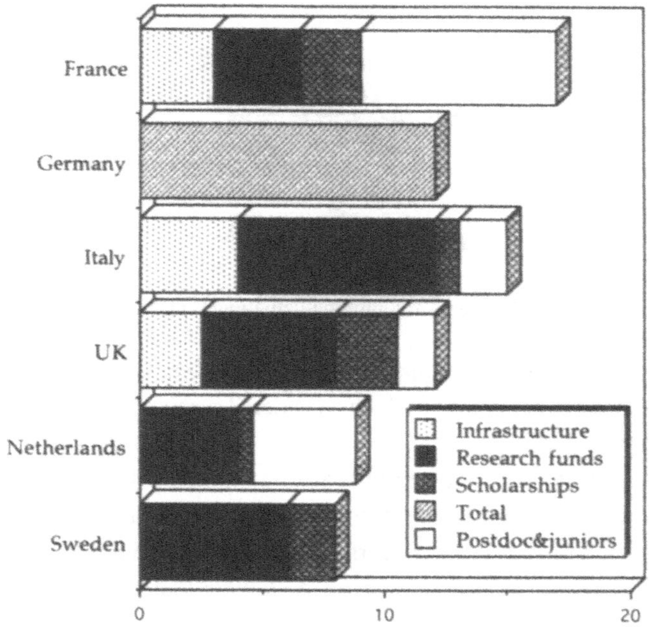

Table 2. Annual public budgets for mathematics (million ECU).

The important size of the French plan results from a reform started in 1988. This reform aimed at correcting an alarming situation characterized by an historic drop in graduates in mathematics (down by a half in ten years) and a difficulty in awarding the 75 postdoctoral grants that were then available. The high number of postdoctoral grants is meant to eliminate one of the main weaknesses of the previous system, that is the absence of a transition between the dissertation and the application for an assistant professorship position[10]. Table 3 shows that the

[10] J.-P. Raoult "La formation par la recherche : l'étape 1989", *Matapli*, 199, pp. 60–64.

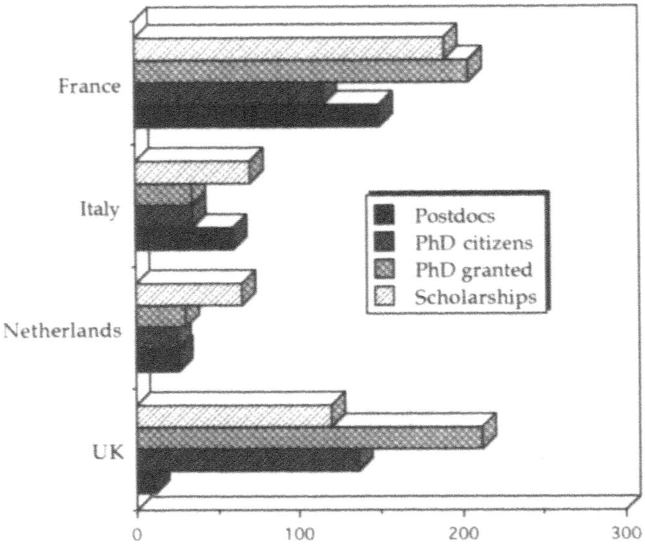

Table 3. Numbers of annual grants for young mathematicians (1990 estimates).

British system suffers severely from this bottleneck effect. This was particularly stressed by the *Kingman Report*[11], which considered it necessary to have a flux of 140 post docs per year.

In general, we do not have any information on the evolution of funding in the various countries. Let us simply mention that the efforts accomplished in France during the last four years concerning training through research have been accompanied by a significant increase in research funds that more than doubled in the same period.

If we assemble all the data from our sample, which concerns more than 10,000 mathematicians for a global population of around 280 million (this is comparable to the American mathematical community), the global annual budget is 85 million ECU. This is well below the $133 million of federal funds in the United States (1989 data) and consequently lower than the revised objective of the *David Report* II[12] which is $250 million per year. This sample however shows that globally Europe is close to the objectives of the David plan concerning young researchers: funding for 1,000 graduate students and 400 postdocs in stationary regime. It thus seems that Europe lags behind on research funds, even taking into account the

[11] *Op. cit.*

[12] *Renewing U.S. Mathematics: a Plan for the 1990's*, Notices AMS, 37, 1990, 813–837.

fact that part of the American individual grant money serves as a complement for the salary of professors (we estimate this sum at $25 million for the 1,900 senior researchers with federal grants).

This diagnosis must not hide the heterogeneity of European situations and ought to be confirmed by deeper analyses. We can fear in particular that the European countries absent from our sample and without a strong mathematical tradition have even lower resources. If we consider the indicator of average financial support, we are still far from 3000 13 ECU per researcher generally considered as the target for mathematics (against twice as much for the experimental sciences). Let us mention in passing that according to information presented by John Chadam, a Canadian colleague who participated in the round table, this average ratio is more or less reached in Canada where the NSERC (Natural Sciences and Engineering Council) has a very active policy that goes beyond the implementation of the goals of the American *David Report*.

In order to establish a topology of research actions financed by public funds in the various European countries, we may distinguish between:

(a) *Support funds*

These funds typically represent more than half of the budgets and often take the form of individual contracts on proposals presented by researchers (responsive funding of the British SERC, individual grants programmes of the German DGF, and so on). This form of individual financing requires an enormous amount of evaluation work that is performed by peer group committees, usually composed of elected or appointed members. During the round table, Professor Carlo Cercignani stressed the tendency of such committees to sprinkle funds, at least in Italy, in order to satisfy their constituencies. He emphasized the importance of serious *a posteriori* evaluation of the quality of the research financed.

France has a different system based on the funding of teams and not individuals. This system encourages the formation of critical masses and facilitates the harmonization of fund allocations corresponding to research and doctoral programmes. Funding is based on four year contracts at the end of which the team is reevaluated.

This type of team financing is part of the recommendations of the *David Report* as an alternative to individual financing, which is traditional in small science. Individual financing by its very nature is able to cover a broad spectrum of themes, but it seems to be less adapted to today's conditions. There is a need to accommodate a massive influx of young researchers and to concentrate efforts within and among universities on certain complex or priority themes. Team financing seems to be more adapted to this need.

(b) *Project based incentives*

These funds include public financing to encourage the structuring of research networks on the national level (research groupings of CNRS in France, RVSM regional groupings of departments of mathematics in the Netherlands) as well as on the European level (*cf.* §3).

The most significant and most exemplary of such programmes are the *Sonderforschungsbereiche* (SFB, special collaboration programmes) of the German DFG, which group together researchers from one university, sometimes in association with neighboring institutions, on a long term (12 to 15 years) interdisciplinary scientific project.

These priority programmes are still bottom-up initiatives since they are selected among proposals submitted by the universities which contribute their personnel and equipment to the projects.

The centers of gravity thus created receive on the average 1 to 1.5 MECU per year from the DFG. The financing is taken up by the Federal Ministry of Education for 75% and by the Länder for 25%. Far-reaching projects can thus be undertaken, including the training and accomodation of young researchers and intense international cooperation[13].

The "Theoretical Mathematics" SFB thus functioned from 1969 to 1985 at the University of Bonn and was the origin of the *Max Planck Institut für Mathematik* founded in 1982. In 1986 there were 3 SFBs in mathematics among the 160 programmes in all disciplines:

— approximation and optimization in applied mathematics (Universität Bonn), followed by the SFB nonlinear PDEs,
— geometry and analysis (Universität Göttingen),
— stochastic mathematical models (Universität Heidelberg).

In 1989, a fourth SFB was added to the list:

— discrete structures in mathematics (Universität Bielefeld).

The British SERC has recently introduced long term thematic initiatives, called *SERC Initiatives*, with an annual funding above 1 MECU for nine years. These top-down initiatives are decided by the Science Board of the Council. There are currently four initiatives in mathematics:

— complex stochastic systems
— nonlinear systems
— image processing
— computational sciences.

Similar top-down projects are in the works, notably in the Netherlands and France, where CNRS has installed an "Interactions of Mathematics" committee.

[13] V. Massow, *Organization and Promotion of Science in the Federal Republic of Germany*, Inter Nationes, Bonn, 1986.

In Italy, CNR has a project for a special five year programme *Progetto Finalizzato Matematica per la Tecnologia e la Societá.* Feasibility studies for this project identified six themes:

- — data, signal and shape analysis
- — mathematical methods for new materials and advanced technology
- — mathematical methods in fluid dynamics and molecular dynamics
- — mathematical methods for complex industrial systems
- — mathematical methods for social and economic systems and environment
- — mathematical methods for chemistry, medicine, biology and ecology.

These projects include industrial cooperation and should receive financing of several MECU per year... if the Italian public deficit reduction programme allows for it.

(c) *Infrastructure funds*

These funds correspond to the indispensable effort to obtain computer equipment for departments of mathematics, which increasingly need high performance graphic workstations for all areas of research. We have tried to indicate an order of magnitude when available. It should be noted that the next generation of microcomputers should deliver this computing power at low prices.

If the prodigious increase in computing power of workstations and microcomputers has made them a tool well adapted to mathematical research, we must not forget that supercomputers still constitute a stimulating area of research. The efficient use of supercomputers requires the development of new mathematical methods. There is a need for intensive computing capabilities, vectorial and massively parallel, that can be used by mathematicians in association with physicists, chemists and specialists of the earth and universe sciences.

(d) *Other*

A non-negligible fraction of research funds is used to organize colloquia and summer schools, to participate in international meetings, to accommodate foreign colleagues and publish preprints. This specificity simultaneously reflects the modesty of global resources and the importance of international communication in mathematical research.

3. European cooperation

These intensive exchanges between researchers constitute a very rapid means of communication. A new idea is made known to a circle of specialists within a few days but will take another year or more to be published. Hence the strong tendency among mathematicians to establish and maintain links with centers of excellence in their domain that are unequally distributed throughout the world.

It is well known that it is extremely difficult to assess the flow of exchanges passing through this dense network of spontaneous cooperation among researchers. We can in any case bet that they are strongly polarized towards the United States, in mathematics as well as in other disciplines, since the United States dominates the scientific communication and has critical research masses with a strong power of attraction.

Without being suspected of advocating European autarchy, we might nonetheless think that the potential for excellence in European centers and their proximity should induce a rebalancing of exchange in favor of Europe. This objective is often present in the international cooperation projects of public research agencies and it is reinforced by incentives from the European Community.

Historically, European research structures have initially favored large facilities in the fundamental science, then large thematic programmes. Since the end of the 1980s, there have been complementary initiatives better adapted to the small sciences such as mathematics. The programmes of the Commission of the European Communities just mentioned are twofold: training and research.

The programmes of the Community concerning mobility of students and educators (ERASMUS, TEMPUS for Eastern Europe) are the subject of a detailed analysis in the "Degrees Harmonization and Student Exchange Programmes" round table report to which we refer the reader. The COMETT programme for associations of university and industry in technological training is evoked in the "Mathematics and Industry" round table. Let us simply recall here that there are less than 40 ERASMUS projects in mathematics in a total of 1,800, corresponding to an estimated 1 MECU financing, that is less than 1% of the global budget of the programme.

We have concentrated on the SCIENCE programme of stimulation of international cooperation and necessary exchanges of researchers in Europe. After an experimental phase in 1983–1984 and a first STIMULATION initiative in 1985–1986, the SCIENCE programme reached its full thrust in 1988 with a budget of 167 MECU over five years. In 1992, the HUMAN CAPITAL & MOBILITY programme takes over, with an approved budget of 488 MECU for the period 1992–1996 with the following repartition:

— 220 MECU for research training fellowships

— 200 MECU for European research networks

— 13 MECU for Euroconferences

— 55 MECU for researchers' access to large equipment.

Starting from the fact that European research is too widely dispersed and that the size of our research centers is too small to organize in-depth projects or very specialized seminars on new subjects, the SCIENCE programme tried to stimulate the creation of European research networks on scientific cooperation projects proposed by the teams themselves. These bottom-up projects also include training through research with post-doctoral fellowships and a mobility aspect with visiting professorships.

As we said in the introduction, the SCIENCE programme devoted 10 MECU to 40 mathematical projects, or 6% of its global budget. We present a more detailed account in Table 4, in which we used a possibly arbitrary thematic classification that seemed relevant to us. We set the "big" projects apart, as well as the Euromath and Space Plasmas Simulation projects which appeared to be atypical in terms of their duration, budget and participating institutions. We then distinguished a group of projects qualified as "pure mathematics" and several groups of application oriented projects, stochastic modeling, computer science, mechanics, nonlinear dynamical systems and engineering:

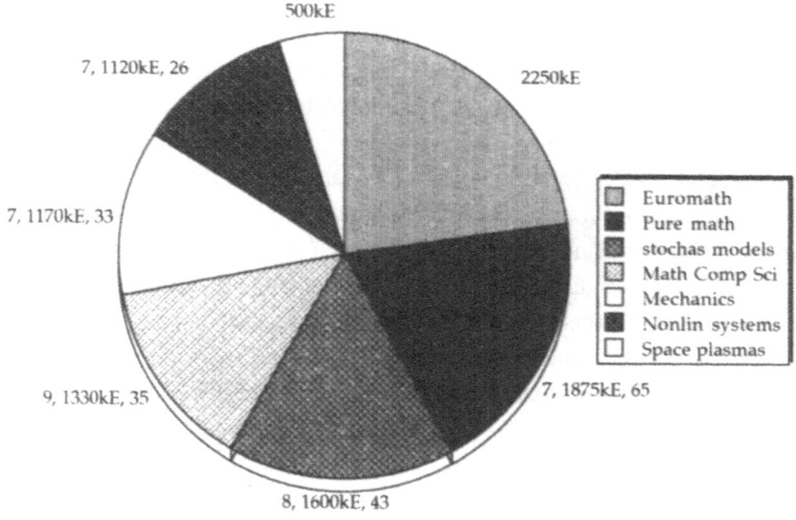

Table 4. SCIENCE projects. (1988–1992): number of projects, funding in kECU, number of participating teams.

For each group, we indicate the corresponding budget, the number of projects and the number of participating teams. The "pure mathematics" project on K-theory has the largest number of participants with 18 teams and a budget of 500 kECU over three years. There are also a few small projects with two or three participating teams. The average number of participating teams is between five and six. This is the optimal size retained for the new HUMAN CAPITAL

& MOBILITY programme, with the condition that the teams must come from at least three different countries.

The projects selected represent only part of the currently hottest or most promising subjects, a non-exhaustive but illustrative list of which is given in the *David Report* II[14]. They represent a little more than 200 research teams, which is still low compared to the European research potential. It would be good if the budgetary increase of the new HUMAN CAPITAL & MOBILITY programme could result in a multiplication of opportunities; this presupposes that mathematicians massively submit quality proposals. We are thus led to the question of programme proposals and their evaluation which were under criticism during the round table discussion. The administrative inertia in proposal processing was heavily criticized. It seems that, paradoxically, the scientific contents of proposals is not sufficiently taken into account. The evaluation criteria are not clear and seem superficial as well as redundant since research teams are also evaluated at the national level. It is worth noticing that the projects often overlap with the above mentioned national thematic initiatives, especially in applied mathematics, and that they probably have already been evaluated.

We have also estimated the annual financing for mathematics from SCIENCE contracts by country, based on the geographical distribution of the 225 participating teams. This division is given in Table 5, which is to be compared with the demographic data of Table 1 and the budgetary data of Table 2.

We may conjecture that the lack of national funding in Great Britain prompted British mathematicians to participate more actively in the programme. The participation of Southern European countries is relatively weak, partially due to a lack of contact with Northern Europe or a lack of experience in proposal.

Before drawing conclusions on European cooperation, we must mention several national initiatives that should fully benefit from the Euro Conference facet of the new HUMAN CAPITAL & MOBILITY programme. These are institutions that regularly host European mathematicians for meetings and congresses or around a given theme: the *Centre International de Rencontres Mathématiques* (CIRM) in Marseille, France, the *Mathematisches Forschungsinstitut* in Oberwolfach and the *Max Planck Institut für Mathematik* in Bonn, Germany, the summer schools of the *Centro Internazionale Matematico Estivo* (CIME) and of the *International Center for Theoretical Physics* (ICTP) in Trieste, Italy.

Other internationally oriented projects are about to begin. They retain the principle of the American NSF for its mathematics research institutes in Berkeley and Minneapolis, namely the organization of research programmes on a given theme for a semester with a good number of visiting participants. The most ambitious of these initiatives has recently been decided in Cambridge. The *Isaac*

---

[14] *Renewing U.S. Mathematics: a Plan for the 1990's*, Appendix B, Recent Research Accomplishments and Related Opportunities, Notices AMS, 37, 1990, 904–1004.

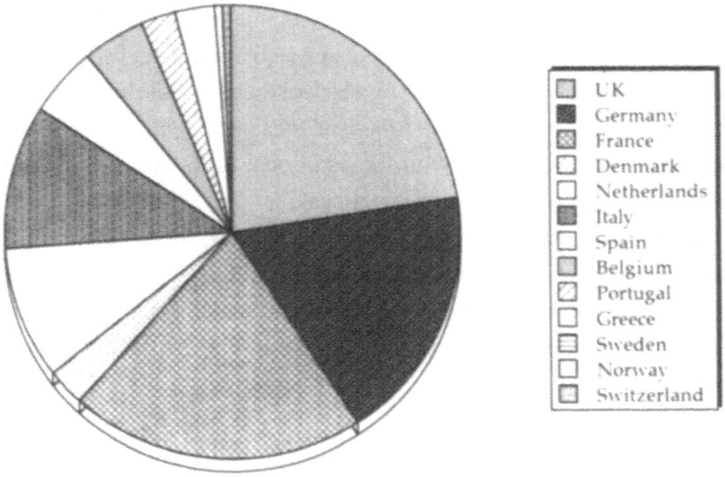

Table 5. Division of SCIENCE funds by country.

Newton Institute will start with an initial budget of 6 MECU over five years from public and private, British and international, sources.

Let us also mention on a more modest scale the *International Centre for Mathematical Sciences*, a joint venture of Heriot-Watt and Edinburgh universities, the project for a *Centre Émile Borel* at the *Institut Henri Poincaré* in Paris, and so on. All these initiatives include training for young researchers. They should offer opportunities for visiting scholars and post doctoral fellows from Eastern Europe and the ex-USSR, while some centers such as ICTP or ICMS are also open to developing countries.

In the special area of supercomputers, we must mention the foundation in Toulouse of the *Centre Européen de Recherche et de Formation Avancée en Calcul Scientifique* (CERFACS), with an annual budget of 5.5 MECU from French and Community sources. CERFACS welcomes researchers, post docs and doctoral candidates from all over Europe.

This general survey of European cooperation is far from exhaustive and we still need to gather more information. For example, our analysis of community action did not include mathematical projects funded by the large thematic technological programmes such as ESPRIT II, RACE, BRITE/EURAM or FUSION. Such common university and industrial projects include a component of mathematical modeling and numerical simulation that sometimes involve mathematicians.

We will return to this point further on.

II. The Objectives of a European Science policy

In light of our analysis of the situation of mathematical research in Europe, it appears that there is a potential for quality distributed throughout various countries, some elements of which are threatened by a lack of means. This diagnosis is similar to that of the *David Report* for the American system in spite of the differences in organization and control and with the additional difficulty of the absence of harmonization among national science policies.

However, if the *David Report*'s objective, namely, the doubling of federal resources over ten years recommended in the mid-1980s, has only been partially achieved, it seems difficult to transpose it in the climate of budget austerity prevailing in the Europe of the 1990s, unless we can find particularly convincing arguments.

1. Priority action for mathematics

Today, mathematics is essential in all branches of science and technology. It also increasingly pervades all aspects of the economy. Our industrialized societies will need more and more mathematical tools and more and more young people who have been trained to use them. This is the basic argument of the David and Kingman reports and the message of the *Mathématiques à Venir* colloquium as well. These documents contain numerous precise examples that illustrate the dynamism and fecundity of mathematical ideas, as well as the fundamental role they have played in the scientific and technological development over the last fifty years[15].

It becomes clear in this perspective that a decline of mathematics caused by a prolonged degradation of material conditions would eventually threaten the other scientific disciplines. Thus a priority effort is required to restore the necessary balance.

Such an objective needs to be supported by a comparative analysis of the various types of financing for different disciplines in each country (aid for training of young researchers, research funds, equipment). It is after having performed such a thorough analysis that the Ministry of Education in France decided to reallocate its research funds in favor of mathematics: 7.6% in 1990 against 4.5% in 1988[16]. This decision was based on an evaluation of teams in collaboration with CNRS

---

[15] This communication effort to make our discipline better known deserves to be pursued and amplified, notably through the *WMY 2000* initiative (World Mathematics Year 2000) sponsored by the International Mathematical Union.

[16] *Rapport de la Direction de la Recherche et des Études Doctorales*, Ministère de l'Éducation Nationale, de la Jeunesse et des Sports, Paris, 1990.

and we take the opportunity to emphasize to what extent the coordination of the various public institutions in France was crucial in deciding and giving impetus to a continuing policy for mathematics.

As we have shown in the first section, this priority aims at disarming in time the "demographic implosion bomb" demonstrated by the age pyramid of European mathematicians in order to ensure renewal for the years 2001–2010. We must attract enough young talents to the field of mathematics by multiplying doctoral fellowships and postdoctoral grants. We must also create career incentives through an appropriate policy of recruiting, promotions as well as incentives at the senior level to continue and direct research. The various budgetary measures thus need to be coherent. We must also take into account the necessity of entirely renewing the pyramidal system of teaching mathematics, both in terms of personnel and methods, for the sake of the education of future generations.

Efforts to catch up must also concern research funds strictly speaking, whose insufficiency has been underscored, the now indispensable computer equipment and the technical personnel that goes with it, as well as libraries.

2. Encouraging the interactions of mathematics

The examples we referred to above show that the current vitality of mathematics is particularly apparent at the interfaces. These interfaces may reside inside the discipline itself where the discovery of more general structures or more powerful methods have set forth new links and have led to the unification of branches that were previously isolated. There is also a growing awareness about the role of applications that increasingly erases the distinction between "pure" and "applied" mathematics. Other sciences nowadays make use of sophisticated and often quite recent mathematical tools, such as nonlinear models in biology or stochastic models in finance. Conversely, several examples of spectacular mathematical discoveries have their roots in other sciences: fractals, wavelets, structure of spaces in dimension 4 and so on[17]. Computer science was born fifty years ago out of the work of a few mathematical geniuses and a few engineers. It still stimulates the need for new mathematics while opening up new directions and new methods of mathematical exploration.

The main impact of computer science and of the explosion in computing capabilities it is undergoing resides in the rapidly growing development of numerical simulation of complex phenomena and systems. This development has profoundly altered the nature and bearing of the mathematics used in industry and services. The intensive modeling effort is thus the source of three areas of interaction for mathematics:

---

[17] These examples are taken from "Interactions des mathématiques" by C. Basdevant and J.-P. Bourguignon, in *Rapport de prospective du CNRS* (1989).

— interaction with computer science through algorithms
— interaction with other sciences through modeling
— interaction with industry and services.

Such fruitful interactions flourished in the United States in the 1980s, due notably to the support of the Department of Defense which allocated $40 million a year to academic mathematical research. The programmes of DARPA (Defense Advanced Research Projects Agency) included the efficient use of supercomputers[18], computational algorithms, dynamical systems, turbulence and signal compression. In the meantime, mathematical modeling played a crucial role in the industrial development of critical technology in micro-electronics or aerospace, again with the support of defense contracts.

There is nothing similar in Western Europe[19], where military support for mathematics, even if it exists in France or Great Britain, is of much more modest dimensions.

In any case, the reduction of defense budgets calls for new strategies. It is particulary urgent for mathematicians to look for new partners in society, in order not to suffer from the same close ties with the military that were established in the United States during and after World War II.

Technological development strategies of the European countries are based more on objectives for improvements in standards of living (health, transportation and, in the near future, environment) and on the quest for competitiveness in the civilian goods market. These strategies offer opportunities, notably in the framework of the large thematic research programmes at the national, intergovernmental or community level. Mathematics has an important role to play in these long-term programmes to nourish and develop the enormous modeling effort they require. Let us simply mention here the problems of climatic changes or the transportation of aircraft of the future. Furthermore, we point out that the "unreasonable efficiency" of mathematics, as the physicist Eugene Wigner said, often leads to high profitability of the initial low investments in terms of the efficiency of the solutions obtained.

Enhancing the interactions of mathematics thus appears to be an especially profitable objective. It requires the creation of interdisciplinary teams and an intensification of modeling efforts in the framework of long term thematic research programmes.

3. Developing links with industry and services

We have already mentioned this objective in our discussion on the interaction

---

[18] Following the recommendations of the Lax Report: *Report of the Panel on Large-Scale Computing in Science and Engineering,* P.D. Lax Chair, NSF, 1982.

[19] To the exception of CERFACS mentioned above, but civilian funded.

of mathematics. We have chosen to elaborate on it because, beyond the dynamism of research, it also concerns two important points: the increase in resources and the adaptation of training programmes.

As the industrial participant to our round table stated in a thought-provoking fashion, industry is not interested in mathematics... but in its utilization in models that lead to improvements in the technical or economic efficiency of industry. The number of mathematicians in industry, either openly declared or "undercover", varies according to branches but still does not reflect the actual importance of mathematical tools. The situation is rapidly evolving due to several success stories, the progressive extinction of the NIH (Not Invented Here) syndrome still present in certain companies and the appearance of an interface culture that facilitates the dialogue between scientists and industrial technical management.

In effect, international competition has constrained industry to engage in permanent technological innovation and forced it to be constantly on the watch for new ideas and to develop relations with exterior research teams.

Even large European firms generally have only local contacts in their country of origin, or possibly through their representatives abroad. They feel the need for a network of relationships with the research world that is more global and better organized. Hence their interest in community programmes such as ESPRIT (computer science and microelectronics), BRITE-EURAM (industrial technology and new materials) or RACE (telecommunications). Industry is paying for half of these programmes, the other half being paid by the EEC. In order to develop links with industry, mathematicians should thus participate in these large technological programmes, for example, in specific basic research facets such as those already existing in ESPRIT II. Another possibility could be research projects in collaboration with industry on pluridisciplinary and fairly long term objectives, initiated in a bottom-up fashion by the partners themselves in a spirit close to that of the EUREKA programme. They should be upstream research-oriented rather than downstream product-oriented.

The other important aspect of the relationship between the mathematical community and firms concerns the scientific training of the future management of industry and services. All analyses agree in foreseeing a strong growth in the demand for young mathematicians on the part of industry. Industry already welcomes trainees and doctoral candidates in its laboratories. We must take this future competition into account in the training plans and maintain a flux of Ph.Ds sufficient to simultaneously ensure the renewal of the academic community and the needs for mathematical expertise by enterprises.

We must, however, also realize that young mathematicians will need other qualifications in their industrial career. A young researcher in mathematics, in general, has no opportunity to be trained in team work, project and instrumentation constraints and budget management, contrary to a doctoral candidate in physics who works in a laboratory. The point is not necessarily to set up a double curriculum including accounting and management courses, but rather to take the measure of the changes in behavior that this recent evolution will bring about. This is also an area where the creation of pluridisciplinary teams can help this transformation and bring more adaptivity and an aptitude to change subjects that mathematicians still often lack. If there is room for very specialized scientists in European industry, the industrial structure is such that it is more difficult for them to change companies in order to improve their career or to adapt to strategic changes than it is in the United States.

More generally, the penetration of mathematical models in every sector creates a strong demand for young graduates in various disciplines with "numerical" qualifications and a mathematical background sufficiently solid to take advantage of modern modeling techniques. What is at stake here for the mathematical community is culturally and scientifically important. It not only implies an increase in teaching duties, but also a profound revision of the goals and methods of mathematical teaching.

4. Reinforcing European cooperation

In our presentation of community programmes such as HUMAN CAPITAL & MOBILITY, we have already explained the importance of cooperation between European teams in order to efficiently organize in-depth research on new subjects. We also showed how the installation of research networks serves the objectives of promoting of doctoral programmes and welcoming Eastern researchers (TEMPUS).

We have just indicated in the discussion on links with industry how the large community programmes aimed at reinforcing the scientific and technological basis and the competitiveness of European industry should give some room to mathematics and encourage interactions.

However, the objectives of European cooperation do not boil down to obtaining more subsidies from Brussels. The action of the Community obeys a few simple rules such as:

— avoiding redundancy and reinforcing cooperation and synergies,

— avoiding having several do what could be better done alone.

This last rule is a common sense variant of what is customarily called the "subsidiary principle".

In the case of mathematics, this means that there is no question that the EEC monopolizes science policy by defining, for example, the optimal size of the mathematics community or by substituting itself for state intervention. We have nonetheless seen that the difficulties of the mathematics community are not solved by national research policies such as those in Germany or Great Britain. In this case, we may imagine that the European Community could play an *incentive* role in initiating the convergence of national policies on objectives that are more appropriate for the development of mathematics, such as those defined in the present section.

This incentive role implies that new mechanisms be set up that would probably be less cumbersome than existing ones. For example, one possibility is that the Commission commit itself to systematically supporting exchange costs (research trips, seminars and so on) of certain European cooperation projects that are conducted within the framework of national programmes (we gave a few examples in Section I). Such procedures would be similar to the *concerted initiatives* of the EEC and would result in less evaluation work since they would naturally rely on national evaluation. They could contribute to the creation of a real common ground for research and should then succeed in mobilizing mathematicians efficiently.

III. Conclusions

The present report tried to present a diagnosis of the situation of mathematics in Europe. The diagnosis showed a large population of professors and researchers whose potential is reduced by the dispersion of efforts and the scarcity of resources. The development of this population appears to be threatened in the near future by an alarming aging problem and by the frequent lack of national policies that are appropriate for the needs of mathematics.

The strategic character of mathematics for the progress of other sciences and for practically all areas of technology is a strong argument in favor of a priority effort for the discipline. Without falling into excessive technocracy, we must nonetheless pay attention to the cost-efficiency ratio in order to define priorities and ensure the coherence of such a policy for mathematics. With these questions in mind, we were led to identify a series of objectives:

— ensure the renewal of academics and researchers,

— encourage interactions of mathematics,

— develop relations with industry and services,

— reinforce European cooperation.

These objectives should be integrated into any policy for research and training.

We must be cautious in assessing future prospects. This is especially true in mathematics where important future applications will probably emerge from unexpected and unforeseeable combinations of methods. It is thus not desirable to concentrate resources too narrowly on special areas and privileged research centers. It seems to us that the best strategy for taking advantage of the variety and potential for creativity of the system consists in bottom-up initiatives in which proposals come from the base and are stimulated by flexible mechanisms that indicate priorities.

Starting from this analysis, we can try to draw a sketch of proposals for a programme. These proposals are intended for national and European decision-making centers, as well as for the scientific community and its representatives. The mobilization of the scientific community is essential in the implementation of an authentic European scientific policy for mathematics.

(1) We have stressed the importance of attracting a sufficient number of new talents to doctoral studies in mathematics in order to ensure the renewal of an aging mathematical population in the years 2001–2010 and also to meet the still largely latent demand for expertise in numerical modeling in industry and services.

It is thus important to encourage students directly through the allocation of doctoral fellowships, but also by improving the attractiveness of academic careers through a corresponding effort in recruiting and post doctoral grants. We must also encourage senior researchers to continue their work actively, to develop interactions with other disciplines and to train and direct younger researchers.

This renewal effort is particularly urgent in certain countries such as Germany and Great Britain. It needs to be continued in countries such as France where it has already begun. It is important to set up coherent programmes concerning all aspects of academic careers. From this point of view, we have mentioned the interest of allocating funds to teams, thus financing at the same time the researchers, post docs, doctoral candidates and part of the equipment.

(2) The creation of European networks in the framework of the former SCIENCE community programme and now of the HUMAN CAPITAL & MOBILITY programme constitutes an interesting opportunity due to their double function of research-training and host of postdocs and visiting scholars, notably those coming from the East. The support for mathematical projects represented 6% of the SCIENCE programme. It deserves to be increased in order to express a real priority for this key discipline and to better cover the wide spectrum of promising themes that demonstrate the current vitality of mathematics.

(3) We have drawn attention to the particularly fruitful role of interaction in the recent advances and future development of mathematics. The vigor brought about by the integration of various branches of the discipline and the explosion in computing capabilities quicken the pace of research and spread the use of mathematical models in all scientific and technical fields.

The large programmes of scientific and technological research should thus give some room to mathematics. In effect, mathematics can nourish these programmes and help in the areas of modeling and numerical simulation where mathematics is required. We refer notably to the creation of interdisciplinary teams on long term basic research themes or even in the framework of industrial cooperation on mid-range projects.

We must thus convince national and community public authorities and potential industrial partners of the efficiency of a continuing effort in favor of basic research that include mathematicians. This research is needed to achieve actual breakthroughs in the large scientific programmes and in the development of critical technology.

(4) The previous arguments show the necessity of mobilizing and the need for coordinating the support for mathematics from various funding sources. This problem, involving various agencies, is to be found in every country as well as in Brussels. We must strive to formulate concerted initiatives with national partners and to envision new projects in the framework of the community programmes.

We thus propose that the European Commission appoint an official representative for the mathematical sciences to play this role of coordination and to implement incentive procedures capable of mobilizing and bringing national policies closer together in favor of mathematics. This representative should also encourage bottom-up initiatives coming from researchers themselves, give more weight to evaluation procedures already performed on the national level, simplify the establishment of contracts and more generally give as much flexibility as possible in the use of community resources.

(5) The implementation of a coherent European policy requires an in-depth knowledge of the state of the educational/research system in mathematics in the various countries. It is thus desirable that the mathematical communities of these countries establish or update surveys on the state of their departments of mathematics, working conditions, available funds and make an assessment of their doctoral programmes and European collaborations. They should also take the recommendations from the other round tables into account concerning the image of mathematics in the general public, the renewal of the teaching of mathematics on all levels, student exchange programmes, collaboration with developing countries, library networks and so on. All these points are objectives for a real European policy.

Acknowledgments. I would like to express my gratitude to all participants in the round table who helped in its preparation and who enriched the discussion with their contribution:

Carlo CERCIGNANI (Politecnico di Milano, Italy)

Christopher LANCE, General Secretary of EMS (University of Leeds, UK)

Hans MUNKHOLM (Odense University, Denmark)

Patrick VALENTIN (ELF, France).

I would also like to thank Claude JABLON, Scientific Director of the ELF-Aquitaine group, who was unable to participate in our session, for the financial support of his company.

I am very grateful to all colleagues who were willing to spend some time answering the questionnaire I had sent to the different mathematical societies in Europe. For the information provided, I would like to thank:

Guiseppe ANICHINI, General Secretary of the Italian Mathematical Union (University of Modena, Italy)

R.Y. SHARP, General Secretary of the London Mathematical Society (University of Sheffield, UK)

Peter SJGREN, President of the Swedish Mathematical Society (Chalmers University, Göteborg, Sweden)

N.M. TEMME (CWI, Amsterdam, the Netherlands)

I also received useful information from J. BENAR (Prague), Zoltan BUCZOLICH (Budapest), Jan BUTKIEWICZ (Warsaw), Luc LEMAIRE (Brussels) and Robert KÖNIGS (DFG, Bonn). I express my gratitude to them all.

My knowledge would have remained even more incomplete without the precious help of Hans FÖLLMER (University of Bonn) and Jean-Pierre RAOULT, *chargé de mission* for mathematics at the French Ministry of Research who made his rich documentation available to me. I also benefitted from information passed on by Claude BASDEVANT (Ministry of National Education, Paris) and Jean-Pierre FERRIER (CNRS, Paris). I can only reiterate my appreciation.

My thoughts on European policy were enriched by stimulating discussions with Jean-Pierre BOURGUIGNON (Ecole Polytechnique, Paris and President of Société Mathématique de France). I am very grateful to him as I am for the encouragements from David CRIGHTON, President of the Mathematics Committee of SERC (Cambridge University, UK).

This work would probably never have been completed without the friendly encouragements and assistance of my colleagues from the Organizing Committee of the Congress, Fulbert MIGNOT, François MURAT and Bernard PRUM, during the preparation of the Congress... and even after.

The usual caveats are particularly in order. None of the persons quoted above is responsible for the errors contained in the present report that result partially from necessary simplifications and from an attempt at synthesis of the data. The opinions expressed and proposals put forth reflect solely the ideas of the author and do not commit either the other organizers of the Congress or the institutions that supported it. On the contrary, the opinions and proposals herein are meant to incite reactions, comments and criticisms that will be extremely welcome.

Round Table K: July 8, 1992

COLLABORATION WITH DEVELOPING COUNTRIES

by Pierre Bérard
Institut Fourier, Université Grenoble 1, Grenoble (France)

The round table *Collaboration with Developing Countries* was held on Wednesday July 8, 1992, in the Amphithéâtre Descartes of the Sorbonne.

Long before the round table, invited speakers were approached and asked to make written contributions. It was also suggested that they might invite interested colleagues to contribute written remarks as well. At the same time, European mathematical societies were requested to provide reports on collaboration projects in their own countries. The invited speakers met shortly before the congress both to coordinate their presentations and to prepare recommendations for future undertakings.

The round table was organized as follows: Presentations were first given by M. S. Narasimhan, A. O. Kuku, R. Labarca and A. Verjovsky, after which there were questions and comments invited from the floor. Then, A. Pelczar, J. S. Martins, J. Eells and J.-M. Lemaire were heard. followed by a few remarks from P. Bérard, and again, further questions and comments were elicited from the floor.

M. S. Narasimhan gave several reasons why developing countries need to promote research in mathematics. He also made comments drawn from his own experience in India and as chairman of the Commission on Development and Exchanges of the International Mathematical Union. A. O. Kuku reported on the current difficulties of African countries regarding mathematics and made specific proposals to European countries. R. Labarca reported on the existing collaboration between European countries and Latin America. He then made very precise proposals for improved collaboration between Latin America and European countries concerning training and research in mathematics, as well as support for libraries. A. Verjovsky reported on ICTP and proposed a fellowship programme supported by the European Community to fund visits by European mathematicians to ICTP and by ICTP staff to European centers. A. Pelczar commented on his experience collaborating with Africa. J. S. Martins reported on collaboration programmes between Portugal and developing countries, as well as on the recent project for an international center for mathematics (CIMAT). J. Eells gave an account of ICTP and ICMS. He insisted on mathematics in industry and included suggestions for the future. J.-M. Lemaire talked about ICPAM. He pointed out the importance of electronic mail in breaking isolation and allowing access to scientific information. He added comments on the possible role of Europe.

This report is based on the written contributions by the invited speakers, mathematical societies or individuals, and on the taped proceedings of the round table. These contributions do not attempt to give an exhaustive description of the problems in developing countries. This report does show, however, that the question of collaboration with developing countries deserved the attention it received at the first European Congress of Mathematics.

One of the goals of the round table was to make recommendations for future undertakings. Recommendations given in Section A have been drawn from the written contributions, from discussions during the preliminary meeting as well as from the round table itself. We hope that these recommendations will serve as guidelines for *ad hoc* committees to implement concrete plans, and, in particular, on the European level.

Sections C through F contain the presentations made during the round table in the order in which they were given. Some speakers only sketched their point of view during the discussion and amplified their remarks in more detail in written contributions. These contributions appear in the appendices. Appendices I and II also contain reports which were not discussed during the round table. Information on ICMS, ICPAM, ICTP and CIMAT are to be found in Appendix III.

Acknowledgements: The organizer of the round table wishes to thank the Organizing Committee as well as UNESCO-SER (Paris) for their financial support. He also wishes to thank all colleagues who participated in the round table or who sent comments and contributions.

TABLE OF CONTENTS

I. Recommendations for Future Undertakings 349
II. List of Invited Speakers 353
III. Contributions by Invited Speakers, I 353
IV. Contributions from the Floor, I 365
V. Contributions by Invited Speakers, II 367
VI. Contributions from the Floor, II 372
Appendix I. Reports by Mathematical Societies 374
Appendix II. Individual Reports 381
Appendix III. Information on ICMS, ICPAM, ICTP, CIMAT 394

I. Recommendations for Future Undertakings

This section contains a summary of the main recommendations made by round table participants.

General observations

(1) The international community or organizations cannot be a substitute for the developing countries, i.e., their governments, decision makers or mathematicians, in fostering the development of mathematical research. Nor can they put a stop to the brain drain of young scientists.

(2) What the international community or organizations can do is to help train, educate and support mathematicians in developing countries. The specific circumstances of a given mathematical community, however, should be taken into consideration. Support should only be awarded to those communities who meet high standards regarding merit and scientific quality. There cannot be double standards.

(3) As far as possible, any support granted should be free of bureaucracy and awarded directly to individuals or local groups. The activities of grantees must be submitted to periodical scanning.

(4) This control could be made by *ad hoc* scientific committees who also receive the applications in the first place.

(5) Developed countries, particularly in Europe, could try to encourage developing countries to invest in basic science by offering a means of alleviating their debt in exchange for investment in basic science: **debt for science** (e.g., see below).

Future undertakings

1. An awareness of the importance of mathematics
2. Training
3. Research
4. Communications, libraries and scientific documentation

Comments

1. An awareness of the importance of mathematics

As mentioned above, the development of mathematical research is strongly dependent on the support of local governments. Mathematicians in developing countries and their colleagues in developed countries must combine their efforts in making governments aware of the importance of mathematics in all its aspects, cultural, educational, technical and scientific. Efforts must also be made by developing countries to stop the brain drain of young scientists, which implies good working conditions, reasonable salaries and career opportunities.

2. Training

2a. Master's level training

In many developing countries, there is a dire lack of mathematics teachers. Therefore, it is important to create and sustain training programmes at the Master's degree level. The first goal must be to train enough teachers. The second is to select gifted students to continue on for a Ph.D either in a local research center or abroad.

Governments of developing countries need to provide the basic means for developing such programmes. The international community could help by providing scientific advice and professors to run the courses.

2b. Doctoral and postdoctoral training

In order to help alleviate brain drain, see point 1, it is important that talented students from developing countries be able to study for their doctorate at advanced universities abroad and then return to their home countries. Easy access to post-doctoral fellowships on a regular basis is important for recent Ph.Ds so that they might keep in touch with ongoing research. They also provide a motivation to return to their home countries.

It is equally important to keep doctoral students in mathematics at universities in developing countries. It would be good if these students could benefit from regular visits by reputable mathematicians. They should also have the possibility of spending six months to a year in advanced research centers abroad during their studies.

3. Research

It is important to develop high quality local research groups. This can be best achieved by concentrating on a few subjects and grouping the researchers themselves in only a few centers (so that a critical mass be reached). Such research groups would benefit from regular visits by reputable mathematicians. Their members should have the opportunity of visiting advanced centers all over the world. Collaboration between research groups in developing countries should be encouraged.

Developing countries should also be encouraged to organize regional or international conferences so that their researchers might meet with their colleagues and keep abreast of recent trends in their fields of interest.

4. Communications, libraries and scientific documentation

One of the main difficulties for mathematicians working in developing countries is isolation and lack of contact with active research groups of reasonable size. Lack of access to research libraries is also an important problem. It would be ideal of course if we could construct, maintain and develop regional libraries. This goal, however, seems unlikely to be achieved in the near future due to lack of funds and a supportive infrastructure. We should therefore seek alternative solutions.

4a. Identify mathematical centers with an already existing library who could receive books and journals either through individual donations and support or book donation programmes organized by international institutions.

4b. Connect mathematical centers in developing countries with reference libraries in order to facilitate access to scientific information. This supposes appropriate financial support.

4c. Persuade reputable publishers to sell books at highly reduced rates or to publish cheap editions. This could be limited to specific basic textbooks. Books could also be donated by publishers to some libraries identified in advance by *ad hoc* committees.

4d. Promote access of developing countries to electronic mail networks because electronic mail is a very efficient means of communicating mathematical materials, sending TEX files or digitalized data.

Comments

Concerning points 1, 2a and 4d, international organizations and, in particular, the European Mathematical Society, could help by addressing recommendations to governments, to the European Community, and so on.

Concerning points 2b and 3, these organizations could help by creating *ad hoc* committees to screen applicants for fellowships and grants and to make periodic checks on the programmes involved.

Concerning points 2a, 4a and 4d, international organizations and EMS could help choose centers to implement the proposed programmes. Concerning point 4d, they could help select books and libraries and approach publishers.

Specific programmes: general comments

The proposed programmes should have a sufficient duration, for example 5 years for research or library programmes and up to 10 years for educational programmes. This will allow time for their impact to be analyzed.

Preferably the programmes should be organized on a small scale basis, i.e., individual to individual, center to center, and so on, and as exempt from bureaucracy as possible.

Support should be awarded by *ad hoc* committees, according to merit and scientific quality. These same committees should also periodically evaluate activities.

A. *Visiting professorship programme*. This programme would provide travel expenses for research visits to centers, living expenses of the visiting mathematician being paid by the center. This funding should apply to mathematicians visiting research centers in developing countries as well as to mathematicians from developing countries visiting a center in a developed country: see points 2b and 3.

B. *Training professorship programme.* This programme would provide travel expenses and salaries to mathematicians working in a training curriculum (e.g., a Master's degree programme) for an extended period of time, e.g., at least six months. This programme should apply to mathematicians from both developing and developed countries. The curricula should be specified in advance: see point 2a.

C. *Ph.D fellowship programme.* This programme should provide travel and living expenses for students from developing countries studying for Ph.D studies in a quality university: see point 2b.

D. *Visiting doctoral student programme.* This programme is intended to support travel and living expenses for doctoral students from developing countries to study in high quality universities for six months to a year to prepare their dissertation: see point 2b. There must of course be some agreement between the institutions involved. One possibility is to have two advisors, one on each side. Another possibility is to have the home institution participate in some European research network.

E. *Postdoctoral visiting programme.* This programme would offer travel and living expenses to recent Ph.Ds to make a three month visit per year over a four year period to high quality research centers: see points 2b and 3.

Note: Programmes C–E could also apply the other way around: for example, recent Ph.Ds from developed countries could be offered fellowships to visit research centers in developing countries, and so on.

F. *Support to conferences.* This programme would provide support to conferences organized in developing countries, especially grants to cover travel expenses of invited speakers and participants from other developing countries: see point 3.

G. *Associate professorship programme.* This programme would offer associate professorships to mathematicians from developing countries for three months a year on a five-year basis, including travel expenses and salary. These professorships should be offered in leading universities: see point 3.

H. *Book donation programme.* As outlined in points 4a and 4c.

Final comments

It is clear that the European Mathematical Society could promote projects such as those described above on the European level. In doing so, we must remain realistic and choose one or two points, then develop these points on a small scale before proceeding further. The EMS could encourage the European Community to use part of the funds in the Human Capital and Mobility programme, for example 10%, to link mathematicians from developing countries to the European Research Networks. European mathematicians belonging to such networks could also try to find further funds within their own countries to link up with colleagues from developing countries.

Certainly, EMS should create a sub-committee to pursue these matters.

II. List of Invited Speakers

— J. Eells, Chairman Mathematics Department, ICTP—International Centre for Theoretical Physics—Trieste, Italy and University of Warwick, UK

— A. O. Kuku, University of Ibadan, Nigeria. President African Mathematical Union

— R. Labarca, Universidad de Santiago de Chile, ELAM: Escolas Latino Americanas de Matemática

— J.-M. Lemaire, EMS and CIMPA—Centre International de Mathématiques Pures et Appliquées—Nice, France

— J. S. Martins, Departamento de Matemática, Coimbra, Portugal

— M.S. Narasimhan, President of the CDE—Commission on Development and Exchanges—of the International Mathematical Union

— A. Pelczar, The Jagiellonian University, Krakow, Poland

— A. Verjovsky, Mathematics Department, ICTP— International Centre for Theoretical Physics—Trieste, Italy

III. Contributions by Invited Speakers, I

M.S. Narasimhan*

Why is research in mathematics important for developing countries? While most people agree that education and research in science are necessary for economic and cultural development in developing countries, there are some who doubt whether research in mathematics, or for that matter any research in any basic science, is necessary for a developing country. The experience of agricultural development in India for example shows that the contribution by pure scientists is very important and not at all negligible. Even if you simply want to adapt a given technology to the needs of a country, you need basic science to accomplish it. It cannot be done just like that. Also, at a time when there is a lot of talk about intellectual property rights, and so on, it is obvious that Third World countries should see to it that they have an infrastructure in basic sciences.

---

* M. S. Narasimhan is President of the Commission on Development and Exchanges of the International Mathematical Union (CDE–IMU) and vice-president of the International Center for Pure and Applied Mathematics (ICPAM–CIMPA) in Nice (France). Beginning October 1st, 1992, he became Chairman of the Department of Mathematics of the International Center for Theoretical Physics (ICTP) in Trieste (Italy). The following text has been transcribed from his notes and from the tape recorded during the round table.

More specifically, mathematics is the vehicle of exact scientific thought and provides the basic language for understanding nature. Because of this, it is clear that the scientific basis of a country will not be solid unless mathematics is well developed. As everybody knows, such highly advanced mathematics is being used in all kinds of so-called applied subjects that without mathematicians a country will never understand what is going on and be completely lost.

There is another aspect which I may call a psychological and cultural aspect for promoting research in mathematics. International achievement in the sciences at the highest level gives the general public confidence in the capacities of a country. It is important to notice that you can have a high level of achievement in mathematics in a developing country with a very modest financial investment as opposed to many other sciences. To give an example, in India the fame of Sriniva Ramanujan gave tremendous confidence in the intellectual potential of the nation to the people at large. In fact, you can trace the development of scientific research in India in this century to the catalytic effect of Ramanujan's work. Many of the great scientists in India who received Nobel prizes and so on were indirectly influenced by Ramanujan. This is something that is little understood but is absolutely important. Again I emphasize that it is probably easier to succeed in mathematics with much less investment than in any other field. That is one more reason why research in mathematics is important for developing countries.

Obviously the major efforts to develop mathematics should come from the developing country itself. Governments must make a firm commitment to develop mathematics and to lend support to mathematicians. Mathematicians themselves in developing countries need to create mathematical groups.

It is true that international institutions and the international mathematical community can play an important role, but it can only be a supporting role. They cannot simply create mathematics in Third World countries, it has to be done by the people who are living in Third World countries and who of course should draw upon the support of the international mathematical community and international institutions.

I will use the example of India to show how this can be done. We are fortunate to have very good support even in pure mathematics. In fact, about eight years ago the government of India set up an organization called NBHM (National Board for Higher Mathematics). The NBHM is only for mathematics and as far as I know, no other country has a funding organization only for mathematics. What has the Board done? It has set up approximately four regional libraries (more are forthcoming), while lending support to single groups of mathematicians. There is also a visiting professorship programme to invite foreign mathematicians (travel, stay in India and travel in India are all met by the Board). The NBHM also gives support to mathematicians to travel to conferences. More importantly, I think, there are very attractive scholarships for students, even at the high school level, to pursue study in mathematics. It suffices for someone to say "I want to study mathematics", and there will be some support. The studies need not specifically lead to research.

There are other projects as well. Having some special organization whose primary purpose was the development of higher mathematics may be one way that some countries can try. I must emphasize here that the contribution of the Indian government has been very substantial. In principle we can even create new research institutes. We have not so far done this for other reasons but in principle this is possible. And this is for mathematics in all its aspects, applied or not.

How can international organizations or mathematical community help mathematicians or mathematics in developing countries? They can help in the formation of talented young mathematicians and in sustaining the formation of groups as well by providing them with means of continuous contact. Again I will give an example from India and more precisely from the Tata Institute of Fundamental Research. When efforts were being made some forty years ago to build up a school of mathematics at Tata, we were extremely fortunate in having a stream of eminent French mathematicians visiting the Institute and initiating students in modern fields of mathematics. Some of these mathematics students also spent time in France as postdoctoral fellows. There is no exaggeration in saying that these contacts played a crucial role in the development of mathematics at the Tata Institute and as a consequence in India. This gives a concrete example of how outside support can aid in the formation of mathematicians and in sustaining contacts so that they are able to continue on.

Let me conclude by saying a few words about the Commission on Development and Exchanges (CDE). This is an organization set up by the International Mathematical Union to help mathematics and mathematicians in developing countries. I mention here some of the activities of this commission. One is awarding travel grants for mathematicians from developing countries to visit any center of mathematics anywhere in the world, or for mathematicians from developed countries to visit centers of mathematics in developing countries. This is our first programme and it is supported by UNESCO.

A second activity is support for conferences which are held in developing countries. Thirdly, the CDE has selected various groups of good mathematicians in developing countries and has given these groups direct support, for example funds they can use to invite people, travel or buy books. This is something that we have tried and which has been quite successful.

The funds at the disposal of CDE are limited but it is amazing how useful they are. A thousand dollars may not seem much, but it makes a big difference for the persons CDE supports. We get the funds from the International Mathematical Union (IMU), from UNESCO, from important contributions by the International Council of Scientific Unions (ICSU) these last two years and also from mathematical societies of different countries. Our experience shows how even with limited funds at one's disposal visible effects can be achieved.

M. S. NARASIMHAN
Mathematics Department
International Center for Theoretical Physics (ICTP)
P.O. Box 586
34 100 TRIESTE (Italy)

A. O. Kuku*

Problems facing mathematical research and education in Africa

Among the various problems preventing rapid progress in mathematical research and education in Africa are the following:

(1) *Inadequacy of journals, books, teaching and research facilities.* It cannot be overemphasized that scientific research can only be productive with adequate journals, books and facilities. Most mathematical research in Africa is carried out in universities where libraries, teaching and research facilities are less than adequate. There is an acute shortage of good textbooks and teaching aids at all levels. Books, where available, are so expensive that students cannot afford them. Consequently, students do not understand the first principles of a subject. They become disillusioned and especially in the mathematical sciences come to hate the subject.

(2) *Poor preparation and shortage of teachers in the mathematical sciences at all levels.* Most of the mathematical sciences departments in the universities are inadequately staffed in quality and quantity. The relatively few lecturers on the spot carry heavy teaching loads to the detriment of their research and teaching efficiency. In the schools and teacher training colleges, many professors do not have specialized degrees in the subjects they teach. Thus the quality of instruction is adversely affected at all levels.

(3) *African mathematicians work in relative isolation.* Mathematicians are spread thin in disciplines all over the continent. Research groups by disciplines are virtually nonexistent in most of the universities. It is difficult to reach critical mass and to have people work together in only one place. Regular communication with their colleagues abroad is an uphill task. Right now most universities in Africa do not have electronic mail, faxes and so on. Where fax or telex exist, it is still not possible to do mathematics by telex and there is a limit to what you can do with fax due to expense. Until recently when regional and subregional scientific organizations in Africa became active, there was very little interaction

---

* A. O. Kuku is President of the African Mathematical Union. The following text contains his written contribution before the round table. It has been slightly edited in accord with the tape made during the round table.

between African mathematicians, especially because contacts abroad have been along colonial lines, French, English, Portuguese, and so on. The African Mathematical Union is doing its best to alleviate the situation.

(4) *Inadequate financial support for research in mathematics.* Mathematical research is inadequately funded throughout Africa. This cannot be overemphasized because many of the countries treat mathematical research as a marginal activity. With problems of foreign exchange common to many African countries, importation of books, journals and equipment have become astronomically expensive. The burden of debt in most African countries has also made developmental effort impossible since the meager resources available are used to reimburse debts.

(5) *Prospective mathematicians are attracted to more lucrative professions.* Records available in African universities indicate that most of the talented students these days opt to go into professional programmes like medicine, law, engineering, and so on, because there is no money in mathematics. Consequently, mathematics departments have to content themselves with the left-overs. This compounds the problem of producing the required critical mass of mathematicians.

What can the European mathematical community do to help?

(1) *Improvement of teaching and research facilities.* The European mathematical community could help, in collaboration with ICSU (International Council of Scientific Unions), TWAS (Third World Academy of Sciences), IMU-CDE (International Mathematical Union – Commission on Development and Exchanges) and all other interested organisations to improve research and teaching facilities in the form of recent books and journals. Reputable publishers should be contacted to sell books to African countries at highly reduced rates. The devaluation of the currency in most African countries has made it difficult for many libraries and individuals to purchase current books and journals. Also, EMS (European Mathematical Society) should join hands with organizations mentioned above in their massive book and journal donation programmes for the developing countries.

(2) *Need for centers of excellence in Africa.* There is a dire need to develop good centers for mathematical research in Africa. The best way to help is not just to bring isolated researchers to Europe to spend many years and get a degree (and then maybe not go back). There should be good centers on the continent itself. The African Mathematical Union has already identified some eighteen universities and centers that could be developed all over the continent. For this programme to materialize, mathematical societies in Europe should sponsor visiting professors to these centers to give lectures, seminars, and so on on a regular basis. Also, graduate students from African centers must spend a period of six months to a year in centers of excellence in Europe before finishing their thesis in African universities. These centers could be identified by EMS beforehand. One important component of combatting brain drain from Africa to the developed countries is to

ensure that mathematical research and training flourish on the African continent. One way to achieve this is to de-emphasize students studying abroad for a long time to work on degrees in European universities. A comprehensive programme of visits by African postgraduate students to centers abroad would go a long way toward alleviating the situation. There must also be a programme encouraging mathematicians from African countries to visit centers abroad in order to improve their research capabilities.

(3) *Need for improvement in communication.* African countries should be accommodated into the "Euro-Math" communication programme whenever possible. If possible, EMS should initiate a preprint/abstract service that would provide information on current mathematical literature. This service should not be directed only at the European mathematical community but also for the developing countries.

(4) *Need for political pressure in Europe to alleviate the burden of debt in African countries.* The developmental effort of developing countries is being seriously hampered by their spending their meager resources on reimbursing debts. It would help if EMS would join its voice in the current crusade to find ways of alleviating the burden of debt in developing countries.

Specific proposals

What can we propose on the European level that is not feasible for individual countries?

Experience over the years indicates that individual European countries would rather strengthen their ties with their former colonies while providing token support to other developing countries. Moreover, the current practice of many African countries looking up to their former colonial masters for support in various directions—technical, scientific, economic—has hampered South-South cooperation along these lines. Thus, looking at African problems, or indeed at problems in developing countries, from a (global) European point of view is likely to lead to realistic solutions for the global benefit of the continent.

Among the various areas where global solution to the problems are desirable are the following:

(1) *Visiting professorships to Africa and placement of research mathematicians in centers in Europe.* This is most likely to be effective on the European level. For one thing EMS is likely to know what is best in Europe and who is most appropriate in what field, taking the various European countries into consideration.

(2) *Book/journal distribution to developing countries.* This could also be organised through ICSU and so on on the European level especially as regards approaching reputable European publishers to ask them to sell books and journals at highly reduced rates or, in some cases, to give them free of charge to some identified developing countries. Many of these publishers will have much more respect for

requests coming from joint efforts than for those from individual countries. Also, fund-raising for the actual distribution of books and journals will be more effective on the European level.

(3) *Global campaign for debt relief for developing countries.* This campaign can be done most effectively on the European level.

Aderemi O. KUKU
Mathematics Department
University of Ibadan
IBADAN (Nigeria)

R. Labarca*

I would like to emphasize the kind of support our countries in Latin-America need to further their scientific development. I think it is essential to have co-operation programmes to obtain, maintain, develop and benefit from scientific knowledge in our countries.

It is worth pointing out that the development of mathematics in our countries is not homogeneous. This depends very much on the interest of the government in developing basic sciences, on the economical and political situation, and also on the existence of people with political and academic authority who are interested in developing this area.

It would be very useful, in order to make the EC (European Community) cooperation fruitful for our countries, to prepare a report on the current state of development of mathematics in the region. During the preparation of this document, we should take the advice of well-known mathematicians and from prestigious institutions.

What are European countries doing to aid in our scientific development?

At the moment there are several initiatives sponsored by European institutions and countries, most of them in the form of cooperation agreements and scientific exchange programmes, which have proved to be very important. Some of these initiatives have been significant for mathematics and I would like to mention them here:

French cooperation: A considerable number of Latin American students have obtained their doctorates in France through cooperation agreements. Many of them have returned to their home countries and on occasion return to France for brief visits to continue their research. Similarly, French mathematicians have visited Latin America and have involved themselves in the activities of local research teams.

---

* R. Labarca has been very much involved in the organization of ELAMs (Latin American Schools of Mathematics). The following text essentially reproduces his written contribution.

A large number of conferences, workshops and schools have taken place in Latin America and have proved to be very useful in improving the research activities of local groups. In addition, French cooperation has included the contribution of advanced textbooks and journals to improve libraries.

Agreements with CNRS: The leading countries in mathematics in Latin America, Brazil and Mexico, have benefited from agreements with CNRS (Centre National de la Recherche Scientifique – France) for at least fifteen years, thus enabling them to increase the research activities of their more advanced teams. Argentina, and more recently, Chile, have had the opportunity of participating in exchange programmes with CNRS.

Research cooperation with Latin America within the scope of the French military service: This programme provides a continuous flow of young French mathematicians to Latin America, providing highly qualified lecturers for graduate programmes and giving rise to first rate joint research programmes. Such cooperation has been particularly beneficial to Argentina, Brazil, Chile, Mexico, Uruguay and, I believe, to France as well.

German cooperation: Cooperation agreements have been established mainly by the DAAD (Deutscher Akademischer Austauschdienst). A number of Latin American students have obtained their Ph.Ds in Germany on related programmes through such agreements. The most effective programmes consisted in sending German mathematicians to Latin America for two- or three-year periods to collaborate in undergraduate and graduate programmes. The DAAD and the GMD (Gesellschaft für Mathematik und Datenverarbeitung) have agreements which include brief visits in Germany as well as support for conferences, laboratory equipment, donations to libraries (advanced textbooks and journals) and postdoctoral fellowships. We must also mention the A. von Humboldt Foundation which has a programme of scientific exchange which continues to provide support for researchers involved in the programme even after their departure from Germany.

British cooperation: These agreements have been established by the British Council and include doctoral studies, brief visits in Great Britain and support for conferences. A visitors programme has recently been established with the support of private foundations. In general, British cooperation has been rather limited.

Spanish cooperation: A cooperation programme has recently been established with Spain. It is rather limited, but is expected to grow.

Cooperation with the Italian government and the International Centre for Theoretical Physics in Trieste: An important cooperation programme has been established by the Italian government, the International Atomic Energy Agency and UNESCO, through the International Centre for Theoretical Physics. This cooperation has been particularly valuable in less developed countries. Students and professors have participated in courses and advanced workshops, collecting information on the latest developments in mathematics. Scientists and researchers have likewise participated in periodical meetings (e.g., on dynamical systems), thus keeping abreast with the latest developments in their field. They have also carried out various research programmes. As a matter of fact, ICTP has played an important role in association with prestigious local institutions, especially IMPA, the internationally praised Latin American center for mathematics, in forging the establishment of excellent research groups in Brazil, Chile, Mexico, Uruguay, Venezuela, as well as in Portugal.

The ICTP has a small postdoctoral fellowship programme, which attracts much interest because it enables fellows to pursue various activities in Europe.

Finally, mention must be made of the agreements between the Brazilian and Peruvian Mathematical Societies, to encourage the development of science in Peru. This agreement has been supported by ICTP.

Some ideas about our needs

The first necessity is the promotion and execution of a permanent policy to support scientific activities in Latin American countries. In the light of our experience, the support policy must be translated into projects which will permit the multiplication, preservation and application of basic scientific knowledge for the benefit of our countries. In relation to the above, I wish to express the following:

a) In 1967, a Master's degree programme was established at the Universidad Técnica del Estado in Santiago de Chile. Its primary aim was to improve the education of mathematics lecturers and help them reach a Master's degree level. Between 1967 and 1978 (the final year of the programme) about 250 students completed this programme and some of them pursued doctoral studies. About fifty received a Ph.D and some thirty-five are working in Chile at present.

This programme, together with other initiatives, dramatically improved the teaching of mathematics in Chilean universities and secondary schools. It was successful as a result of the commitment of local authorities and due to the funding granted by the Ford Foundation for fellowships, remuneration for American mathematicians on long-term visits and support for libraries.

I cite this example because I think there are many countries in the region that urgently need to improve the competence of their mathematics lecturers. They also need modern, up-to-date textbooks, journals and computer equipment. In these countries there is a number of mathematics researchers who would benefit from basic scientific exchange because their own funds are very limited. Some Latin American countries do, however, have some means at their disposal to provide their teachers with at least a Master's degree. In any case, there are several areas in mathematics that are not developed at all or where there is only a very reduced number of researchers. In these areas it is absolutely essential to send students abroad for their doctorate degrees.

b) Natives of some countries who do have a doctoral degree ought to be encouraged to return to their own countries (e.g., Colombia and Peru), including young scientists living in the northern hemisphere or in other countries of the continent (e.g., Argentina, Brazil, Chile, Uruguay). The Argentinean and Peruvian examples are dramatic.

c) Another basic necessity is a permanent policy to support the preservation of qualified human resources in Latin American countries. A continuous emigration to the northern hemisphere or within the region is caused by the unstable economical and political conditions in our countries.

d) We require larger and better libraries with recent textbooks and at least one hundred titles in scientific journals.

e) Finally, Latin American industry needs the creation of a highly qualified center for the applications of mathematics to the production of new technologies and new products.

What should the European Community contribution be?

First, I want to make some comments concerning the execution of any cooperation programmes for Latin America:

a) They ought to be free of bureaucracy.
b) Applications for these programmes must be independent of local research councils.
c) The decisions must be discussed with authorized scientific committees.
d) Scientific exchange should be on the basis of quality and merit on an international level.

Concerning specific programmes

1. Human Capital and Mobility: Add extra funds to enlarge the EC programme "Human Capital and Mobility" (or create a new one) in order to allow for the inclusion of the following countries: principally Argentina, Brazil, Chile, Mexico, Uruguay and Venezuela (refered hitherto as ABCMUV countries). The following aspects should be included in this programme:

— *Doctoral and postdoctoral fellowships:* This programme should provide fellowships for researchers at the doctoral and postdoctoral levels to work in a host institution for a period of twelve to thirty-six months (doctoral level) and of six to twelve months (postdoctoral level). In this programme, EC nationals would apply in an ABCMUV institution and ABCMUV nationals in an EC institution.

— *Research networks:* Such networks should involve at least two EC and two Latin American teams. The purpose is to develop research in a specific direction.

— *Support for conferences:* This programme would enable EC nationals to participate in conferences in Latin America and ABCMUV nationals in European conferences.

2. Mathematics library-journals programme: The aim of this programme is to contribute to the scientific development of Latin America by completing a collection of basic journals in mathematical science. This programme would provide complete series of the main European journals for the ABCMUV countries. This could be done according to the following: three collections in Brazil distributed by the Brazilian Mathematical Society, two collections in Mexico distributed by the Mexican Mathematical Society; two collections in Argentina distributed by the Argentinian Mathematical Union; one collection in Chile distributed by the Chilean Mathematical Society; one collection in Uruguay at the Universidad Nacional de la República and one collection in Venezuela distributed by the Venezuelan Mathematical Society.

3. Short-term visit programme: This programme would provide the possibility for researchers from EC countries to visit Latin American institutions and vice-versa.

4. EC fellowship programme: This programme would provide twelve fellowships at the Ph.D level and six fellowships at the postdoctoral level for Latin American nationals to apply to ABCMUV institutions.

5. EC Scientific aid programme: This programme is an EC contribution for the establishment of two long range Master of Science programmes in mathematics. The first one to be addressed to Central America nationals and the second one to South America nationals, not in the ABCMUV countries. One aim of this programme is to select young people who would be able to obtain a doctorate, either in Latin America or in Europe, and then must return to their home country. The other is to provide systematical improvement of the competence of lecturers in mathematics. Both programmes ought to be applicable for ten years or more throughout Latin America with basic support conditions such as an appropriate building, a basic library (with modern textbooks and journals), professors involved in the programme with a solid formation and committed to performing well.

The European contribution could consist of fellowships, the employment of European and Latin American professors for one or two years, the maintenance of a basic library and a number of fellowships for bright students to pursue doctoral studies either regionally or in Europe in accordance with previously proposed programmes (see 1 and 4 above).

6. Research Centre for Applied Mathematics in Industry: The aim of this programme would be to provide partial financial support for the creation, in one of the ABCMUV countries, of an International Centre for Applied Mathematics in Industry. The Centre's aim will be to give assistance to Latin American industries for problems related to new technologies and the production of new products.

As a rule in the above propositions: The Latin American governments and/or institutions involved in these agreements ought to contribute significantly and continuously to these programmes.

Acknowledgements: I am indebted to many Latin American mathematicians for several comments and suggestions concerning the present report. Special thanks are due to: R. Miatello (Argentina); J.L. Barbosa, C. Camacho, F. Cardoso, L.A. Medeiros, N. Maculan, J. Palis Jr., A. Simis and K. Teneblat (Brazil); R. Bamón, M. Blázquez, J. Billeke, R. Correa, S. Martínez, S. Navarro, S. Plaza, R. Rodriguez and R. Rebolledo (Chile); J. Lewowicz and R. Markarian (Uruguay).

Rafael LABARCA
Departamento de Matemática
Universidad de Santiago de Chile
Casilla 307 - Correo 2
SANTIAGO DE CHILE (Chile)

A. Verjovsky*

Let me very briefly explain what the International Center for Theoretical Physics (ICTP) is. It is located in Trieste, Italy. ICTP was founded about 25 years ago by A. Salam, Nobel laureate. Essentially, the center is financed by the Italian government (up to 90%), due to a policy of generosity and farsightedness. The center is clearly beneficial for Italy as well and quite recently the Italian Parliament has passed a law that provides stable financial support for the ICTP. A mathematics section was created six years ago and we now have a staff of around 25 visitors from the Third World. By visitors, I mean postdoctorates or mathematicians. ICTP is therefore a unique place. To the best of my knowledge, there is no other place that offers the opportunity for mathematicians from third world countries to spend one or two years as postdoctorates or visiting researchers. A.O. Kuku has already mentioned isolation as a recurrent problem in the third world. The center plays an essential role in breaking up this isolation. ICTP has a very good library, although it still needs to be expanded. ICTP plays an essential role in identifying and recognizing centers of excellence—no easy matter—in particular with the help of young fellows. Consequently, we now have a growing directory of names and institutions.

Let me now move on to concrete proposals. In order to reinforce the permanent staff of ICTP, it would be a great help if the European Community could provide ICTP with visiting European scientists (exact number to be determined) who would spend from one to three months at ICTP.

---

* A. Verjovsky comes from Mexico. He is presently senior scientist at the Department of Mathematics of the International Center for Theoretical Physics (ICTP) in Trieste (Italy). The following text has been transcribed from the tape recorded during the round table.

The presence of such scientists would be very important for the ICTP staff, especially if they were working in areas of interest to our staff. It would also be very important if there were European centers open to visits by young mathematicians from ICTP. These young people would then have the opportunity of visiting other places during their stay at ICTP. This would be financially feasible since ICTP already covers travel expenses from the developing country of origin to Italy.

Alberto VERJOVSKY
Department of Mathematics
International Center for Theoretical Physics (ICTP)
P.O. Box 586
34 100 TRIESTE (Italy)

IV. Contributions from the Floor, I*

One conferencee raised the question of Eastern European countries. It is a well-known fact that many of these countries are facing severe problems. Being a European matter, this question was not discussed further although it is of great importance.

One conferencee criticized the "narrowness of the mathematical themes represented at ICTP or Tata Institute" and also questioned the *a priori* high standards set by these centers.

As M.S. Narasimhan noted, there is some disagreement here. He himself thinks that it is necessary to have high standards. P. Bérard mentioned the example of IMPA in Brazil which first chose to limit the range of its topics of interest in order to develop strong groups. J. Eells said that there are two aspects concerning ICTP. Concerning *Conferences*, he thinks that the center should invite all serious proposals (with specific topics and specific names) and that the variety of themes could be broadened. Concerning the *Visiting Staff* (25 fellows visiting for one or two years), Eells said that the topics will have to be more concentrated in order to achieve what has been done at IMPA or at Tata and what is currently being done at the CIMAT (Centro de Investigacion en Matematicas, Mexico) that is to develop strong groups. He also said *we cannot have double standards.*

Prof. Hazewinckel conveyed the following message: "If you want to help, do it yourself. If you have a good idea for a course (say applied harmonic analysis) work it out, give names and work with existing centers like ICTP." He also suggested donating books or journals that we do not use any more either to ICTP (cf. book donation programme) or through direct personal contacts in developing countries. He also suggested asking ex-colonial governments for support for concrete proposals.

---

* Comments were invited from the floor on the above contributions. The following text has been transcribed from the tape recorded during the round table.

N.D. Tri, President of the Vietnamese Mathematical Society, recalled that in 1960 Vietnam was able to invite up to ten French mathematicians each year to give lectures and that since 1982 the priority areas for cooperation between Vietnam and France do not include basic sciences anymore. He insisted that it is very important to develop mathematical research in developing countries. There are several mathematical centers in Vietnam (Mathematics Institute, Institute of Computer Science, Department of Mathematics in Hanoi, Da Nang and Ho Chi Minh City, and so on). Among the difficulties which these centers face, Tri mentioned two. The salary of a full professor at top level is presently $60 a month and secondly, in 1992 all institutes had to cut off their subscriptions to scientific journals from foreign countries due to lack of funds. Tri said that it is important to be able to send both young as well as confirmed mathematicians to centers of excellence abroad. Vietnam has been doing so thanks to the help of the international community. Tri also insisted however that it is important to organize international conferences in Vietnam so that many Vietnamese mathematicians can interact with foreign mathematicians. Two such conferences will be organized in 1993. Tri required the help and support of the international mathematical community to make it possible for Vietnam to organize such conferences on a regular basis.

Prof. Dineen (Ph.D under the late L. Nachbin, Brazil) recalled a key factor in the development of third world mathematics: the problem of the successful third world mathematician. If he is really successful he can either get a job, say in Berkeley or New York, and is then lost for the third world, or stay at home and may very well loose touch with mathematics. This is a real dilemma. Dineen suggested that some of the major European universities establish permanent professorships for three months a year for mathematicians from the third world: they would have a place to go every year and there is a lot of prestige for someone who has a professorship, for example, in Paris. The key for the development of mathematics in the third world is to allow mathematicians both to keep in touch with their field and to keep in touch with their home country and help local doctoral students.

Prof. Rofman mentioned the difficulties that are encountered when hardware equipment has to be chosen by a developing country. Unless carefully chosen this could cause an enormous waste of money. He suggested to have an *ethics committee* to help in such cases.

Al-Amrani reported on mathematics in Morocco, cf. Appendix II.

Prof. Schweitzer, PUC, Rio de Janeiro, Brazil, called the attention of the conferencees to the fact that IMPA, the leading mathematical center in South America, is seriously menaced by the present situation in Brazil. Schweitzer asked for political influence from the Brazilian government, e.g., debt for science. He also insisted that this is not merely a financial problem, but also the way first-rate institutions are treated.

Prof. Schiffels made comments on the report he wrote on behalf of the German Mathematical Society, *cf.* Appendix I.

V. Contributions by Invited Speakers, II

A. Pelczar*

I think that African societies should be aware how important mathematics is. This seems like a trivial statement, but it is not entirely clear that those in power are aware of this. Mathematics is a fundamental element of general culture as well as education, but this fact is not always acknowledged. Our role should be to support all those in developing countries who try to inform their own authorities about the importance of mathematics. There are some exceptions, like Nigeria, but there is a general lack of knowledge concerning mathematics. There are even secondary level schools in Africa that do not teach mathematics. It is our duty to help establish conditions which will improve education in these countries regarding mathematics.

There is also the question of doctoral students. There are two possibilities: one is to invite doctoral candidates to important centers in Europe or elsewhere, the other, for example, is to send professors to African countries. My opinion is that we should do both but that perhaps the first option is to send visiting professors and perhaps a few good graduate students can return with the visiting professor. This might be a good beginning for collaboration at a higher level. I have been in Nigeria and a student then came with me for a Ph.D at Krakow and the Ph.D was finished in Nigeria.

Concerning education in primary and secondary schools, it is of course necessary that teachers be evenly distributed throughout the whole country. As for research, a few good research centers is the best approach. I am pleased that Nigeria has now established such a center. Some governments think that if they have five good researchers, these researchers should be sent to five different universities in the country. This is not a good policy. Europeans can make suggestions and give advice, eventually providing support to help create good centers of mathematics. This is a practical suggestion for both sides, for our friends in Africa but also for Europeans.

A. PELCZAR
The Jagiellonian University
Institute of Mathematics
30-059 KRAKOW (Poland)

---

* A. Pelczar is Rector of the Jagiellonian University in Krakow, Poland. The following text has been transcribed from the tape recorded during the round table.

J.S. Martins*

A) The Portuguese collaboration in mathematics has been scarce, so far. According to information from the universities and the "Gabinete para a cooperação com os países de língua portuguesa" (Office for Cooperation with Portuguese Speaking Countries), the collaboration in the last two years consisted almost exclusively of some graduate level courses given by Portuguese professors at "Escola de Formação de Professores" (Teachers Training School) in Praia, Cape Verde.

There are, at the moment, about twenty professors of mathematics in Portuguese universities who have taught or studied in pre-independence universities in Angola and Mozambique. The well-known political difficulties in these African countries have, until now, upset the schedule and the achievement of further undertakings, as well as preventing the training of an adequate staff and the stabilization of these universities.

Undergraduate students from Angola, Mozambique, Cape Verde and Guinea-Bissau, mostly with a very weak background, are studying in several Portuguese universities for a degree in mathematics, engineering, and so on. Many of them have Portuguese grants. Two Rumanian students are in Lisbon universities, one working for a M.Sc., the other for a Ph.D.

Some mathematics books and other publications have been sent free, mainly to Mozambique University.

In some areas of the mathematical sciences, Brazil can no longer be considered a developing country. There have been Portuguese professors in Brazilian universities and some Brazilian professors in Portuguese universities, either for short visits or for longer stays, from six months to two years.

B) The Portuguese Mathematical Society presented a proposal from the Mathematics Department of the University of Coimbra to the European Mathematical Society for an international center of mathematics, CIMAT. Intended to be a center similar to CIMPA, ICTP and ICMS, one of its purposes is the collaboration with developing countries. Due to linguistic affinities, the easiest collaboration would be with Portuguese or Spanish speaking countries. (Obviously, English and French are spoken throughout the Portuguese and international community!) Mathematicians from all nationalities may be invited to cooperate. The Center will be established in Coimbra as a joint venture by almost all the Portuguese universities. The organizing committee foresees its legal creation in the near future and intends to raise funds for its activities. Up to the moment, and unlike the patronage of French and Italian Governments, permanent sponsorship, although requested, has not been granted by the Secretaries of State for Science and for Cooperation.

The main objectives for collaboration include:

---

* The following text reproduces the report written by J.S. MARTINS on behalf of the Portuguese Mathematical Society and presented at the round table.

— Advanced courses for teacher training, summer schools and workshops in Portugal or abroad (mainly in developing countries),
— hosting students for M.Sc. or Ph.D, as well as mathematicians, from developing countries,
— sending specialists for short visits or longer stays at centers in developing countries,
— providing mathematical literature for university libraries in developing countries.

J.S. MARTINS
Departamento de Matemática
Universidade de Coimbra
Apartado 3008
3000 COIMBRA (Portugal)

J. Eells*

The ICTP has a programme of now approximately 25 postdoctoral students who come for one or two years, sometimes longer. They come because they are of high quality and that is probably going to have in the immediate future an additional condition, high quality in areas which have to fit together and those particular areas might change from time to time. There are two aspects at ICTP: the conference center and the research center, which makes it more than a conference center. Concerning the conference center, let me re-issue an invitation to all of you to submit sensible, detailed proposals for meetings to be held in 1996–1997–1998 (we are committed that far ahead). We need new ideas just as one needs new blood.

Concerning third world development, we have continually heard about the need for and possibility of increasing financial support for mathematics in industry. This is one of the purposes of the new center (ICMS, International Center for Mathematical Sciences) in Edinburgh (UK). The first meeting was held this year in Edinburgh. ICMS is a conference center and there is a plan to have postdoctoral fellows just as we have at ICTP. The next programme of ICMS is on materials and this is more oriented toward mathematics in industry than the first meeting, which had to do with mathematical physics and gauge theory. We also have an increasing interest in mathematics in industry in Chile. This was meant to say, first, that not all centers are doing the same thing; second, that there is a basic need for enlarging over what we have done in the past decade, and third, that any cooperation between future centers in mathematics that can be made, should be made.

---

* J. Eells has been Chairman of the Mathematics Department at the International Center for Theoretical Physics (ICTP) in Trieste (Italy) until September 1992. He is a member of the Board of the International Center of Mathematical Sciences in Edinburgh (UK). The following text has been transcribed from the tape recorded during the round table.

Let me now turn to what seems to be a very important point of this meeting today, i.e., suggestions we have to convey to the European mathematical community. We need fellowships. This is very important. We need travel support for Europeans to go to third world countries and vice versa. The centers could be a big help in this regard. For example, ICTP has a large number of people (80 this year) who spend some time at the center and then go off to other various institutes.

I would like to insist on the importance of having a directory of mathematicians, especially those working in third world countries, so we could keep track of researchers who are still active. Let me mention the example of Sudan where there were over 30 mathematicians in 1975 who are now scattered throughout the world. I suggest that EMS (European Mathematical Society) should be able to establish a secretariat for that purpose. The present World Directory of Mathematicians does not fully meet this problem.

It is also important that information concerning conferences should be centralized and that conferences, large and small, should be announced two years in advance to avoid overlap and to allow invited lecturers to attend.

J. EELLS
Institute of Mathematics
University of Warwick
COVENTRY CV4 7AL (United Kingdom)

J.-M. Lemaire*

Let me first give some indications on ICPAM–CIMPA (International Center for Pure and Applied Mathematics–Centre International de Mathématiques Pures et Appliquées). It has been said several times that there are "centers like ICTP or ICPAM". Let me begin by saying that ICPAM differs from ICTP by an order of magnitude of 40, both with respect to funds and staff. ICPAM is a small center and it has no research component. In this respect, it is very different from ICTP. A research institute should concentrate on a few given subjects, as already stressed above, while a conference center or a service organization, which is what ICPAM is for now, can organize activities in various fields according to the demand, in this case from developing countries. In this respect, one should point out that ICPAM organizes conferences in developing countries. I think this really helps in various respects. As Professor Narasimhan already mentioned, it gives confidence to the people. ICPAM helps choose the lecturers, organize the meeting and also provides funds. We have already held such conferences in Egypt, Nigeria, Chile and China.

I frequently heard that the situation is very bad in developing countries, that times are difficult, working conditions impossible, and so on. I feel that in some

* J.-M. Lemaire has been Délégué Général of ICPAM– CIMPA, International Center for Pure and Applied Mathematics, in Nice (France). The following text has been transcribed from the tape recorded during the round table.

instances local governments in developing countries could help much more than they do. I know of at least one African country whose Finance Minister is a mathematician and maybe his government could help more than it actually does. Let me say something positive and mention a counterexample. We had a conference in Yaounde. I attended it as an ICPAM representative; ICPAM supported the conference financially. This conference lasted ten days and *every single day* during the conference there were many journalists asking questions: "What are you doing?" "What is mathematics?" "What is it good for?" Well, I have certainly never seen that many journalists at this first European Congress of Mathematics! Through the media, information on and interest in mathematics can reach the government, the decision or policy makers. If you organize a conference in your country, you can use this kind of thing to convey your message about mathematics.

I would also like to insist on the importance of electronic mail. It is a cheap and efficient means of breaking up isolation. Much has been said about establishing regional libraries in the third world. This involves large amounts of money and so far nothing has been done. Electronic mail or fax could be used to provide access to information, e.g., to large libraries to obtain photocopies. The idea would be to establish an e-mail network so that every mathematician could have access to information. Concerning the problems of libraries, the University of Bordeaux (France), in collaboration with ICPAM, provides bibliographic information— so far by ordinary mail, e-mail might make it even more efficient—to some African countries. The e-mail issue leads to the problem previously alluded to by Professor Rofman, concerning hardware equipment and the proposed ethics committee. For example, it might be advisable to be careful in choosing a network, as some are being used less and less by the scientific community.

We cannot provide developing countries with all they need. Only they can meet their basic needs. We can however provide them with up-to-date information about mathematical research and the mathematical environment.

Concerning Europe, I think it is the natural setting to bring together all the individual initiatives. I think we should coordinate our activities: ICPAM, ICMS, ICTP and ourselves are going to do so more and more. Take the e-mail business, for example. This is not a national issue because it is an international network. No single European country can take the initiative. Each country can provide funds of course, but the coordination, the scientific organization needs an overall direction. The same comments apply to databases and, for example, Euromath.

What we should do at EMS is to advocate that some room be made for the developing countries and that we take the necessary steps so that those means remain available to developing countries.

J.-M. LEMAIRE
C.I.M.P.A.
1 Avenue Edith Cavell
06 000 NICE (France)

Closing Comments by P. Bérard

Some fifty years ago, Hermann Weyl showed his concern toward us *the next generation.* We too have a duty to the next generation, in particular, toward those in developing countries. Before quoting Hermann Weyl, let me mention the following comment made by Professor A. Simis from Brazil (his contribution appears in Appendix II).

A. Simis: *The active mathematical groups in developing countries need to be treated along the policy lines of CDE *, i.e., merit and high standards. The last thing they need is a third-world paternalistic view lacking funds!*

H. Weyl: *We do not claim for mathematics the prerogative of a Queen of Science; there are other fields which are of the same or even higher importance in education. But mathematics sets the standard of objective truth for all intellectual endeavors; science and technology bear witness to its practical usefulness. Besides language and music, it is one of the primary manifestations of the free creative power of the human mind, and it is the universal organ for world-understanding through theoretical construction. Mathematics must therefore remain an essential element of the knowledge and abilities which we have to teach, of the culture which we have to transmit to the next generation. Only he who knows what mathematics is, and what is its function in our present civilization, can give sound advice for the improvement of our mathematical teaching.* (Source: from ETH Collection 1944, cited in Gesammelte Abhandlungen, Springer-Verlag, 1968.)

VI. Contributions from the Floor II*

A colleague from Moldavia called our attention again to the difficulties in the countries of the ex-Soviet Union. There is a lack of information, of means of communication (fax, electronic mail) and of hard currency to buy books or subscribe to international journals. He thinks that there should be a discussion on these dying countries. This last point was supported by Y. Kosmann-Schwarzbach. Prof. Hazewinckel supported these comments. He insisted that we do not throw away computers, faxes, and so on, and that we donate them (directly to some person or institute and not through the surviving bureaucracies) to Eastern European countries or developing countries. A conferencee from Iran mentioned that his institute, as well as others in Iran, would be very pleased to invite mathematicians from the ex-Soviet Union (in particular those from countries around the Caspian sea) and that Iran could pay them to come. He invited any such mathematician to get in touch with institutes in Iran.

---

* Commission on Development and Exchanges of the International Mathematical Union.

* The conferencees were invited to make comments on the preceding contributions. The following text has been transcribed from the tape recorded during the round table.

Prof. Rodrigues from Portugal suggested that a network of European centers like ICTP, ICPAM, and so on, be established. J.-M. Lemaire agreed. He added that there are already friendly relations between ICTP and ICPAM. J.-M. Lemaire also recalled that the European Community programme *Human Capital and Mobility* unfortunately does not involve developing countries. He said that we should tighten links between Europe and the developing countries and stated that there is much concern in developing countries that Europe might now invest only in Eastern Europe at the expense of developing countries. Y. Kosmann-Schwarzbach insisted that it is a very rewarding experience organizing ICPAM schools.

One conferencee from Tunisia mentioned that he had seen very little advertising about the European Congress and lamented this fact.

Lemaire mentioned the organizational reasons for this and explained that the European Congress is primarily concerned with European affairs. Concerning Tunisia, the question is whether or not Tunisia should apply to the European Mathematical Society, because of its European links, through its own mathematical society, assuming it exists, or rather apply to the African Mathematical Union for geographical reasons or perhaps to both.

APPENDIX I

Contributions made by Mathematical Societies

The German Mathematical Society (DMV)*

The following is a report on Germany's collaboration with developing countries. I have made a survey of the mathematical departments and research institutes in Germany. I proposed to take as a definition of "Developing Country" the list of the OECD (Organization for Economic Cooperation and Development). The OECD list is very large, and includes countries such as Brazil and China, for example.

According to the answers I received there are many academic links to developing countries. Some of these are partnerships existing for at least 5 to 20 years, some of which are sustained by a variety of German mathematical departments. The partners of such large undertakings are in Brazil, Columbia, Ethiopia, Thailand and Tunisia. A second category could be the links that have existed for 2 to 5 years; most of them are meant to continue. Such links are supported by one or several German mathematicians. Often there is some explicit project, programme or contract. The partners are in Algeria, Argentina, Chile, the People's Republic of China, Indonesia, Cuba, South Korea, Kuwait, Madagascar and Zimbabwe. A third category might be the many links between single German mathematicians and single foreign universities where contacts are for periods ranging from one month to two years.

The financial and organizational promotion of the links is supported by various German governmental or public institutions. First, there are the Federal Ministries of Foreign Affairs, of Economic Cooperation, of Education and Science, and of Research and Technology. Second, there are the Science Administrations of the Bundes-Länder, which are responsible for our universities. Third, by commission of the Bund or the Länder, or in cooperation with them, there are special institutions such as the Alexander von Humboldt Foundation (AvH), the German Academic Exchange Service (Deutscher Akademischer Austauschdienst, DAAD), the German Society for the Advancement of Scientific Research (Deutsche Forschungsgemeinschaft, DFG), the German Society of Technical Cooperation (Deutsche Gesellschaft für Technische Zusammenarbeit, GTZ), the Volkswagen Foundation and several other institutions. The DAAD, for example, has programmes for long term lecturerships (4 months to 5 years) and short term lecturerships (1 to 3 months), which are given to Germans to teach in developing countries. There are also grants for students and young scientists from these countries. The programme "Postgraduate Studies with Significance for Developing Countries" of the

---

* The following text has been prepared by G. Schiffels, on behalf of the Deutsche Mathematiker Vereinigung (DMV). Full details of his report have been published in the internal journal of the DMV (April 92).

DAAD supports the curriculum of "Industrial Mathematics" at the University of Kaiserslautern. The Gesellschaft für Mathematik und Datenverarbeitung (GMD) organizes a project of collaboration with Brazil, which has existed for 20 years.

I will now present some suggestions and questions, which are partially based on information from my colleagues. Some institutes and departments of some developing countries have all the personnel and equipment required for collaboration on an equal basis. But in most cases, collaboration also requires aid to the participants in developing countries. This may involve:

— training of research and junior scientists
— teaching in the different curricula
— providing literature and computer equipment
— long-standing cooperation, even helping in administration.

In several developing countries mathematicians are also very isolated from each other because their fields of interest are too varied. Perhaps there should be a planning and decision-making process to diminish the random generation of too many areas of interest. Contacts with neighboring countries or universities are still often meager. Which fields of research and teaching are estimated to be most important for the development of culture or industry in a given country? Which curricula should be strengthened or reformed in relation to the labor market in the country? Often there is too much emphasis on the training of future professors. Some of my colleagues have expressed the opinion that long-standing cooperation is the most important and effective, for example, for developing new curricula or for helping junior scientists. There should be more such "systematic"cooperation and it should be better supported by the political agencies.

G. SCHIFFELS
Fakultät für Mathematik
Universität Bielefeld
Postfach 8640
4800 BIELEFELD 1 (Germany)

The Jànos Bolyai Society*

1. Hungarian mathematicians in developing countries. Numerous Hungarian mathematicians have spent some time teaching in developing countries, usually for at least one year each. It must be noted that lately the number of such visits has decreased. Most of these visits occurred between the mid-sixties and late seventies. Some of the visiting mathematicians succeeded in launching talented students and colleagues into successful mathematical careers.

* The following text has been sent by Pham Ngoc Anh on behalf of the Jànos Bolyai Mathematical Society (Hungary).

2. University students from developing countries in Hungary. Hungarian universities have been receiving students of mathematics from many countries, most of them from Vietnam, Mongolia, Cuba and some African countries. Judging from the experiences of the past ten years, one can say that a major concern has been the weak background that these students received in high school (with the exception of students from Vietnam). Despite this, they still obtained their university diplomas, very often without proper mathematical knowledge. It is difficult to judge from here whether such practices helped their countries of origin or not. The relatively high standards in the Hungarian universities made these students' studies rather painful. They avoided personal contacts with fellow students as much as with their professors, so that, once back home again, they were not able to make use of any help offered by the Hungarian Mathematical Society. After their return they usually stopped having any mathematical contacts with Hungary.

3. Postgraduate studies of students from developing countries in Hungary. The university diploma offered by Hungarian universities is roughly equivalent to an M.Sc. degree. After obtaining it, there are three further levels of scientific degrees: the first is called "university doctorate", the second is the degree of "candidate of sciences", finally the third, the highest level is what is known as the degree of "doctor of sciences". Scientists coming to Hungary from developing countries for a scientific degree on a basis of state agreements were, as a rule, enrolled in the second degree level, i.e., that of a candidate of sciences. However, many of them were below even the standard university level. Still, with the exception of a very few extreme cases, in the end they obtained their degrees. A common feeling is that a major factor obstructing foreign students in Hungary is the lack of a good command of the Hungarian language, although with a proper knowledge of the English language (which very few of the candidates had) it would be quite possible to do research successfully. This deficiency can hinder the work of even talented students. This may be the reason why these candidates, like students coming to get their first university degrees, have no further contacts with Hungarian mathematicians after their return.

During the last 10 years the number of students who have received the degree of candidate of sciences was: 8 from Egypt, 2 from India, 5 from Cuba, 22 from Vietnam and 11 from Mongolia. Presently, the number of postgraduate students is 1 from Egypt, 4 from Iraq, 1 from Libya, 4 from Vietnam and 5 from Mongolia. Most of them are fully financed by Hungary. These expenses consist of two major parts. The first component is living expenses of the students, which is approximately $3400 per person per year. The second part of the funding is what one could call "research expenses". These expenses are usually built into the budget of the institution of the candidate, so it is more difficult to give a numerical estimate for these expenses.

Other arrangements—like exchanges between academies or universities as well as practical studies organized by the Ministry of Education—also exist, which make it possible for scientists from developing countries to spend from a few months to one or two years in Hungary. It is difficult to judge the value of these visits. Unfortunately, it seems that people visiting Hungary within such frameworks were not always chosen according to their professional merit.

4. Recommendations

(i) It should be arranged on a European level so that every year at least 5 to 10 mathematicians are sent to developing countries for periods from a couple of months up to one year. In the long run sending mathematicians seems to be the most economical and efficient way for Europe to help those developing countries which already have an established university system with a relatively high mathematical educational standard. These visiting mathematicians could help determine the mathematical needs of the host countries, find the most efficient methods of cooperation, and they could also take part in selecting students and scientists to be sent for further education abroad. A good example for this is the visit of Grothendieck to Vietnam in 1967. At that time the impact of his visit was not clearly visible. However, looking back on this event now, it is clear that it gave a lot of support and inspiration to the development of the Vietnamese Mathematical Society.

(ii) It would be worth considering whether the students coming from developing countries, with a few exceptions, shouldn't be sent to a teacher training college first. Then, after a year or six months based on their results, some of them could continue their studies at the university level. The advantages of such a system would be twofold. First, it may offer a more refined education, better adapted to the real needs of the sending countries and also avoiding the difficulties that these students would face in an institution that is insensitive to their problems. Secondly, it may help the development of the very talented students by helping them in the adjustment process. Finally, it could be financially advantageous to make the unsuccessful candidates' stay shorter.

(iii) According to the present Hungarian practice, the sending country decides which candidates should come to Hungary for studies or research. This practice seems to be rather unfortunate. It would therefore be very helpful if the mathematical knowledge of the potential candidates could be tested through a standard (Hungarian or even European) testing process. The host institutions would then have the final say in whom to accept for studies at that institution.

(iv) The "library conditions" in developing countries are usually very poor, with almost no foreign books and journals available to local scientists. A coordinated action would be needed to help these countries to receive the necessary scientific information. Support from major publishing companies, universities and research centers would be essential in this matter.

PHAM NGOC ANH
Mathematical Institute
Hungarian Academy of Sciences
H-1364 BUDAPEST, Pf. 127 (Hungary)

The London Mathematical Society*

This report describes some cooperation programmes involving (1.) The Royal Society, (2.) The Overseas Development Administration (ODA) and the British Council.

1. The Royal Society. The Royal Society is an independent learned society, self-governing under a Royal Charter, for the promotion of the natural sciences, including mathematics and all applied aspects, such as engineering and medicine. It encourages both national and international activities in a similar way to national academies overseas. Its aims include, among other things, the encouragement of scientific research and the promotion of its application, and the promotion and facilitation of international scientific relations.

The Society's annual expenditure is around $15 million. The greater proportion of this derives from a Parliamentary Grant-in-aid administered by the Society.

The Royal Society has a long history of encouraging scientific contact between scientists (including mathematicians), technologists and engineers in the United Kingdom and Africa, Asia and Latin America. It has agreements with partner organizations in 10 countries, namely, Argentina, Brazil, Chile, Egypt, India, Korea, Mexico, Pakistan, Philippines and Venezuela. There are also funds to support visits to countries in Africa, Asia and Latin America with which it has no formal agreements.

Opportunities exist for visits in each direction for scientists of postdoctoral or similar level in the natural sciences, including mathematics. In most cases the sending side pays the international fares of its nationals and the host side their local costs to an agreed tariff. Each agreement has an annual quota expressed in person-months and/or in balancing sums of sterling on the Society's side and in the currency of the agreement partner.

These arrangements provide funds to support *Study Visits and Fellowships*. The purpose of *Study Visits* is to establish or develop personal contacts, to exchange ideas and information, to give lectures and where relevant to acquire new techniques. *Fellowships* support visits by younger British scientists of postdoctoral

---

* The following report has been written by R.Y. Sharp, Council and General Secretary of the London Mathematical Society.

status who wish to undertake research overseas in the countries of Latin America, South East Asia, the Indian Subcontinent and Africa, for periods of up one year.

In addition, proposals for long term visits (six to twelve months) to the United Kingdom can be considered under the Royal Society Developing Country Fellowships Scheme, the aim of which is to enable scientists of proven ability working in developing countries or newly industrialized countries to pursue research, learn new techniques or undertake other forms of study in the natural and applied sciences (including mathematics) by working with colleagues in the United Kingdom. In January 1991, the value of a Developing Country Fellowship amounted to a monthly allowance of $650, with provision for a small number of subsidiary visits to academic centers.

In addition to the above scheme, the Royal Society operates a Developing Commonwealth Fellowships Scheme (for scientists, including mathematicians, resident in developing Commonwealth countries (other than India) wishing to visit the UK or to visit a developing Commonwealth country other than their own) and operates exchange agreements between the UK and USSR, Bulgaria, Czechoslovakia, Hungary and Poland, Romania, Yugoslavia.

I have not managed to discover the order of financial support for mathematics through these programmes.

2. The Overseas Development Administration (ODA) and the British Council. The Overseas Development Administration (ODA), part of the Foreign and Commonwealth Office, is the British Government Department responsible for Britain's aid programme, which in 1990-91 amounted to over $1,500 million. One of the many areas in which the ODA concentrates a small part of its support for education projects in developing countries is in the training of scientists and technologists. Much of the day-to-day running of the ODA's education projects is carried out by the British Council, with its extensive network of overseas offices.

The ODA provides a Higher Education Grant to the Committee for International Cooperation technics through small-scale link arrangements and other activities in support of their counterparts in developing countries. The fund is managed by the British Council, and about $2 million per annum is used for operational purposes.

Further funding is available through the Low Priced Books Scheme, currently supported at the level of $1.2 million per year. This scheme enables British publishers to make available subsidized student editions of popular British textbooks through booksellers in a selected number of developing countries. The books have a distinctive binding, and are aimed at post school students who would otherwise not be able to afford to own copies of key texts. The scheme has just been revised to give it a sharper focus to concentrate on poorer countries, and to cover subjects closely related to ODA subject priorities.

One further major source of aid funds for educational purposes is the ODA's contribution to the core funding for the British Council (some $23 million in 1988-89). This is used by the Council for a variety of purposes consistent with the aid programme, including the provision of British Council libraries and branch libraries, scholarships and lecture tours for developing countries. I understand that the British Council was involved in the placing in 1990-91 of 55 students from developing countries in 25 different UK higher educational institutions to follow mathematics courses.

The Committee for International Cooperation in Higher Education has a brief from the ODA which covers all developing countries. The Committee provides advice to British Council Representatives overseas on the establishment of links between higher education institutions in developing countries and the higher education sector in Britain.

A substantial number of British university and polytechnic departments operate link arrangements with their counterparts in developing countries from Algeria to Zimbabwe. Links funded by the ODA's Education Division through the Higher Education Group of the British Council operate over virtually all scientific and technological disciplines involving British university and polytechnic departments. According to the British Council, there were about 12 links of the above type in mathematics that were active at 30 August 1991, and several of these had been active for a few years. The countries involved were Brazil, China, India, Korea, Lesotho, Malaysia, Mexico, Thailand, and Zambia. The purposes of the links varied from research collaboration in applied mathematics or mathematical education to curriculum development, the upgrading of computing facilities or the improvement of mathematical instructions in the developing country. The expenditure on these links for the financial year 1990-91 varied from $2,000 to $11,000. Each link supports a number of visits in either direction to further the aims and objectives of that link.

More substantial links, funded directly from the ODA's technical cooperation programme, exist between a range of British institutions and their counterparts in developing countries.

R.Y. SHARP
Department of Mathematics
University of Sheffield
SHEFFIELD S3 7RH (UK)

APPENDIX II

Individual Reports

A. Al-Amrani

Contemporary mathematical activities in Morocco

Our goal here is to inform the interested reader about current mathematical activities in Morocco. We aim to give an account of the situation, rather than to express an opinion, insofar as this is possible. We will essentially concentrate on research and training in the university. The author wishes to emphasize this topic only deals with mathematics done by Moroccan mathematicians* in Morocco.

History. The first faculty of sciences in Morocco was created in Rabat in 1958. Up to 1970, it was related to the *Faculté des Sciences* of Bordeaux who had, for example, the right to inspect the examinations taking place in Rabat. It is worth recalling that from 1912 to 1956, Morocco was a French protectorate. In fact, scientific university teaching in Morocco can be traced back to 1940, when it depended on the famous *Faculté des Sciences* of Algiers. The second faculty of sciences in the country was not founded until 1978 in Marrakech. From this date on, faculties of sciences were built all around the country in Fès, Oujda, Casablanca, Tétouan and so on. All these faculties have a mathematics department. The *Société Mathématique du Maroc* was founded in Rabat in 1990. Actually, this is not the first mathematical society. It is a successor of an association that was born in the 1970s, but which had long since ceased to exist in practice.

Training. To the best of our knowledge, the first *doctorat d'État* dissertation in mathematics obtained by a Moroccan was defended in France towards the end of the 1960s. At the beginning of the 1970s, three other Moroccans followed suit, still in France. About ten years later, a Moroccan woman was awarded the same title in Paris in applied mathematics. The first *thèses de troisième cycle* entirely prepared in Morocco under the direction of Moroccan mathematicians were defended in 1978.

Towards the end of the 1970s, due to scientific cooperation agreements between the United States and Canada on the one hand and Morocco on the other, American and Québécois universities opened their doors to Moroccan assistants and lecturers seeking a Ph.D. In fact, three or four Moroccans had already obtained a Ph.D in the United States. These doctorates were accepted as equivalent to the Moroccan *doctorat d'État.* From this date on, a growing number of lecturers from Moroccan faculties who already had a French or Moroccan *thèse de troisième cycle* left for the United States or Canada in order to prepare a Ph.D. The average length of their stay was four years. Coming back home, they were first appointed

* who were most often trained abroad.

assistant professors then, four years later, full professors by seniority, like their predecessors who had been trained in France or elsewhere (Spain, for example).

This does not mean that France was entirely abandoned by Moroccan graduate students. A certain number of *doctorats d'État* were completed in France during the 1980s. The first *doctorat d'État* prepared in Morocco was defended in 1987 at the faculty of sciences of Rabat.

The institution, a few years ago, of a single dissertation in France caused a peculiar situation for the training of Moroccan doctoral candidates in France. In contrast with other Ph.D degrees, such as those awarded by American, Belgian or Canadian universities, the French *nouvelle thèse* is not accepted as equivalent to the Moroccan *doctorat d'État* by local authorities. This is due to historical reasons made clear by what was said before, but also due to political reasons. Actually, there are no scientific grounds for this refusal when other Ph.D degrees are automatically accepted.

In any case, this is why a certain number of dissertations that had been prepared in French departments in the last few years were defended in Morocco (Rabat, Fès and so on) as local Moroccan *doctorats d'État*. It is quite possible that in some cases there are other reasons for this phenomenon.

Let us point out that the Moroccan university hierarchy is as follows: assistant, lecturer (*maître-assistant*), assistant professor (*maître de conférences*) and full professor. The first rank is currently becoming obsolete. To become a lecturer, a French or Moroccan *thèse de troisième cycle* is required, or a French *nouvelle thèse*. To become an assistant professor, a French or Moroccan *doctorat d'État* or a foreign Ph.D (other than the French *nouvelle thèse*, of course!) is required.

Research. In our opinion, the question of doctoral preparation by graduate students (*thèses de troisième cycle*) or by lecturers (*doctorats d'État*) is of prime importance regarding the mathematical research done in Morocco. In effect, the work accomplished in these dissertations or in relationship to them constitutes the bulk of Moroccan mathematical output. The desire to obtain a doctoral degree, and later a faculty position, is very strong among an overwhelming majority of students. Besides, the tradition of the master instilling knowledge in his disciples goes back several centuries in Morocco. This is still a strong factor in the modern mentality.

It is rare that a Moroccan academic who already has a *doctorat d'État* or an equivalent degree—most often prepared abroad—will continue to do significant research once back in Morocco. Given the current state of affairs, this would add nothing to his career. He receives no material or even *moral* incentives to do research. However, in the last few years, exceptions to this rule are becoming more numerous. Some professors, either alone or in collaboration with foreign colleagues, have continued their research after coming back to Morocco.

Currently, there are around 500 mathematicians who hold positions in the faculties of sciences, *Écoles Normales Supérieures* and engineering schools. The distribution is approximately 45 professors, 70 assistant professors (who will become professors after four years of teaching) and 385 lecturers. There are around 30 professors, 30 assistant professors and 85 lecturers in the *Faculté des Sciences* and various schools in Rabat. The activities of these mathematicians include the direction of dissertations, the organization of mathematical meetings and publishing.

Conclusion

It is clear that for at least fifteen years a certain amount of research in mathematics has developed in Morocco. There have been attempts to form research teams, more or less successfully. Unfortunately, these initial efforts are not well supported by governmental authorities. The necessary means for developing research are not forthcoming. The library of the department of mathematics of the faculty of sciences in Rabat, the most important in the country, is dying due to lack of maintenance and supplies. Teaching in computer sciences, when it exists, is more theoretical than anything else due to lack of computers. A Moroccan academic most often pays for his professional trips out of his own pocket, whether it is within the country or outside it. No budget is allocated for research from which a mathematician could benefit. Even if the state signs a cooperation agreement with another country, the agreement is not respected. The example of the *actions intégrées* with France is well known: trips by Moroccans to France were never reimbursed and the duration of visits by the French to Morocco was shortened...
Sources: 1) Author's 10 years experience in Rabat. 2) Newsletters 1 and 2 of the *Société Mathématique Marocaine*. 3) Publications from the faculty of sciences at Rabat: annual reports of research departments (1980–1989), index of theses and dissertations (1952–1988).

A. AL-AMRANI
Département de mathématiques
Université Louis Pasteur
67084 STRASBOURG Cedex, France

R. Correa

On the development of applied mathematics in Chile

The end of the 1960s saw the appearance of mathematical engineers in the Engineering School of the University of Chile. Such engineers are specialized in applied mathematics and find their place amidst interdisciplinary teams to help work on difficult problems. This was the beginning of a trend that would lead, in the ensuing years, to the development of one of the most flourishing scientific teams in Chile.

The founder of the mathematics department in this school was Professor

Mellado. He profoundly altered the composition of the teaching team that was in charge of training engineers to adapt it to this new line of development. Mellado found inspiration in the applied mathematics taught in French engineering schools, in particular, at ENSIMAG in Grenoble, headed at the time by Professor Gastinel. Scientific cooperation with French applied mathematicians was firmly established and played a crucial role in the development of applied mathematics in Chile. In an initial phase, Chilean universities received important donations of books and benefitted from the enthusiasm of numerous military or civilian *coopérants**. They contributed to the training of the first Chilean researchers who were able to continue their graduate studies in France due to French government grants.

A systematic training plan for researchers was set up. With the support of French cooperation, which we mentioned above, and of other sources, young Chilean researchers went to France and the United States to pursue their doctoral studies. This plan was a success; most researchers came back home to continue their research and teach in Chile. The support of the authorities of the Faculty of Physical Sciences and Mathematics of the University of Chile was critical. This support supplied the complementary resources needed to let researchers stay abroad long enough to complete their training and to provide them, upon their return, with a suitable framework for their scientific interests. In the beginning, researchers were trained in the classical areas of applied mathematics: numerical analysis, probability theory, statistics, optimization and theoretical computer science. The return of these researchers created an extremely favorable context for the development of local research in applied mathematics. They induced young people to enter a scientific career, thus ensuring the continuation of this endeavor. This is why today more than 150 mathematical engineers have graduated from the Engineering School of the University of Chile. About 50 of them have a doctoral degree and about 20 will have one soon. Of the 50 doctorates, five were in mathematical economics, five in mechanics or electricity, five in pure mathematics, five in computer sciences and the rest in the more conventional areas, numerical analysis, probability theory and optimization. Approximately half of them obtained their degree in France and the other half in the United States. Nowadays, a large number of Chilean researchers have achieved international recognition, as can be seen from their publications.

Researchers from the Mathematics Department of the Faculty of Physical Sciences and Mathematics of the University of Chile have the opportunity of spending two month stays in reputable centers of the Northern hemisphere. This is handled within the framework of various agreements for scientific cooperation signed with international agencies, the French government and the European Com-

---

* French draftees have the opportunity of doing their military service as *coopérants* in a foreign country, where they can teach, advise, cooperate with research teams and so on. There are also civilian *coopérants*.

munity. Furthermore, they can benefit from grants by the Chilean governments that are allocated through national project competitions.

The Mathematics Department of the Faculty of Physical Sciences and Mathematics of the University of Chile is currently the most developed applied mathematics center in Chile. It has reached a mature stage with twenty or so established researchers. An average of fifteen first-rank researchers come to visit every year for about three weeks to develop common research. The department organizes a Franco-Chilean Congress of Applied Mathematics, with important local and regional attendance, as well as a Mathematical Physics Congress every other year. Similarly, a CIMPA school in pure and applied mathematics is organized every two years in collaboration with the University of the Frontier (Temuco, Chile). The department has research projects that attracted the interest of Digital Corporation, so that researchers were provided with latest technology computers. This gave the department computing facilities that are extremely well adapted to the needs of its research. The high quality research environment attracts the best students from the Engineering School, one of the oldest in Latin America. The new mathematical engineers have popularized the use of mathematics for solving practical problems in the country's industrial milieu. It is remarkable that before 1980 there were almost no publications in applied mathematics in Chile. Now, there are around thirty applied mathematics publications per year. The Journal of Applied Mathematics—Revista de Matematicas Aplicadas—has reached its thirteenth volume and is most probably the oldest applied mathematics journal in Latin America. Its editorial board features internationally renowned researchers, which ensures a certain level of quality.

A doctoral programme was created three years ago at the Mathematics Department of the Faculty of Physical Sciences and Mathematics of the University of Chile. This programme is the crowning of all the efforts made by several persons to develop applied mathematics in Chile. There are about ten students enrolled in this programme. We are now starting to develop strong relations with industry, either through direct agreements or through common projects.

Our main task is currently to create a Center for Industrial Mathematics whose aims would be to promote mathematics as a major tool in industry and to become a training and research center in applied mathematics that would be a shining light for all Latin America. Creating such a center presupposes very close scientific and technological relations with similar foreign centers. An endeavor of such extent would need, in addition to local support, an important scientific and financial aid from international development agencies and governments of developed countries.

Rafael CORREA
CONICYT
Canada 308 Casilla 297-V
SANTIAGO 21 (Chile)

Paul-André Meyer

Mathematical Cooperation between France and China

The "Centre Sino-Français" at Wuhan

The mathematical cooperation between China and France is a special one since there exists a Chinese-French mathematical institute located at Wuhan. It is the last remaining institute of a much wider project of a French university in China, proposed in 1980 by President Giscard d'Estaing. All departments collapsed early on except mathematics and a smaller department of French civilization. This Institute has a Chinese director, at the present time J.R. Yu, and a mixed French and Chinese teaching staff, particularly, a resident French professor of mathematics (for the last five years, Guy Roos from the University of Poitiers), and two teaching assistants who usually are young Ph.Ds in mathematics or computer science, performing their "Service National" (military service) in the French international cooperation projects. France also provides a resident teacher of scientific French. Four to six additional lectures a year are given by visiting mathematicians, who stay from three weeks to three months. The aim of the institute is to provide undergraduate and graduate mathematical instruction to locally selected Chinese students. The best of these students are selected to go to France for a doctorate (4 to 8 fellowships per year are provided for such students, paid equally by Chinese and French funds). These students are supposed to return to Wuhan after getting their doctorate and to participate in the teaching and research activity of the Centre.

Practically all the teaching at the undergraduate level ("premier cycle" and part of "second cycle") is now done by the Chinese staff, unfortunately in Chinese, which thus delays the process of learning western languages. There is officially no graduate school in Wuhan, but in practice students have some contact with the first part of the French "troisième cycle", only the so-called DEA remaining to be done in a French university in one year. The level of these students is quite comparable to that of the best graduate students in France.

The activity of the Centre is not only teaching. It also organizes high level research seminars with nationwide announcements, as well as summer schools and symposia. Up to now these activities have been located at Wuhan, a pleasant campus on the East Lake with good lodging facilities, although the climate is only pleasant in spring and early fall. The university is highly regarded among Chinese universities. It has an outstanding library. The choice of Wuhan however was forced by China on reluctant French diplomats, who preferred to establish their institutes at Shanghai and have never really been reconciled to Wuhan. I personally think that Wuhan has allowed for a much closer contact between the Chinese and the French than Beijing or Shanghai where the French institute would have encountered more political difficulties.

The Wuhan project depends on the French Ministry of Foreign Affairs, although its general orientation is set by a Franco-Chinese commission, which during the years following Tian An Men could not operate properly. It certainly is one of the major French cooperation projects outside the traditional French cultural areas in Africa and Indochina. The cost of the operation is considerable, especially the salaries of the permanent staff. Since 1991, the French Ministry of Education has assumed the burden of the resident professor's salary. This is the beginning of an acknowledgment that the Centre is a kind of university institute abroad. On the other hand, this has only been done on a provisional basis and it is not clear whether the whole operation will last or not.

In practice, two years ago the universities of Wuhan and Orsay signed an official agreement and the Ministry of Foreign Affairs delegated a global amount of money to the University of Orsay. In turn, the President of Orsay has nominated a scientific commission that makes a number of immediate decisions in agreement with the residing professor: choice of visitors, organization of symposia, installation and maintenance of computer hardware and software and also, unfortunately to a limited extent, scientific relations with China as a whole. This represents a lot of administrative work. Although this work is performed with the help of the Service des Relations Extérieures at Orsay and several of my colleagues have invested a lot of time, thought and effort, it still is a heavy load on a group of people all of whom have "something else to do".

A few difficulties. Mathematical cooperation with China has a firm basis at Wuhan and the achievement over ten years has been considerable. Still, the situation is far from satisfactory.

1) The amount of money spent on Wuhan alone is nearly the entire budget for China. This seems unreasonable. The system is very rigid and does not easily allow for the financing of visits elsewhere (although sometimes a short visit to Wuhan pays the travel expenses for a long visit elsewhere...). The situation is even worse the other way around: how to finance the stay in France for Chinese mathematicians from other places.
2) The status of the Centre within the university itself is not clear. Given the general poverty of Chinese institutions, the Centre is considered a kind of privileged body in Wuhan. Although M.Y. Qi, the current president of the university and himself a mathematician, has done his best to smooth things over, the Centre is looked upon without much sympathy. On the French side, the situation is not much better. The project is under frequent attack, often from people who have never been there. It lacks stability. Its staff is hired for one or two years contracts. The general budget is submitted to the yearly fluctuations of foreign affairs policy. The presence of a resident staff has in fact been questioned every year since I started following events.
3) The main complaints concern the overall balance of the teaching. There have been several failures among the students. France complained that 1) students are selected on a basis that is too restricted to Hubei and the neighboring

provinces and 2) that the selection of students sent to France has not always been free of political interference, although the presence of a residing professor has kept this within acceptable bounds. Nonetheless, there has been a large proportion of excellent young mathematicians from Wuhan. The Chinese complaints are that too many students do not return to China, although some have, in spite of the difficult conditions on their return. Anyway, the status of the students is likely to change, since China would prefer them to complete a Chinese doctorate instead of a French doctorate.

Prospects. The future of mathematical cooperation with China depends on two things. A reasonable preservation and development of the investment already made in Wuhan and the development of parallel systems and links with other universities. In both cases, there may be a role for European participation. As far as Wuhan is concerned, this participation might take the form of organizing meetings, summer schools and symposia on a European basis.

On the other hand, our aim has been to involve other organizations in the project, for instance, the École Normale Supérieure of Paris, the École Polytechnique, the Ministry of Research and Technology and CNRS. The main object should be to promote scientific exchange, particularly at the postdoctoral level. A general agreement between CNRS and the Academia Sinica is supposed to involve mathematics as well, but it has proved inefficient possibly for two reasons. One is that the Academia Sinica does not represent Chinese science as a whole and gives priority to its own people and its own preferred areas, in which mathematics ranks only second or third. In addition, France is supposed to provide funds for the stay of Chinese visitors and China is supposed to cover travel expenses, but it is too poor to pay its share. A European exchange project with China would be very welcome, possibly taking the form of links on a given mathematical subject between European and Chinese universities.

Paul-André MEYER
Département de Mathématiques
Université Louis Pasteur
67084 STRASBOURG Cedex (France)

R.J. Miatello

Comments on possible collaboration of Europe with developing countries

Perhaps the most important problem in doing research work in a developing country is having to work in almost complete isolation. Often there is no one with whom to discuss a problem or a specific technical question related to it. The availability of e-mail has been rather helpful in this direction but still the problem persists. It would be most helpful to have visitors coming on a regular basis who give lectures on present-day research and also a perspective for the near future on the main directions in a particular field. This, I feel, is essential so that the most promising young students see a future for themselves in staying or eventually coming back to their country.

The so-called brain drain is a bitter reality. Our country, for instance, has lost most of its best researchers in mathematics and physics during the past two decades. In our department, we are working hard to change this, but we encounter difficulties of many kinds. From our own experience, we feel that what is most positive in this is the contact between our young people with visitors from some of the main centers of research. We have about two or three visitors a year for a period of around one month each. This has provided both academic and moral support. The main difficulty in maintaining these visits is the cost of air fare to Argentina, which in the case of European countries comes close to US$ 2,000. This sum for a visit of six weeks is unobtainable for most countries. At present, agreements exist with individual European countries which can and are being used. In 1990 we had W. Meyer from Münster, a well known differential geometer, visiting Córdoba via an agreement with Germany (DAAD). Argentina paid for the living expenses and Germany covered travel expenses.

These agreements should be extended if possible. Something of this sort at the European level could help in (a) diminishing the bureaucracy involved, (b) making the visit of an expert available to more than one country or department at a time, for instance, by giving priority to joint applications which would benefit several departments. In our case, we have had rather close contact with the University of Chile and IMPA (Rio de Janeiro) for several years, but we should increase this interaction even more in the future. Along these lines, perhaps it is unrealistic, but still worth trying, to obtain air fare rebates (say to 2/3 of the value) from airlines.

Another issue of the same nature would be to increase the opportunities for research visits from a developing country to a European country. Both air fare and living expenses are a barrier. A common example is that of a young doctorate who returns to his home country. His chances of being able to travel periodically to the northern hemisphere are very slim, in particular, because he has not yet built a curriculum vitæ. A consequence is that often young people wait several years after their doctorate before returning home and then eventually decide not to come back at all.

Another problem is the cost of scientific journals. This is often too expensive for developing countries, especially during particularly bad economic years. This has caused frequent interruptions in the delivery of journals, gaps in the series and discouragement on the part of many researchers and students. To work to obtain a special rebate price from publishers for libraries of developing countries would be most helpful and is perhaps an attainable goal. I do not see that this would mean such a great loss for a publisher. In particular, we have been significantly hurt by the increase in price of Springer-Verlag journals of mathematics (for instance

Inventiones Math., US$ 1,800 a year, *Math. Annalen*, *Math. Zeitschrift*, US$ 1,400 a year).

These are the main aspects regarding the expansion of scientific research in our department and from the point of view of mathematics, which has the advantage not to require chemical products or as much sophisticated new equipment as experimental science does, apart from computer equipment. To sum up, I feel that (a) enhancing the interaction of researchers and (b) enhancing the availability of information, through journals and other means, are important areas where the European community could provide great help. Such help is already available and is being provided by individual countries but could be made much more efficient on the European level by speeding up the procedures and announcing widely new programs of visits or joint meetings.

Finally it should be said that it is understood that for aid to be profitable and a factor of change, it is essential that the developing country share the costs and efforts of moving forward. If aid is provided to many departments who are not making a genuine effort to improve, then funds would be wasted and whole programs would lose their support.

Most of the points mentioned above are the obvious ones but could perhaps be a starting point for other newer and more daring ideas for interaction among the countries. This is a very worthwhile and challenging test for the near future.

Roberto J. MIATELLO
FAMAF, Univ. Nac. Córdoba
5000 CÓRDOBA (Argentina)

A. Simis*

The following comments are the natural product of my experience as a researcher in a developing country, but they also unavoidably reflect conversations held on various occasions with colleagues in developed countries and, last but not least, my previous experience as a member of the CDE (1986–1990).

1. What Are Developing Countries?

I know that the correctness of this terminology is not under debate. Nevertheless, it is clear that some clarification is in order if one means to have a realistic and effective policy. I am sure the members in this debate are well aware of the differences between countries such as Argentina, Brazil, Chile, on one side, and other countries in very poor regions of Africa. Understanding such differences may help grasp why the needs of such countries, no matter how economically

* A. Simis has been a member of the CDE (Commission on Development and Exchanges of the International Mathematical Union). He frequently makes references to this commission but it is clear that his comments apply to any organization sharing the same goals, namely to foster mathematical research in developing countries.

different, are all needs of developing countries. Sometimes, a country is classified under "developing" not because it lacks the principal resources to stand up to the modern challenge, but rather due to relative political and social instability. This is not equivalent to lacking modern resources to face up to that challenge. As a corollary, when discussing future policy or guidelines in this debate, a few distinctive approaches ought to be implemented if anything at all is expected to be done. Under the uncomfortable risk of sounding "separatist", I will take the liberty of subclassifying developing countries into two groups, called I and II for the mere sake of the discussion. It will become clear that such subclasses reflect not only the present level of general socio-economic development of the countries, but also and, sometimes, mainly their level of mathematical sophistication and self-financing.

2. Developing Countries, Group I

I hereby give, if I may, a concise tentative list of "members" naturally falling in group I. In alphabetical order: Argentina, Brazil, Chile, China, India, Mexico, South Korea, Vietnam. This is a short list which may be increased, since I certainly have made omissions. The common feature among the above countries is a minimum stability that assures some reasonable uniform research activity, manpower and education. I am, of course, better informed about details in Latin American countries that in the remaining ones in the group, so we must be prudent here. Thus, for example, South Korea may be much better off in high tech production than Brazil or Argentina, but perhaps lacks the mathematical tradition of the institutions of the latter as well as their average output quality, by which I mean the output that results from internal research investment rather than the performance of individuals outside their country of origin.

I believe that, as a rule, the current need of Group I is mainly a *permanent credit line* to be opened to them by developed countries. And I mean it as literally as possible. Here are some highlights regarding this suggestion.

1) It is unclear what the exact format of such a procedure might involve. For example, it might involve only government resources from developed countries or also from private foundations, financing institutions, banks, and the like.
2) The main objective of this credit line would be to help out a country in the group whose normal investments in mathematical research came to a sudden halt due, for example, to political change. This would give active mathematicians in the problem country some relative security for the natural flow of their work. In particular, they could rid themselves of the recurring émigré syndrome.
3) The credit flow should stop as soon as the native research financing agencies resume their normal activities.
4) The credit line would not come to the rescue of the whole mathematical community in a problem country but rather to established or clearly emergent

groups of researchers, following the traditional policy of CDE guided by usual high standards.

5) The members of CDE should have access to every *in loco* evidence they need in order to reach a verdict as to the emergency situation in a particular country and as to what groups might be salvaged.
I believe this could be very effective and free of bureaucratic obstructions.

3. Developing Countries, Group II

This group comprises the vast majority of the remaining developing countries. Further subdivision might be appropriate to encompass the different needs of these countries. The magic password here seems to be *massive financing*, from the very basic topic of mathematical libraries through effective research manpower to direct investment in physical infrastructure. Here are some highlights:

1) Obviously, there is not enough money anywhere presently to finance research *ab initio* in all countries in the group, so one suggestion is to carefully select a few countries as testing grounds for the policy.
2) Output would be expected from the test countries as soon as they found a measure of self-sufficiency with the appropriate follow-up by the CDE members.
3) The process would be slow and at best results would show up only in the long run. The CDE should nominate special subcommittees for the periodic evaluation of local scientific activities.
4) The same high standards as before should be applied by CDE as to what groups should continue to receive massive support.
5) The support should be focused, action compact and free of bureaucratic obstructions.

I am, of course, assuming that none of this heavy financing is feasible within the normal national budget of the countries of Group II. True as it seems as a rule, some extra care has to be exercised when considering countries such as Iraq, Kuwait, Saudi Arabia which have special characteristics. Also, some thought has to be given as to the classification of countries such as Israel and South Africa which in many respects are developed countries.

4. General Suggestions

a) Found a Committee for Advancing Research (COREA) within the European Mathematical Society to help foster mathematical activities in developing countries. The First European Congress of Mathematics seems to be a most appropriate occasion.
b) Search means of establishing foundational grants within the European framework to assist developing countries with the aforementioned support.

c) Make operational a procedure under which important publishing houses (such as Springer, Birkhäuser, Dunod, and so on) would print cheaper editions of books and journals, say, at half or one third the regular price, so that libraries and/or individuals in developing countries could afford acquiring the main publications.

d) Find a means to overcome the permanent deterioration of buying power of local currency in developing countries due to rising inflation and an unequal international trade balance, so as not to interrupt subscriptions and bookshelf updating.

e) A similar mechanism as in the preceding item which would allow for a reasonable recycling of computer equipment for mathematical purposes.

5. Concluding remarks

It is my personal belief that some of the above suggestions can be implemented with the active support of CDE through the prestige enjoyed by its members in the mathematical community at large and, in particular, in the European mathematical community.

Clearly, a clear-cut aggressive action is to be taken by the selected local groups in developing countries in order to annihilate the usual native obstructions for attaining the discussed goals.

What is generally clear is that the active mathematical groups in developing countries need to be treated along the policy lines of CDE, i.e., merit and high standards. The last thing they need from the European community is a third world paternalistic view lacking funds!

Aron SIMIS
Departamento de Matemática
Universidade Federal da Bahia
40 210 SALVADOR, Bahia (Brazil)

APPENDIX III

Centers in Collaboration

This Appendix contains data concerning centers which organize some collaboration schemes with developing countries. The data have been drawn from information issued by these centers that reached the round table organizer through a poll conducted on the occasion of the round table.

I.C.T.P.

The International Center for Theoretical Physics (ICTP) is part of the International Atomic Energy Agency (IAEA, Vienna-Austria) and of the United Nations Educational, Scientific and Cultural Organization (UNESCO, Paris-France). Its Director, since its creation in 1964, is Professor A. Salam. The ICTP is financed by the IAEA, by UNESCO and primarily by the Italian government for its ongoing activities. Some programmes receive funds from other countries*. The main objectives of the ICTP are:

a) to help foster the growth of advanced studies and research in physical and mathematical sciences, especially in the developing countries,
b) to provide an international forum for scientific contacts between scientists from all countries.
c) to provide facilities to conduct original research to his visitors and associates, principally from developing countries.

The first programme relating to mathematics was held in August 1971. The programmes in mathematics were then consolidated. From 1971 onwards, courses were held every year and workshops, conferences and symposia were gradually added to the list of activities. Although some research in mathematics was being done at ICTP before 1986 (mainly in collaboration with the International School of Advanced Studies ISAS-SISSA of Trieste, Italy), the research in mathematics at ICTP started on a firm basis in September 1986, under the direction of J. Eells (University of Warwick, UK) and A. Verjovsky (Mexico).

In addition to the mathematicians in residence (around 30, the main themes of research being geometry of PDE's, global variational theory, complex analysis, dynamical systems, differential geometry, topological methods in analysis, complete integrability), the research group includes visiting mathematicians, postdoctoral fellows, research students and associate members**. Mathematicians from industrialized countries are welcome to spend their sabbatical leave at ICTP. The

---

* cf. *Mathematics at the International Center of Theorerical Physics, ICTP Trieste-Italy 1989* for more details

** Associate members are selected scientists from and working in developing countries who have the opportunity of making three visits to ICTP—from 6 weeks to 3 months—in a six-year period. In 1989, 65 of them were active mathematicians.

mathematicians from ICTP collaborate with their colleagues from ISAS-SISSA. The mathematical activities include regular seminars, lectures, workshops and colleges.

The *Office of External Activities of ICTP* sponsors summer schools, workshops and seminars organized in developing countries either by ICTP or by scientists from the developing countries themselves. The *Book Donation Programme* aims at providing selected libraries or research centers in developing countries with scientific books and periodicals or unused surplus of scientific equipment. The ICTP also concluded *agreements with universities*, institutes and departments for them to send junior scientists for 40 to 120 days every year.

The activities of ICTP are advertised through by now well-known posters and leaflets (general scientific programme, individual workshops, schools or colleges). Some limited financial support is available to mathematicians from developing countries. As far as mathematics is concerned, the ICTP currently offers a limited number of *postdoctoral fellowships* (one year) and of visiting mathematicians *scholarships* (3 to 9 months).

I.C.T.P.	Telephone: (40) 22.401
Strada Costiera, 11	Cable: CENTRATOM
P.O. Box 586	Telex: 460392 ICTP I
34136 TRIESTE, Italy	Telefax: (40) 22 41 63
	root@ictp.trieste.it

C.I.M.P.A.-I.C.P.A.M.

The Centre International de Mathématiques Pures et Appliquées (International Center for Pure and Applied Mathematics, CIMPA-ICPAM) was founded in Nice, France, in 1978. The objectives of the CIMPA are:

1) to promote mathematical research and higher education in developing countries (both in mathematics and computer science),
2) to provide high level training to mathematicians from developing countries,
3) to improve cooperation between mathematicians of industrialized countries and developing countries (both North-South and South-South collaboration,)
4) to collect, publish and distribute mathematical documentation in developing countries.

The main sponsors of CIMPA are the French Ministry of Education, the French Ministry of Foreign Affairs, the City of Nice, UNESCO, the Royal Society and the University of Nice Sophia-Antipolis.

The activities of CIMPA are as follows:

a) organization of international schools of mathematics and computer sciences in Nice, other French universities or in developing countries,
b) publishing scientific works (in particular documentation used in the courses)
c) collecting, producing and distributing scientific documentation in the service of developing countries

d) financing grants for trainees who attend mathematics or computer science courses.

The CIMPA circulates leaflets calling for the courses that are organized.

Telefax: (40) 22 41 63

CIMPA
1, Avenue Edith Cavell
06000 Nice France

Telephone: (33) 93 53 18 43
Telefax: (33) 93 81 73 48
cimpat@taloa.unice.fr
Telegrams: CIMPANICE

I.C.M.S.

The International Center for Mathematical Sciences (ICMS) was established in Edinburgh (UK) as a joint venture by Edinburgh and Heriot-Watt universities to be a center of excellence. Its activities began in 1991. Participation by mathematicians from the developing world in these activities is welcome. The aims of the ICMS are:

1) to create an environment in which mathematical sciences will develop in new directions,
2) to encourage and exploit those areas of mathematics that are of relevance to industry and commerce,
3) to promote international collaboration with particular reference to mathematicians working in the developing world.

The activities of the ICMS will consist of:

a) Research programmes and workshops, aimed at stimulating the chosen topic by bringing together researchers with different backgrounds and different perspectives, who might otherwise not have the opportunity to work together,
b) Courses of various types for wide audiences, including courses for postgraduates, and courses aimed at participants from industry and commerce.

The ICMS received start-up support from Edinburgh and Heriot-Watt universities, the London Mathematical Society, the Edinburgh Mathematical Society and Standard Life Insurance. It has embarked on a campaign to raise funds in order to insure its development for the next years.

I.C.M.S.
Mathematics Department
Heriot-Watt University
Riccarton Edinburgh EH14 4AS
Scotland UK

Telephone: (31) 451 32 50
e-mail: icms@cara.ma.hw.ac.uk
Telefax: (31) 451 32 49

C.I.M.A.T.

The international center for mathematics (CIMAT: Centro Internacional de Matemática) is being organized and will be located in Coimbra (Portugal). One of its aims will be cooperation with developing countries.

Round Table L: July 9, 1992

MATHEMATICAL LIBRARIES IN EUROPE

organized by P. Barrat,[1] G. Sureau,[2] L. Zweig[3]

Université Paris VII[1], Université Paris XI[2], Ecole Normale Supérieure[3]

Paris (France)[1], Orsay (France)[2], Paris (France)[3]

Note to the Reader

The texts included in the Proceedings of the round table *Mathematics Libraries in Europe* have been assembled from the record made on July 9, 1992. These texts do not constitute an exact transcription of the tape, but are rather highlights of the discussions and presentations. In particular, we have changed the order of a few sections and have tried to arrange them by theme.

The round table was made possible by a grant and the encouragement of the organizers of the First European Congress of Mathematics and of the *Sous-Direction des Bibliothèques.*

In particular, we wish to thank the following persons who were willing to participate in the round table:

- Laszlo Babai (Budapest, Chicago)
- Domenico Bogliolo (Rome)
- Catriona Byrne (Springer-Verlag)
- Stanislaw Goldstein (Lodz)
- M. Guillaume (Clermont Ferrand)
- Pilar Ibarrola (Madrid)
- Sergej B. Kuksin (Zurich, Moscow)
- Hervé Le Crosnier (Caen)
- Ay Ong (Amsterdam)
- Ms. Ovadieva (Sofia)
- Dietmar Petroll (Oberwolfach)
- Samuel M. Rankin III (A.M.S. Providence)
- Françoise Renzetti (Grenoble)
- Agnès Szendrei (Szeged)
- Dan Stoica (Iasi)
- Bernard Tessier (Paris)
- Fleming Topsoe (Copenhagen)
- David Tranah (Cambridge University Press)

We do not overlook all the anonymous participants in our thanks.

The Organizers

October 1992

Introduction

The compilation and analysis of the questionnaire sent out throughout Europe a few weeks before the Congress allowed us to define the main problems of libraries. In light of these results, the round table oriented its presentations and debates around three themes.

— Financing of libraries

— Scientific publishing, problems of documentation

— New technology, the computerization of libraries, new media and problems of personnel.

About the questionnaire (for the results, see Appendix 1)

In January 1992, 200 questionnaires were sent to target libraries on the one hand, and on the other hand, around a hundred were sent to European mathematical societies to be distributed among their own mathematics libraries.

72 responses were obtained, which seems meager yet promising, considering that it is the first time that such a survey has taken place and that a directory of European mathematics libraries strictly does not yet exist.

The large number of responses from French mathematics libraries (27) is due to the existence of a French mathematical network. Everybody knows everybody and specialized libraries are perfectly well known. It should be noted that some Western European libraries participated in the round table even though they had not responded to the questionnaire.

15 themes were proposed for consideration:

— Financing of libraries
— Consistency of financing
— Acquisitions' policy and the development of collections
— What do mathematicians have to say in these policies
— Exchange policy
— Interlibrary loans
— Computerization of libraries
— Cataloging norms
— Gray literature
— Ancient books
— Library personnel
— Increase in the cost of documentation
— New technology: CD-ROM, databases...
— Administrative structures
— Relationships with scientific publishers

The majority of responses made it apparent that financing of libraries and a consistent source of funding was the major preoccupation.

The second point raised was personnel. Whether in large libraries or small, the lack of personnel is flagrant. Moreover, the qualifications of the personnel are also often lacking. Is it necessary to recall that the libraries are the laboratories of mathematics departments and that a laboratory does not go without qualified personnel.

A third subject of concern stems from the fact that 75% of libraries are computerized, some very recently. Thus the problem of norms and new technology preoccupies a fair number of librarians and users. Right now, all institutions are faced with a worrisome phenomenon, the prodigious increase in the volume of published documentation and the inflation in prices. This is an important theme that was retained for discussion and comments.

Who responded to the questionnaire?

It is very interesting to see that the libraries who responded are above all libraries whose diversity and volume reflect the scientific activity of the mathematics department to which they are attached. As we see in Tables 1 and 2 in Appendix 1, the libraries possess on the average 200 to 250 current periodicals and from 10,000 to 30,000 works. They are principally aimed at readers in mathematics departments, researchers and graduate students. Obviously the quantity and diversity of documentation depends in a large part on the number of researchers and on the number of research areas in which the departments are active.

The number of readers in the libraries varies from 100 to 1,000. We should note that academic or university libraries who responded to the survey and who sometimes have interesting mathematical collections were not able to determine which of their readers were in mathematics for administrative reasons.

Similarly, these academic or university libraries cannot distinguish specific collections and show up on peaks in histograms 1 and 2, Appendix 1 (1,000, 10,000 or 100,000), that do not correspond solely to mathematical collections.

Remarks on budgets

It was very important to know the means of financing for these libraries. The responses underlined the financial difficulties of the majority of the libraries, whether in the East, or in the West, in such countries as Germany, Italy and Holland. The majority of these institutions have budgets that are unfortunately fixed in the sense that their budget remains constant or decreases. As a result of documentary inflation, the libraries have to resort to cancelling subscriptions (43% have already had recourse to such cancellations). Only the most important libraries have up to now escaped the axe. French libraries, it should be noted, have varied sources of financing. For this reason, they are perhaps suffering the least and, in any case, are cancelling the least subscriptions.

To avoid cancelling subscriptions, some libraries are diminishing their book orders, sometimes drastically. Others have had to stop binding their periodicals.

Mathematicians are very close to "their documentation"

Mathematicians play a non negligible role in the acquisition policy for books and periodicals in their libraries. In a majority of cases, this policy is decided by a committee of mathematicians.

In most libraries, mathematicians have free access to the shelves. Almost all libraries lend their books (all of the collection or only part of it) to their researchers, but a good number of libraries, especially in France, do not lend their periodicals.

80% of libraries have exchange policies. This policy gives some oxygen to several institutions, especially in Eastern Europe. The circulation of "gray literature" is intense and increasing, with an important scientific impact on the community of mathematicians. This circulation grows all the more as "modern means" tend to become widespread.

75% of libraries are now computerized, most of them for cataloging and lending, a certain number also for acquisitions and management and only a few for periodicals. Computerization started slowly about ten years ago. Its pace has quickened the last five years. For example, among the 36 Western European libraries which answered the questionnaire, only 6 are not yet computerized. In France, only another 6 are not computerized out of a total of 27.

Computerization, however, has often been rather anarchic, if we may say so, as each library decided on its own software and norms. Nonetheless, the creation in France of an "informal network" has led to an attempt at normalization: 3/4 of libraries use the same software and follow the same norms.

If, in answer to informal questions, some people mention problems of work space, most insist particularly on financing and personnel problems.

If this questionnaire was helpful in defining some of the problems and focusing the discussion, it is clear that we must consider it as an initial contact and that a more comprehensive survey of all mathematics libraries in Europe would be useful.

FINANCING OF THE LIBRARIES

The Situation of Libraries in Russia

by S. Kuksin

Moscow (Russia) and ETH-Zentrum, Zürich (Switzerland)

The problems mentioned here are identical to those of the former Soviet Union, including the Baltic Republics, the Ukraine and Moldavia. Russia is in very serious economic trouble and certainly will not recover during the next 3 to 5 years.

The main problems are the following:

— Academic and scientific institutions and libraries are very short of funds. Subscriptions to foreign journals are now being cancelled and new books are no longer being purchased. This is the situation in Moscow and St. Petersburg, and the problem is even more serious in smaller centers.

— Europe and the United States are in the process of helping Russian libraries by sending them journals and books; however the mailing problem has not yet been overcome. There has to be special agreements with the mathematical societies that are presenting an aid program for the use of express mail, diplomatic channels and Russian mathematicians visiting foreign universities. New solutions must also be found.

— What is actually needed is for young Russian mathematicians to have access to the dozen or so of the leading mathematical journals to give them a picture of modern mathematics.

— The second problem concerns books. Libraries did not obtain many books from western publishers. Of those that had been previously obtained, many of them had been translated into Russian. The translations were then used in other countries of eastern Europe. Now very few foreign books are being translated and this is a serious problem.

— Another very serious problem concerns photocopy machines. Photocopying in Russia has always been problematic, although it is worse now because even if there is a machine, there is no paper, no toner, etc.

Help is urgently needed for our universities and our mathematical centers for books and journals! Thank you.

The Situation of Libraries in Poland

by L. Goldstein

Lodz University, Institute of Mathematics, Lodz (Poland)

The purpose of these notes is to present the situation in libraries in Poland from the point of view of mathematicians. To give an impression of the situation in Polish libraries, we shall try to answer a list of simple but important questions, namely:

1. How easy is it to get to the library?
2. Is it possible to find a book or a journal we are looking for in the library?
3. How easy is it to actually access the book or journal, i.e., how long do we have to wait for it?
4. Is it possible to borrow a book or a journal?
5. How easy is it to make a copy of an article we have just found?

6. Is interlibrary cooperation helpful?
7. How easy is it to obtain information on the appearance of new books and periodicals, and of articles in journals?
8. What is the impact of new technologies?
9. What are the main problems and what are the prospects for improvement?

In order to be able to answer these questions, I shall briefly present the organization of higher education in Poland and its consequences for the organization of libraries.

There are about 10 universities and many educational institutions having the status of a university, such as technical universities or medical academies. There is usually one main university library, gathering books and periodicals on all possible subjects. Strictly mathematical libraries can be found in mathematical institutes and departments. There is also one outstanding library of the Mathematical Institute of the Polish Academy of Science. The answers to the above questions depend very much on the kind of library we consider.

Most of the time it is easy to reach the local mathematical library, but the situation is further complicated when more than one mathematics institute exists in a city, which often happens. Since the universities are usually dispersed all over the city, the main university library is sometimes not very conveniently situated.

Main libraries normally have a large amount of books, but they are usually not those a mathematician is looking for. The situation is much better in local mathematical libraries, where even if a book is not in the library, it is often possible to both order and obtain it within a few months.

The situation with periodicals is much worse. Some libraries offer quite a number of different titles, but few are of real importance, and many important ones are missing. Moreover, it is difficult to find a complete set of any particular journal — many issues or even years are missing. The only notable exception is the Library of the Polish Academy of Science. The situation differs from place to place, from good in Warsaw, acceptable in Wroclaw and Kracow, to very bad in other places.

Accessing a book starts with finding it in one of the good but traditional catalogues and continues with asking a library employee to obtain it. The actual waiting time for a book or journal varies from half a minute in a local library to half an hour or longer in the main university library. Only newly-arrived periodicals can be found on the generally accessible shelves.

It is relatively easy to borrow a book or a journal, but in the main library it might mean waiting until the next day. Of course, many books and periodicals are very hard to obtain once they have been borrowed. Getting a book from another library requires a lot of patience.

We are usually well informed on the appearance of new books and periodicals by means of catalogues and newsletters. On the other hand, information on forthcoming articles is very difficult to obtain. Also there are few places subscribing to Mathematical Reviews, Zentralblatt für Mathematik or current mathematical publications.

The process of "informatization" of the libraries has just begun. It is hindered by the lack of standards. There is a growing number of mathematicians with a computer in their room, and networks are quickly expanding. On the other hand, there is little chance of using on-line database systems offered by some information centers in Western Europe because of lack of money.

The main problem is of course money or rather lack of it. This is responsible for the very poor accessibility to journals. Another problem is the doubling of resources connected with the existences of many rival institutions of the same kind in one city. Exchange of periodicals with educational institutions abroad is too centralized to be of great help for provincial (but not necessarily small) universities. The universal application of the consecutive numbering of the books makes it very difficult to put them on shelves for easy access.

The main university libraries are financed centrally. The local libraries' existence depends on the flow of money from grants from members of the institutes. This makes their future very uncertain.

The Financial Situation of Mathematical Libraries in Hungary

by A. Szendrei

Jozsef Attila University, Bolyai Institute, Szeged (Hungary)

In Hungary there are six libraries with a significant mathematical collection; two general libraries: the central library of the Hungarian Academy of Sciences and the central library of the Technical University of Budapest, and four mathematical libraries: the library of the Mathematical Research Institute of the Hungarian Academy of Sciences and the mathematical libraries of three universities (in Budapest, Debrecen, and Szeged). It should be noted that the mathematical libraries of the three universities are not completely independent of the central libraries of those universities.

The mathematical libraries in Hungary have similar problems; therefore, although I have information mostly on the library in Szeged and the library of the Mathematical Research Institute in Budapest, I think these two will reflect the general situation.

Financing

The financing of the mathematical libraries in Hungary comes mainly from government sources, and the real value of this financing has been strictly decreasing for at least a decade. Just to illustrate this with figures for the last two years: in 1991 the inflation rate was 36% while the financing of these institutions was increased nominally by 5%; this year the inflation rate is expected to be 27%, and it has already been announced that no nominal increase is planned in financing for 1993.

We have some other sources as well:

1) Research grants. The grant system was introduced in Hungary in 1986. Primarily, the applicants for these grants are research groups. The source I am referring to is that research groups having grants offer a certain (not low!) percentage of their grant money for the library of their institutes. (Consequently, they have less money for such expenses as travelling and inviting guests.)
2) TEMPUS. The TEMPUS program allows the Eastern partners in the projects to buy books to raise the level of teaching activities covered by the projects.
3) Occasional donations by foundations and institutions in Western countries, and donations by individuals.

Unfortunately, there is a problem with the latter two sources: in most cases sources from Western countries are not allowed to be spent on ordering journals (even for a limited period of time).

How do we try to cope with the situation?

Periodicals. Because of rising prices and insufficient financing we have been forced, from time to time, to cancel subscriptions of several journals. This process started almost a decade ago, so that now we have reached a point where new cancellations would have a severe effect on current research. The most painful loss to the library in Szeged was in 1989 when we were obliged to cancel the *Crelle Journal*, which we had possessed since its very beginning in 1823. Cancelling subscriptions was coordinated so that each important journal could be found in at least one mathematical library in Hungary.

Since 1991, the Hungarian export-import company for cultural goods, which sold Hungarian journals at an unfairly low price and at the same time was disinterested in negotiating prices of books and journals it bought for the Hungarian libraries, has lost its monopoly. As a consequence, in that year, Western journals cost our libraries less than the previous year. However, it was during this year that the price of journals from Eastern countries rose sharply. It is rather surprising, but in many cases a journal from the former Soviet Union costs practically the same as its English translation. Therefore it is very probable that Hungarian libraries will switch to the English translations, and at the same time will be forced to cancel some of them.

Because of the financial difficulties, exchange of periodicals is very important for us. The mathematical libraries in Hungary have extensive exchange programs, and we are interested in widening them.

Books. I think it is fair to say that the number of books acquired by the mathematical libraries in Hungary is far below the needs. However, if we want to make a year-by-year comparison, the picture is more complex, and depends on how many funds were available from what I called extra sources that can only be spent on books. For example, in the library of the Mathematical Research Institute, the number of books being acquired in recent years has been decreasing, while at the library in Szeged there was a slight increase, thanks to TEMPUS.

The library in Szeged also keeps the books that are reviewed in the mathematical journal of the University of Szeged (*Acta Sci. Math.*), and these are a very valuable contribution to the collection of books.

Services in libraries. All mathematical libraries in Hungary are in old buildings and there is no room for expansion; in some of them the reading room is very small. Copying is available in all libraries; however, in most cases demand for the copiers far exceeds their capacity.

In the Szeged library, there is a computerized catalogue of books where the decimal and the AMS classifications are also included. The whole process was carried out without any extra fund. In the library of the Mathematical Research Institute, computerization is in the making.

Summary

The financial situation in mathematical libraries in Hungary is urgent, although it is certainly better than in many other East European countries. So far, we have been able to survive with a few sacrifices, thanks to the generosity of colleagues and some foreign institutions. However, we have rather gloomy prospects for the near future.

I would like to conclude with some suggestions on how our problems could be eased.

1) We would welcome offers for exchange of journals, as well as superfluous copies of books that some libraries might have.
2) It would be a great help if publishers regularly sent us their yearly catalogues — even if we cannot afford to buy the books each time. We would also be grateful if publishers sent us a free copy of each new publication that we request for a book review.
3) The restriction of Western projects or donations to the effect that the fund they provide cannot be used for ordering journals ought to be reconsidered. Furthermore, the COST program does not seem to allow spending money on books or journals, which is bad news for us.
4) The financial aid from EC or Western governmental sources intended for Hungarian libraries usually ends up in the main general libraries. Perhaps the European Mathematical Society could try to see to it that a portion of

these sums be transferred to the special mathematical libraries, maybe using coordination of the Hungarian Mathematical Society.

Discussion

Financing of libraries

The financing problems facing Western European libraries are less dramatic than in the East. Nonetheless, the results of the questionnaire show that 43% of libraries in the West have cancelled subscriptions in the last two years (cf. Appendix 1).

The situation in France is slightly better because French mathematicians organized themselves in order to obtain multiple funding from various sources.

In Germany, a large number of universities and mathematics departments are developing interlibrary loans and cooperation between neighboring centers in order to achieve savings on the subscriptions.

In Italy, there are also financial problems concerning computerization as well as personnel. Quite a few centers are forced to draw from their research funds in order to maintain their libraries and subscriptions have had to be cancelled.

Traditionally, the exchange of periodicals between universities was a means of lowering financial pressure. In this way, in spite of the lack of currency, Eastern European countries were able to obtain journals. This allowed their mathematicians to keep in touch with the West through their libraries and stimulated their creativity, even if the exchanged periodicals were of varying value. It so happens that more and more Western centers are now unable to exchange the periodicals created by their mathematicians because these journals are no longer published by university presses but rather by commercial publishers. Such journals are totally inaccessible to poor countries and are very hard to buy when budgets are strained. Concerning journals published in the East, material problems disrupt their publication.

Another cause of financing problems is the steadily rising cost of documentation, combined with a serious documentary inflation induced either by the appearance of new journals or by the increase in the number of volumes of existing ones.

SCIENTIFIC PUBLISHING AND PROBLEMS OF DOCUMENTATION

There are several studies on the subject: the most recent ones are those of the European Mathematical Society and the American Mathematical Society (*Notices* **vol. 36** no. 9, 1989, pp. 1193–1198). Another study concerning scientific periodicals in general appeared in *Nature* **vol. 341**, 9/28/1989 issue, by John B. Merriman.

Documentary Inflation and the Rising Cost of Journals

by C. Byrne

Editor, Springer-Verlag, Heidelberg (Germany)

I would like to make a few remarks on some aspects of mathematical, and more generally, scientific publishing, in particular, on mathematical publishing and its interactions with libraries. It is certain that the problems you encounter as librarians or as scientists concerning publications correspond to problems in the publishing business.

Year after year, we observe an increase in the volume of published documentation. It is true for books, it is also true for journals since new journals are created every year. I checked into *Zentralblatt für Mathematik* from 1972 to around 1991: 21,749 books have been recorded. This only concerns books, which means that, even if we do not take into account the acceleration in publications, at least 1,000 books are published every year. This is enormous. And in the last few years journals have become increasingly numerous as well. Of course, it is easier to keep track of new journals that appear than old ones that disappear—this happens too—but I think nonetheless that the total number is increasing.

What are the causes of these phenomena which constitute a real problem for you as well as for us? In the last ten years, we have observed a decrease in the sales of each published book and a decrease in the number of subscriptions to each journal. As the number of copies of each book printed diminishes and as the circulation of each journal is reduced as well, the fixed charges in the price of each book or journal are obviously increased. Therefore, the prices go up even though we know that libraries have fewer and fewer readily available funds.

I want to point out here that in mathematics we do not depend on libraries, not even in the main, I think. We insist on publishing for individuals, which is obviously becoming harder and harder due to the rise in prices. Most mathematicians think that once a book has been published and that a certain number of copies have been sold to libraries, the publication is written off and the rest is just profit. This is not the case. Sales to libraries, of which we are assured from the beginning, only number a few hundred. Beyond this are sales to individuals or libraries which order afterwards. In any case, the sales to libraries are not sufficient to write off publication costs.

The reasons for this explosion in publication volume and prices reside in a large part in the community of mathematicians itself. The subject is complex, mathematics itself splits up into more and more specialized branches and one needs books and journals to be able to work. At the same time, these books and journals find fewer and fewer readers.

Concurrently, however, there is a need for information from the other branches of the discipline. We are in an era of interactions between branches of mathematics that formerly appeared to be disjointed but no longer are. Mathematicians feel a strong need to become informed about other areas of their subject. There is thus a very definite demand for books which give their readers information about neighboring areas. We must say that such sales decrease much less.

The decrease in sales and circulation is not uniform. It corresponds to the needs and development of mathematics. There is also a sociological reason for the fact that the number of books published today is larger than what it was 20 or 30 years ago. Whereas in the past publishing was somehow élitist, it is now accessible to more mathematicians. This development has had good as well as bad effects, namely a proliferation of publications which is perhaps not sufficiently under control. This is where, in my view, an appeal to mathematicians themselves should be made.

All the books we publish are previously read by mathematicians who give us their advice on these manuscripts. These referees have more or less amounts of time in which to do the work, the manuscript is more or less interesting to them depending on the case. I would like to make an appeal to mathematicians not to approve the publication of a manuscript on the sole ground of technical quality, not to be content with saying "this is good, it is good mathematics, the rest is none of my business". I think that every mathematician who feels responsible in some way for his community, and in particular for the life of his library, must also wonder whether there really is enough people and libraries for which this particular book is necessary and indispensable. Are the articles published in journals sufficient? Is communication through other means sufficient? Is this book really necessary? Will it have any readers? We ask our referees this question and it is often hard for them to give an answer. But as we always consult several persons, their points of view complement each other and often converge. My impression is that quite often mathematicians do not think much about this aspect when they give advice on a submitted manuscript.

I would like to talk about publishing in general, but I will say a few words on Springer's policy. I looked at what we have been doing since 1984 and I saw that the number of published books stayed relatively stable without any increase. I am only talking about English language books since German language books are primarily textbooks and only concern the German-speaking community. There are of course fluctuations, which is quite natural, taking into account that the publishing of a book cannot be programmed, the publication date depending essentially on the author. The figures for 1992 are rather similar to those for 1984–1985. We see no reason whatsoever for increasing the number of books we publish per year. We have also been faced for several years with rising publication costs. This rise is due to several factors. The manufacturing costs for a book represent only a small part of the production, distribution, diffusion and sales costs. It is nonetheless the part of costs that is easiest to influence and we did that as much as possible: we buy the typography and filmsetting (...) abroad, in particular in Asia, to reduce manufacturing costs maximally. We print a lot in the United States because the dollar is low. Obviously, this only represents a limited part, but we are able to manage this part a little better. Unfortunately, our influence on distribution costs is more limited.

Concerning journals, Springer publishes around twenty mathematical journals. Most of them were created in the 70s and are representative of the splintering of mathematics. Three new journals were created between 1980 and today. Two of them present new interfaces between mathematics and computer science and the third one was recently published in New York to serve as an interdisciplinary forum for mathematicians, engineers and physicists. We want the number of our journals to remain stable at all costs. We reject on the average one journal per month. We are not looking for ideas to create new journals. In fact, we often reject such ideas. When these proposals for new journals reach us, we do not always reject them immediately. We try to work seriously for the persons who propose a journal, so we ask around for advice. However, the consensus is almost always that, even if the journal has genuine scientific value, it is currently impossible to launch a new journal. From an economic point of view, the libraries just cannot afford to subscribe.

I thus repeat my appeal to mathematicians. Does mathematics really need to keep splintering? Do we have to create more and more small journals intended for ever narrower groups of mathematicians? Is this good for the community of mathematicians? Such journals will never reach a very high circulation. They can be excellent, of the best quality, have the best editorial board you can imagine, but they will never have as many subscribers as classical journals have.

Our classical journals have lost an average of almost one hundred subscribers per journal during the last ten years. One journal stayed stable, another lost almost two hundred, the average is around one hundred. This is a lot when you think that without exception their circulation is below one thousand. Ten years ago, there were circulations above one thousand. There are none today.

The table of Appendix 2, taken from an article in *Nature (vol. 341, 1989, 28/9) by John B. Merriman* describes eloquently the difficulties new journals meet to reach a sufficient number of subscriptions. The four curves show that the trend is nothing but increasing: one observes that a scientific journal created between 1982 and 1986 can barely reach the number of subscribers after five years that older journals had after one year. I tried to draw a small curve (it is the dotted curve, the second one from the bottom of Appendix 2) that looks exactly like the other curves: it shows the total of four mathematical journals launched between 1984 and 1988. We see precisely the same trend as for other scientific journals.

Let me ask mathematicians what is the advantage of publishing articles in a journal that only have one, two or three hundred subscribers. Are they sure their article will reach all the people that might be interested? It seems to me that something can be done on the level of existing journals. Indeed, every time a new journal is proposed, existing journals are criticized: they are not sufficient or they are not good for this or that reason, or they are not what they should be. Isn't there still time to do something here, to stabilize these journals that do have subscriptions and very faithful subscribers? We know only too well how a librarian is loathe to cancel a subscription. An existing subscriber, generally speaking, is as faithful as possible.

Questions touched upon during the discussion on documentation costs

Increase in the number of volumes

If during the 70s libraries were still able to face the increase in the number of volumes of a journal, this is no longer the case today. When in the course of a few years a journal goes from one volume to several, this amounts to taking a new subscription for many libraries. Some commercial publishers, well aware of how reluctant librarians are to cancel a subscription, excessively increase the number of volumes and thus put some libraries in difficulty. A few people suggest that the scientific editors are primarily responsible for this situation.

Mathematicians should pay more attention to their library and should be more critical about acquisitions. Only mathematicians can bring about a change in the mathematical publishing market.

Proliferation of new journals and marginal cost of extra copies

What to choose? Either increase the number of volumes of existing journals or create new journals?

A publisher who increases the number of volumes of a journal, even by adding new sections, is not taking much risk since the journal is already on the market. On the contrary, the launching of a new journal should only occur when a new subject emerges that is not represented in existing journals. Why is it then that some publishers enter such a risky market knowing in advance that the number of copies sold will be extremely low for many years?

It is worthwhile recalling that the marginal cost of production of extra copies of a journal is considerably less than the subscription price of the same journal. According to some individuals, this cost would only represent 10 to 18% of the subscription price. It is possible that if commercial publishers would lower the price of their journals they would see an increase in their circulation.

According to other opinions, this is contradicted by the fact that a new journal of high quality, launched for a relatively low price about five years ago, has not even reached 200 subscribers today.

Additionally, these figures are extremely small in terms of publishing. The production costs are not the only ones that count. In fact, they have rather been decreasing in the last few years, due to delocalization of typography and typesetting towards countries where these activities are less expensive, and so on. When dealing with international diffusion, other factors come into play that could help explain the high subscription prices.

However, is it really necessary that publishing houses send so many representatives to visit mathematicians in their universities, only to obtain the vague promise that if they fancy writing a book, they will publish it here rather than there? All this contributes to the final price.

Low-cost production?

Cheap new editions exist in certain countries such as India. Would it be possible to let other countries such as Eastern European countries benefit from them?

A necessary condition for cheap new editions is that it be preceded by a full-price edition. Some buyers must have paid the full price.

Do mathematical publications have to be very sophisticated and thus cost so much?

Conclusion

Scientific publishing is a kind of publishing that is not always easy, but is certainly profitable given the prices and the number of new jounals that are issued. On the other hand, mathematics certainly is not a "popular" subject which explains its low circulation.

NEW TECHNOLOGIES AND COMPUTERIZATION

Computerization of Spanish Libraries

by Pilar Ibarrola

(Universidad Complutense, Facultad de Matematicas, Madrid (Spain))

The computerization of Spanish libraries is proceeding in a rather disorganized fashion with a lack of instruction made worse by the situation of the National Library. Public libraries are slowly getting computerized and the changes in choice of software bode ill for future problems. The situation is better in universities and in the CSIC (Superior Council for Scientific Research). Existing networks are the following:

I *DOBIS-LIBIS network*. This includes the universities of the Basque Country, Grenada, Navarre, Oviedo, Salamanca, Seville and the UNED (National University for Correspondance Courses). These institutions all use the DOBIS-LIBIS system and the hub of the network is located in Oviedo with a common catalogue.

II *REBIUN university libraries network.* This includes about twelve university libraries using various software (VTLS, LIBERTAS, ABSYS, SABINI, DOBIS-LIBIS...) Its objectives are interlibrary loans and a common catalogue. All these libraries are currently connected together via a CD-ROM except those using DOBIS-LIBIS.

III *CSIC libraries network.* This is composed of more than 80 libraries using the ALEPH system.

IV *SLS Libertas network.* This is currently being constructed (Madrid, Majorca, Malaga...) These libraries have a possible connection with the British JANET network.

V The National Library uses an in-house system called ILIADA and is compiling a catalogue (ARIADNE) of 8,000,000 documents.

VI Finally, the thematic network DOCUMAT groups together 17 university mathematics libraries in addition to those of the CSIC. The objectives of this network are: (1) To compile a common catalogue for mathematical journals that exist in Spanish libraries. (2) To choose and buy new journals to be added to the collection of the various libraries (decisions made during one annual meeting). (3) To spread mathematical information through the exchange of abstracts, brochures or other appropriate means. (4) To exchange documents between members of the network. To send off of photocopies with compensatory accounts (clearing). (5) To install the Math-Sciences database. (6) To prepare the future on-line connection between libraries, given that although all participating institutions are computerized, they are using different systems. The connection will be effected in the framework of the IRIS program. The common catalogue will reside in the Digital VAX 631 of CSIC which will be the central computer. Each library will be connected with the VAX through IBERPAC in X25. The network will be connected with the British Library through ICYT (CSIC information and documentation center).

The French Experience and Norms

by Pierre Barrat

(Université Paris VII, Département de Mathématiques, Paris (France))

The mathematics libraries in France have joined a so-called "network", which is more accurately described as a pressure group. Financial problems are at the root of the meetings held by librarians in mathematics. They next tried to unify their computerization, as much as possible. In January 1985, it was decided to select a common software to avoid anarchic choices. A cataloging standard was defined. This standard evolved in time to get closer to the MARC norm of the Library of Congress of the United States.

The experience of French libraries is in this respect relatively interesting as almost 3/4 of them are computerized with compatible and exchangeable files. A certain number of these catalogues have been merged to create an on-line file that can be read, for example, from any Minitel in France. The elimination of duplicates was a slow and painful work performed by André Voutier. This file only concerns books because in France there is a computerized national catalogue for periodicals, concerning all subjects and all kinds of libraries.

The French example, without being perfect, is nonetheless instructive. It shows how important the choice of a cataloguing standard is that is the same for everyone and that it is probably simpler to use the same software. All libraries that want to get computerized and are planning to create a network should adopt common norms, and if possible an international norm such as MARC (for example) so as to avoid the errors of disorganized and anarchic computerization. This project could be the subject of another meeting in the future.

Discussion

Digitalization and copyright problems

The new technologies represent a great advance for libraries because they make computerized catalogues possible, often in common, but the final goal is not cataloguing. It is rather the mathematical text. It is not enough to have a computerized library and be able to exchange files. It is also necessary that the mathematician who has tracked down such and such a book in this or that library be able to obtain the text.

A text can be photocopied and sent by fax. This supposes that someone has made the photocopy and thus that there is supplementary personnel. This also involves raised telephone expenses. It would be interesting to make a comparison in the running costs: photocopy and fax on the one hand, digitalization of a text (as is foreseeable in the near future) and e-mail on the other.

Digitalization poses the problem of copyright. The same information is going to be preserved for a number of years and be used ten or twenty times. This represents a real competition for commercial publishers. Many juridical questions remain unsolved: for example, in the digitalization projects of the Bibliothèque de France as well as in those of supplying electronic documents (FOUDRE project). Currently, the only texts in which digitalization is legally permissible are preprints.

Speed and cost of the installation of new technologies in publishing

There are two factions:

— The first one insists on the speed of technological evolution. Its advocates think that in few years, journals will be available on screen at no cost and will be printed on the spot. There are already several questions to publishers: when are they going to publish a computer file containing the contents of forthcoming journal issues, so that libraries can distribute them automatically to their readers? Is it the case that electronic publishing will eliminate all distribution costs? What is the real cost of e-mail: is it as cheap as people seem to think it is?

— The second one emphasizes that this evolution will take a good generation, because the conflict of interests between scientists and commercial publishers must be solved beforehand. Publishers and scientists must work together on the problems involved, concerning the cost of realization and management of an electronic journal (is it really going to lower production costs and thus the price of journals) as well as copyright questions.

Recommendations

The recommendations that result from the Mathematics Libraries in Europe round table concentrate essentially on five points.

(1) Continuity of library financing, incentives to find multiple funding sources

It is essential to stress the importance of mathematical documentation. These documents (journals and books) keep their topicality for a long time. They are frequently consulted and represent the result of the research of mathematicians. The library is at the same time one of the inspiration sources for mathematicians and the place where they work, in contrast with other scientists. This is the reason why it is necessary for each mathematics department to have a library. The library serves as laboratory.

The continuity in documentation is fundamental: stable, long-term funding is of the essence. Finding various sources of funding might provide a solution to this problem.

(2) Problems of personnel in libraries

It is of the utmost importance to call to the attention of authorities understaffing and/or inexperienced personnel. Given the nature of library work, professional qualifications are indispensable and made all the more necessary by the appearance of new technologies. A laboratory cannot function correctly without qualified personnel.

(3) Aid to Eastern European libraries

A first possibility could be to let the European Mathematical Society manage a fund intended to buy a certain number of periodicals. The management of this aid must be carefully studied: choice of target institutions, distribution circuits (make sure that the periodicals reach their destination within a reasonable delay), and so on.

Another solution, or a parallel one, could be to effect a differentiation in subscription prices depending on the country. In particular, the sums that various institutions may release to help mathematics libraries in the East could be used to obtain reduced subscription rates from publishers, for example.

(4) Installation or development of European mathematical networks

The questionnaire sent before the round table showed that 75% of all libraries are more or less computerized. A more comprehensive study of the computerization of European mathematics libraries could serve as a basis for presenting a serious case to the European Community asking for a "thematic" subsidy.

(5) Relationship of mathematicians with the publishing business

Mathematicians have an essential role to play. Faced with documentary inflation, they should consider seriously all creation of new journals, be more critical concerning editorial boards and participate actively in the choice of acquisitions in their libraries.

Finally, we recommend a second round table on all these subjects and possibly others.

APPENDIX 1
Results of the Questionnaire

A survey of mathematics libraries in Europe was started at the beginning of 1992. This survey was intended to define the various problems for which the libraries share concern. Around 200 questionnaires were sent to libraries on the one hand, and on the other hand, around a hundred were sent to European mathematical societies to be distributed among their own mathematics libraries.

We received 36 answers from Western Europe, 9 from Eastern Europe and 27 from France.

We are well aware that for European countries outside France, this is just a beginning, albeit a promising one. Concerning France, the network of French mathematics libraries was helpful in targeting addressees and collecting a large number of answers primarily from mathematics departments.

The order of preference of proposed themes was explicit enough and helped us to define the orientations of the round table (financing, documentation cost, documentary inflation, computerization and new technologies, library personnel).

Collections and readers

Tables 1 and 2 give an idea of the quantity of documents possessed by various libraries.

The number of readers in research libraries depends on the size of their department, and their educational level corresponds to at least four years of university studies. On the contrary, university libraries did not differentiate readers according to their discipline and their educational level is more varied.

Users almost always have access to the shelves and can borrow almost all books. Periodicals can be borrowed in half of European libraries (but only in one third of the French libraries).

Funding

There is a significant difference in financing among European libraries. French libraries receive funds from very diverse sources, whereas libraries in the rest of Europe are essentially dependent on a unique source (government, mathematics department or institution).

If most libraries have a regular funding (either good or bad), they are all concerned with their diminishing purchasing power in the face of documentary inflation and rising publication prices.

Several solutions are suggested in the face of this situation. If some are resigned to their fate, others consider looking for patrons. Many libraries try first to compensate for this situation by decreasing book purchases through a vigilant acquisition policy and the development of coordination with other libraries. Above all, everyone is starting to reflect on the causes of this inflation (43% have already cancelled subscriptions).

On the average, libraries devote one third of their budget to book purchases and two thirds to subscriptions. The binding budget is around 5 or 6%, but unfortunately this only concerns a small number of libraries.

Libraries that have been computerized often benefited from specific funds.

Documentary policy and interlibrary relations

In all libraries, most acquisitions are decided either by a committee of mathematicians, or by a mixed committee of mathematicians and librarians.

If 80% of the libraries surveyed are developing exchanges, 40% admit, with the exception of Eastern European countries, that this has little impact on their budget, although for 46% the scientific impact is significant.

What is exchanged? Mostly preprints and periodicals and few publications in series.

60% of the libraries take part in interlibrary loans. Most libraries have essentially local (city or region) connections in the form of common catalogues and a supply of photocopies.

In several places, acquisitions are coordinated to increase documentary coherence and also as a savings measure.

New technologies

75% of the libraries are computerized, 40% of which were computerized about 5 years ago. Computerization has been slow but steady and its pace has increased since 1985.

The most commonly computerized function is cataloguing, then acquisitions and loans. The machines most widely used are PCs or Macs. In big libraries, the main frame is the main server.

In most countries, computerized cataloguing norms are diverse, often local, and the same is true for software. France stands apart with a more unified situation: one software is dominant and the MARC standard is widely used.

More than 50% of libraries do not have a CD-ROM yet and 50% (not necessarily the same 50%) have no access to databases.

Almost 40% belong to a national network and slightly more than 30% to a thematic network.

Questions raised

The financial problem is essential for all. Apart from this, the main difficulties are caused by personnel problems and sometimes a lack of space.

In some cases, a blatant lack of personnel prevents normal operation of the library. In other cases, lack of qualifications produces the same effect. Without exception, all institutions raise the issue of lack of personnel.

In view of the results of this survey and of the discussions of the round table of the European Congress of Mathematics, it seems that many subjects deserve an in-depth study. European mathematical documentation is still not well known and needs a more comprehensive approach.

The generalization of new technology should bring the various institutions closer to each other. A network would probably constitute a precious help, even if it is informal in the beginning. Coordination of standards could only reinforce relations and exchanges. More frequent meetings should enhance existing ties and lead to common projects.

Besides, this "future network" would be a more appropriate framework in which convincing arguments could be found to substantiate financial requests necessary for the smooth functioning of libraries. Indeed, we can only stress once more that libraries are the essential tools of mathematicians.

APPENDIX 2
Statistical Data

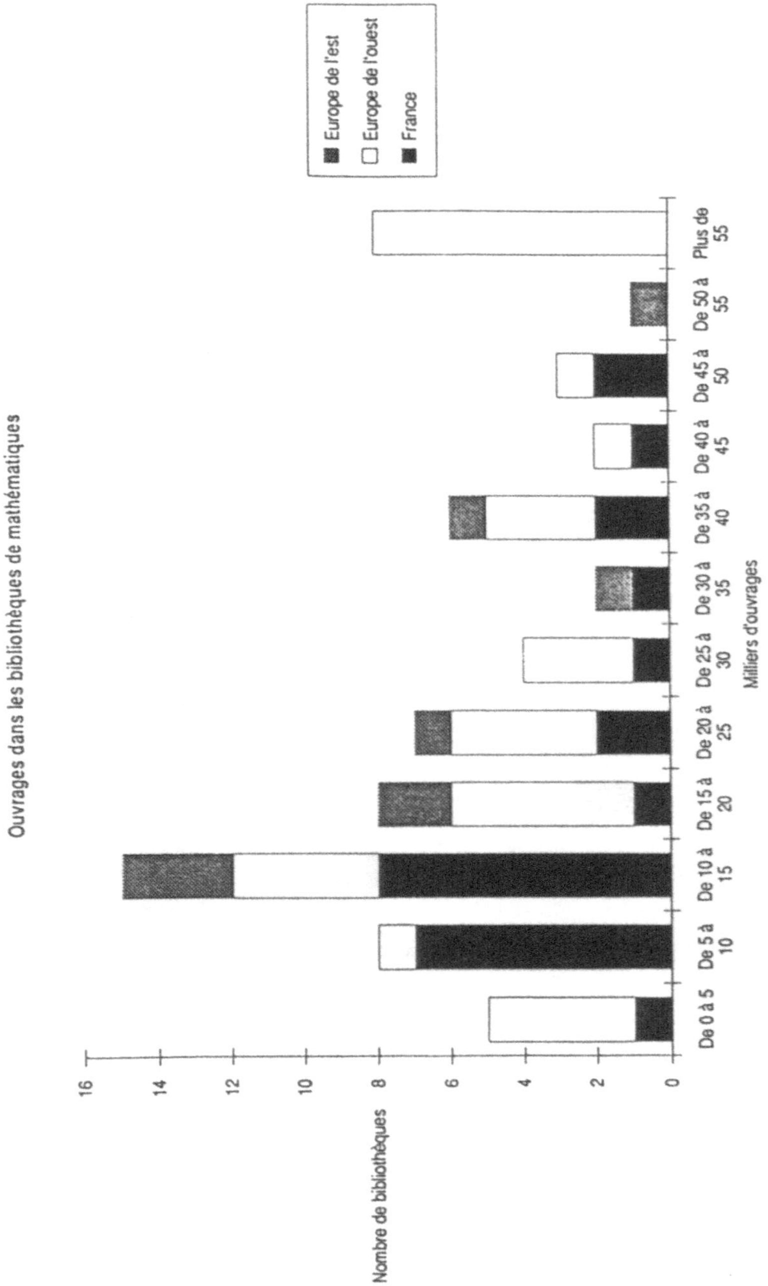

Tableau 1. Ouvrages dans les bibliothèques de mathématiques

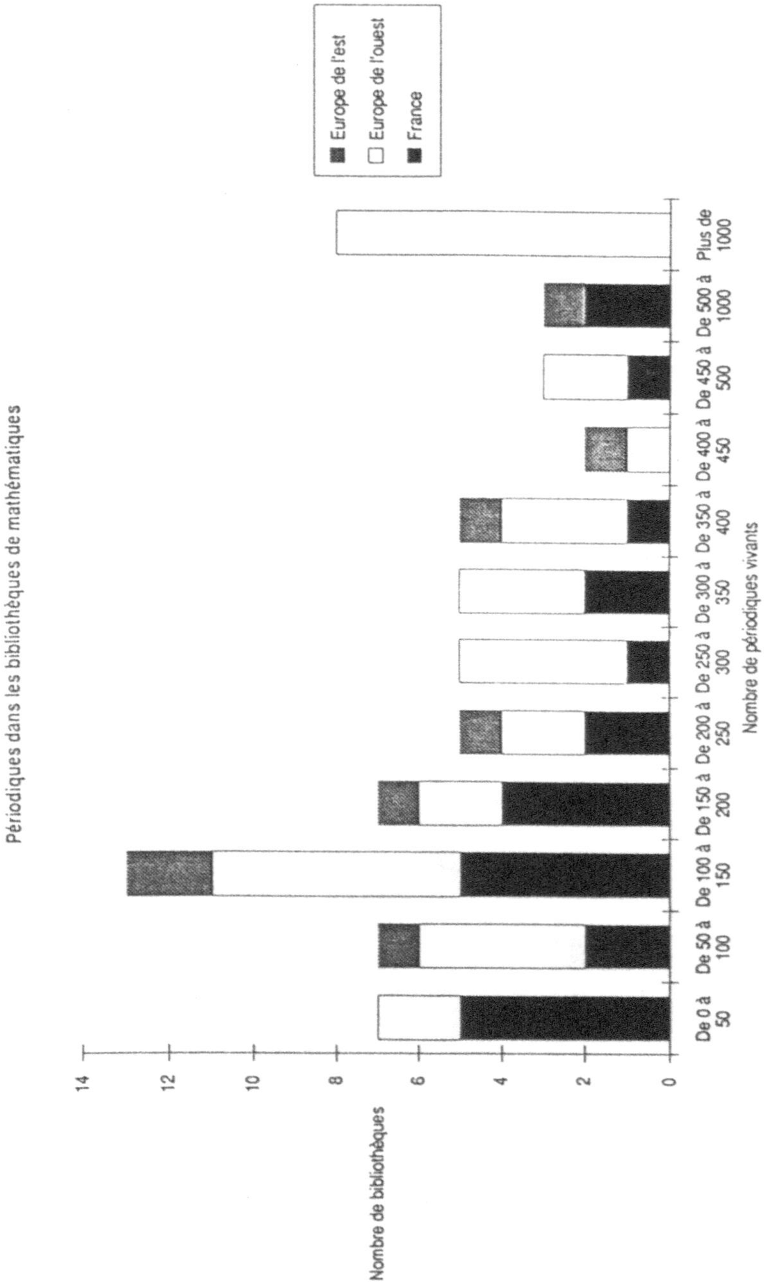

Tableau 2. Périodiques dans les bibliothèques de mathématiques

	Abonnements supprimés			Suppressions prévues		
	OUI	**NON**	**sans réponse**	**OUI**	**NON**	**sans réponse**
Europe Ouest	18	14	4	16	15	5
Europe Est	6	3	--	5	4	--
France	7	19	1	5	21	1
Total	**31**	**36**	**5**	**26**	**40**	**6**
		43 %			**36 %**	

Tableau 3. Suppressions d'abonnements

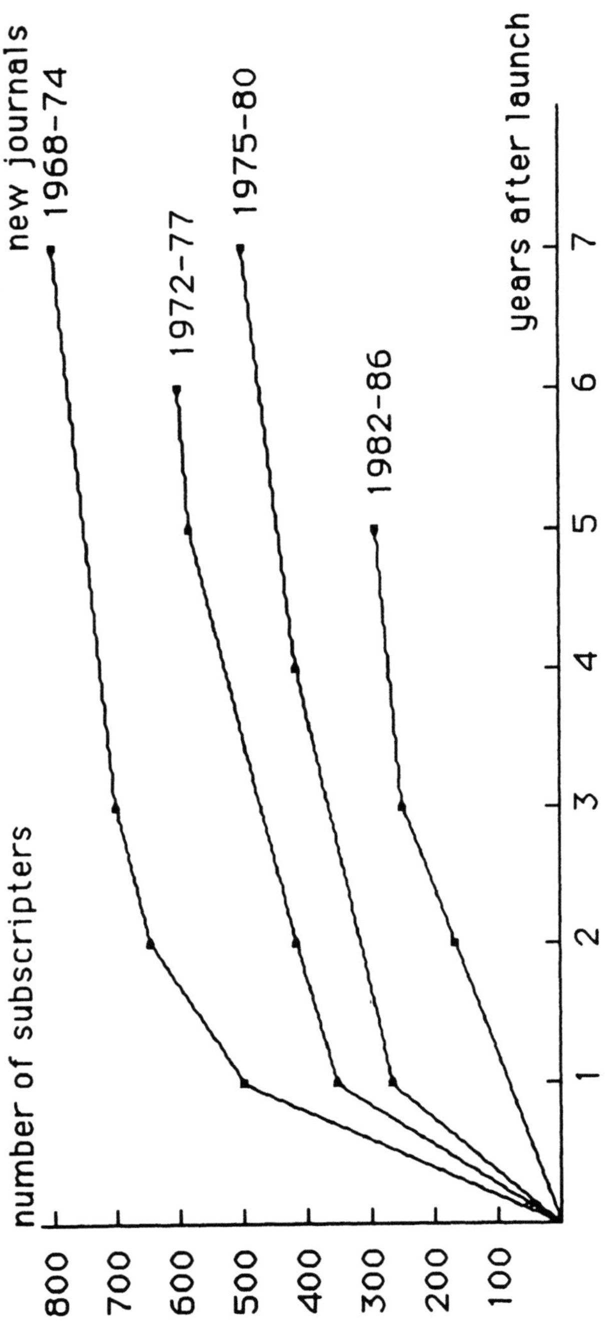

Table 4. Decline in circulation figures for new journals. (From: A. Stevens, *Learned Publishing*, **2**, 1–7, 1989)

APPENDIX 3

Manufacturing Costs

by C. Byrne
Editor, Springer-Verlag

The manfacturing costs for a book—typesetting, filmsetting, paper, ink, printing and binding costs—only represent a small part of the price observed in bookstores. The large part of this price is due to distribution costs, which include rebates for the bookseller or wholesale retailer (forty percent of the list price on average at Springer). Distribution costs also include advertising, storage, mailing and documentation costs (in our own catalogues, in the "Books in Print", at the Library of Congress, and so on). It is a fact that the maintenance of a distribution network is becoming more and more costly, all the more so, if this network is international.

In an effort to compensate for this situation, we are trying to decrease manufacturing costs. This is still possible, especially through relocalization of the services we purchase. For more than ten years, we have been purchasing a lot of typography in Asia, where prices are around fifty percent or less of Western European prices for a comparable quality. The current low value of the dollar enables us to buy a large part of the printing and binding in the United States. However, Asia is the place to look for new opportunities for top quality work at lower prices. An analogous situation has already developed in other industries and it is clear that it is detrimental to European industry. Specialized typography as in mathematics or chemistry is slowly heading for extinction in Europe.

Let us give a few figures. Globally, the manufacturing costs in European publishing have increased by twenty-percent in the last five years and by fifty-five to sixty percent in the last ten years. This twenty percent increase in costs should normally have resulted in an increase of thirty percent or more in list prices. An increase that the market (i.e., libraries and scientists) would have considered exaggerated. We were able to stay well below thirty percent due to the measures outlined above.

We must add to delocalized typography the impact of typography done by the authors themselves. It is a blessing for scientific publishing that, in this period of restrictive budgets and circulations, $\TeX$ processing has been almost universally adopted. In effect desktop publishing can produce significant typesetting cost savings, if certain standards are met. In this regard, we refer to the *Notices of the AMS* 1989 article by our colleague M. Gilchrist.

APPENDIX 4
Electronic Publishing: A Few Remarks

by C. Byrne
Editor, Springer-Verlag

A publisher can define his role as an information distributor. More precisely, his role is to collect information, or have it collected, to organize and present it in a fashion that corresponds to the needs of the users. Publishers also have to make this information available on the market and publicize its contents and means of acquisition.

In the case of paper medium, publishers possess a profound expertise and centuries-old experience in all these areas. However, concerning electronic media, the publishers' competence is limited to the early stages. They must either subcontract, which presupposes a good knowledge of available possibilities, or collaborate with one or several partners. This division of work will remain indispensable in the future as a means of staying abreast of technical advances. A publisher tries to meet the needs that are taking shape in the market, that is, to follow the demand and not the opposite. "Publishers are product-makers. They have no experience as market-makers".

Up to now, publishers have participated in the compilation of electronic catalogues and bibliographies, on disk, on-line and recently on CD. Since norms have been established and since the prices of CD-drives have reached an affordable level, CD-ROM-based products are now meeting with a very encouraging success. CD-ROMs are already used as complements for textbooks, allowing for interactive work. Springer-Verlag has announced several products of this type for next year. It goes without saying that these projects are joint efforts with the research teams of Apple and SUN in order to make the products available to a broad public at reasonable prices.

Let us recall that what is currently technically possible is still far beyond what is economically feasible on the scientific market scale. This is why publishers seem to be rather slow in developing and distributing electronic publications. Investments are very large and even long-term profitability is far from certain.

Mathematicians often ask us to replace our paper journals with electronic journals distributed by e-mail, in the hopes of lowering subscription prices. This proposition amuses all those who have received e-mail bills! E-mail costs are actually high, and if they are met through public funding in most research institutions in Western Europe or in the United States, no one knows how long this is going to last. Anyway, publishers have to meet their own costs. You must add to these costs a complete internal or external structure of technical control to make sure that norms are scrupulously respected for each article and to correct by hand any such "manuscripts", if the need arises, to ensure error-free publication. Computer scientists who already publish a considerable number of electronic journals know this well.

During the first European Congress of Mathematicians, to be held in Paris from July 6 to July 10, 1992, a round table will be held dealing with the European mathematical libraries. Would you be so kind to fill out the questionnaire, so that the organizers of the round table will be in a better position to evaluate the type of problems facing the various mathematical libraries.

1. Do you wish to participate in the round table?

☐ yes ☐ no

(The round table will probably be held July 9)

If not, is this due to financial problems ? ☐ yes

If yes, please get in touch with us very promptly :

Pierre BARRAT
Bibliothèque de mathématique-recherche
Tour 56, 4ème étage
4 place Jussieu
75252 Paris Cedex 05 France
Tél : (33) (1) 44 27 37 23
Fax : (33) (1) 44 27 78 18
e-mail : barrat@mathp7.jussieu.fr

Geneviève SUREAU
Université Paris-Sud
Bibliothèque mathématique
Bâtiment 425
91405 Orsay France
Tél : (33) (1) 69 41 70 51
Fax : (33) (1) 69 41 62 21
e-mail : sureau@junon.matups.fr

Liliane ZWEIG
Bibliothèque de mathématiques
Ecole Normale Supérieure
45, rue d'Ulm
75230 Paris cedex 05 France
Tél : (33) (1) 44 32 31 44
Fax : (33) (1) 44 32 20 80
e-mail : zweig@dmi.ens.fr

Address

2. Name of library : ....

3. Address : ....
....

4. Telephone : ....

5. Fax : ....

6. E-mail address : ....

7. Library head : ....

8. Type of organization :
☐ academic
☐ mathematics department
☐ research institute
☐ other (please describe)

Holdings

9. What collections, other than mathematics, does the library contain ? :
a...
b...
c...
d...

10. Number of books : ....

11. Number of bound periodicals : ....

12. Number of active subscriptions to periodicals : ....

13. Other documents (special collections, ancient books, etc...) : ....
....
....

Library users :

14. Number : ....

15. Level :
☐ professors, researchers
☐ doctoral students
☐ graduate students
☐ undergraduate students
☐ first year students
☐ others

16. Do your readers have free access to bookshelves ? : ☐ yes ☐ no

17. Can journals be borrowed ? : ☐ yes ☐ no

18. Can books be borrowed ? : ☐ yes ☐ no

19. Who can borrow books ? :
    - ☐ all users
    - ☐ some
    - ☐ none

20. Do you participate in an interlibrary loan system ? : ☐ yes ☐ no

Financing

21. Specify the different sources of your financing (in percentage if possible) :

    a. your institution : %
    b. mathematics department : %
    c. governement : %
    d. public research institute : %
    e. private research institute : %
    f. other (please specify) : %

    Total : 100 %

22. Are your resources regular ? : ☐ yes ☐ no

23. Are they subject to sudden fluctuations ? : ☐ yes ☐ no

    If yes, please describe : ....
....
....
....

Spending

24. Percentage of your budget for :

    a. journals : %
    b. books : %
    c. binding : %
    d. computers : %
    e. other (specify) : %

    Total : 100 %

Acquisitions

25. They are decided upon by :
    - ☐ Librarian
    - ☐ Panel of mathematicians
    - ☐ Other (specify)

26. Do you acquire any materials by exchange with other institutions : ☐ yes ☐ no

27. What do you supply in an exchange program ? :
    ☐ preprints ☐ journals ☐ serials

28. What do you receive in return ? :
☐ preprints ☐ journals ☐ serials

29. Importance for your savings through exchanges :
☐ important ☐ medium ☐ not important

30. Scientific impact on users :
☐ important ☐ medium ☐ not important

Modern technologies

31. Is your library computerized ? : ☐ yes ☐ no

32. Since when ? : ....

33. For which functions ? :
☐ catalogue
☐ journals
☐ loan
☐ acquisitions
☐ finances

34. Software :
☐ local
☐ national
☐ international

Specify : ....
....
....

35. Type of computer :
☐ PC
☐ Station (ex. Sun,Apollo)
☐ Mini (ex. Vax...)
☐ Mainframe

36. Do you belong to one/many network(s) by subject ? : ☐ yes ☐ no
Specify : ....
....

37. Do you belong to one/many national network(s) ? : ☐ yes ☐ no

38. What format do you use for the machine readable catalog ? :
....
....

39. Do you have CD-ROM ? : ☐ yes ☐ no
Specify : ....
....

40. Access to databases ? : ☐ yes ☐ no

Specify : ....
....

Possible Problems

Price rises :

41. Did you recently have to cancel subscriptions ? : ☐ yes ☐ no

42. Do you intend to cancel subscriptions ? : ☐ yes ☐ no

43. What do you intend to do to face the rapid increase of mathematical publishing ? :
....
....
....

44. Library staff : principal problems : ....
....
....

45. Administrative changes in the structure of your library :
....
....
....
....

Relations with libraries in your country :

46. Do you share a/many collective catalogue(s) ? : ☐ yes ☐ no

If yes, specify : ....
....
....

47. Could you describe your library's relations with other national or international libraries ? :
....
....
....

48. Specific points you wish to discuss :
....
....
....

SURVEY

In the list below, which topics seem to you most important (1,2,3,...in order of priority) ?

- ☐ Library financing
- ☐ Regularity of financing
- ☐ Acquisitions policy
- ☐ Mathematician's role in library policy
- ☐ Exchanges
- ☐ Interlibrary loan
- ☐ Computers in library
- ☐ Catalog format
- ☐ Preprints
- ☐ Ancient books
- ☐ Library staff
- ☐ Price rises
- ☐ New technologies
- ☐ Administrative structure
- ☐ Relations with scientific publishers
- ☐ Other (specify)

Round Table M: July 6, 1992

MATHEMATICS AND ECONOMICS

by Bernard Cornet[1]

Université de Paris I, Paris (France)

1. Introduction

The aim in organizing a round table on the subject "Mathematics and economics" within the European Congress of Mathematics was to show the role of modelization in economics (and more generally in social sciences), to present the type of questions which are asked together with some of the mathematical tools that are used. Modelization in economics is clearly distinct from modelization in physics, mechanics or engineering, sciences which are more familiar to mathematicians. Experiments play a different role and often its absence in social sciences leads to abstract axiomatized models to analyze subjects such as competition, conflict, cooperation, reputation...

Three main themes have been selected to present the interface mathematics – economics. The first one, market (or price) theory, also called general equilibrium theory, is the more detailed formalized model in economics and is generally considered a model of reference. The second one is the theory of financial markets, which can be considered as an integral part of the previous one but has been developed as an autonomous field during the past twenty years. Finally, the last theme deals with game theory and is deeply linked to economics from its origins even if its actual and potential applications go far beyond economics.

In his book *Eléments d'économie politique pure*, written in 1874, Leon Walras presented the first mathematical analysis explaining that the quantities of goods

[1] This text resumes the discussions of the participants of the round table of the European Congress of Mathematics which was held in Paris on July 6, 1992. All the participants have made a huge effort to synthesize the progress of the theory in their field. The remaining errors should only be credited to the author whose task was just to write in a unified way the different presentations; helpful comments and suggestions on the editing were made by Joseph Abdou, Jean-Marc Bonnisseau, Monique Florenzano, Pascal Gourdel, and Nicole El Karoui. The participants of the round table were: Robert Aumann, The Hebrew University of Jerusalem; Bernard Cornet, Université de Paris I and Ecole Polytechnique; Gérard Debreu, University of California, Berkeley, visiting the Laboratoire d'Econométrie de l'Ecole Polytechnique; Egbert Dierker, Universität Wien; Darrell Duffie, Stanford University; Nicole El Karoui, Université de Paris VI; Jean-Michel Lasry, Université de Paris IX and Caisse des Dépots et Consignations, Paris; Jean-François Mertens, Université Catholique de Louvain, and Sylvain Sorin, Université de Paris X.

bought by consumers, their labour supply, the quantities of goods produced or used by the firms and the prices observed on the different markets could be interpreted as an equilibrium situation among a large number of agents having conflicting interests. It was, however, only during the thirties that the question of the existence of such an equilibrium was explicitly posed by Abraham Wald (1935). This question was solved in a mathematically correct way by Kenneth Arrow and Gérard Debreu (1954). The book *Theory of value* by Gérard Debreu presents the synthesis of the work during this period, which together with the work by Tjalling C. Koopmans, John Nash, John Von Neumann, and Oskar Morgenstern lay the foundations of mathematical economics.

If the works in the fifties essentially used the reasoning of general topology, properties of convex sets and the fixed point theorem of Kakutani, the developments of Walrasian theory over the past thirty years lead to the use of more sophisticated mathematical tools. The existence of equilibria in more general models goes far beyond the use of Brouwer and Kakutani theorems and may involve techniques from algebraic topology or differential topology. In Section 3 we shall present existence results of equilibria in the particular but important case of exchange economies (without production) which are formally defined in Section 2.

Competition between agents implicitly embodies the fact that every particular agent is negligible; this fact is rigorously formalized by considering a measure space of agents without atoms, a fundamental example being the unit interval, as first considered by Aumann (1964). The study of "infinite economies", one of the main fundamental results being the equivalence theorem between the core and the set of Walras equilibria, started the extensive use of measure theory in economics and initiated developments of new mathematical techniques such as the integration of correspondences. The core of an exchange economy is defined in Section 9 which also presents the core equivalence theorem.

The determinedness of equilibria, counting their numbers, the study of the geometric properties of the set of equilibria, etc... were the points of departure for the use of differential topology; these techniques and, in particular, Sard's theorem or transversality results allowed us to define constructive ways to reach an equilibrium, a subject which was first tackled by Herbert Scarf with combinatorial techniques. These questions are presented in Section 5 where an algorithm is defined to reach equilibria.

The Arrow–Debreu model is still in the problematics of recent works which try to embody in the theory new aspects such as risk and uncertainty, incomplete markets, nonconvexities and imperfect competition, infinite economies, etc.... For the interested reader, we refer to the recent *Handbook of Mathematical Economics*, Vol. IV, (1991) edited by Hildenbrand and Sonnenschein.

If Finance is an integral part of economic theory and can be analyzed in the general equilibrium framework (cf., for example, the article by Darell Duffie

(1991) in the Handbook), over the past twenty years it had to answer specific problems which led to the development of Finance as an autonomous field.

This field challenges the mathematician in several ways. First, on the theoretical side, with the complexity of the modelization of the basic problems (no-arbitrage opportunities, optimization of the wealth or the risk inherent to an asset, or the return of a portfolio...). But also on a practical side, with the proximity existing between the models and their immediate use or the original way market took possession of models, particularly, Black and Scholes formula.

Finance theory uses tools coming from different branches of mathematics. As early as 1900, Louis Bachelier introduced Brownian motion to modelize the price of actions. Since then, stochastic calculus remains a major and important tool which will be discussed in Section 8. Modelization in Finance which started extensively in the seventies with the works by Markowicz, Merton, Black and Scholes, also makes an important use of techniques from optimization theory, partial differential equations and statistics. In Section 6, a basic model of financial markets is presented in a two-period economy. The next section gives some consequences of the important no-arbitrage assumption in terms of asset pricing. In Section 8, we shall present some specific aspects of financial markets in a continuous time framework.

The aim of Game Theory is to study situations of conflict and cooperation. The conceptual basis and the mathematical formalism of Game Theory started with the book of John Von Neumann and Oskar Morgenstern, *Theory of Games and Economic Behavior*, written in 1944. The reference to economics in the title shows its importance as a field of application of game theory; one can in fact consider that the strategic approach to economics dates back to French economists Cournot and Bertrand in the nineteenth century. But applications of game theory can also be given in other fields such as biology or social sciences. For example, the notion of evolutionary equilibrium allowed us recently to clarify the problem of selection of species.

Von Neumann and Morgenstern essentially distinguish two main approaches to modelize a game. The strategic approach (also called the non-cooperative approach) considers agents (players) acting only according to their own interest and following the rules of a game which are known to everybody. The coalitional approach (also called cooperative approach) allows the agents (players) to form coalitions and tries to determine the probable issues which take into account alliances and thus the power of the coalitions. A fundamental program of research in Game Theory is still to bridge the gap between cooperative and non-cooperative approaches. We shall present in Section 9 a fundamental example of cooperative games in an economic setting: the core of an economy. We shall also state the so-called "equivalence theorem" between the core and the set of Walras equilibria, the latter being a typical non-cooperative concept. In Section 10, we shall briefly expose the basic notions of non-cooperative games.

The mathematical techniques traditionally used in Game Theory cover equilibrium theory, convex analysis and fixed point theorems (Brouwer and Kakutani). It appears that the main concepts of Game Theory have a semi-algebraic nature and recent works use geometric techniques (from homology theory) to study, for example, strategic stability.

2. Exchange Economies

We[2] consider an *exchange economy* with a finite number, ℓ, of goods, a set A of consumers (agents) and no producers for the sake of simplicity (for the general model we refer to Debreu (1959)). In the following, the set A will either be finite, or $(A, \mathcal{A}, \nu)$ will be a measure space. In the latter case, $\mathcal{A}$ is the σ-field of coalitions and ν is a finite, positive, countably additive measure defined on $\mathcal{A}$. If $\nu(A) = 1$, the measure $\nu(E)$ of a coalition $E \in \mathcal{A}$ is interpreted as the fraction of the totality of agents in E. We further assume that the measure space $(A, \mathcal{A}, \nu)$ is complete, i.e., every subset of every set of measure zero belongs to $\mathcal{A}$. The characteristics of an agent $a \in A$ are his (her) *preference relation* $\preceq_a$ on the closed positive cone R^ℓ_+ of R^ℓ and his (her) *initial endowment* $e(a)$ in R^ℓ_+. The strict preference relation $\prec_a$ is defined in a standard way as follows: $x \prec_a y$ means that $[x \preceq_a y$ and not $y \preceq_a x]$. We assume that the binary relation $\preceq_a$ is complete, reflexive, transitive and satisfies the assumptions of:

closedness: *the set* $\{(x, y) \in R^\ell_+ \times R^\ell_+ \mid x \preceq_a y\}$ *is closed; and*

monotony: *for every* x, y *in* R^ℓ_+ *such that* $x < y$, *then* $x \prec_a y$.

We denote by $\mathcal{P}$ the set of preference relations on R^ℓ_+ satisfying the above assumptions. Thus, the set of agents characteristics is the Cartesian product $\mathcal{P} \times R^\ell_+$ on which we define a topology as follows; since a preference relation in $\mathcal{P}$ is a closed subset of $R^{2\ell}$, we endow $\mathcal{P}$ with the topology of closed convergence (cf. Hildenbrand (1974)) and $\mathcal{P} \times R^\ell_+$ is endowed with the product topology. The measure space structure on $\mathcal{P} \times R^\ell_+$ is derived from its topology in a standard manner by considering the Borel σ-field of that topology. An (exchange) economy $\mathcal{E}$ specifies for each agent $a \in A$, his (her) characteristics $(\preceq_a, e_a)$ in $\mathcal{P} \times R^\ell_+$; formally, the economy $\mathcal{E}$ is a mapping from A to $\mathcal{P} \times R^\ell_+$. If $(A, \mathcal{A}, \nu)$ is a measure space, we further require that $\mathcal{E}$ be measurable and that the endowment mapping e from A to R^ℓ_+ be integrable. We shall say that $\mathcal{E} : \mathcal{A} \to \mathcal{P} \times \mathcal{R}^\ell_+$ is a finite (resp. measureable) economy if A is finite (resp. if $(A, \mathcal{A}, \nu)$ is a measure space and the above assumptions are satisfied).

---

[2] If x and y are two vectors in R^ℓ, $x \geq y$ (resp. $x > y$, resp. $x \gg y$) means that $x_h \geq y_h$ for every coordinate $h = 1, \cdots, \ell$ (resp. $x \geq y$ and $x \neq y$, resp. $x_h > y_h$ for every h) and we denote by $x \cdot y = \Sigma_{h=1}^{\ell} x_h y_h$ the scalar product of R^ℓ. We let $R^\ell_+ = \{x \in R^\ell \mid x \geq 0\}$ and $R^\ell_{++} = \{ x \in R^\ell \mid x \gg 0 \}$.

An *allocation* for the economy $\mathcal{E}$ specifies, for each agent $a \in A$, his (her) consumption vector $x(a) \in R^\ell_+$; formally, x is a mapping from A to R^ℓ_+ that we require to be integrable if $(A, \mathcal{A}, \nu)$ is a measure space.

We now can define the equilibrium notion due to Walras (1874) for finite and measurable economies. If A is finite (resp. $(A, \mathcal{A}, \nu)$ is a measure space), a pair (x^*, p^*) of an allocation x^* and a price vector p^* in R^ℓ is a *(Walrasian) equilibrium* if:

(i) for every $a \in A$ (resp. for almost every $a \in A$) $x^*(a)$ is the best element for $\preceq_a$ in the budget set: $\{y \in R^\ell_+ \mid p^* \cdot y \leq p^* \cdot e(a)\}$; and
(ii) $\Sigma_{a \in A} x^*(a) = \Sigma_{a \in A} e(a)$ (*resp.* $\int_A x^* d\nu = \int_A e d\nu$).

We denote by $W(\mathcal{E})$ the set of (Walras) equilibria allocations of the economy $\mathcal{E}$. An allocation that satisfies only (ii) is called a state of the economy.

3. Existence of Equilibria

The first question to be considered is the existence of such an equilibrium. At first, proofs of existence of an economic equilibrium were obtained via a fixed point theorem of the Brouwer type or the Kakutani type, a technique which was not available in Walras' time. In the next section, we shall present the heuristic but interesting argument, due to Walras, which amounts to counting the number of equations and the number of unknowns and to notice that equality of the two "implies" the existence of an equilibrium. The first satisfactory answer to the existence problem was given by Wald (1935) and the result stated below is a particular case of a general existence theorem of Arrow and Debreu (1954) and McKenzie (1954).

Theorem 1. *Let $\mathcal{E}$ be a finite economy. Then there exists an equilibrium (x^*, p^*) of $\mathcal{E}$ such that $p^* \gg 0$ under the following additional assumptions:*

(C) *for every $a \in A$ and every $x \in R^\ell_+$ the set $\{y \in R^\ell_+ \mid y \succeq_a x\}$ is convex;*
(S) $\Sigma_{a \in A}\, e(a) \gg 0$.

The next result guarantees the existence of an equilibrium when $(A, \mathcal{A}, \nu)$ is a measure space of agents. The space A said to be atomless if, for every $E \in \mathcal{A}$ with $\nu(E) > 0$, there exists $S \subset E$ such that $S \in \mathcal{A}$ and $0 < \nu(S) < \nu(E)$. A standard example is the unit interval with Lebesgue measure.

Theorem 2. *Let $(A, \mathcal{A}, \nu)$ be an atomless measure space of agents, then there exists an equilibrium (x^*, p^*) of $\mathcal{E}$ such that $p^* \gg 0$ if we further assume that* $\int_A e d\nu \gg 0$.

This result, due to Aumann (1966) [see also Hildenbrand (1974)] relies, first, on a fixed point argument as in the finite case, and secondly, on Lyapunov's theorem on the range of an atomless finite dimensional vector measure. We notice that no convexity assumption on the preferences is made in Theorem 2.

4. Excess Demand Functions and Stability

We consider a finite exchange economy $\mathcal{E} : A \to \mathcal{P} \times R^\ell_+$ and we suppose that $\Sigma_{a\in A} e(a) \gg 0$, together with the strict convexity of the preferences of every agent a, i.e.,

(SC) *let x, x' be in R^ℓ_+ such that $x \neq x'$ and $x \preceq_a x'$ then, for every $t \in]0, 1[$, one has $x \prec_a (1-t)x + tx'$.*

Then, for every agent a and every price vector $p \in R^\ell_{++}$, there exists a unique element $d_a(p)$, called demand (vector) of consumer a, which maximizes the preferences $\preceq_a$ in his (her) budget set $\{y \in R^\ell_+ \mid p \cdot y \leq p \cdot e(a)\}$.

We then define the *excess demand function* (of the economy) $F_\mathcal{E} : R^\ell_{++} \to R^\ell$ by

$$F_\mathcal{E}(p) = \Sigma_{a\in A} d_a(p) - \Sigma_{a\in A} e(a).$$

We notice that $p^* \in R^\ell_{++}$ is an equilibrium price of $\mathcal{E}$ if and only if

$$F_\mathcal{E}(p^*) = 0.$$

In other words, p^* is a solution of a system of ℓ equations (since $F_\mathcal{E}(p)$ belongs to R^ℓ) with ℓ unknowns $p_1, \cdots, p_\ell$. As we shall see in the next lemma, the homogeneity property allows us to fix one variable, say p_ℓ, to remain only with $\ell - 1$ unknowns. But Walras noticed that, in the above system, the equations are not independent since, for every $p \in R^\ell_{++}$, one has $p \cdot F_\mathcal{E}(p) = 0$. Hence, we can equivalently consider the system of the first $\ell - 1$ equations with the first $\ell - 1$ unknowns, a counting argument though unsatisfactory, which led Walras to suggest that the existence of an equilibrium would prevail ... several years before Brouwer's fixed point theorem was discovered.

It is worth noticing that, under the above assumptions, $F_\mathcal{E}$ satisfies the following properties (cf. Debreu (1982) for the proof of the result).

Proposition 3. *(i) (positively homogeneous) $F_\mathcal{E}(\lambda p) = F_\mathcal{E}(p)$ for every $\lambda > 0$, and every $p \in R^\ell_{++}$; (ii) (continuity) $F_\mathcal{E} : R^\ell_{++} \to R^\ell$ is continuous; (iii) (Walras' law) $p \cdot F_\mathcal{E}(p) = 0$ for every $p \in R^\ell_{++}$; (iv) $F_\mathcal{E}$ is bounded below; (v) (boundary condition) if $(p^q) \subset R^\ell_{++}$ converges to $p^0 \in \partial R^\ell_+$, $p^0 \neq 0$, then $\|F_\mathcal{E}(p^q)\| \to +\infty$.*

Let S be the unit sphere of R^ℓ and $S_{++} = S \cap R^\ell_{++}$. From the homogeneity property of $F_\mathcal{E}$, the restriction of $F_\mathcal{E}$ to S_{++} characterizes the excess demand function $F_\mathcal{E}$ and we shall denote them identically in the following. Thus the excess demand function $F_{\mathcal{E}} : S_{++} \to R^{\ell}$ of the economy $\mathcal{E}$ can be geometrically represented, as in Figure 1, by a continuous vector field on the submanifold S_{++}.

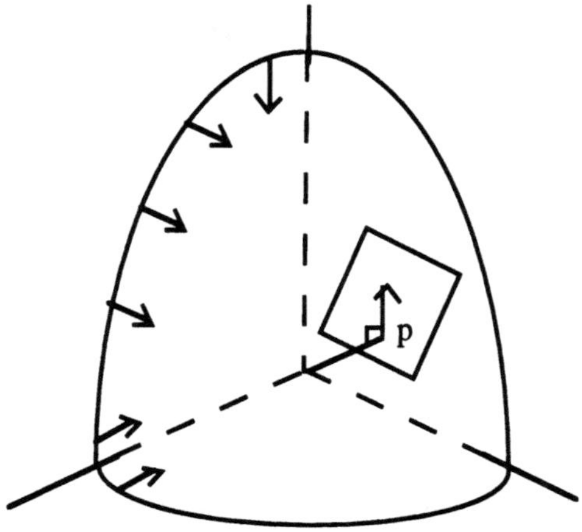

Figure 1.

One further checks that the above boundary condition *(v)* can be rigorously interpreted by saying that "if p is close to the boundary of S_{++}, then the vector field $F_{\mathcal{E}}$ points inward S_{++}". This remark is the point of departure of another existence approach which uses tools from differential topology. It allows us also to give more information about local uniqueness of equilibria, the number of the set of equilibria, etc.. . . a subject to which we refer to Dierker (1982) and Smale (1982) for surveys.

Our purpose is now to give some insight into the "stability" of equilibria. One century ago, Walras proposed that outside of equilibria, one should increase the price of goods for which demand exceeds supply and decrease the prices of other goods. The so-called Walras' *tâtonnement process* can be mathematically formalized by the following differential equation

$$\dot{p}(t) = F_{\mathcal{E}}(p(t)).$$

During the sixties and beginning of the seventies, a huge amount of literature was devoted to the study of the stability of Walras' tâtonnement process. As we shall see below, it appears to be very unstable and allows "bad" behaviour such as the one in Figure 2, with a unique unstable equilibrium and a periodic orbit.

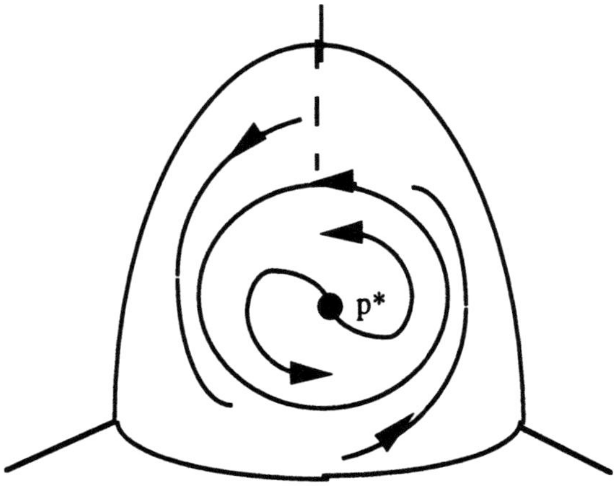

Figure 2.

This fact is a consequence of the following result, also of interest for itself, which shows that Walras' law and the continuity property both characterize excess demand functions of finite exchange economies (up to some technical properties on the boundary of the simplex). For the proof of the result due to Sonnenschein, Mantel and Debreu, we refer to Shaffer and Sonnenschein (1982).

Theorem 4. *Let* $F : S_{++} \to R^{\ell}$ *be a continuous mapping such that* $p \cdot F(p) = 0$ *for every* $p \in S_{++}$. *For every* $\varepsilon > 0$, *there exists a finite exchange economy* $\mathcal{E} : A \to \mathcal{P} \times R^{\ell}_{+}$ *such that card* $(A) = \ell$, *for every* $a \in A$, $\succeq_a$ *is closed, monotone, strictly convex,* $e(a) \gg 0$ *and* $F(p) = F_{\mathcal{E}}(p)$ *for every* $p \in S_{\varepsilon}$, *where* $F_{\mathcal{E}}(p)$ *is the excess demand of* $\mathcal{E}$ *and* $S_{\varepsilon} = \{x \in S \mid x_h \geq \varepsilon$ *for every* $h\}$.

This result, which can be seen as a very negative one, raises two important questions on which some work is still in progress. The first one, raised by S. Smale, is to reconsider the tâtonnement process by defining more sophisticated ways to study disequilibrium adjustments. Smale proposed to consider a global version of the Newton method, which is also of interest to solve nonlinear systems of equations and is constructive (see the next section for this question). The second question, for which we refer to recent work by Hildenbrand, tries to study distributions of agent characteristics which allows us to recover some properties of excess demand functions.

5. Computation of Equilibria

As we already said, the proof of the existence of an equilibrium uses a variant of Brouwer's fixed point theorem, hence it is not constructive. Algorithms for the computation of economic equilibria were initiated by H. Scarf. They first used combinatorial methods and more recently S. Smale proposed a differential equation, called the "global" Newton method, whose solutions generically converge to an economic equilibrium.

Hereafter, to reach an equilibrium, we define an algorithm which is inspired by the work of Scarf and Eaves. We consider a finite exchange economy $\mathcal{E}$ with an excess demand function $F : S_{++} \to R^\ell$ satisfying the properties of Proposition 3.

Let $L = \{1, \cdots, \ell\}$ and Σ be the simplex of R^ℓ, i.e., $\Sigma = \{p \in R^\ell_+ \mid \Sigma^\ell_{h=1} p_h = 1\}$.

We define, for every $p \in \Sigma$, the set

$$\Lambda(p) = \{h \in L \mid F_h(p) = \max_j F_j(p)\} \text{ if } p \notin \partial\Sigma,$$

$$\Lambda(p) = \{h \in L \mid p_h = 0\} \text{ if } p \in \partial\Sigma.$$

We recall that $p^* \in R^\ell_{++}$ is an equilibrium of $\mathcal{E}$ if and only if $F(p^*) = 0$ and we notice (using Walras'law) that this equation is satisfied if and only if

$$\Lambda(p^*) = L.$$

One easily sees that one can choose a mapping $\lambda : \Sigma \to L$ such that $\lambda(p) \in \Lambda(p)$ for every p satsifying the following two properties (α) and (β). We first say that a set $M \subset \Sigma$ is completely labeled, if $\lambda(M) = L$, and we recall that a facet of a k-dimensional simplex M, is a $(k-1)$-dimensional face of M.

(α) *The set of vertices of* Σ *is completely labeled.*
(β) *No facet of* Σ *is completely labeled.*

We define the extension $\lambda : R^\ell_+ \setminus \{0\} \to L$ of the mapping λ (the two being denoted identically, in the following, for the sake of simple notations) by letting $\lambda(q) = \lambda(\pi(q))$ where π is the projection into Σ from 0.

We also define the regular triangulation $\mathcal{T}$ of R^ℓ_+ above Σ, as in Figure 3.

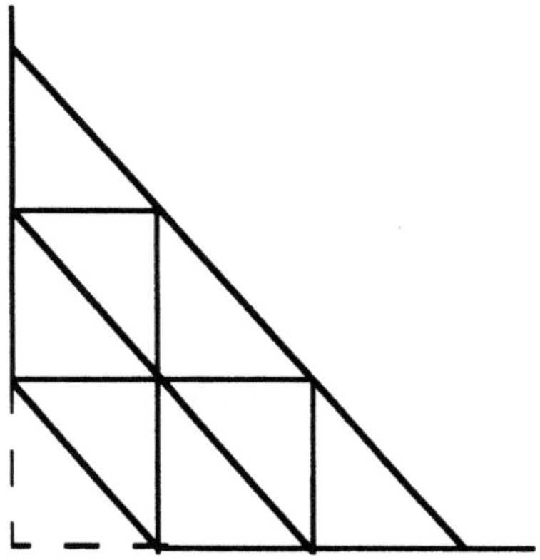

Figure 3.

We now construct a sequence (t_n) of $(\ell-1)$-simplexes in $\mathcal{T}$ with completely labeled vertices and satisfying the property that adjacent $(\ell-1)$-simplexes are facets of a ℓ-simplex in $\mathcal{T}$.

We start from Σ by taking $t_0 = \Sigma$. Clearly, there is one ℓ-simplex in $\mathcal{T}$ of which Σ is a facet and there is one $(\ell-1)$ simplex adjacent to Σ with completely labeled vertices. Property (β) implies that an $(\ell-1)$-simplex with completely labeled vertices cannot be contained in ∂R^{ℓ}_{+}.

Given an $(\ell-1)$-simplex with completely labeled vertices different from Σ, it has two adjacent $(\ell-1)$-simplexes with completely labeled vertices. If t_n belongs to the sequence, one of these two adjacent $(\ell-1)$-simplexes has been used to get to t_n; thus one can choose t_{n+1} to be the other one.

One then shows that (i) the sequence (t_n) never returns to a previously used $(\ell-1)$-simplex, (ii) that it never terminates, and (iii) it moves away from 0.

Let s_n be the projection of t_n into Σ from 0. Then it has completely labeled vertices and diam $s_n \to 0$. The compactness of Σ implies that there exists a subsequence of s_n which converges to p^*, clearly $p^* \notin \partial\Sigma$ and p^* is an equilibrium price of $\mathcal{E}$.

6. Financial Markets in a Two-period Economy

To capture both time and uncertainty in the simplest way, we consider a model with two periods ($t = 0, t = 1$) in which one of S states of nature ($s = 1, \cdots, S$) occurs at time $t = 1$. The main results can be extended to a stochastic process over many time periods.

We again consider a finite exchange economy with L physical goods, and a finite set A of agents having preferences $\preceq_a$ and initial endowments $e(a)$. In this section, however, we shall focus on the individual behaviour of an agent without considering the interactions between all the agents and the resulting equilibrium equations. The main new feature is now the notion of financial markets which provide instruments to redistribute income across time (and also across the states). Hereafter, we consider Arrow's security model which allows us to present the main features of a financial market equilibrium.

There are n securities (assets) in the model; an asset j ($j = 1, \cdots, n$) is characterized by a vector $d_j = (d_{js}) \in R^S$, where d_{js} represents the claim (dividend) in units of account if state s prevails at time $t = 1$. Thus, for each agent, a portfolio $\theta = (\theta_j) \in R^n$ of securities gives him at time $t = 1$ a dividend $\theta \cdot d_s = \Sigma_{j=1}^{n} \theta_j d_{js}$ if state s prevails (where $d_s = (d_{js}) \in R^n$).

We denote by q_j the price of security j, that is, the price to be paid at time $t = 0$ to get the dividend d_{js} at time $t = 1$ if state s prevails. Hence $q = (q_j) \in R^n$ is the security price system.

We now present the idea of *no-arbitrage*, also called the "absence of free lunch", a basic concept of finance. Heuristically, it says that to receive a positive dividend at time 1 at every state s, one has to pay something at time $t = 0$. Formally, the security pricing $(q, d_1, \cdots, d_n)$ is said to satisfy the *no-arbitrage property* if

$$(NAP)\left\{\begin{array}{l}\text{for every } \theta = (\theta_j) \in R^n \\ \Sigma_{j=1}^n \theta_j d_j > 0 \Rightarrow \theta \cdot q > 0; \text{ and} \\ \Sigma_{j=1}^n \theta_j d_j = 0 \Rightarrow \theta \cdot q = 0.\end{array}\right.$$

We recall that, for a vector $x \in R^S$, the notation $x > 0$ means that $x \geq 0$ and $x \neq 0$.

7. Asset Pricing

We shall now give some consequences of the no-arbitrage property. As a consequence of the so-called "theorem of the alternative" (or Farkas' lemma), one deduces the following state pricing result.

Proposition 5. *Assume that the no-arbitrage property holds, then for every state $s = 1, \cdots, S$, there exists $\pi_s > 0$ such that $q_j = \Sigma_{s=1}^S \pi_s d_{js}$ for every $j = 1, \cdots, n$, or equivalently,*

$$q = \Sigma_{s=1}^S \pi_s d_s.$$

Moreover, the state price system $\pi = (\pi_s) \in R^S$ is uniquely determined if and only if the vector space spanned by $\{d_j \,|\, j = 1, \cdots, n\}$ is R^S.

We shall now give an important property of the asset price system π which will be shown to be the expected value of its dividend, up to a normalization to be defined below. The underlying measure, which is directly related to the vector (π_s), is called the *risk neutral probability*.

To define the normalization, we first consider some portfolio (if any) $\bar{\theta} = (\bar{\theta}_j)$ having a dividend $\Sigma_{j=1}^n \bar{\theta}_j d_j \gg 0$. Assuming that (d, q) satisfies the no-arbitrage property, we deduce that $q \cdot \bar{\theta} = \Sigma_s \pi_s \Sigma_{j=1}^n \bar{\theta}_j d_{js} > 0$. We can therefore normalize prices and dividends, respectively, as follows:

$\hat{q}_j = q_j / q \cdot \bar{\theta}, \quad j = 1, \cdots, n;$

$\hat{d}_{js} = d_{js} / [\Sigma_{j=1}^n \bar{\theta}_j d_{js}], \; j = 1, \cdots, n, \; s = 1, \cdots, S.$

The pair $(\hat{q}, \hat{d})$ satisfies also the no-arbitrage property. Thus, from Proposition 5, there exists a state price vector $\hat{\pi} \in R_{++}^S$ such that

$$\hat{q}_j = \Sigma_{s=1}^S \hat{\pi}_s \hat{d}_{js} \; j = 1, \cdots, n.$$

We notice that, for the portfolio $\bar{\theta}$, we have $\bar{\theta} \cdot \hat{q} = 1$ and $\Sigma_{j=1}^n \bar{\theta}_j \hat{d}_{js} = 1$ for all s; this remark and the above equations imply that $\Sigma_{s=1}^S \hat{\pi}_s = 1$ and $\hat{\pi}_s > 0$

for every s. Thus we can interpret $\hat{\pi}$ as a vector of probability assessments of the states. Considering an element in R^S as a random variable on $\Omega = \{1, \cdots, S\}$ into R, endowing Ω with the σ-algebra $\mathcal{F}$ consisting of all subsets, we define on $(\Omega, \mathcal{F})$ the probability measure Q in a standard way, as follows:

$$Q(E) = \Sigma_{s \in E} \hat{\pi}_s \text{ for every } E \in \mathcal{F}.$$

From above, we now deduce that the normalized asset price system is the expectation under Q of the normalized dividends, that is, formally,

$$\hat{q}_j = E^Q(\hat{d}_j) \;\; j = 1, \cdots, n.$$

The measure Q is called the risk neutral probability or an equivalent martingale measure by Harrison and Kreps (1979) who extended the above framework to a continuous-time setting. This case, which will be developed in the next section, involves mathematical techniques which come either from probability theory (Brownian motion, diffusions possibly with jumps, martingale theory, Monte Carlo simulations...) or from partial differential equations (parabolic equations, numerical analysis...).

This analysis can also be extended to multi-period models; pricing formulas, like the famous Black and Scholes' one, follow from the existence of a risk neutral probability π. Another natural generalization is to allow transaction costs. In discrete time this leads to the use of Bellman dynamic programming and in continuous time to the use of the Hamilton–Jacobi–Bellman non linear partial differential equation and stochastic control of diffusions (possibly with jumps).

8. Financial Markets: Continuous Time

In his seminal paper, published in the *Annales de l'Ecole Normale Supérieure* in 1900, Louis Bachelier introduced the main principles of modern finance:

— the use of probability theory in the valuation of what he defines as "premia", and which are exactly options;

— the introduction of the Brownian motion to model the dynamics of the underlying process and the importance of the Markov property;

— the absence of arbitrage opportunities, recognized through the existence of both buyers and sellers at different quoted prices.

In his terms: "Le marché ne croît à un instant donné ni à la hausse, ni à la baisse du cours vrai". Hence, taking position in liquid markets is a fair game.

Around the same period, Einstein in 1905 defined formally Brownian motion, the main properties of which were studied by Wiener in 1923. Both were working, having in mind applications in the area of physics; it is of some interest to point out the major differences which appear when applications are given in finance theory. The setting for uncertainty is a filtered probability space $(\Omega, [0, T], \{\mathcal{F}_t; t \in [0, T]\}, P)$; the σ-algebra $\mathcal{F}_t$ represents the information available at time t.

A typical use of Brownian motion is signal theory where the observed signal $dX(t)$ is supposed to be the sum of a deterministic signal and a Brownian perturbation (white noise). Formally,

$$dX(t) = h(t)dt + \sigma dW(t),$$

where W is a standard Brownian motion in R^q. The main goal in this case is to identify the drift $h(t)$, observing $X(t)$.

Similarly, in Finance, the price process $S(t)$ of assets at time t is shown to satisfy the following stochastic differential equation:

$$dS(t) = S(t)[\mu^S(t)dt + \sigma^S(t)dW(t)],$$

where $\mu^S(t)$ is the instantaneously stochastic return (claim, dividend) at time t of the asset, $\sigma^S(t)$ is its volatility and W is a standard Brownian motion as previously.

To simplify the presentation, we also suppose that there exists in the model a non-risky asset with price $S^0(t)$ at time t. In the absence of risk, it is a standard result that $S^0(t)$ is the solution to the following ordinary differential equation

$$dS^0(t) = S^0(t)r(t)dt,$$

where $r(t)$ is the short term interest rate.

In this continuous time model, the no-arbitrage property implies the following relationship between the risky and the non-risky asset returns. Formally, there exists a stochastic R^q-valued process $\lambda(t)$, the so-called risk premium vector, such that $\mu^S(t) = r(t) + \sigma^S(t)\lambda(t)$ almost everywhere, that is, $\lambda(t)$ is the price that the market assigns to the noise $W(t)$ and which does not depend on the asset S under analysis.

The following result shows, as in the discrete case, the existence of a risk neutral probability measure Q satisfying the following properties.

Proposition 6. *There exists a probability measure Q, equivalent to the objective probability P defined by*

$$\frac{dQ}{dP} = \exp(-\int_0^T \lambda(s)dW(s) + (1/2)\int_0^T |\lambda(s)|^2 ds),$$

and we define a Q-Brownian motion as

$$\hat{W}(t) = W(t) + \int_0^t \lambda(s)ds.$$

(a) Then the normalized price process of the assets, $\hat{S}(t) = S(t)/S^0(t)$, satisfies the following stochastic differential equation:

$$\frac{d\hat{S}(t)}{\hat{S}(t)} = \sigma^S(t)d\hat{W}(t),$$

and it is a martingale for the risk neutral probability measure Q and the filtration $\mathcal{F}_t$, that is

$$\hat{S}_t = E_t^Q(\hat{S}_T),$$

where $E_t^Q(\cdot) = E_t^Q(\cdot|\mathcal{F}_t)$.

(b) Let $\hat{H}_T$ be a contingent claim which is a terminal normalized value of the portfolio $(\theta(t))$, then the normalized price process $\hat{c}(t)$ of $\hat{H}_T$ is given by

$$\hat{c}(t) = E_t^Q(\hat{H}_T),$$

and $(\theta(t))$ is called the hedging portfolio of the claim $\hat{H}_T$.

In Finance Theory, the main problem is to identify the volatility parameter $\sigma^S(t)$, which fully characterizes the dynamics of the asset prices. Consequently, the appropriate tools had to be designed to address the estimation of the parameters of stochastic processes and this has generated a renewal of interest in these methods: volatility estimator, ARCH models...

The accuracy of the model has been and is questioned; in particular, because of the thickness of the distribution tails, Mandelbrot and Levy processes are also considered.

For the pricing, hedging of portfolio in a multi-period setting, sophisticated probabilistic tools became necessary: Itô's formula, Girsanov's theorem, Bessel processes, Markov processes...

Bachelier concludes. "Le marché, à son insu, obéit à une loi qui le domine, la loi de probabilité".

These questions are very important for practitioners (financial institutions but also firms) involved in projects requiring borrowing in the future or dealing with foreign currencies. For them, the crucial problem is to properly use the new financial instruments (futures, options, swaps) in order to build the appropriate Hedge.

Stochastic modelling provides a unified framework to price and to measure the sensitivities of the different instruments.

9. The Core of an Economy

The core of an economy consists of the states of the economy which cannot be blocked by a group (coalition) of agents. We shall see that the core always contains the set of Walras equilibria and that these two sets coincide when the influence of every agent is negligible, a property rigorously formalized by considering an atomless measure space of agents. This "equivalence theorem" gives a deep insight into the role of coalitions in price theory.

We now formally define the notion of the core of an exchange economy $\mathcal{E} : A \to \mathcal{P} \times R^{\ell}_{+}$. If A is a finite set of agents (resp. $(A, \mathcal{A}, \nu)$ is a measure space of agents), a coalition E is a subset of A (resp. E is an element of $\mathcal{A}$); an allocation $x : A \to R^{\ell}_{+}$ is said to be *attainable* (or a state) if

$$\Sigma_{a \in A} x(a) = \Sigma_{a \in A} e(a) \text{ (resp. } \int_A x d\nu = \int_A e d\nu).$$

The coalition E is said to *block (or improve upon)* an allocation x if there exists an attainable allocation y such that

$$y(a) \succ_a x(a) \text{ for every a } \in A \text{ (resp. for almost every } a \in A).$$

The *core* of the economy $\mathcal{E} : A \to \mathcal{P} \times R^{\ell}_{+}$, denoted by $C(\mathcal{E})$, is the set of all attainable allocations of $\mathcal{E}$ which no coalition can block.

The following result has an important economic interpretation although its proof is mathematically trivial.

Proposition 7. *Let $\mathcal{E}$ be a finite economy; then $W(\mathcal{E}) \subset C(\mathcal{E})$.*

The above result and the existence theorem of Walras equilibria immediately yield the following result on the non-vacuity of the core.

Theorem 8. *Let $\mathcal{E}$ be a finite economy; then the core $C(\mathcal{E})$ is nonempty under the assumptions of Theorem 1.*

The main result, however, states that the two concepts of Walras equilibria and of core allocations are equivalent when every individual agent is negligible. Formally:

Theorem 9 (Aumann). *Let $\mathcal{E}$ be a measurable economy such that $(A, \mathcal{A}, \nu)$ is an atomless measure space of agents and $\int_A e d\nu \gg 0$, then $W(\mathcal{E}) = C(\mathcal{E})$.*

For the proof of this result, we refer to Hildenbrand (1974).

10. Non Cooperative Games

We first present one of the most popular kind of games: those which are played by two players in many steps as, for example, the game of chess. At each step only one player makes a move among a finite set of available possibilities (actions). At the end of the game, either one player is declared winner and the other loser (and the loser has to pay one unit to the winner), or the play is a draw.

This class of games is referred to as the two-person zero-sum games with perfect information. In this type of game it is natural to ask which one of the two players will win. The answer to this question is known as Zermelo's theorem. Formally:

Theorem 10 (Zermelo). *In a two-person zero-sum game with perfect information, either player one can force a win, or player two can force a win, or both sides can force at least a draw.*

The idea of the proof relies on the important fact that only one player moves at each step and that each player has a full observation of the past moves.

This is no longer the case when the players are supposed to play simultaneously, that is, perfect information is lost. Consider, for example, the following game of "matching pennies" in which the payoff of player 1 is given in the following table.

Player 2

		Strategy 1	Strategy 2
Player 1	Strategy 1	1	-1
	Strategy 2	-1	1

The interpretation is that each player shows the other player one side of a coin. If both players show the same side, then player 1 wins. If not, player 2 wins.

The reader should easily convince himself that there is no "winning strategy" in any naive sense for any of the two players. The term of "winning strategy" will now be defined in a formal and general setting.

A game in *strategic (or normal) form* is a list $G = (A, (S_a)_{a\in A}, (g_a)_{a\in A})$, where A is the set of agents (called players hereafter), for every $a \in A$, S_a is the set of *strategies* of player a, and $g_a : \Pi_{a\in A} S_i \to R$ is the *payoff function* of player a.

The game G is "played" as follows. Each player a chooses a strategy $s_a \in S_a$ and the choices are done simultaneously by all the players. Then player a receives the payoff $g_a(s)$, where $s = (s_a)_{a\in A}$ is the list of strategies of all the players. In the following, we shall say that the game G is finite if A is finite (and for the sake of simplicity we assume that $A = \{1, \cdots, n\}$) and if all the sets S_a are finite.

A *two-person zero-sum game* is a strategic game G such that $n = 2$ and the sum of the payoff functions of the two players is equal to zero, that is, for all strategy pairs s, one has $g_1(s) + g_2(s) = 0$. In other words, what is won by one player is lost by the other one.

The game of chess and indeed every game with perfect information can be put in a strategic form. As for chess, a strategy of a player is the full description of his action policy along the whole game including what he should do at nodes he would never reach. As described above, the number of strategies of a player clearly increases exponentially with the size of a game.

A two-person zero-sum game G is said to have a value v if player 1 has a strategy $s_1^* \in S_1$ that guarantees him a payoff greater than or equal to v, that is, $g_1(s_1^*, s_2) \geq v$ for every $s_2 \in S_2$, and if player 2 has a strategy $s_2^* \in S_2$ that guarantees him that his loss is less than or equal to v, that is, $v \geq -g_2(s_1, s_2^*) = g_1(s_1, s_2^*)$ for every $s_1 \in S_1$.

The couple $(s_1^*, s_2^*) \in S_1 \times S_2$ is called an *equilibrium* of the game and one easily checks that it satisfies the two above inequalities if and only if it is a *saddle point* of g_1 in the sense that

$$g_1(s_1, s_2^*) \leq v = g_1(s_1^*, s_2^*) \leq g_1(s_1^*, s_2), \quad \forall s_1 \in S_1 \ \forall s_2 \in S_2.$$

Clearly, the game of "matching pennies" fails to have a value in the above sense. This failure is rooted in the discrepancy between what a player can force (as a leader) and what he can defend (as a follower).

We shall now see that the game of "matching pennies" has in fact a value in a weaker sense which is equal to 0, by allowing the players to "mix" their strategies, a fundamental idea first introduced by Von Neumann and Morgenstern. Indeed, let player 1 choose strategy 1 with probability 1/2 and strategy 2 with probability 1/2; this guarantees him an expected (average) payoff $(1/2)1 + (1/2)(-1) = 0$. The same strategy for player 2 also guarantees him a loss of 0. Hence, 0 is a value of this game when we allow "mixed" strategies, a notion that we now formally define.

Let $G = (\{1, 2\}, (S_1, S_2), g = g_1)$ be a two-person zero-sum game; the *mixed extension* of G is the two-person zero-sum game:

$$\hat{G} = (\{1, 2\}, (\hat{S}_1, \hat{S}_2), \hat{g}),$$

where $\hat{S}_a$ is the set of probability distributions on S_a and, for every $\mu_1 \in \hat{S}_1$, $\mu_2 \in \hat{S}_2$, $\hat{g}(\mu_1, \mu_2)$ is the expectation of the random variable g with respect to the product probability $\mu_1 \otimes \mu_2$, that is, in a finite setting,

$$\hat{g}(\mu_1, \mu_2) = \Sigma_{s_1 \in S_1, s_2 \in S_2} \mu_1(s_1)\mu_2(s_2)g(s_1, s_2).$$

The elements in $\hat{S}_a$ are called *mixed strategies* to distinguish them from the elements in S_a which are called *pure srategies.*

We now can state the fundamental theorem of zero-sum games.

Theorem 11 (Von Neumann-Morgenstern). *Let G be a finite two-person zero-sum game; then its mixed extension $\hat{G}$ has a value.*

Actually, this result can naturally be extended to games with compact Haussdorff sets of strategies and to a continuous payoff by defining appropriately its mixed extension.

We end this section by coming back to the general case of a game in strategic form $G = (A, (S_a)_{a\in A}, (g_a)_{a\in A})$. In this case, the natural extension of the value and of the associated equilibrum notion (s_1^*, s_2^*) defined previously is given as follows.

A *Nash equilibrium* point of G is an n-tuple $s = (s_a)$ of strategies in $\Pi_{a\in A} S_a$ such that, every player a has no profitable deviation, that is, $g_a(s) \geq g_a(s_1, \cdots, s_{a-1}, t_a, s_{a+1}, \cdots, s_n)$ for every $t_a \in S_a$.

The mixed extension of the game G is defined similarly as in the zero-sum case to get the following result.

Theorem 12. *The mixed extension of a finite game in strategic form has a Nash equilibrium.*

Actually, the finiteness of the game is not essential nor does one need that the game be a mixed extension of some finite game to get a Nash equilibrium. Indeed, one can prove the following result.

Theorem 13. *Let $G = (\{1, \cdots, n\}, (S_a), (g_a))$ be a game in strategic form and assume that for every $a = 1, \cdots, n$*

(i) *S_a is a non-empty, convex, compact subset of a Hausdorff locally convex real topological vector space;*

(ii) *the mapping $g_a : s_a \to g_a(s)$ is quasi-concave, for every $(s_1, \cdots, s_{a-1}, s_{a+1}, \cdots, s_n) \in \Pi_{j\neq a} S_j$;*

(iii) *the mapping $g_a : S_1 \times \cdots \times S_n \to R$ is continuous.*

Then the game G has a Nash equilibrium.

References

[1] Aumann, R.J., "Market with a continuum of traders", *Econometrica* **32** (1964), 39–50.

[2] Aumann, R.J., "Existence of competitive equilibria in markets with a continuum of traders", *Econometrica* **34** (1966), 1–17.

[3] Arrow, K.J. and G. Debreu, "Existence of an equilibrium for a competitive economy", *Econometrica* **22** (1954), 265–290.

[4] Arrow, K.J., "Le rôle des valeurs boursières pour la répartition la meilleure des risques", *Econométrie*, Centre National de la Recherche Scientifique, Paris, 1953, 41–48. Translated as "The role of securities in the optimal allocation of risk-bearing", *Review of Economic Studies* **31** (1964), 91–96.

[5] Bachelier, L., "Théorie de la spéculation", *Annales Scientifiques de l'Ecole Normale Supérieure* **17** (1900), 21–86.

[6] Black, F. and M. Scholes, "The pricing of options and corporate liabilities", *Journal of Political Economy* **81** (1973), 637–59.

[7] Cournot, A., *Recherches sur les principes mathématiques de la théorie des richesses*, Hachette, Paris, 1838. Translated as: *Research into the Mathematical Principles of the Theory of Wealth*, Kelley, New-York,1960.

[8] Debreu, G., *Theory of Value*, Wiley, New York, 1959.

[9] Debreu, G., "Existence of competitive equilibrium". In *Handbook of Mathematical Economics, vol. II,* Arrow and Intriligator (eds.), North Holland, Dordrecht, 1982.

[10] Dierker, E., "Regular economies". In *Handbook of Mathematical Economics, vol. II,* Arrow and Intriligator (eds.), North Holland, Dordrecht, 1982.

[11] Duffie, D., *Security Markets, Stochastic Models*, Academic Press, San Diego, 1988.

[12] Duffie, D., "The theory of value in security markets". In *Handbook of Mathematical Economics, vol. IV*, W. Hildenbrand and H. Sonnenschein (eds.), North Holland, Dordrecht, 1991.

[13] Einstein, A., "On the movement of small particles suspended in a stationary liquid demanded by the molecular-kinetic theory of heat", *Annales Physik* **17** (1905).

[14] Harrison, J-M. and D.M. Kreps, "Martingale and arbitrage in multiperiod securities markets", *Journal of Economic Theory* **20** (1979), 381– 408.

[15] Hildenbrand, W., *Core and equilibria of a large economy*, Princeton University Press, Princeton, 1974.

[16] Hildenbrand, W., "Core of an economy". In *Handbook of Mathematical Economics, vol. II*, Arrow and Intriligator (eds.), North Holland, Dordrecht, 1982.

[17] Hildenbrand, W., and H. Sonnenschein, *Handbook of Mathematical Economics*, vol. IV, North Holland, Dordrecht, 1991.

[18] Koopmans, T.C., *Three Essays on the State of Economic Science*, McGraw Hill, New York, 1957.

[19] Kakutani, S., "A generalization of Brouwer's fixed point theorem", *Duke Mathematical Journal* **8** (1941), 457– 459.

[20] McKensie, L.W., "On equilibrium in Graham's model of world trade and other competitive systems", *Econometrica* **22** (1954), 147–161.

[21] Nash, J.F., "Equilibrium points in N-person games", *Proceedings of the U.S.A. National Academy of Sciences* **36** (1950), 48–49.

[22] Von Neumann, J. and O. Morgenstern, *Theory of Games and Economic Behavior*, 3rd ed, Princeton University Press, Princeton, 1954.

[23] Scarf, H.E. (with the collaboration of T. Hansen), *The Computation of Economic Equilibria*, Yale University Press, New Haven, 1973.

[24] Scarf, H., "The computation of equilibrium prices: an exposition". In *Handbook of Mathematical Economics, vol. II*, Arrow and Intriligator (eds.), North Holland, Dordrecht, 1982.

[25] Shafer, W. and H., Sonnenschein, "Market demand excess demand functions". In *Handbook of Mathematical Economics, vol. II*, Arrow and Intriligator (eds.), North Holland, Dordrecht, 1982.

[26] Smale, S., "A convergent process of price adjustment and global newton methods", *Journal of Mathematical Economics*, **3** (1976), 107–120.

[27] Smale, S., "Global analysis and economics". In *Handbook of Mathematical Economics, vol. II*, Arrow and Intriligator (eds.), North Holland, Dordrecht, 1982.

[28] Wald, A., "Über die eindeutige positive Lösbarkeit der neuen Produktionsgleichungen", *Ergebnisse eines mathematischen Kolloquiums* **6** (1935), 12–20.

[29] Walras, L., *Eléments d'économie politique pure*, Corbfaz, Lausanne, 1874. Translated as: *Elements of pure economics*, Irwin, Chicago, 1954.

[30] Zermelo, E., "Über eine Anwendung der Mengenlehre auf die Theorie des Schachspiels". In *Proceedings of the Fifth International Congress of Mathematicians, vol. II*, Hobson, E. and A. Love (eds.), Cambridge University Press, Cambridge, 1912, 501–504.

Table Ronde O : 8 juillet 1992

MATHÉMATIQUES ET CHIMIE

par Edgar Soulié
Commissariat à l'Energie Atomique,
DSM-DRECAM-SCM, Centre d'Etudes de Saclay, Gif-sur-Yvette (France)

La chimie est à la fois une vaste discipline scientifique et une industrie qui contribue de façon décisive à façonner la société actuelle par les produits bruts ou demi-finis qu'elle livre aux industries de transformation et les produits finis qu'elle élabore.

La table ronde ne pouvait rendre compte de toute la diversité des relations entre mathématiques et chimie. Notre ambition plus modeste fut d'apporter un témoignage sur quelques aspects de ces relations, à travers les contributions de quelques chimistes, travaillant dans des entreprises ou des institutions de caractère académique, et de deux mathématiciens qui se sont intéressés à des problèmes qui se posent en chimie.

Sans les outils informatiques, ordinateur et périphériques associés (écran de visualisation graphique et imprimante), les mathématiques ne pourraient guère orienter, faciliter ou accélérer les recherches et les applications en chimie. Le lecteur devra garder à l'esprit l'intervention éventuelle des mathématiques lorsque l'intuition et l'expérience (difficile à décrire) de l'expert ont été mises à profit pour discerner quels sont les théories, modèles ou schémas explicatifs les plus pertinents a priori.

Aux éléments et aux corps purs à l'échelle macroscopique (la nôtre) correspondent respectivement les atomes et les molécules à l'échelle microscopique. En chimie, on s'intéresse à la fois aux propriétés des molécules et à leurs transformations les unes dans les autres par échange d'atomes : les réactions chimiques. L'existence des réactions chimiques appropriées conditionne la possibilité d'obtenir les molécules. Le développement fulgurant de la chimie repose donc largement sur les découvertes de réactions chimiques produisant les molécules désirées.

Le plus souvent, plusieurs réactions chimiques ont lieu dans le même milieu; pour prévoir l'évolution de ce milieu, on modélise ces réactions par des équations et des équations différentielles exprimant la conservation des éléments, les relations entre les concentrations d'espèces impliquées dans une même réaction et les vitesses d'évolution de ces concentrations. P. Hanusse montre sur un exemple comment on entreprend, avec une stratégie appropriée, la réduction de la complexité du modèle représenté par ces équations différentielles. Cette réduction de la complexité n'a pas pour seul intérêt de faciliter les calculs, elle concourt surtout à dégager les traits essentiels du comportement du système.

On pourrait croire qu'il ne s'agit que d'un problème académique, à la frontière entre cinétique chimique et mathématiques. Or cette difficulté se présente aussi pour le système chimique de la haute atmosphère (un milieu raréfié), impliquant l'ozone, dont tout un chacun dit maintenant se préoccuper, les oxydes d'azote (émis notamment par les véhicues automobiles) et les chlorofluorocarbones utilisés dans les circuits réfrigérants et comme moyen de dispersion des aérosols.

Alors que de nombreuses réactions chimiques ont lieu en phase liquide, d'autres réactions impliquent la phase gazeuse et deux ou plusieurs phases solides. La nature des produits de la réaction peut dépendre des conditions de température et de pression du milieu où la réaction se produit. Pour comprendre et prévoir quelles espèces chimiques seront formées et sous quelle(s) forme(s) elles se trouveront, le chimiste (ou le métallurgiste) ont besoin de s'appuyer sur la thermodynamique. Cependant, les données thermodynamiques expérimentales et les diagrammes de phase ne forment pas naturellement un ensemble cohérent et compatible avec les lois de la thermodynamique, en raison des erreurs inhérentes à toute mesure. Avant d'utiliser ces données pour piloter un réacteur chimique (l'endroit où l'on conduit la réaction chimique), il est nécessaire, comme l'indique C. Bernard, de les "mettre en cohérence", c'est-à-dire de les "rectifier" de telle façon qu'elles satisfassent aux relations prévues par les lois de la thermodynamique. Ce problème peut être formulé mathématiquement comme un problème d'optimisation non linéaire avec contraintes. L'aptitude de l'industrie à fabriquer le matériau requis avec la pureté requise en utilisant l'installation disponible dans les meilleures conditions économiques dépend entre autres de la qualité des valeurs finales obtenues lors de cette mise en cohérence.

Pour analyser les informations chimiques disponibles, les méthodes statistiques peuvent être complétées par d'autres méthodes comme l'expose J. Gasteiger, la reconnaissance de formes et les réseaux neuronaux. Ces derniers sont adaptés à l'investigation de systèmes pour lesquels on ne dispose pas d'un modèle prédictif et suscitent l'intérêt tant en chimie qu'en médecine par exemple.

Alors qu'à notre échelle, le chimiste observe, manipule, étudie des corps purs et mélanges, il a besoin de se représenter les molécules pour analyser ses observations et mesures, pour avoir l'intuition de nouvelles réactions. Depuis la découverte de la diffraction des rayons X par Max von Laue au début du siècle, l'arrangement tridimensionnel des atomes a été déterminé pour de nombreux composés cristallisés. La détermination de la structure géométrique des molécules s'est cependant heurtée à deux difficultés majeures :

1) il s'avère parfois difficile, voire impossible, de faire cristalliser certaines substances, ou d'obtenir un cristal de taille suffisante (dimensions de 50 à 250 micromètres)

2) Alors que la théorie de la diffraction fait intervenir des amplitudes qui sont des nombres complexes, on ne mesure que des intensités qui sont les carrés des modules des amplitudes.

G. Bricogne expose comment l'intuition chimique et des méthodes statistiques raffinées (mises en oeuvre sur ordinateur) permettent de surmonter la difficulté de l'information manquante appelée "problème des phases" en cristallographie.

La structure cristalline d'une molécule fournit une information irremplaçable sur laquelle s'appuie le raisonnement chimique. Elle indique notamment la symétrie du site occupé par chacun des atomes de la molécule. Or la symétrie joue aussi un rôle décisif pour expliquer les propriétés physiques, en particulier les spectres tels que spectre vibrationnel, spectre d'absorption électronique, spectre de résonance magnétique nucléaire, spectre de Mössbauer, spectre photoélectronique, etc. M. Golfier montre à quel point la théorie de la représentation des groupes est féconde pour prévoir et ensuite interpréter les spectres observés (y compris des spectres observés pour des molécules interstellaires). La théorie des groupes, combinée avec les éléments de la chimie quantique s'avère également un moyen puissant pour comprendre la réactivité chimique comme M. Golfier le montre sur des exemples.

La compréhension des spectres comme celle de la réactivité chimique s'appuie d'autre part sur la chimie quantique, que présente M. Allavena, et qui est intimement liée à la théorie de la représentation des groupes.

La mécanique quantique introduit la description de la liaison chimique par une "fonction d'onde" qui est solution de l'équation de Schrödinger. Cette équation aux dérivées partielles — que l'on cherche à résoudre pour connaître la probabilité de trouver un électron dans une région donnée de l'espace (à l'échelle de l'angström !)— ne peut être résolue numériquement que par une succession d'approximations.

De façon approchée, on décrit cette fonction d'onde, qui dépend des coordonnées de plusieurs électrons, comme un produit de fonctions d'onde monoélectroniques. Et les fonctions monoélectroniques sont elles-mêmes approximées par des combinaisons linéaires de fonctions d'onde atomiques. Ces combinaisons dépendent essentiellement de la symétrie et certaines propriétés qualitatives des fonctions d'onde peuvent être déduites de la théorie des groupes avec des calculs simples — alors que d'autres propriétés exigent de longs et délicats calculs. L'espace mathématique ayant la puissance du continu est nécessairement remplacé en pratique par un ensemble fini de fonctions.

De nombreux problèmes d'unicité et de convergence peuvent donc se poser. L'expérience accumulée par les quanto-chimistes depuis une quarantaine d'années indique souvent quel niveau d'approximation doit être choisi pour obtenir un résultat significatif, selon la propriété recherchée. Lorsqu'ils existent, les spectres expérimentaux servent de référence à cet égard. La chimie quantique permet souvent de prédire le comportement des molécules non seulement dans leur état "normal" de plus basse énergie, qui est stable, mais encore dans des états excités transitoires, aux durées de vie parfois très courtes, qui sont observables avec les techniques résolues en temps.

La chimie quantique permet aussi d'aborder la différence des propriétés dans le coeur et à la surface de la matière; car la connectivité des atomes de surface diffère de celle des atomes situés à l'intérieur d'un échantillon par ailleurs homogène. Les quanto-chimistes sont devenus des utilisateurs importants des ordinateurs "vectoriels" avec lesquels ils ne peuvent cependant pas traiter les problèmes posés à propos des grosses molécules. Peut-être les ordinateurs "parallèles" permettront-ils prochainement d'aborder les calculs de structure électroniques de molécules complexes comportant de nombreux électrons autrement que de façon semi-empirique. Les progrès de la chimie quantique dépendent beaucoup de gros logiciels dont l'utilisation intelligente exige un effort important d'apprentissage.

En "chimie de base" et en métallurgie, l'industrie livre des produits relativement peu nombreux en très grandes quantités. A l'inverse, en pharmacochimie, on produit des quantités réduites de substances chères essentielles à notre santé. Qu'est-ce qui fait la différence entre une molécule ayant une activité pharmacologique et une autre qui lui ressemble, mais est dépourvue de toute activité pharmacologique? Sans attendre que les réponses aux questions de ce genre aient été trouvées, les laboratoires de l'industrie chimique et pharmaceutique ont recherché empiriquement les molécules les plus actives pour en faire des médicaments. G. Moreau présente une heuristique très employée utilisant la corrélation entre structure (géométrique) et activité (pharmacologique). A cette approche empirique, on préfère maintenant une approche utilisant la minimisation de l'énergie ou une autre approche basée sur l'application de l'équation fondamentale de la dynamique (transposée de notre échelle à l'échelle de l'atome!). En raison de la complexité des molécules intéressant la pharmacologie et la biochimie, on rencontre de nombreux minima secondaires de l'énergie qui rendent incertaine la détermination de la configuration d'énergie minimale — celle que la nature adopte le plus souvent.

L'importance de cette détermination a amené deux mathématiciens, S. Gomez et D. Romero, à étudier et appliquer deux algorithmes bien adaptés à la recherche d'un minimum global, l'algorithme du recuit simulé et l'algorithme du tunnel. Le premier est bien connu pour les problèmes d'optimisation combinatoire, que l'on retrouve dès lors que l'on recherche un minimum global parmi des minima locaux.

Le second est appliqué à la détermination de la géométrie de petis agrégats, des intermédiaires entre l'atome isolé et la matière à l'échelle macroscopique.

Alors que les mathématiques traditionnelles décrivaient les propriétés de courbes rectifiables et de surfaces dont les sections planes sont des courbes rectifiables, on rencontre dans la nature, et en particulier en chimie, des phénomènes que l'on représente beaucoup mieux en faisant appel aux objets fractals systématiquement explorés par B. Mandelbrot. A. Le Méhauté et F. Héliodore abordent la géométrie fractale et les équations différentielles d'ordre non entier en chimie physique. On peut prévoir que des modèles fractals seront de plus en plus introduits dans les modèles des phénomènes de diffusion, d'adsorption de la matière, comme dans les réactions chimiques en catalyse hétérogène.

Remerciements. Les organismes suivants ont apporté leur soutien à la Table Ronde Mathématiques et Chimie par un don. Qu'ils en soient profondément remerciés.

— Direction de la Chimie du Centre National de la Recherche Scientifique (C.N.R.S.)
— Direction des Sciences de la Matière du Commissariat à l'Energie Atomique (CEA)
— Société Rhône-Poulenc
— Société Solvay

Edgar Soulié
Commissariat à l'Energie Atomique
DSM-DRECAM-SCM, Centre d'Etudes de Saclay
F-91191 Gif-sur-Yvette Cedex, France

Mathématiques et modélisation en dynamique chimique

par Patrick Hanusse
CRPP/CNRS – Université de Bordeaux I, Pessac (France)

Les chimistes, qui pratiquent une science expérimentale, sont conduits à utiliser des notions, des outils, des méthodes appartenant à de très nombreuses branches des mathématiques et de l'informatique. Bien évidemment, cette largeur ne peut qu'entraîner un manque de profondeur, et parfois un manque de recul, conduisant souvent à l'utilisation de "boîtes noires", que ce soit en terme d'outils ou de méthodes. Un chimiste ne peut cependant pas se permettre de trop investir dans la dimension mathématique de ses problèmes sans prendre le risque de se marginaliser.

Le chimiste doit donc se tourner vers les mathématiciens, les intéresser à ses problèmes. Cependant, les mathématiciens ont leurs objectifs et leur langage propres, qui ne correspondent pas toujours à ses besoins et possibilités. Les problèmes des chimistes sont souvent d'une complexité trop grande et échappent aux possibilités des mathématiques, rendant aléatoires les investissements que pourrait y faire un mathématicien, sans courrir, lui aussi, un risque de marginalisation. C'est pourtant dans les "marges" que des progrès nouveaux peuvent être réalisés. Il faut cependant reconnaître que de nombreux exemples de collaborations fructueuses existent.

Pour dégager des perspectives nouvelles de développement, il apparaît important de tenter de définir une base conceptuelle commune qui permette, en quelque sorte, de "mathématiser" ce qui ne l'est pas encore. Nous allons tenter d'illustrer cette idée dans ce qui suit.

Comme dans de nombreux domaines de la science s'opposent deux points de vue que l'on peut sommairement dénommer *réalisme* et *universalité*. Il y correspond, d'une part un *langage spécifique* et spécialisé, qui s'est constitué pour des raisons pratiques et historiques, d'autre part un *langage générique*, qui est généralement celui des mathématiques.

Les chimistes utilisent différents langages spécialisés pour décrire leurs observations. C'est le cas du modèle moléculaire, niveau de description que, pour des raisons de formation, le chimiste se sent obligé, consciemment ou non, de considérer comme plus "vrai" que tout autre. Il ne constitue pas toujours le bon niveau de description pour l'étude de phénomènes qui représentent en fait des comportements ou des classes de comportements physiques globaux, des structures universelles indépendantes des spécificités de bas niveau du domaine.

Nous allons montrer sur un exemple comment on peut passer d'une description "réaliste", au sens du spécialiste, mais complexe, à une description, sinon universelle, du moins plus "mathématique". Définir une stratégie pour réduire ainsi la complexité et découvrir les structures universelles sous-jacentes, nous semble une voie possible pour jeter les bases d'un nouveau rapprochement entre une discipline comme la Chimie et les Mathématiques.

Un exemple: la modélisation en dynamique chimique (1,2)

Pour modéliser le comportement dynamique d'un système chimique, le chimiste écrira un *schéma réactionnel* réaliste, (de son point de vue, il y aurait là fort à dire), mais complexe, comportant beaucoup de variables et beaucoup de paramètres, comme par exemple le schéma qui fait intervenir 13 espèces (dimension de la trajectoire) dans 33 processus (incluant les échanges dans un Réacteur Continu Agité) et comporte 21 paramètres (voire p. 457).

A chaque espèce, chimique telle que Br^- ou BrO_3^-, est associée une concentration notée $[Br^-]$, $[BrO_3^-]$. Les concentrations évoluent au cours du temps sous l'effet des réactions chimiques, décrites sous la forme "$A + B => C$" indiquant que A et B réagissent pour former C. La vitesse v d'une telle réaction est de la forme $v = k[A][B]$, où k est une constante de vitesse, paramètre associé à ce processus, $[A]$ et $[B]$ les concentrations des espèces correspondantes. Plus généralement, la vitesse d'un processus réactionnel est proportionnelle au produit des concentrations des espèces réagissantes. Une telle réaction contribue à la vitesse d'évolution de $[A]$, $d[A]/dt$, par un terme "$-v$", de même pour $[B]$, et par un terme "$+v$" pour la concentration $[C]$. Ainsi, au schéma réactionnel ci-dessus, qui met en jeu 13 espèces, sera associé un système de 13 équations différentielles non-linéaires du premier ordre. La réaction étant ici conduite dans un "Réacteur Continu Agité", on considérera le système comme homogène. Dans le cas con-

CSTR KO Br-, BrO_3^-, H^+, ArOH, H_2O

BrO_3^-	$+ Br^- + 2H+$	<=>	$HBrO_2 + HOBr$
$HBrO_2$	$+ Br^- + H^+$	=>	2HOBr
HOBr	$+ Br^- + H^+$	<=>	$Br_2 + H_2O$
BrO_3^-	$+ HBrO_2 + H^+$	<=>	$2BrO_2. + H_2O$
$2HBrO_2$		=>	$BrO_3^- + HOBr + H^+$
BrO_2	+ ArOH	=>	$ArO. + HBrO_2$
BrO_2	+ BrArOH	=>	$BrArO + HBrO_2$
Br_2	+ ArOH	=>	$BrArOH + Br^- + H^+$
HOBr	+ ArOH	=>	$BrArOH + H_2O$
BrArO.	+ ArO.	=>	$Dh + Br^-$
BrArOH	+ ArO.	<=>	BrArO. + ArOH

Exemple de schéma réactionnel complexe mettant en jeu 13 espèces chimiques (Br^-, BrO_3^-...), tel qu'il est fourni à un système informatique de simulation. La première ligne spécifie que la réaction a lieu dans un Réacteur Continu Agité dans lequel les espèces indiquées sont injectées avec un flux KO. Suivent les processus réactionnels de la forme " A + B => C", indiquant que les espèces A et B réagissent pour former l'espèce C. Un flèche simple indique que la réaction a lieu vers la droite seulement. Il existe donc 15 processus réactionnels, 5 processus d'entrée et 13 processus de sortie du réacteur, soit 33 au total. Ce modèle présente 21 paramètres: 15 constantes de vitesses, le flux d'entrée KO et les 5 concentrations des entrées.

traire, la diffusion des espèces devrait être prise en compte et on obtiendrait un système de "Réaction-Diffusion".

En règle générale, un tel système d'équations différentielles ne peut être étudié analytiquement. La trajectoire du système devra être calculée numériquement. Il faudra rechercher les états stationnaires, points fixes, singularités d'un tel système, étudier leur stabilité, tracer un diagramme d'état (de grande dimensionnalité), simuler le comportement. Pour ce faire, il faudra utiliser la théorie de la stabilité, la théorie des bifurcations, des formes normales, des catastrophes, l'analyse numérique, etc.

Les solutions globales ne peuvent pas être facilement répertoriées. Dans un système fermé, à cause de la conservation des espèces chimiques, on sait seulement que la trajectoire est bornée. Plusieurs attracteurs peuvent coexister, points fixes, cycles limites, tores denses, attracteurs étranges. Le choix de la condition initiale a donc un grande importance. En faisant varier les divers paramètres, on cherchera à déterminer les points de bifurcations d'où se déploient les divers attracteurs.

On sait que le comportement de tels modèles est, heureusement, caractérisé par une dimensionnalité effective beaucoup plus basse que celle de l'espace dans lequel a été formulé le problème. Dans le cas contraire, rien ne serait vraiment possible, "compréhensible" au sens étymologique même. Ainsi le schéma réduit suivant se comporte "quasiment" comme le schéma complexe précédent.

Cette réduction de complexité que pratique parfois "à la main" le chimiste manque cependant singulièrement de bases théoriques, en-dehors des situations simplificatrices relativement évidentes. Le plus souvent, la description obtenue n'est pas perçue comme constituant la réalité ultime du système étudié, la présence d'une universalité ou généricité, mais comme un pis-aller (on est bien obligé de faire des simplifications, des approximations). Ces pratiques devraient pourtant rapprocher le langage du chimiste de celui des mathématiciens, à condition cependant que l'on puisse définir une stratégie générale. Voici comment nous avons procédé sur cette exemple.

X	=>	Y
X + Y	=>	
Y	<=>	2 Z
2 Y	=>	
Z	=>	W + Y
W	=>	1/2 X

Schéma réactionnel réduit ne comportant plus que 4 espèces dans 7 processus. Le comportement de ce schéma est quantitativement quasiment identique à celui du schéma complet. Le principe de la méthode de réduction est décrite dans le texte.

Lorsque l'on observe le comportement du schéma réactionnel complet dans une situation typique, en effectuant un simulation numérique, on constate que la forme du comportement, dans le cas présent un oscillation de type cycle limite, est considérablement plus simple, d'un point de vue morphologique, que le schéma réactionnel qui lui donne naissance. En d'autres termes, le schéma réactionnel, pris dans son ensemble, est un objet complexe, comportant beaucoup de termes, beaucoup de relations. Potentiellement, la trajectoire du système dans l'espace de dimension 13 déjà mentionné pourrait prendre des formes très complexes. Il n'en est rien. La forme des séries temporelles donne l'intuition qu'une description compacte, en "quelques mots", mais pertinente, devrait pouvoir être construite.

Est-il possible de mettre en relation la simplicité morphologique du comportement avec un modèle mathématique, un schéma réactionnel simplifié, de complexité comparable? Pour ce faire, il faut d'abord caractériser la forme de la trajectoire de façon quantitative.

Si l'on veut extraire l'essentiel du comportement, il faut extraire l'essentiel de la forme. Pour ce faire, nous pratiquons un analyse hiérarchique de la trajectoire dans l'espace à 13 dimensions. Le principe en est simple. La description la plus grossière de la trajectoire est un segment joignant le premier et le dernier point de la série temporelle. Le degré de précision de cette description peut être mesuré par la distance la plus grande de l'arc de trajectoire au segment ainsi défini. Le segment initial sera alors remplacé par les deux segments formés au moyen du point de la trajectoire le plus distant. En répétant récursivement l'opération sur chaque segment, on obtient un description polygonale hiérarchique de la trajectoire, en réalité ici, du cycle limite.

Ce faisant, on construit un arbre binaire qui code les partitions de chaque segment quand la résolution augmente et dont la profondeur est liée à la précision que l'on choisit. Dans un contexte plus général, nous avons dénommé "dendrone" ce type de structure arborescente (4-7). A quelle résolution faut-il arrêter la décomposition ? La réponse à cette question est fournie par la structure de l'arbre lui-même. Lorsque l'on mesure le nombre de noeuds de cet arbre en fonction de la résolution, ce qui traduit le nombre de segments entrant dans la description à cette résolution, on observe que cette fonction présente un plateau très marqué. La présence de ce plateau est le signe qu'il existe une description stable ou persistante, qui reste inchangée dans un large intervalle de résolution. C'est la description morphologique "essentielle" que nous recherchons. Elle ne comporte que 5 segments. Notons qu'il est essentiel d'opérer dans l'espace complet. Ce résultat est à la fois surprenant et attendu. Il traduit la simplicité de la forme et la dimensionnalité effective réduite de la dynamique. Il existe donc, au cours du cycle limite, 5 phases essentielles dans cette dynamique.

Pour obtenir le schéma réactionnel réduit présenté plus haut, nous pouvons alors analyser, globalement, mais également au cours de chacune des phases ainsi définies, les liaisions qui existent entre les diverses variables et les divers processus que définit le schéma complet. On peut ainsi remarquer les processus négligeables, ceux qui sont en équilibre, détecter les dépendances entre processus, définir les processus équivalents, etc. Il n'est pas possible de décrire ici davantage cette étape (pour plus de détails, voir (3)). Le schéma réduit que l'on obtient ne fait plus intervenir que 4 variables dans 7 processus. Son comportement est pratiquement quantitativement identique à celui du schéma complet. Il décrit la dynamique globale au cours du cycle limite. En examinant la contribution de chaque processus au cours de chacune des phases identifiées, on peut en interpréter le comportement.

Remarquons que la réduction ainsi effectuée permet de garder une formulation de type chimique, c'est-à-dire ayant la forme d'un schéma réactionnel. Il s'agira néammoins en général d'un pseudo-schéma faisant intervenir des pseudo-espèces chimiques. La relation entre espèces "vraies" et pseudo-espèces peut ne pas être simple. En général, les chimistes acceptent ce genre de modèle, car il ressemble encore à un schéma réactionnel. Disons qu'il utilise le même type de langage que celui auquel ceux-ci sont habitués. On pourrait cependant pousser plus loin la réduction, mais dans ce cas on perdrait la forme chimique du modèle, pour obtenir un système d'équations différentielles. A l'extrême, s'agissant d'un cycle limite, nous pourrions obtenir la forme normale de Hopf correspondante ou un déploiement pertinent de celle-ci.

Nous voyons sur cet exemple qu'il existe toute une hiérarchie de descriptions, de modèles, qui vont du "réalisme" spécifique du domaine au "générisme" propre aux mathématiques, mais surtout qu'il semble possible, au moins dans certains cas, de définir une procédure explicite qui permette de construire cette hiérarchie.

Ce type d'approche n'est pas spécifique de l'exemple présenté ci- dessus. Il semble possible de définir une stratégie assez générale, basée sur la description hiérarchique d'une observation de nature quelconque, qui permette d'extraire les divers niveaux pertinents de description ainsi que le langage adapté à chaque niveau. On peut citer l'analyse de données complexes, d'objets fractals, d'images (4-7).

Cette perspective nous semble de nature, dans certains cas, à favoriser la découverte de structures universelles indépendantes du domaine, donc à rendre possible une "mathématisation" de phénomènes divers qui relèvent encore d'une description spécifique. Certains problèmes résisteront encore longtemps, d'autres pas. Cependant, il convient de ne pas négliger les obstacles que l'on peut qualifier de culturels, ceux qui tiennent à l'attachement de chaque spécialiste à un ou des langages qui lui sont familiers, et qu'il finit par considérer comme plus vrais que d'autres, en tout cas plus parlant pour lui. Le concept de compréhension ne s'exprime pas de la même façon pour tout le monde, même si toute entreprise de théorisation consiste précisément à unifier et rendre universel ce qui apparaît, au départ, comme spécifique et particulier. S'attacher à élaborer une telle méthodologie de réduction de complexité nous paraît de nature à rapprocher Mathématiques et Chimie.

Références

[1] Ch. Vidal et H. Lemarchand, *La réaction créatrice*, Hermann, Paris, 1988.

[2] P. Bergé, Y. Pomeau et Ch. Vidal, *L'ordre dans le chaos*, Hermann, Paris, 1988.

[3] P. Hanusse, "Analyse morphologique de trajectoires en dynamique chimique", *J. Chim. Phys.* **84** (1987), 1315–1327.

[4] P. Hanusse, Ph. Guillataud, "Object detection and identification by hierarchical segmentation", *Lecture Notes in Computer Science* **427**:217 (1990), Springer-Verlag, Heidelberg, New York.

[5] P. Hanusse et Ph. Guillataud, "Sémantique des images par Analyse Dendronique", *Actes de RFIA-91*, AFCET (ed.), Paris 1991.

[6] P. Hanusse et Ph. Guillataud, "Dendronic Analysis of pictures, fractals and other complex objects". In *Fractal geometry and computer graphics*, Encarnaçao et al. (eds.), Springer-Verlag, Heidelberg, New York, 1992.

[7] Ph. Guillataud, "Contribution à l'Analyse Dendronique des images. Application à l'analyse d'images de phénomènes physico-chimiques". Thèse Université de Bordeaux I, 1992.

Patrick Hanusse
CRPP/CNRS
Avenue Schweizer
F-33600 Pessac, France

Mathématiques et science des matériaux

par Claude Bernard
Laboratoire de Thermodynamique et Physico-Chimie Métallurgiques,
E.N.S.E.E.G. – Domaine Universitaire, Saint-Martin d'Hères (France)

Je me propose modestement de faire ici un bilan de l'intervention des mathématiques dans un domaine de la chaîne de conception et d'optimisation des procédés d'élaboration des matériaux.

Pour bien élaborer un matériau, *il vaut mieux bien* le connaître : diagramme d'équilibre et propriétés thermodynamiques. Et, si l'on souhaite réaliser une simulation thermodynamique du procédé, préalable à la mise en oeuvre du premier prototype, *il faut très bien* le connaître.

Si cette "bonne connaissance" n'est pas déjà disponible dans une banque de données, elle est à établir. Ce qui veut dire qu'il faut mettre en cohérence deux grandes familles d'informations expérimentales trop souvent disjointes : les informations qui concernent la topographie, des équilibres entre phases et les informations relatives aux grandeurs thermodynamiques en présence, qu'elles soient intégrales, partielles ou dérivées.

Pour expliciter la méthode de mise en cohérence, quelques rappels sont nécessaires.

L'état macroscopique d'un système est défini par un certain nombre de variables d'état, telles que la pression P, le volume V, la température T... et les nombres de moles des espèces chimiques constituant le système. Ces variables ne sont pas indépendantes, elles sont reliées entre elles par une équation d'état. La thermodynamique nous apprend l'existence d'autres fonctions d'état moins évidentes. En effet, par rapport à des variables d'état, par exemple P et T, il est parfois possible d'identifier une propriété du système qui s'exprime comme variation infinitésimale

$$\delta A = f(P, T)\delta T + g(P, T)\delta P.$$

Lorsque la somme des δA le long d'un chemin ne dépend que de l'état initial et de l'état final, on obtient une fonction A de T et de P que l'on appelle fonction d'état. En d'autres termes, si A est une différentielle totale, à savoir si $\partial f/\partial P = \partial g/\partial T$, il existe alors une fonction A déterminée par les conditions $\partial A/\partial T = f$ et $\partial A/\partial P = g$. Par exemple, A peut être l'entropie S (δS étant $\delta q/T$ où δq est la variation de la quantité de chaleur). [1, 2].

Suivant les contraintes imposées à un système, la composition à l'équilibre de celui-ci est obtenue pour le minimum d'une fonction d'état déterminée. Si par exemple la température et la pression sont maintenues constantes, la fonction à minimiser est alors l'enthalpie libre de Gibbs du système G. Pour les systèmes condensés, cas des solutions par exemple, pour lesquels le produit $P.V$ (pression.volume) est habituellement négligeable,

$$G \simeq -kT \ln Q$$

où Q fonction de partition de la solution considérée s'écrit :

$$Q = \sum_i g_i \exp(E_i/kT)$$

où E_i est l'énergie du système dans l'état i, k est la constante de Boltzmann, et g_i le facteur de dégénérescence de l'état i.

Il existe plusieurs autres manières de décrire la fonction G. Des méthodes relativement directes comme la description polynomiale et le modèle de Redlish–Kister [3], ou des méthodes plus complexes comme la méthode variationnelle thermodynamique basée sur le système de référence HSY (Hard-Sphere-Yukawa) [4]. Ce système est caractérisé par un ensemble de sphères dures chargées interagissant par des potentiels de Yukawa : les trois paramètres — σ, le diamètre de sphère dure, ϵ la force du potentiel d'ordre au contact et X, la constante d'écrantage du potentiel — sont accessibles par une variante de la technique variationnelle de Gibbs-Bogolioubov. Cette technique est particulièrement intéressante pour certains alliages car il suffit de connaître l'énergie totale ; elle permet d'atteindre simultanément les densités d'états électroniques, les grandeurs thermodynamiques de mélange et les facteurs de structure partiels.

Concernant les grandeurs thermodynamiques de certains alliages liquides, une bonne approximation consiste à utiliser le réseau de Bethe [5] comme support géométrique, c'est-à-dire la description de l'ordre chimique local (fluctuation de composition) à partir d'interactions de paires effectives variant avec la composition, et une approximation entropique la méthode variationnelle des amas (paire ou triangle ou tétraèdre) dite CVM d'un réseau cubique à faces centrées désordonné. L'intérêt d'une telle démarche est qu'elle permet de traiter les parties solide et liquide d'un diagramme de phases avec le même degré d'approximation, condition nécessaire à l'étude des équilibres hétérogènes qui nous préoccupent.

Une fois décrite la fonction à minimiser, pour chacune des phases, les coefficients devront en être optimisés en tenant compte de toutes les informations expérimentales.

Comme les grandeurs expérimentales font souvent intervenir des équilibres entre phases, il existe des relations implicites entre les coefficients de celles-ci. Les relations à traiter pour minimiser la somme des carrés des erreurs ne sont pas linéaires et nécessitent le recours à la méthode de Gauss-Newton.

Problèmes

Les difficultées rencontrées au cours de l'optimisation sont liées à des problèmes de convergence qui peuvent être dûs à l'incompatibilité de certaines mesures entre elles (si tel résultat expérimental est exact, tel autre ne peut être que faux) ou à l'existence de minima secondaires de la fonction. On rencontre également des problèmes d'unicité de solution qui tiennent peut-être à la nature de certains modèles.

D'autres difficultés surgissent au cours de l'étape de modélisation de la fonction à minimiser. Une mention spéciale sera faite pour ce qui concerne les problèmes liés à la dégénérescence (multiplicité) des états de configuration, problèmes sous-jacents dans tout modèle de solution [1].

En effet, une présentation générale de l'outil thermodynamique permettrait de dégager l'importance des modèles statistiques dans les domaines de *l'interprétation*, de *l'estimation* et de la *prévision* et le terme clef de l'établissement de toute formulation des grandeurs thermodynamiques de mélange des alliages à partir de la physique statistique, est l'estimation de la dégénérescence des états de configuration $g(E)$. (Ceci peut être déjà mis en évidence par la présentation des modèles les plus simples) [6]. Or, la simulation par la méthode de Monte Carlo, qui conduit, elle, à la valeur exacte de $g(E)$, montre l'importance des conséquences qui s'en déduisent :

- allure des courbes d'entropie et d'enthalpie ;
- interprétation des transitions ordre-désordre.

Il est donc fortement souhaitable de rechercher une méthode analytique qui se rapproche le plus possible des résultats obtenus par calculs numériques.

Pour conclure, un exemple d'utilisation des données optimisées précédemment montrera l'impact de celles-ci sur la simulation d'un procédé expérimental. Soit un réacteur de dépôt chimique en phase vapeur, c'est-à-dire un tube à parois froides au centre duquel un substrat chaud va être balayé par un mélange gazeux. Connaissant la pression dans le tube, la température du substrat, les débits de gaz injectés, et en faisant l'hypothèse du système isolé et à l'équilibre, il est possible de calculer la nature et les proportions des phases déposées sur le substrat, ce qui nous donne un diagramme de dépôt (Figure 1) [7].

Actuellement le couplage de cette approche purement thermodynamique avec les équations de diffusion et de flux est en cours, il doit permettre de tenir compte de la variation de composition des dépôts le long du substrat.

Les résultats tels que ceux présentés sur la Figure 1 sont obtenus par minimisation de la fonction énergie libre totale du système en prenant en compte les contraintes linéaires liées aux divers bilans atomiques. Cela peut être réalisé soit par un algorithme du type Newton appliqué à la méthode des multiplicateurs de Lagrange, soit par un algorithme de minimisation directe (gradient conjugué,

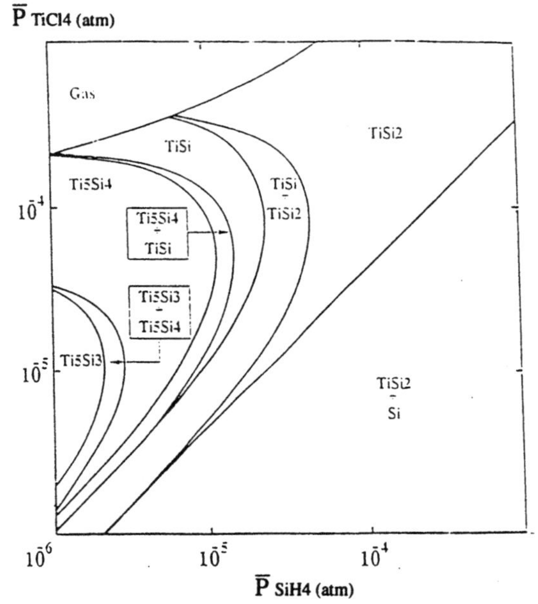

T = 1073 K

Pt (SiH4 + TiCl4 + H2) = 3.3 10^{-3} atm

Figure 1. Diagramme calculé, donnant, en fonction des pressions de vapeur de $TiCl_4$ et de SiH_4 introduites dans le réacteur, la nature des phases solides déposées dans chaque domaine à partir d'une phase vapeur $TiCl_4$, SiH_4, H_2 dans des conditions de pression totale et de température fixées [7].

Hooke et Jeeves [8]). Nous préférons une méthode exploratoire parce qu'elle permet de mieux contrôler les contraintes d'inégalité (non négativité des nombres de moles) et nous utilisons en particulier l'algorithme de Hooke et Jeeves parce qu'il permet d'explorer les variables indépendamment les unes des autres, et de ne calculer que la partie de l'énergie qui dépend de cette variable. Ce type de méthode nous a permis de résoudre des problèmes très complexes (200 variables) sans avoir à fournir de valeurs approchées pour les inconnues contrairement aux autres méthodes. Là également, nous éprouvons des difficultés à optimiser les paramètres de la convergence.

Ainsi, on peut constater que dans ce domaine très restreint de l'optimisation des données et des procédés, les mathématiques sont fort utiles. Bien que les méthodes auxquelles on fait appel soient relativement simples, de nombreux problèmes sont rencontrés, et l'aide de mathématiciens est certainement indispensable à leur résolution.

Références

[1] C.H.P. Lupis, *Chemical Thermodynamics of Materials*, Elsevier, New York, 1983.

[2] G. Bruhat, A. Kastler et R. Vichnievsky, *Thermodynamique*, Masson, Paris, cinquième édition, 1962, pp. 113–121.

[3] O. Redlich, A.T. Kister, *Industr. Eng. Chem.* **40**: 345 (1948) .

[4] A. Pasturel, J. Hafner, P. Hicter, *Phys. Rev. B* **32**: 5009 (1985).

[5] L.M. Falicov, F. Yndurain, *Phys. Rev. B* **12**: 5664 (1975).

[6] C. Bichara, J.P. Gaspard, J. Cl. Mathieu, *J. Chem. Phys.* **89**: 7 (1988), 4339-4345.

[7] R. Madar et C. Bernard, *J. Vac. Sci. Technol. A8* **(3)** (1990), 1413–1421.

[8] R. Hooke et T.A. Jeeves, *J. Assoc. Comp. Mach.* **8**: 212 (1961).

Claude Bernard
Laboratoire de Thermodynamique et Physico-Chimie Métallurgiques
E.N.S.E.E.G.–Domaine Universitaire
F-38042 Saint-Martin d'Hères Cedex, France

Knowledge Acquisition in Chemistry

by Joachim Johann Gasteiger

Computer-Chemie-Centrum, Universität Erlangen-Nürnberg, Erlangen (Germany)

Most chemical phenomena are too complex to be put into mathematical equations that can be solved. Chemical knowledge is gained to a large extent from individual observations and inductive learning by deriving general rules from a series of data. Data analysis is therefore of paramount importance in chemistry.

Statistical and pattern recognition methods are playing an increasing role in the analysis of chemical information. Flexible and robust methods are very much desired. This is underscored by the rapid acceptance of neural networks for chemical applications [1,2]. Quite a few problems in chemistry have to deal with the relationships between the structure of a compound and its physical, chemical or biological properties. The approach taken to find such relationships is outlined in Figure 1.

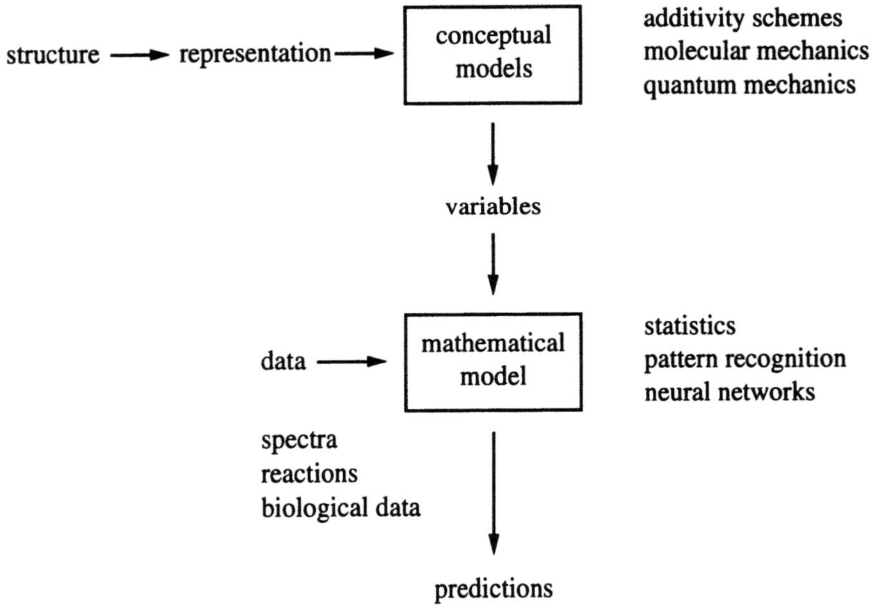

Figure 1. The building of models to derive relationships between chemical structure and data.

The endeavor to derive chemical knowledge from individual data is supported by the increasing availability of chemical information in databases. More than 13 million compounds have been stored, databases of several millions of chemical reactions are available, hundreds of thousands of spectra of chemical compounds have been stored in computer readable form together with information on the structure of those compounds.

Chemical Structures

A major problem in the analysis of chemical information is the representation of chemical structures [3]. Chemists have created a rich and internationally accepted language for representing the structure of molecules: the structural formula. A structural formula is essentially an undirected, weighted graph. Graph theory therefore plays a major role in the computer representation and manipulation of chemical structure information (Figure 2).

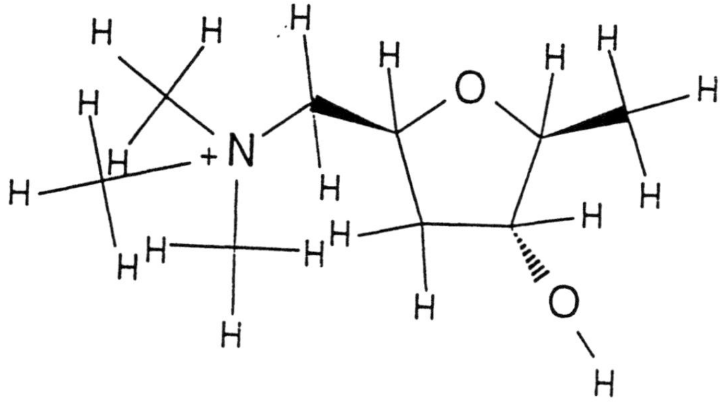

Figure 2. Structural formula of muscarine, the poison in the mushroom fly agaric. The wedge and dots indicate stereochemical information.

Graph theoretical methods are used to derive a unique representation of a chemical compound, an indispensable requirement for storing informations on compounds in databases. Furthermore, the perception of rings, of molecular symmetry, of aromaticity, and the coding of stereochemical information rest on graph theory.

However, with the increasing sophistication of our insight into chemical structures, the limitations of a structural formula are increasingly felt. Many applications in chemistry and biology have to deal with the three–dimensional aspects of chemical structures (Figure 3).

In fact, the representation of chemical structures has developed to a point that details of the three–dimensional electron distribution in molecules have to be taken into consideration. The graphical representation of such three–dimensional structures has matured to a high degree of sophistication (Figure 4). However, methods for extracting the important features of the three–dimensional electron distribution, for comparing molecules and for defining similarity at such a high level of structure representation are just being explored. Figure 5 shows a two–

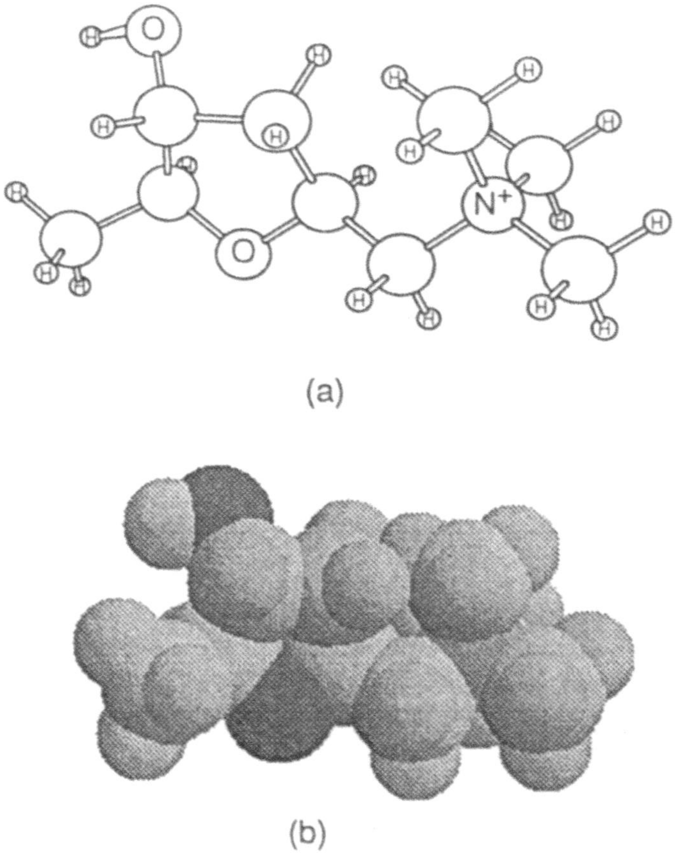

Figure 3. Ball and stick model (a) and space–filling–model (b)

dimensional map obtained by a Kohonen neural network [4] from the molecular electrostatic potential on the van der Waals surface of muscarine (cf. Figure 4) [5].

In essence, the Kohonen neural network allows a non–linear projection of the three–dimensional formation of a property on a molecular surface onto the surface of a torus that can be converted into a two–dimensional plane [5].

Chemical Reactions

Chemical reactions are usually represented by the structural formulae of the starting materials and products of a reaction (Figure 6). The extraction of knowledge from databases of chemical reactions requires finding: (a) the reaction center, i.e., the atoms and bonds directly involved in the reorganization of electrons during a reaction, (b) the required context, i.e., groups of atoms and bonds not directly

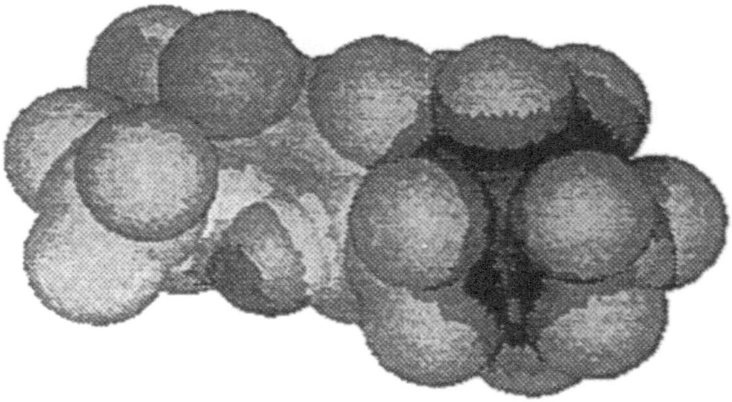

Figure 4. The molecular electrostatic potential on the van der Waals surface of muscarine.

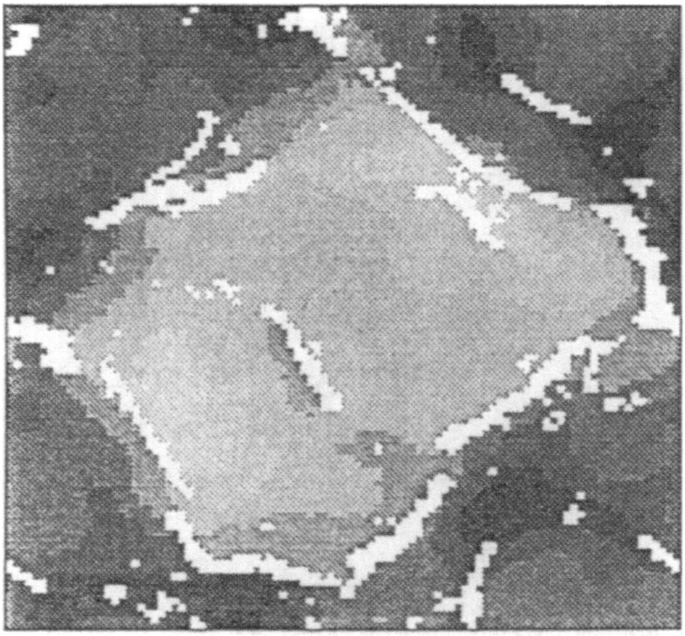

Figure 5. Kohonen map of the molecular electrostatic potential on the van der Waals surface of muscarine.

involved in a reaction but required to make a reaction go, and (c) the variable portion, i.e., those parts of the molecules that may vary in the individual instances of a given reaction type.

Methods for the automatic assignment of reaction centers have been developed. Machine learning techniques for the classification and generalization of chemical reactions based on conceptual clustering methods are being developed [6].

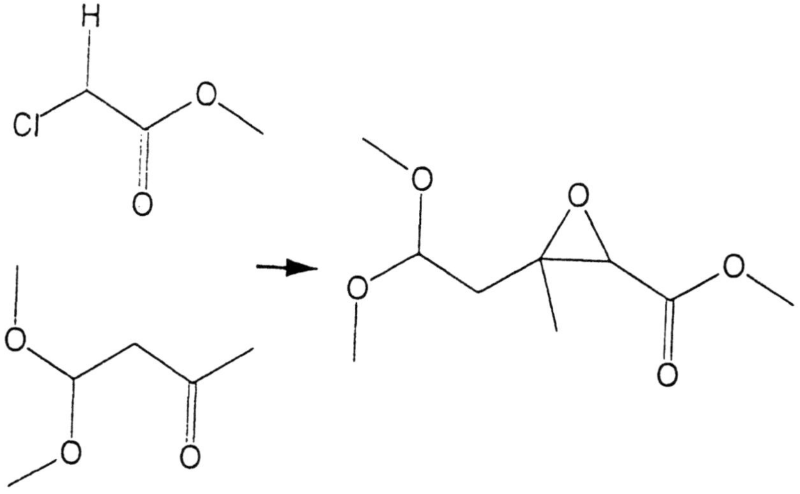

Figure 6. Chemical reaction equation.

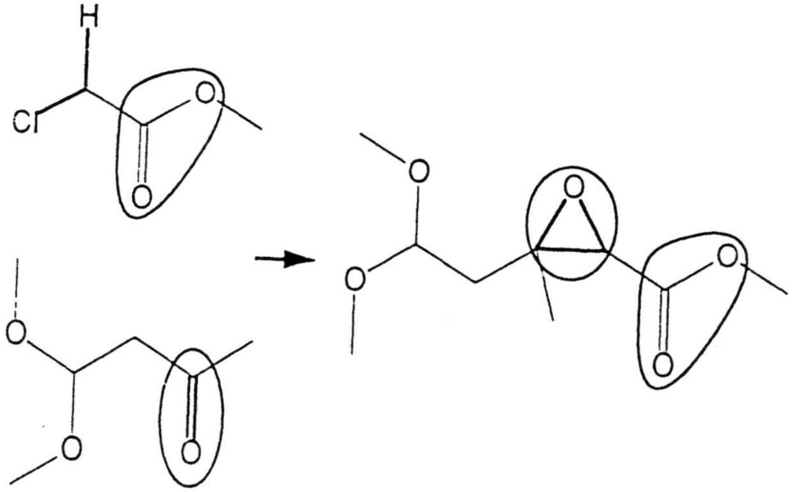

Figure 7. Reaction center (in bold) and required context (encircled) of a chemical reaction.

Spectral Information

Spectra of compounds are rich in information that helps in the elucidation of the structure of compounds. However, the relationships between the structure and spectral data are again highly complex. For example, a mass spectrum records the fragmentation of a molecule after ionization by high energy electron impact.

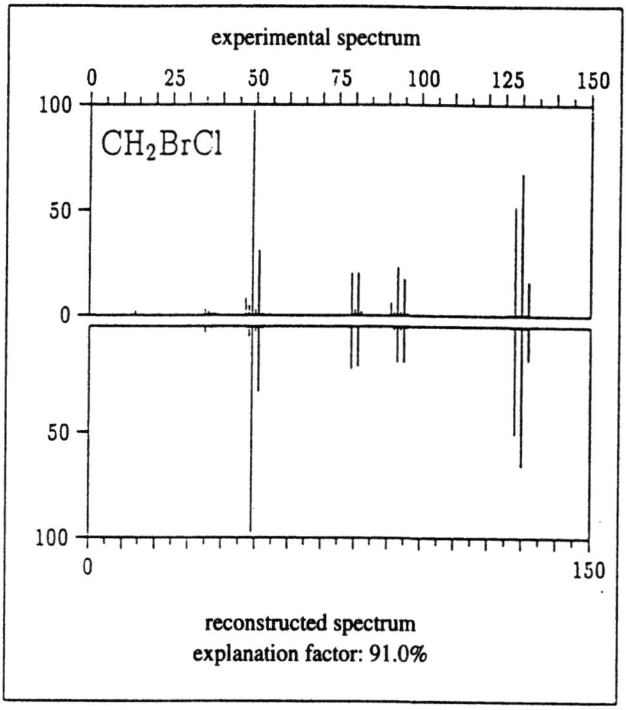

Figure 8. Experimental and reconstructed mass spectrum of CH_2ClBr.

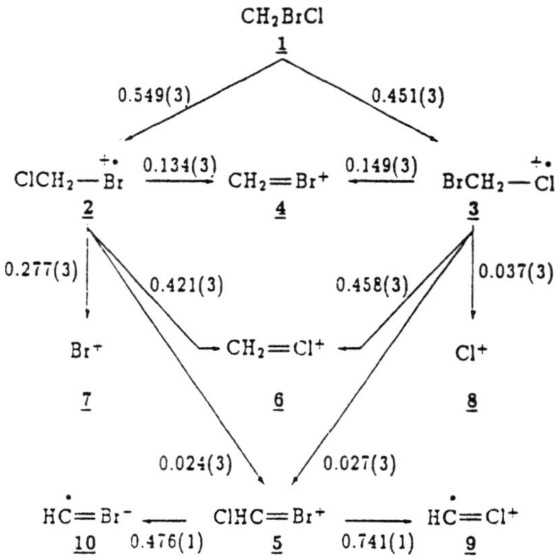

Figure 9. Fragmentation scheme used for explaining the mass spectrum in Figure 8.

The fragmentations and rearrangements occurring in the mass spectrometer can be elucidated by an assembly of conceptual and mathematical models [7]. Figure 8 (preceding page) shows the experimental mass spectrum of an organic compound and that part of the mass spectrum that can be explained by the fragmentation scheme shown in Figure 9.

Clearly, the structures analysed in Figures 8 and 9 are quite simple. More complex organic molecules lead to much more extensive fragmentation schemes that, nevertheless, can still be analysed.

References

[1] Anderson, J. A. and Rosenfeld, E. (eds.), "Neurocomputing: Foundations of the Research", MIT Press, Cambridge, 1989.

[2] Zupan, J. and Gasteiger, J., "Neural Networks: A new method for solving chemical problems or just a passing phase?" *Anal. Chim. Acta* **248** (1991), 1–30.

[3] Warr, W. A. (ed.), *Chemical Structures*, Springer-Verlag, Berlin, 1991.

[4] Kohonen, T., "Self–Organization and Associative Memory", Springer-Verlag, Berlin, 1988.

[5] Gasteiger, J., Li, X., Rudolph, C., Sadowski, J. and Zupan, J., "The Representation of Molecular Electrostatic Potentials by Topological Feature Maps", *J. Am. Chem. Soc.*, May 1994.

[6] Rose, J. R., Gasteiger, J. and Gelernter, H., "Machine Learning Applied to the Hierarchical Classification of Chemical Reactions", *J. Chem. Inf. Comput. Sci.* **34** (1994), 74–90.

[7] Gasteiger, J.; Hanebeck, W., Schulz, K.–P., "The Prediction of Mass Spectra from Structural Information", *J. Chem. Inf. Comput. Sci.* **32** (1992), 264–271.

Joachim Johann Gasteiger
Computer-Chemie-Centrum
Universität Erlangen-Nürnberg
Nagelbachstraße 25
D-91052 Erlangen, Germany

Mathematical Methods for Determining Crystal Structure

by Gérard Bricogne

MRC Laboratory of Molecular Biology, Cambridge (United Kingdom) and LURE, Orsay (France)

The X-ray analysis of crystal structure may be viewed as an inverse problem in which the three-dimensional structure of a crystal is to be reconstructed from the results of an X-ray diffraction experiment. The intensities of the scattered beams which can be measured in such an experiment are essentially the squared moduli of the Fourier coefficients associated to the triply-periodic electron density distribution in the crystal. Unfortunately the corresponding phases cannot be measured, yet they must be restored by some means before the desired picture of the crystal structure can be obtained by Fourier synthesis. This constitutes the "phase problem" of X-ray crystallography.

Because of its extreme practical importance, the phase problem has been under constant attack for eighty years, but there is still no generally applicable method for solving it. Its recalcitrant nature derives from the fact that it is mathematically indeterminate unless sufficient chemical information is brought to bear on its solution to constrain it down to a reasonably unique answer, while the full content of this chemical information cannot be adequately captured by any of the well-developed devices of mathematics until a stereochemical parametrisation of the electron density becomes possible. The central difficulty lies therefore in obtaining an electron density map which can be interpreted by model-building.

A statistical approach to the phase problem was initiated in the early fifties (Hauptman & Karle, 1953) and later developed into successful Direct Methods for the determination of small structures (Giacovazzo, 1980). It views an unknown crystal structure as made up of atoms with known chemical identity but unknown positions, and considers the latter as random, with an initially uniform distribution in the asymmetric unit of the crystal. Structure determination consists in the gradual removal of that randomness. For this purpose, limit theorems of probability theory are invoked to estimate the joint probability distribution of suitably chosen "structure factors" (i.e., triply-indexed Fourier coefficients); formal substitution of the observed amplitudes of the latter into these distributions is then expected to yield conditional joint distributions for the phases, indicating that certain combinations of phase values are more probable than others once the moduli of the structure factors are known.

Traditionally the Edgeworth series (Cramér, 1946) was used to approximate joint distributions (Klug, 1958), but this turns out to be unsuitable for large structure factor amplitudes where methods from "large deviation theory" are needed (Bricogne, 1984; Dacunha-Castelle, 1984). In addition, this series was previously derived under the assumption that the distribution of the random atomic positions is always uniform — a severe handicap in the macromolecular field because of the existence of solvent regions. These two limitations can be overcome simultaneously by using the saddlepoint method (Daniels, 1954) instead of the Edgeworth series; this always yields optimal estimates of joint probabilities involving large amplitudes, and is equivalent to requiring that the distribution of random atomic positions should be updated whenever phase assumptions are made so as to retain maximum entropy under the constraints embodied in these assumptions (Bricogne, 1984). As discussed elsewhere (Bricogne, 1991a) this approach does not amount to a wholesale adoption of the so-called "maximum entropy principle" (Levine & Tribus, 1979; Jaynes, 1983) but instead to a consolidation of the analytical apparatus of direct methods.

This revised statistical analysis of the phase problem leads naturally to a general multisolution strategy of structure determination in which the space of hypothetical phase sets is explored in a hierarchical fashion by building a search tree, similar to those used in game-playing computer programs. Each trial phase set is ranked according to a certain statistical criterion (the log-likelihood gain, or the Bayesian score) which acts as a heuristic function in guiding the subsequent growth of the tree. Further theoretical developments have led to a comprehensive Bayesian theory of the phase problem (Bricogne, 1988) within which all phasing techniques can be treated, with major improvements; and to an extension of these procedures to the use of powder diffraction data where the phase problem becomes a "hyper-phase problem" (Bricogne, 1991b). This approach has been successfully applied to the direct determination of small crystal structures from single-crystal and powder data, to the phasing of electron diffraction amplitudes, and to the ab-initio generation of phases for a small protein (see for instance Bricogne, 1993a).

The areas of mathematics called upon in this field comprise analytical probability theory and the theory of hypothesis testing, in the same vein as classical communication theory (as illustrated by the work of Rice, 1944, 1945); Fourier theory in all its aspects, with particular emphasis on questions related to sampling theorems (see e.g., Bricogne, 1993b); elements of group theory, module theory and integral representation theory (as used in the classification of crystallographic groups by Brown et al., 1978); and in some recent enhancements of the search strategy, the theory of error-correcting codes and of their connections with packing and covering problems (Thompson, 1983; Conway & Sloane, 1988) and with multidimensional discrete Fourier transform algorithms.

References

[1] Bricogne, G., "Maximum Entropy and the Foundations of Direct Methods", *Acta Crystallographica* **A40** (1984), 410–445.

[2] Bricogne, G., "A Bayesian Statistical Theory of the Phase Problem. I. A Multichannel Maximum-Entropy Formalism for Constructing Generalised Joint Probability Distributions of Structure Factors", *Acta Crystallographica* **A44**(1988), 517–545.

[3] Bricogne, G., "The X-ray crystallographic phase problem". In *Maximum Entropy in Action*, B. Buck and V.A. Macaulay (eds.), Oxford Univ. Press, Oxford, 1991a, pp. 187–216.

[4] Bricogne, G., "A Multisolution Method of Phase Determination by Combined Maximisation of Entropy and Likelihood. III. Extension to Powder Diffraction Data", *Acta Crystallographica* **A47** (1991b), 803–829.

[5] Bricogne, G., "Direct Phase Determination by Entropy Maximisation and Likelihood Ranking: Status Report and Perspectives", *Acta Crystallographica* **D49** (1993a), 37–60

[6] Bricogne, G., "Fourier Transforms in Crystallography: Theory, Algorithms and Applications." In *International Tables for Crystallography, vol. B,* for the International Union of Crystallography, U. Shmueli (ed.), Reidel Publ. Co., Dordrecht, 1993b, pp. 23–106.

[7] Brown, H., Bülow, R., Neubüser, J., Wondratschek, H. and Zassenhaus, H., *Crystallographic groups of four-dimensional space*, John Wiley, New York, 1978.

[8] Conway, J.H. and Sloane, N.J.A., *Sphere Packings, Lattices and Groups*, Springer-Verlag, New York, 1988.

[9] Cramér, H., *Mathematical Methods of Statistics*, Princeton Univ. Press, Princeton, 1946.

[10] Dacunha-Castelle, D., "Reconstruction des phases en cristallographie par maximum d'entropie (d'après G. Bricogne)", *Séminaire Bourbaki* **628** (1984).

[11] Daniels, H.E., "Saddlepoint Approximation in Statistics", *Ann. Math. Stat.* **25** (1954), 631–650.

[12] Giacovazzo, C., *Direct Methods in Crystallography*, Academic Press, London, 1980.

[13] Hauptman, H. and Karle, J., *The Solution of the Phase Problem: I. The Centrosymmetric Crystal*, ACA Monograph No. 3, Polycrystal Book Service, Pittsburgh,1953.

[14] Jaynes, E.T., *Papers on Probability, Statistics and Statistical Physics*, Reidel Publ. Co., Dordrecht, 1983.

[15] Klug, A., "Joint Probability Distributions of Structure Factors and the Phase Problem", *Acta Crystallographica* **11** (1958), 515–543.

[16] Levine, R.D and Tribus, M., *The Maximum Entropy Formalism*, MIT Press, Cambridge, 1979.

[17] Rice, S.O., "Mathematical Analysis of Random Noise", *Bell System Tech. J.* **23** (1944, 1945), 283–332, (parts I and II); **24**, 46–156 (parts III and IV). Reprinted in *Selected Papers on Noise and Stochastic Processes* (1954), N. Wax (ed.), Dover, New York, pp. 133–294.

[18] Thompson, T.M., *From error-correcting codes through sphere packings to simple groups*, Carus Mathematical Monographs, no. 21. The Mathematical Association of America, 1983.

Gérard Bricogne
MRC Laboratory of Molecular Biology
Mills Road
Cambridge CB2 2QH, United Kingdom
and
Laboratoire pour l'Utilisation du Rayonnement Electromagnétique
F-91405 Orsay Cedex, France

La théorie des groupes en chimie

par Michel Golfier
Laboratoire de Chimie Organique de Synthèse
Université Paris-Sud, Orsay (France)

Si l'utilisation de la symétrie par les chimistes est ancienne, ne le traitement courant de ses problèmes à l'aide de la théorie des groupes ne remonte guère à plus d'une cinquantaine d'années pour la détermination des structures moléculaires par les méthodes spectroscopiques, et à une trentaine d'années pour la description des liaisons dans les molécules et l'étude de certains mécanismes de réactions. Les ouvrages actuellement disponibles et écrits à l'usage des chimistes développent les applications des groupes ponctuels finis (plus rarement celles des groupes continus) notamment pour les diverses spectroscopies, la formation des liaisons chimiques, l'application, en chimie des complexes, de la théorie du champ cristallin et du champ de ligands, et l'étude du déroulement de la réaction chimique par l'intermédiaire des diagrammes de corrélation et de la "règle de non-croisement".

Ce qui suit décrit, en développant quelques exemples, les principaux emplois des groupes par un "chimiste moyen" qui, sans être un théoricien, a néanmoins besoin d'utiliser la symétrie des molécules de manière fréquente.

1. Représentations linéaires

Les utilisations de la théorie des groupes dérivent essentiellement des propriétés de leurs représentations linéaires. On associe à un espace vectoriel V le groupe $GL(V)$ des isomorphismes de cet espace sur lui-même. Une représentation linéaire d'un groupe fini quelconque est un homomorphisme de ce groupe dans le groupe $GL(V)$. Les représentations sont faites à partir d'espaces vectoriels ordinaires (par exemple l'espace des vecteurs traduisant les déplacements des atomes dans le cas des spectroscopies infra-rouge ou Raman) ou d'espaces vectoriels de fonctions (notamment à partir des diverses orbitales moléculaires s, p et d, qui sont des fonctions décrivant chacune le comportement d'un électron). Au produit de deux fonctions ou au résultat de l'action d'un opérateur sur une fonction, est associé le produit tensoriel des représentations auxquelles appartiennent ces fonctions et cet opérateur. Ce produit tensoriel de représentations joue un rôle important dans les applications.

Ces représentations sont le plus souvent réductibles et les deux problèmes auxquels se trouve confronté le chimiste sont la décomposition d'une représentation réductible en une somme directe de représentations irréductibles, et la détermination de la composante d'une fonction qui appartient à une représentation irréductible donnée. Ce dernier problème est facilement résolu par la technique des projecteurs : le projecteur relatif à une représentation irréductible est un opérateur construit à partir des caractères de cette représentation; appliqué à une fonction appartenant à un espace vectoriel support d'une représentation réductible du groupe de symétrie de la molécule, il en sort la composante qui appartient au sous-espace support de cette représentation irréductible.

Parmi les résultats les plus généraux obtenus, on signalera seulement les trois suivants, qui couvrent la plus grande partie des applications courantes.

2A. Classification des solutions de l'équation de Schrödinger

La propriété de commutativité de l'opérateur hamiltonien H avec tous les opérateurs de symétrie de la molécule, démontrée par Wigner en 1927, entraîne que les solutions de l'équation de Schrödinger appartiennent à des espaces de fonctions supports de représentations irréductibles du groupe de symétrie de la molécule. Ces solutions peuvent donc être affectées d'une étiquette qui est celle de la représentation irréductible à laquelle elles appartiennent et qui caractérise leur symétrie. La dimension de cette représentation irréductible correspond au degré de dégénérescence de ces fonctions, c'est-à-dire au nombre d'entre elles qui sont solutions de l'équation de Schrödinger du système pour la même valeur de l'énergie. La détermination de l'appartenance des fonctions aux représentations irréductibles des groupes de symétrie est indispensable pour le tracé des diagrammes où interviennent ces fonctions.

Application aux complexes des éléments de transition

Les éléments de transition sont caractérisés par la présence d'orbitales d, qui peuvent décrire jusqu'à 10 électrons. Ils forment des ions complexes avec de nombreux donneurs d'électrons appelés ligands (ou coordinats). Les répartitions les plus fréquentes de ces ligands autour de l'ion central correspondent à des structures *octaédriques* (6 ligands aux sommets d'un octaèdre plus ou moins régulier, dont la symétrie maximum est celle du groupe O_h) ou *tétraédriques* (4 ligands aux sommets d'un tétraèdre, symétrie maximum T_d).

La simple considération des modifications de symétrie apportées à la structure électronique de l'ion par la présence des ligands a permis de comprendre un grand nombre de propriétés spectrales ou magnétiques de ces complexes. Pour se limiter aux deux cas ci-dessus, on peut par exemple considérer les ligands comme des charges ponctuelles (ou des dipôles) placées aux sommets d'un octaèdre ou d'un tétraèdre : si les ligands sont identiques, le champ électrostatique qu'ils créent ("champ cristallin") est de symétrie O_h (groupe de l'octaèdre régulier) ou T_d (groupe du tétraèdre régulier). Dans l'ion libre, les cinq orbitales d sont dégénérées et forment la base d'une représentation irréductible de dimension 5. En présence du champ cristallin, il y a une levée partielle de dégénérescence et ces orbitales se partagent en deux groupes qui sont support de deux représentations irréductibles de dimensions respectives 2 et 3 : E_g et T_{2g} du groupe O_h ou E et T_2 du groupe T_d, comme l'indique la Figure 1.

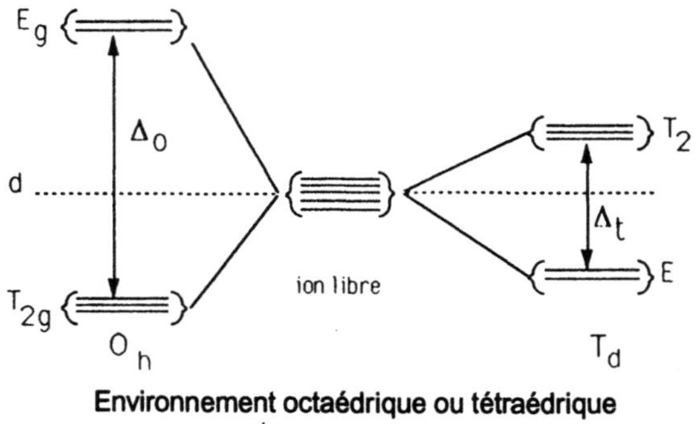

Figure 1.

Le paramètre D, qui caractérise la "force" des ligands, est un paramètre phénoménologique déterminé expérimentalement. On peut tracer les diagrammes montrant l'évolution de l'énergie des états électroniques en fonction de ce paramètre, alors que la symétrie de ces états est conservée lorsque ce paramètre varie (diagrammes de Sugano-Tanabe). La théorie du champ cristallin peut être améliorée en considérant des transferts partiels d'électrons entre l'ion et les ligands (théorie du champ de ligands). Elle permet, notamment par l'intermédiaire de diagrammes de corrélation entre les états électroniques des atomes à plusieurs électrons, l'interprétation de nombreuses propriétés de ces ions complexes (cf. Curie, Cotton, chap. 9) dont la chimie et les applications connaissent un développement considérable depuis une trentaine d'années.

2B. Condition suffisante de nullité d'intégrales ("éléments de matrice")

La théorie des groupes fournit des règles simples pour déterminer une condition suffisante pour la nullité d'intégrales des types : $\int \Psi^* \Theta d\tau$ et $\int \Psi^* F \Theta d\tau$, où F est un opérateur. Les intégrales sont étendues à tout l'espace, dont l'élément de volume relatif à l'ensemble des variables est désigné par $d\tau$. Pour que ces intégrales ne soient pas nulles pour des raisons tenant à leur symétrie (ce qui ne les empêche pas d'être éventuellement nulles pour d'autres raisons) il faut que le produit tensoriel des représentations irréductibles auxquelles appartiennent les fonctions Ψ^* (fonction complexe conjuguée de Ψ), Θ, et l'opérateur F contienne, dans sa décomposition en somme directe de représentations irréductibles, la représentation complètement symétrique. La plupart des applications font appel à ce résultat. C'est notamment le cas des méthodes spectroscopiques.

Applications aux méthodes spectroscopiques

Les diverses spectroscopies correspondent à l'absorption (ou l'émission) d'énergie $h\nu$ transportée par une radiation de fréquence ν. Le système qui absorbe cette radiation subit une transition entre un état d'énergie E_1 (généralement l'état de plus basse énergie) et un état d'énergie E_2. Une condition nécessaire est que $E_2 - E_1 = h\nu$. Selon la valeur de cette énergie, l'absorption conduira à une modification de l'état électronique (spectroscopie ultra-violette (UV)), de l'état de rotation *et* de vibration (spectroscopies infra-rouge (IR) et Raman) ou de l'état de rotation seul (spectroscopie de micro-ondes). S'y ajoutent des conditions liées à la symétrie de l'état initial (qui appartient généralement à la représentation complètement symétrique), à celle de l'état final, et à celle d'un opérateur qui dépend du type de spectroscopie considéré : ces conditions supplémentaires constituent *les règles de sélection*, qui indiquent les transitions "permises" ou "interdites", c'est-à-dire donnant lieu ou non à l'absorption d'une radiation.

Ces règles de sélection découlent des théorèmes sur la nullité des éléments de matrices. Elles traduisent la nécessité que l'intégrale $\int \Psi^*_{final} F \Psi_{initial} d\tau$ ne soit pas nulle par suite de la symétrie des fonctions de l'état initial ($\Psi_{initial}$), de l'état final après absorption (Ψ_{final}) et de l'opérateur F (cf. Herzberg, Fétizon).

L'opérateur F est généralement un opérateur scalaire, vectoriel ou tensoriel. Ainsi pour les spectroscopies Raman et surtout infrarouge, qui sont d'usage courant dans les laboratoires, l'opérateur F est :

— en spectroscopie infra-rouge un opérateur vectoriel μ (trois composantes μ_x, μ_y, μ_z, appartenant aux mêmes représentations irréductibles que les composantes x, y et z d'un vecteur).

— en spectroscopie Raman un opérateur tensoriel de polarisabilité a (six composantes $\alpha_x^2, \alpha_y^2, \alpha_z^2, \alpha_{xy}, \alpha_{xz}, \alpha_{yz}$, appartenant aux mêmes représentations irréductibles que les produits x^2, y^2, z^2, xy, yz et zy).

Une transition est *permise* (ou "active") si $\int \Psi^*_{final} F_i \Theta_{initial} d\tau \neq 0$, pour l'une au moins des composantes F_i des opérateurs μ ou α. Ces règles sont différentes dans l'une et l'autre spectroscopies, mais les transitions ont lieu entre les mêmes états vibrationnels, et fournissent les mêmes fréquences d'absorption (lorsqu'elles sont actives).

Il est long et difficile de calculer les fonctions $\Theta_{initial}$ et Ψ^*_{final} , mais *elles appartiennent à des représentations irréductibles du groupe de symétrie de la molécule*, qu'il est très facile de connaître en décomposant la représentation réductible dont l'espace des vecteurs-déplacements des atomes est le support. Le simple examen du nombre maximum de transitions permises dans l'une et l'autre spectroscopies permet souvent de choisir la structure correcte d'un produit entre plusieurs possibles. Ainsi l'absence de coincidence entre les bandes d'absorption d'un produit observables en infra-rouge, et celles observables en Raman laisse présumer l'existence d'un centre de symétrie : dans un tel cas, les représentations irréductibles sont soit de type g, soit de type u; or les composantes x, y, z sont de type u, alors que leurs produits sont nécessairement de type g : les bandes actives en IR sont interdites en Raman et vice-versa.

L'emploi de la théorie des groupes est indispensable pour l'interprétation des spectres ultra-violets (cf. Herzberg) ou de micro-ondes (cf. Kroto), où elle permet notamment l'interprétation des alternances observées dans l'intensité des bandes de rotation des molécules très symétriques comme N_2 ou l'acétylène.

2C. Diagrammes de corrélation

1) Liaisons dans les systèmes conjugués

L'étude des mécanismes de réaction vise à décrire les changements qui interviennent dans certaines liaisons, entre l'état initial (les réactifs) et l'état final (les produits). La théorie des groupes joue un rôle important dans la description des liaisons chimiques. Pour les liaisons simples ou les liaisons doubles isolées, cette description peut se faire en considérant l'interaction (c'est-à-dire la combinaison) de deux orbitales, une sur chacun des atomes formant la liaison (en utilisant éventuellement pour certains atomes des "orbitales hybrides ", qui sont des combinaisons linéaires des orbitales atomiques rendant mieux compte de la symétrie locale au voisinage de l'atome considéré). Mais pour des systèmes qui, comme le butadiène, contiennent des doubles liaisons séparées par une seule liaison simple (liaisons doubles dites "conjuguées") cette description des liaisons p du butadiène (liaisons provenant du recouvrement latéral de deux orbitales p d'axes parallèles) ne peut se faire sous la forme de deux doubles liaisons localisées (Fig. 2A) mais doit s'étendre à toutes les orbitales p d'axes parallèles (Fig. 2B). Ceci conduit à décrire le comportement des électrons qui participent à ces liaisons à l'aide d'orbitales moléculaires (OM), combinaison de toutes ces orbitales p (Fig. 2B)

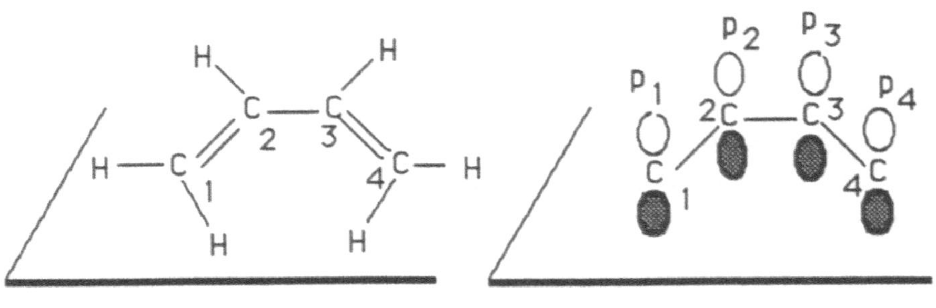

Figure 2.

Le "chimiste moyen" utilise des méthodes d'approximation qui doivent plus à son intuition chimique qu'à la rigueur mathématique. Elles sont imposées par l'impossibilité de résoudre rigoureusement l'équation de Schrödinger, hormis dans quelques très rares cas, et surtout par la nécessité de trouver un traitement général de molécules variées, sans calculs excessifs. C'est le cas de la méthode de Hückel (la plus courante pour un chimiste non théoricien). Elle permet de ramener le traitement complexe d'une molécule à n électrons à celui d'un seul électron décrit par un hamiltonien effectif H_{eff} qui est supposé tenir compte des interactions des autres électrons.

Pour la description des liaisons π du butadiène, le comportement de l'électron $n^\circ\ i$ est décrit par une fonction χ_i , solution de l'équation $H_{eff}\chi_i = E_i\chi_i$.

Dans la méthode de Hückel, les fonctions χ_i sont des combinaisons linéaires de n orbitales atomiques p, chacune centrée sur un atome du système conjugué : $\chi_i = \sum_{k=1}^{n} c_{ik}p_k$ les solutions χ_i sont déterminées en minimisant l'expression de l'énergie :

$$E = \left(\int \chi_i H_{eff}\chi_i d\tau\right) \Big/ \int \chi_i^2 d\tau \ .$$

Ceci aboutit à chercher les solutions d'un déterminant de rang n dont le terme général est de la forme $H_{ij} - ES_{ij}$ et dans lequel apparaissent des intégrales (souvent désignées sous le nom "d'éléments de matrice") qui sont:

$$\int \phi_i H_{eff}\phi_j d\tau = H_{ij} \quad et \quad \int \phi_i\phi_j d\tau = S_{ij} \ .$$

Certains de ces éléments sont considérés comme nuls pour des raisons d'éloignement des atomes i et j concernés. On ajoute les approximations suivantes : $H_{ii} = \alpha$, quel que soit l'atome de carbone du système conjugué. $H_{ij} = b < 0$ si les atomes de carbone i et j sont directement liés, et $H_{ij} = 0$ dans tous les autres cas. $S_{ii} = 1$ (condition de normalisation); $S_{ij} = 0$ dans tous les cas où $i \neq j$.

Mais ces simplifications ne suffisent pas à abaisser le degré du déterminant. Par contre, si l'on exprime les $OM\chi_i$ non en fonction des OA $p_k(k = 1$ à 4, Fig. 2B), mais en fonction de combinaisons linéaires $\chi_i = \sum_{j=1}^{n} c_{ij}\phi_j$ utilisant des **orbitales de symétrie adaptée** ϕ_j qui reflètent la symétrie de la molécule (ce sont des combinaisons linéaires des OA qui appartiennent à des représentations irréductibles de son groupe de symétrie), on aboutit à une factorisation du déterminant qui simplifie la détermination des solutions. Ces orbitales de symétrie adaptée sont facilement obtenues en utilisant la technique des projecteurs.

Il faut souligner qu'on a intérêt à simplifier les problèmes en ramenant un système non symétrique à un système symétrique voisin, quitte à corriger, dans un deuxième temps, les résultats obtenus en considérant comme une perturbation les modifications qui permettent de repasser du système symétrique utilisé pour les calculs au système non symétrique initial.

2) Diagrammes de corrélation. Règles de Woodward-Hoffmann

Les composés analogues au butadiène donnent des réactions thermiques et photochimiques qui ont longtemps intrigué les chimistes. Ils ne s'expliquaient pas leur très grande stéréospécificité : dans le cas d'un butadiène disubstitué par deux groupes CH_3 en position trans par rapport à la liaison centrale (Fig. 3, isomère E,E), la cyclisation photochimique donne un diméthyl- cyclobutène *cis*, dont l'ouverture par une réaction thermique aboutit à l'isomère E,Z, et non à l'isomère E,E; mais ce dernier peut être obtenu par l'ouverture thermique du diméthyl-cyclobutène *trans*, qui résulte lui-même de la cyclisation photochimique de l'isomère E,Z.

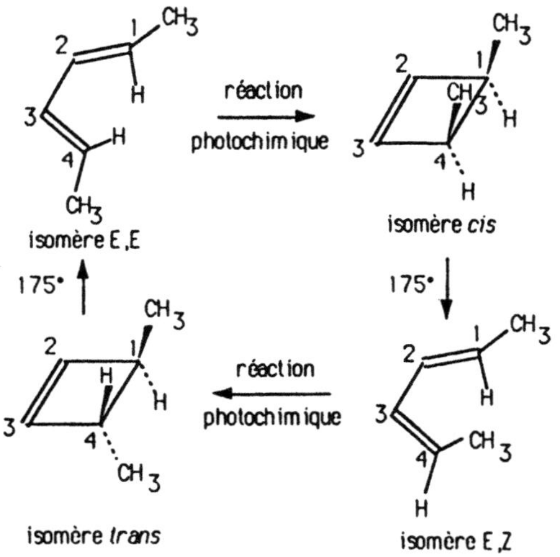

Figure 3.

Les deux réactions thermiques correspondent à des rotations des atomes de carbone C1 et C4 dans le même sens, et les réactions photochimiques à des rotations en sens inverse. Les résultats sont *les mêmes* si l'on considère des systèmes conjugués plus étendus mais contenant un nombre *pair* de doubles liaisons: ils sont *opposés* pour un nombre *impair* de doubles liaisons.

Comme le montre l'exemple précédent, il existe deux types de cyclisations (Fig. 4).

— la **cyclisation disrotatoire**, au cours de laquelle le plan de symétrie P est conservé (groupe C_s) : la rotation en sens inverse des extrémités de la molécule place les deux groupes CH_3 en position *cis* (d'un même côté) par rapport au cycle.

— la **cyclisation conrotatoire**, au cours de laquelle l'axe de symétrie C_2 est conservé (groupe C_2) : la rotation dans le même sens des extrémités de la molécule place les deux groupes CH_3 en position *trans* (de part et d'autre) par rapport au cycle. Les énergies sont les mêmes que dans le cas précédent (en négligeant les faibles différences liées à la position des groupements CH_3).

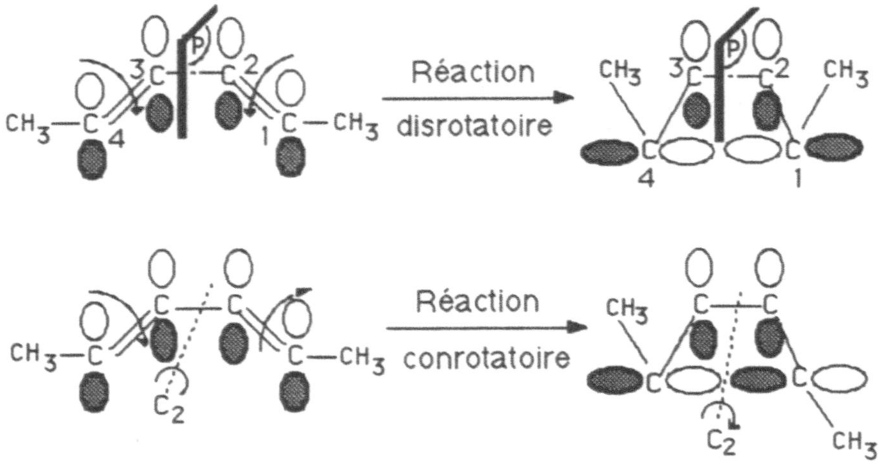

Figure 4.

Cette régularité remarquable a été attribuée par Woodward et Hoffmann (1965) à la symétrie des orbitales moléculaires du système par rapport à l'élément de symétrie de la molécule qui est conservé au cours de la réaction. Elle se traduit dans les **règles de Woodward-Hoffmann**, qui indiquent les réactions permises ou interdites; ces règles s'appliquent à un type de réactions très général, les *réactions péricycliques*, dont les cyclisations ne sont qu'un exemple. Cette interprétation, par l'orientation qu'elle a donné aux recherches chimiques et les idées nouvelles sur la nature de la réaction chimique qu'elle a suscitées, constitue l'avancée théorique la plus importante faite depuis un demi-siècle.

Pour étudier une réaction telle que la cyclisation du diméthyl- butadiène et diméthyl-cyclobutène (ou la réaction inverse d'ouverture du cycle; Fig. 3) on examine à l'aide des diagrammes de corrélation l'évolution des orbitales moléculaires qui subissent de profondes modifications : dans la cyclisation du diméthyl-butadiène précédent, on constate que les orbitales moléculaires χ_i deviennent des orbitales σ ou π de la molécule finale de diméthyl-cyclobutène (*cis* ou *trans*). Le passage du diène au cycle est considéré comme une succession de perturbations qui n'altèrent pas la symétrie des orbitales dérivant par transformation continue d'une orbitale donnée, ou qui tout au plus lèvent la dégénérescence des certaines d'entre elles. Ce principe permet de tracer, pour chaque type de réaction (thermique ou photochimique, disrotatoire ou conrotatoire) un diagramme de corrélation entre orbitales moléculaires du produit de départ et du produit d'arrivée, et de suivre qualitativement l'évolution de l'énergie des électrons décrits par ces orbitales. Pour une réaction thermique, on obtient les deux diagrammes de la Figure 5, construits soit en décrivant l'évolution des orbitales à l'aide d'un paramètre continu, soit en utilisant la règle de non-croisement des orbitales de même symétrie (cf. infra).

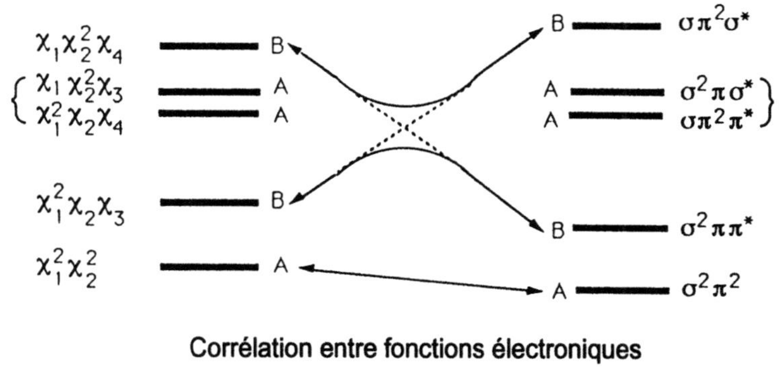

Figure 5.

La réaction thermique part d'un état fondamental de la molécule de diméthyl-butadiène (orbitales χ_1 et χ_2, les plus basses, occupées par deux électrons) et aboutit :

— à un état fondamental (électrons dans les orbitales σ et π du diméthyl-cyclobutène) si la réaction est conrotatoire; l'énergie d'activation est faible et la réaction est **permise**.

— à un état doublement excité (électrons dans les orbitales σ et π^* du diméthyl-cyclobutène) si la réaction est disrotatoire : l'énergie d'activation serait alors plus grande que dans le premier cas et la réaction serait très lente; on dit qu'elle est **interdite**.

C'est donc la réaction conrotatoire qui se produit. La symétrie des diagrammes de corrélation et le principe de microréversibilité montrent que c'est aussi la réaction conrotatoire qui est permise pour la réaction inverse, l'ouverture du diméthyl-cyclobutène en diméthyl-butadiène.

La réaction photochimique (non représentée) part du premier état excité et aboutit au premier état excité du diméthylcyclobutène, si la réaction est disrotatoire, et à un état excité supérieur si elle est conrotatoire: c'est donc la réaction disrotatoire qui est permise.

On aboutit aux mêmes conclusions si, au lieu de considérer les diagrammes de corrélation entre orbitales moléculaires, on considère les diagrammes entre fonctions électroniques totales (Figure 6), que l'on peut tracer à l'aide des premiers. On indique ci-dessous le cas de la réaction disrotatoire, qui permet d'illustrer un résultat important : la règle de non-croisement des lignes de même symétrie.

Les fonctions représentant les états électroniques possibles sont classées selon

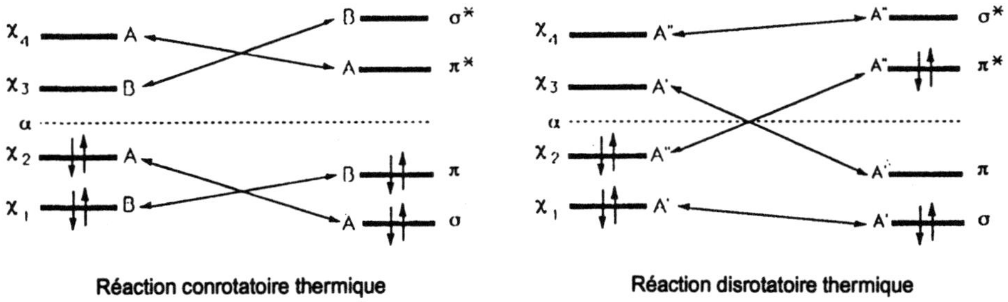

Figure 6.

les représentations irréductibles A ou B du groupe C_2 auxquelles elles appartiennent. La fonction électronique totale est le produit des orbitales moléculaires de tous les électrons. Elle appartient donc à la représentation irréductible produit tensoriel des représentations irréductibles de chacune des OM. Comme une OM occupée par deux électrons apparaît sous la forme ψ_2, qui appartient à la représentation complètement symétrique, on peut se limiter à des diagrammes de corrélation ne faisant intervenir que les OM qui subissent de profonds changements au cours de la réaction. Pour le diméthyl-butadiène, ce sont précisément les quatre OM χ_i provenant de la combinaison des orbitales de symétrie adaptée, et pour le diméthyl-cyclobutène, ce sont les orbitales σ, π, π^*, et σ^*. La symétrie de ces fonctions est déterminée par le produit tensoriel des représentations de chacune des OM composantes. On retrouve le fait que la corrélation se fait entre les états fondamentaux $\chi_1^2\chi_2^2$, et $\sigma^2\pi^2$, ce qui correspond à une réaction thermique. Le diagramme de corrélation des OM (Fig. 5) montre que le premier état excité du butadiène $\chi_1^2\chi_2\chi_3$, devrait évoluer vers l'état $\sigma\pi^2\sigma^*$, et qu'en sens inverse, le premier état excité du cyclobutène $\sigma^2\pi\pi^*$ évoluerait vers l'état $\chi_1\chi_2^2\chi_4$. Tous ces états sont de même symétrie et il existe entre eux, au voisinage du point de croisement, une forte interaction des états intermédiaires qui permet au système de passer de l'un à l'autre (ligne en trait plein) au lieu de suivre le chemin normal (ligne en pointillés). C'est la **règle de noncroisement des lignes de même symétrie**. Les lignes de même symétrie semblent se repousser. Les situations qui, comme celle-ci, correspondent à un *croisement évité*, mettent en jeu des énergies d'activation importantes et sont défavorisées : la réaction conrotatoire à partir de ces états excités (réaction photochimique) est donc interdite. Mais il n'y a pas d'interaction entre les états de symétries différentes et une ligne de corrélation entre deux états A pourrait croiser une ligne entre deux états B.

Conclusion

Dans de nombreux cas, surtout pour des molécules dont la description ne fait intervenir que des orbitales *s* et *p*, l'utilisation de la théorie des groupes n'est pas indispensable : la considération de la symétrie par rapport à un ou deux éléments suffit. Beaucoup de chimistes n'utilisent cette théorie que de manière indirecte en appliquant, souvent sans connaître leur origine, les résultats qui en découlent, comme la règle du "croisement évité". Elle est par contre indispensable dans la chimie des éléments de transition qui repose sur l'existence d'orbitales d plus ou moins remplies, et dans les études spectroscopiques. L'intérêt croissant des chimistes pour la théorie des groupes est marqué par la publication de plus en plus fréquente d'articles pédagogiques dans le *J. Chemical Education.* Il est aussi illustré par l'évolution de l'ouvrage de Cotton intitulé *Chemical Applications of Group Theory*, le plus fréquemment cité dans les publications chimiques. Entre la première édition (1963; traduction française, Paris, Dunod, 1968) écrite, selon la préface, pour le "chimiste moyen", et peu mathématisée, et la 3ème édition (1990), le volume du texte a pratiquement doublé; les aspects théoriques de la représentation linéaire des groupes y sont présentés de façon plus rigoureuse et des indications sont données sur les groupes continus $C_{\infty v}$ et $D_{\infty h}$ (groupes souvent omis dans les ouvrages plus anciens destinés aux chimistes). Des techniques ont été progressivement introduites (opérateurs de projection dont l'usage simplifie l'obtention des combinaisons linaires d'orbitales de symétrie adaptée à celle des molécules étudiées) et des sujets d'application plus ou moins récente ont été développés (applications en chimie des complexes et en cristallographie, utilisation des groupes pour les combinaisons d'orbitales qui traduit l'augmentation de l'importance de la méthode des orbitales moléculaires, par rapport à la méthode des liaisons de valence, malgré un regain d'intérêt récent pour celle-ci).

Le développement actuel des méthodes cristallographiques de détermination des structures moléculaires, l'intérêt suscité par la chimie de composés très symétriques (dérivés du bore, agrégats) et la découverte des agrégats carbonés du type fullerènes (cf. Coulombeau) ne peuvent que renforcer cette tendance.

Bibliographie

[1] Albright, T.A., Burdett, J.K. et Whangbo, M-H., *Orbital Interactions in Chemistry*, Wiley, New-York, 1985. Construction et emploi des diagrammes de corrélation entre orbitales moléculaires. Nombreux exemples montrant la variété et le généralité des problèmes abordables par cette méthode. Pour étudiants de maîtrise et chercheurs.

[2] Cotton, F.A., *Chemical Applications of Group Theory*, 3ème édition, Wiley, New-York, 1990. Ouvrage de référence pour étudiants à partir de la licence et chercheurs.

[3] Coulombeau, C. et Rassat, A., Calculs des propriétés électroniques et des fréquences normales de vibration d'agrégats carbonés formant des polyèdres réguliers et semi-réguliers, *J. Chim. Phys.* **84** (1987), 875–882.

[4] Curie, D., *Champ cristallin et luminescence. Application de la théorie des groupes à la luminescence cristalline*, Gauthier-Villars, Paris,1968. Pour étudiants de 3ème Cycle.

[5] Fétizon, M., Gervais, H.P. et Guichardet, A., *Théorie des groupes et de leurs représentations*, Éditions Ellipses, Paris, 1987. Présentation rigoureuse, à l'usage des mathématiciens, des concepts de la représentation linéaire des groupes en vue des applications en chimie, et plus particulièrement en spectroscopie.

[6] Herzberg, G., *Molecular Spectra and Molecular structure*, I. Spectra of Diatomic Molecules, 2ème éd., Van Nostrand, Princeton, 1950. II. Infrared and Raman Spectra of Polyatomic Molecules, Van Norstrand, Princeton, 1945. III. Electronic Spectra and Electronic Structure of Polyatomic Molecules, Van Nostrand, Princeton, 1966. Ouvrages de références pour les spectroscopies moléculaires.

[7] Jaffé, H.H. et Orchin, M., *Theory and Applications of Ultraviolet Spectroscopy*, Wiley, New-York, 1962. Chap. 4 à 6. Pour étudiants de maîtrise et chercheurs.

[8] Kettle, S.F.A., *Symmetry and Structure*, Wiley, New York, 1985. Présentation à l'usage des chimistes, à l'aide de nombreux exemples traités en détail. La partie purement mathématique (1/3 de l'ouvrage) est traitée en annexe.

[9] Kroto, H.W., *Molecular Rotation Spectra*, Wiley, New York, 1975. Chap. 3.

[10] Schonland, D.S., *La symétrie moléculaire*, Gauthier-Villars, Paris, 1971. Pour étudiants à partir de la licence et chercheurs.

[11] Woodward, R.B. et Hoffmann R., *The Conservation of Orbital Symmetry*, Verlag Chemie, 1970. Développement des trois articles (1965) des mêmes auteurs, qui ont profondément modifié les idées sur la réaction chimique en montrant l'importance de la symétrie dans les interactions orbitalaires. Une étape majeure dans l'histoire de la chimie.

Michel Golfier
Laboratoire de Chimie Organique de Synthèse
Université Paris-Sud
F-91405 Orsay Cedex, France

Mathématiques et chimie quantique

par Marcel Allavena
Laboratoire de Dynamique des Interactions Moléculaires
Université Pierre et Marie Curie, Paris (France)

La référence à la chimie quantique dans le titre de cet exposé souligne la volonté d'attirer l'attention sur le secteur le plus mathématisé d'un domaine scientifique plus vaste, souvent dénommé Chimie Théorique et avec lequel il a été historiquement confondu.

La Chimie Théorique aujourd'hui regroupe des disciplines très diverses. Sans être exhaustif, on peut distinguer au moins quatre pôles : la cinétique chimique, étroitement liée à la Mécanique Statistique (théorie de Marcus (1),...), l'informatique chimique (synthèse assistée par ordinateur,...), les méthodes diagrammatiques, vaste ensemble de règles construites sur la combinaison des orbitales et la topologie des édifices moléculaires (règles de Woodward et Hoffmann, (2)...). Ce troisième pôle est évidemment celui qui est le plus proche de la Chimie Quantique proprement dite. Cette dernière étant définie comme l'ensemble des procédures disponibles pour la résolution de l'équation de Schrödinger associée à un ou plusieurs systèmes moléculaires. Cette définition un peu restrictive risque de faire apparaître notre discipline comme une simple extension de la physique atomique et moléculaire ; certes les liens sont étroits et de nombreuses méthodes de calculs sont empruntées à ces disciplines, mais les objectifs sont autres. La Chimie Quantique a surtout comme objectif l'étude de la transformation des systèmes, plus que l'étude des molécules elles-mêmes ; c'est le résultat de l'interaction entre molécules qu'il convient d'expliquer ou de prévoir.

Pour mieux comprendre les problèmes que pose la Chimie Quantique à ses disciplines voisines, il convient de rappeler sa problématique générale. Pour prévoir la structure d'une molécule ou la transformation d'un système de plusieurs atomes ou molécules (sous l'effet par exemple d'un rayonnement ou des interactions électromagnétiques entre les constituants du système) l'objectif est toujours le même : rechercher la configuration énergétique la plus stable, et cet objectif est atteint en suivant une double démarche. D'abord, déterminer, par résolution de l'équation de Schrödinger, l'énergie totale du système pour une position donnée des noyaux atomiques du système.

Bien entendu, la configuration des noyaux atomiques choisie pour ce premier calcul ne correspond pas nécessairement à la configuration d'équilibre recherchée. Il convient donc de répéter le calcul jusqu'à ce que l'on atteigne le point de plus basse énergie. L'ensemble des points ainsi obtenus sous-tend une surface d'énergie potentielle E(X)*, et la connaissance de cette surface détermine complètement les propriétés du système. La deuxième étape de la procédure générale consiste donc à construire point par point la surface.

Pour la première étape, la Chimie Quantique dispose d'une panoplie de méthodes (3) très diverses qui s'échelonnent des méthodes les plus rigoureuses ne faisant appel qu'aux principes généraux de la Mécanique Quantique et résultant de l'application des méthodes variationnelles ou de perturbation telles que les méthodes Hartree Fock (H.F.) ou d'interaction de configurations (I.C.) jusqu'aux méthodes ad hoc comme les méthodes de la fonctionnelle densité. En outre, existent des méthodes plus frustes, déduites des méthodes rigoureuses ou construites à partir d'opérateurs effectifs, et dont certaines grandeurs sont calibrées sur des données expérimentales (méthodes dites semi-empiriques). A l'exception de quelques cas particuliers comme celui de la molécule H_2, les méthodes disponibles en chimie quantique n'offrent que des solutions approchées de l'équation de Schrödinger. En outre, mêmes ces solutions approchées ne peuvent être mises en oeuvre correctement en raison de la complexité des problèmes abordés. Tel est le cas de la méthode Hartree Fock, une des plus couramment employées, qui décompose l'équation originale associée aux mouvements de n électrons en interaction (équation de Schrödinger indépendante du temps) :

$$H\Psi(r_1, r_2, \ldots, r_n) = E\Psi(r_1, r_2, \ldots, r_n) \tag{1}$$

en une suite d'équations couplées aux pseudo valeurs propres :

$$F\phi(r_i) = E_i\phi(r_i) \tag{2}$$

chacune décrivant le mouvement d'un électron dans le champ moyen des autres, ce qui se traduit par le fait que l'opérateur différentiel F (dit opérateur de Fock) est construit avec les fonctions ϕ associées à chacun des électrons. Mais même dans le cadre de cette approximation, les équations H.F. (2) qui en résultent n'ont pu être résolues à ce jour pour les systèmes polyatomiques. Elles le sont par contre pour les systèmes atomiques qui bénéficient de la symétrie sphérique.

Ce n'est qu'en 1951 (4) qu'une méthode approchée de résolution des équations H.F. a été proposée. Elle consiste à développer chaque solution recherchée ϕ sur une base complète de fonctions connues (les orbitales atomiques) :

$$\phi = \sum C_n \chi_n \tag{3}$$

* L'énergie E est une fonction de l'ensemble des coordonnées relatives (X) des noyaux atomiques les uns par rapport aux autres. La surface se réduit à une courbe si X ne représente qu'une seule coordonnée (molécule diatomique) et à une surface dans R^3, si X représente deux coordonnées (molécules triatomique linéaire). Au-delà on parlera d'une hypersurface.

il en résulte que le système d'équations différentielles (2) se transforme en un système d'équations algébriques linéaires et que la recherche des solutions se ramène à une diagonalisation de matrice. Les solutions ainsi trouvées, ϕ, dépendent très fortement de l'extension des bases d'orbitales atomiques utilisées. Comme pratiquement les bases utilisées sont loin d'être complètes, ce défaut de complétude renforce les défauts intrinsèques de la méthode, et fait l'objet de nombreuses discussions au sein de la communauté des chimistes théoriciens. Il faut admettre que le choix d'une bonne base ressort encore actuellement plus d'un savoir faire artisanal et des disponibilités en ressources informatiques que de la rigueur mathématique. En outre, la résolution des équations (2) exige la mise en oeuvre d'une procédure itérative : choix d'un jeu de coefficients initiaux $C^{(0)}$, puis construction de la matrice F et diagonalisation qui fournit un premier jeu de coefficients $C^{(1)}$ et ainsi de suite. La convergence n'est pas assurée, elle est souvent forcée par l'adjonction d'algorithmes qui amortissent les oscillations des itérations successives.

L'enjeu de la plupart de ces calculs étant de rechercher la configuration des différents atomes énergétiquement la plus stable, on est en définitive ramené à chercher le minimum absolu sur la surface $E(X)$. La recherche des extrema sur une hypersurface soulève encore des problèmes. L'un des plus anciens fut soulevé par Löwdin (5) dans le cadre de l'approximation H.F. Ce problème est le suivant : est-on sûr d'atteindre un minimum stable, lorsque l'on applique la procédure H.F. traditionnelle c'est-à-dire en imposant à la solution de respecter des contraintes de symétrie de spin et d'espace? Le problème des instabilités H.F. ne semble pas avoir reçu à ce jour une réponse définitive.

Si le traitement du problème H.F. à l'aide d'une base de fonctions données ne soulève pas de problèmes majeurs, il n'en est pas de même lorsque l'on veut résoudre par le même procédé l'équation relativiste de Fock-Dirac appliquée aux systèmes moléculaires. Des solutions non physiques et des problèmes de convergences apparaissent, souvent désignés sous le nom "d'effondrement variationnel" ("variational collapse") (6). Ces problèmes sont liés au caractère non borné (inférieurement) de l'opérateur de Dirac, et à l'inapplicabilité de la méthode variationnelle dans un tel cas. Des recherches se poursuivent actuellement (7).

Au-delà de l'approximation H.F., de nombreuses méthodes ont été développées pour dépasser l'approximation du champ moyen qui néglige la corrélation instantanée entre les électrons. La variation énergétique qui en résulte est faible, quelques % de l'énergie totale pour une configuration donnée des noyaux, mais elle peut jouer un rôle décisif lorsque l'on compare des énergies correspondant à des états électroniques ou des configurations géométriques différentes.

Dans cette direction, la théorie des perturbations est un outil privilégié. Les chimistes théoriciens ont bénéficié dans leur recherche des avancées obtenues par les physiciens s'intéressant au problème à N-corps, (par exemple, choix des séries de perturbation qui assurent que l'énergie d'un système croît en même temps que le nombre des particules qui le constituent (size consistency). L'étude de la convergence des séries est plus supposée que démontrée. La complexité des calculs est telle qu'il est difficile de dépasser les quelques premiers ordres de perturbation ($n \leq 4$). Mais contrairement aux méthodes variationnelles (type H.F.), qui assurent que l'énergie exacte est une borne inférieure de l'énergie calculée, en théorie de perturbation la valeur calculée peut approcher la valeur exacte par valeur supérieure ou inférieure suivant l'ordre des perturbations auquel s'arrête le calcul.

D'une manière générale, le degré d'approximation varie suivant la complexité des systèmes considérés : la précision est excellente pour des petits systèmes (entre 10 et 20 électrons); au-delà, elle dépend d'un compromis entre la nature des grandeurs recherchées, les disponibilités en temps calcul, et la taille des ordinateurs disponibles. Des systèmes contenant jusqu'à $\simeq$ 1000 électrons sont traitables, mais on est loin de la précision expérimentale. Nombre de résultats restent donc actuellement à un niveau qualitatif, permettant plus de prévoir des corrélations entre grandeurs que des valeurs absolues. Le problème de la précision en Chimie Quantique est encore sérieux et sa solution n'est pas prévisible dans un avenir proche, sauf si les mathématiciens nous proposent de nouvelles méthodes de résolution des équations aux dérivées partielles. Dans la communauté des chimistes quanticiens, la recherche de nouvelles techniques mettant à profit les ressources des méthodes numériques ou statistiques commence à être explorée. L'essor des méthodes dites de Monte Carlo Quantique mérite d'être mentionné (8). Ces procédés permettent d'atteindre la solution exacte de l'équation d'onde, mais ne s'appliquent encore qu'à des systèmes à nombre restreint d'électrons. Il est intéressant de noter à ce propos que ces recherches soulèvent des problèmes mathématiques qui semblent n'avoir reçu que peu d'attention jusqu'à présent, ainsi en est-il de la conservation (ou non) du nombre de cellules nodales (région où la fonction propre conserve le même signe) d'une fonction propre d'un hamiltonien contenant un paramètre variable (9).

La seconde étape du processus général, construction de l'hypersurface E(x), soulève aussi quelques problèmes. Le premier n'est pas une difficulté d'ordre mathématique mais constitue cependant une réelle pierre d'achoppement de la Chimie Quantique. Il concerne le nombre de points requis pour construire une surface de potentiel convenable, et on retrouve ici à nouveau un problème de précision. Un calcul simple montre que si on se restreint à 10 points par degré de liberté et 2' de temps calcul ordinateur par point, pour une molécule aussi simple que le méthane (CH_4, 5 atomes, 9 coordonnées relatives indépendantes), le nombre de points nécessaires est de 10^9 et le temps de calcul dépasserait plusieurs siècles. Il y a des parades à cette catastrophe, qui consistent à fixer un certain nombre de coordonnées supposées peu mobiles au cours du réarrangement structural examiné. Mais on entre à nouveau dans le domaine des approximations plus ou moins mal maîtrisées. En outre, pour des hypersurfaces compliquées (nombre de degrés de liberté élevé) la recherche du minimum minimorum n'est pas simple. Les méthodes locales (type méthode du gradient) sont couramment exploitées, mais des méthodes plus globales sont souhaitées et le recours aux méthodes de recuit simulé est désormais fréquent. Des progrès sont souhaités dans ce domaine, ainsi que dans l'analyse topologique : caractérisation des extrema, recherche des cols et vallées d'énergie minimale reliant les extrêma, etc...

Une autre caractéristique de la chimie théorique est d'avoir consacré beaucoup plus d'effort à l'étude des états stationnaires qu'aux aspects dynamiques des interactions intermoléculaires.

La résolution de l'équation de Schrödinger dépendant du temps :

$$i\hbar\frac{\partial\Psi}{\partial t} = H\Psi \tag{4}$$

n'a reçu toute l'attention nécessaire que dans le cas relativement simple de l'interaction d'un système moléculaire avec un champ électromagnétique supposé suffisamment faible pour que l'opérateur hamiltonien représentant cette interaction puisse s'écrire :

$$H = H_0 + V(t) \tag{5}$$

et donc que le problème soit traitable en théorie de perturbation. Le traitement aux 1er et 2nd ordres permet de prévoir d'une manière à peu près satisfaisante les probabilités de transition entre les niveaux énergétiques sous l'effet des divers rayonnements. Cependant le calcul des intensités et des formes des spectres reste encore délicat et les initiatives dans ce domaine reviennent souvent aux physiciens (RMN,...).

Cependant les récents développements des spectroscopies résolues dans le temps (enregistrement de spectres sur des intervalles de temps très courts ($< 10^{-12}$ ou 10^{-15} sec.) et des expériences de collisions ont suscité un renouveau des recherches dans ce domaine. L'étude des solutions de l'équation (4), que l'on peut formellement écrire sous la forme:*

$$\Psi(t) = U(t, t_0)\Psi(t_0) = T\exp\left(+\frac{i}{\hbar}H(t')dt'\right)\Psi(t_0) \tag{6}$$

se fait en deux étapes, l'une consistant à trouver une forme spécifique de $\Psi(t)$ et l'autre à suivre sa propagation au cours du temps. On trouvera une revue des

---

* $U(t,t_0)$ est l'opérateur d'évolution qui permet de déduire la fonction Ψ au temps t en fonction de sa valeur au temps t_o. Il s'exprime en fonction d'un opérateur exponentiel précédé d'un opérateur d'ordonnancement dans le temps (T), indispensable en raison de la non commutativité de l'opérateur H pris à des temps différents.

différentes méthodes actuellement disponibles ainsi qu'un résumé de la méthode auto cohérente dépendant du temps (TD SCF) dans les références (10 et 11).

L'interface avec l'informatique est actuellement un domaine très sensible. Là, c'est la Chimie Quantique qui doit s'adapter à un outil dont les performances s'accroissent à un rythme extrêmement rapide. L'évolution des architectures : vectorielle, parallèle, oblige à repenser les codes traditionnels si on veut profiter au mieux des avantages offerts par les nouveaux matériels. En raison de la structure de ses algorithmes, la Chimie Quantique a assez peu bénéficié des avantages de la vectorisation. Or, déjà s'ouvre l'aire du parallélisme qui exigera des aménagements encore plus profonds des logiciels (12), le dialogue entre les équipes de chimistes quanticiens et d'informaticiens sera indispensable sous peine de manquer une étape décisive.

Enfin on ne peut clore ce bref panorama sans évoquer les rapports de la Chimie Quantique avec la Mécanique Classique. Le raccord entre les deux approches se fait chaque fois que l'on aborde des systèmes, ou la dynamique de systèmes dont la densité de niveaux d'énergie avoisine celle d'un continuum.

Au point de vue structure, c'est le cas des très gros systèmes moléculaires tels que les acides nucléiques ou les protéines qui sont traités par les méthodes de la Dynamique Moléculaire, une procédure consistant à résoudre numériquement les équations de Newton établies pour un système de forces qui dérivent de potentiels fournis par la Chimie Quantique (13).

L'aspect dynamique peut être illustré par l'étude des mouvements des noyaux atomiques dans les édifices moléculaires, qui sont décrits en se référant à la théorie des petits mouvements de la Mécanique Classique. Or depuis l'avènement des Lasers, le mouvement des noyaux peut être examiné dans des systèmes fortement excités où les densités d'états d'énergie sont très élevées et les comportements non linéaires. L'étude de ces situations (14) nécessite le recours à des ressources de la Mécanique Classique peu familière aux chimistes (travaux de Poincaré, Kolmogorov, Arnold ...). Enfin les théories du chaos pénètrent peu à peu le milieu des chimistes quanticiens et les discussions sur le chaos quantique dépassent déjà le cadre des cénacles étroits dédiés au "non linéaire".

Née il y a environ 60 ans, la Chimie Quantique s'est développée en empruntant nombre de ses méthodes aux mathématiciens et en s'appuyant sur l'informatique. Aujourd'hui nous constatons qu'en voulant aborder des problèmes de plus en plus complexes, le recours et le dialogue avec ces disciplines doit se poursuivre et devrait peut être même s'amplifier.

Question de Monsieur Bernard Beauzamy, directeur de l'Institut de Calcul Mathématique (ICM), Université de Paris VII.

Vous avez signalé que l'amélioration de la précision des calculs en chimie quantique nécessitait de nouvelles méthodes de résolution des équations aux dérivées partielles. Pouvez-vous commenter ce besoin ?

Réponse de Monsieur Marcel Allavena :

La recherche des solutions de l'équation de Schrödinger (équation aux dérivées partielles) pour les systèmes polycentriques (molécules) conduit à la résolution d'équations intégro-différentielles (les équations de Hartree-Fock) dont nous ne connaissons que des solutions approchées sous forme de développements tronqués de fonctions connues (les orbitales atomiques). Les hypothèses simplificatrices qui réduisent l'équation originale à une équation traitable par les techniques standard et les troncatures des développements sont les principales sources de l'imprécision de nos résultats, imprécision qui croît avec la taille des systèmes.

L'application de l'équation de Schrödinger au traitement des molécules date des années 30 et la résolution pratique des équations de Hartree-Fock n'a été connue qu'au début des années 50 (C. C. J. Roothaan (*Rev. Mod. Phys.,* vol. 23 (1951) p. 69 sqq.).

Il peut être intéressant de reconsidérer ces démarches à la lumière des progrès accomplis par les mathématiciens durant ces quarante dernières années. La lecture de quelques articles clefs sur la molécule H_2 (K. Kolos et L. Wolniewicz, *J. Chem. Phys.,* vol. 41, (1964), p. 3663 sqq. et p. 3674 sqq.) pourrait aider les mathématiciens à se rendre compte des problèmes soulevés par la résolution de l'équation de Schrödinger dans le cas moléculaire.

Références

[1] W. Forst, *Theory of Unimolecular reactions*, Academic Press, New York, 1973.

[2] (a) Y. Jean, F. Volatron, *Les orbitales moléculaires en chimie*, McGraw Hill, New York, 1991. (b) Nguyen Trong Anh, *Les règles de Woodward-Hoffmann*, Ediscience, Paris, (1970).

[3] Pour une bonne initiation aux méthodes traditionnelles de la chimie quantique et un panorama des références indispensables voir: *Methods of electronic structure theory, III*, Henry F. Schaeffer (ed., Plenum Press, New York, 1977. On trouvera des mises au point plus techniques dans les volumes **67** et **69** (1987) de la revue : *Adv. in Chem. Phys.*

[4] C.C.J. Roothaan, *Rev. Mod. Phys.*, **23**: 69 (1951).

[5] P.O. Löwdin, *Rev. Mod. Phys.*, **35**: 496 (1969).

[6] W. Kutzelnigg, *Int. J. Quant. Chem.*, **25**: 107 (1984).

[7] A. Mohanty, E. Clementi. In *Modern Techniques in Computational Chemistry*, E. Clementi (ed.), E.S.C.O.M. Leiden, 1990, p. 693. L'ouvrage dans son ensemble constitue aussi une bonne mise au point sur quelques problèmes actuels.

[8] (a) Monte Carlo Methods in Quantum Problems. Ed. by M.H. Kalos, Reidel, (1982). (b) M. Caffarel, P. Claverie, *J. Chem. Phys.* **88**: 1088 (1988), idem **88**: 1101 (1988).

[9] M. Caffarel, X. Krokidis, C. Mijoule, *Europhysics Lett.*, **20**: 581(1992).

[10] R. Kosloff, *J. Phys. Chem.* **92**: 2087 (1988).

[11] C. Leforestier, R.H. Bisseling, C. Cerjan, M.D. Feit, R. Friesner, A.Guldberg, A. Hammerich, G. Jolicard, W. Karrlein, H.D. Meyer, N. Lipkin, O. Roncero, R. Rosloff, *J. Comp. Phys.* **94**: 59 (1991).

[12] Il peut être intéressant de consulter à ce sujet, le numéro de la revue "*Theor. Chim. Acta*", (**84** (1993), 255–473) entièrement dédié au calcul parallèle en chimie physique.

[13] W.F. Van Gunsteren, H.J.C. Berendson, Angew. *Chem. Int.*, Ed. Engl., **29**: 992 (1990).

[14] L'étude des "modes locaux" est un bon exemple de ces développements. On peut trouver un résumé simple de ce problème dans : B.R. Henry, *Acc. Chem. Research* **10**:207 (1977).

Marcel Allavena
Laboratoire de Dynamique des Interactions Moléculaires
Université Pierre et Marie Curie
4, Place Jussieu, Tour 22
F-75252 Paris Cedex, France

Mathématiques et chimie en recherche pharmaceutique

par Gilles Moreau
Centre de Recherches Roussel UCLAF, Romainville (France)

Pour beaucoup de scientifiques, la recherche pharmaceutique et les disciplines qu'elle met en jeu, chimie, pharmacologie, sciences de la vie en général, sont aux antipodes des mathématiques. Le mathématicien René Thom est même allé jusqu'à dire, récemment lors d'une émission à la télévision, je cite : "... *déjà, la chimie ça branle dans le manche...*". Cette comparaison pittoresque avec un outil dont le manche est mal fixé, donc imprécis, mauvais, traduit sans détour cette opinion. La chimie et la recherche des médicaments, disciplines confrontées à la réalité, sont cependant pleines de bon sens et extrêmement rationnelles, et elles n'ignorent pas les mathématiques quand elles peuvent servir.

Une molécule une formule

Pendant longtemps, les chimistes n'ont perçu une molécule qu'à travers la formule développée (structural formula en anglais) qui définit mais ne décrit pas physiquement une molécule. Malgré leur apparence de hiéroglyphes, les formules développées sont des êtres totalement mathématiques. Une formule développée est d'abord un graphe qui indique quels sommets sont directement liés à quels autres sommets. On pourrait montrer que toutes les informations présentes dans une formule peuvent être exprimées comme des propriétés des sommets. On peut considérer le graphe comme un espace constitué d'un nombre fini de points, les sommets, dans lequel la métrique est simplement le plus court trajet séparant deux sommets, compté en nombre d'arêtes. Les diverses propriétés attachées aux sommets sont donc distribuées dans cet espace et on peut définir et calculer une fonction d'autocorrélation de cette distribution, par analogie avec ce qu'on appelle fonction d'autocorrélation en traitement du signal : $F(\delta t) = \int f(t) f(t + \delta t) dt$. Les composantes $V^{(k)}$ du vecteur autocorrélation sont définies ainsi :

$$V^{(k)} = \sum_{i \leq j}^{(k)} f(a_i) f(a_j)$$

où $f(a_i)$ désigne la propriété considérée associée à l'atome a_i. La somme S porte sur toutes les paires d'atomes a_i, a_j séparées par la même distance k. Ainsi une formule est transformée en un vecteur facile à manipuler. Certaines questions restent posées, par exemple : existence de graphes distincts ayant la même fonction d'autocorrélation ; transformation inverse, etc. Plus généralement la structure de graphe se prête à de nombreux calculs, leur portée pratique reste cependant très limitée.

Approche rationnelle

Aujourd'hui, pour concevoir un médicament, on commence par étudier les mécanismes biologiques qui sont impliqués dans la maladie que l'on désire combattre. Ainsi, on pourra identifier des cibles moléculaires sur lesquelles on espère agir au moyen de molécules appropriées. Cette idée est fondée sur le fait que les éléments de ces mécanismes sont souvent des enzymes et leurs substrats, des récepteurs biologiques et leurs hormones ou leurs médiateurs. Des récepteurs biologiques ce sont, par exemple, les "opsines" qui captent les photons qui arrivent sur notre rétine ; des récepteurs analogues existent partout dans nos cellules pour capter des molécules étrangères (goût, odeur, etc.) ou des molécules endogènes, hormones ou médiateurs qui assurent la communication entre diverses fonctions.

Prenons l'exemple d'une hormone : sa fixation sur son récepteur se fait avec une très grande affinité et entraîne une série d'évènements biochimiques et finalement une réponse. Un médicament capable d'antagoniser cette hormone sera une molécule ayant une affinité comparable pour le récepteur mais dont la fixation n'entraînera pas la réponse physiologique. Le récepteur étant occupé, l'hormone ne pourra pas y accéder, c'est ce que schématise la Figure 1.

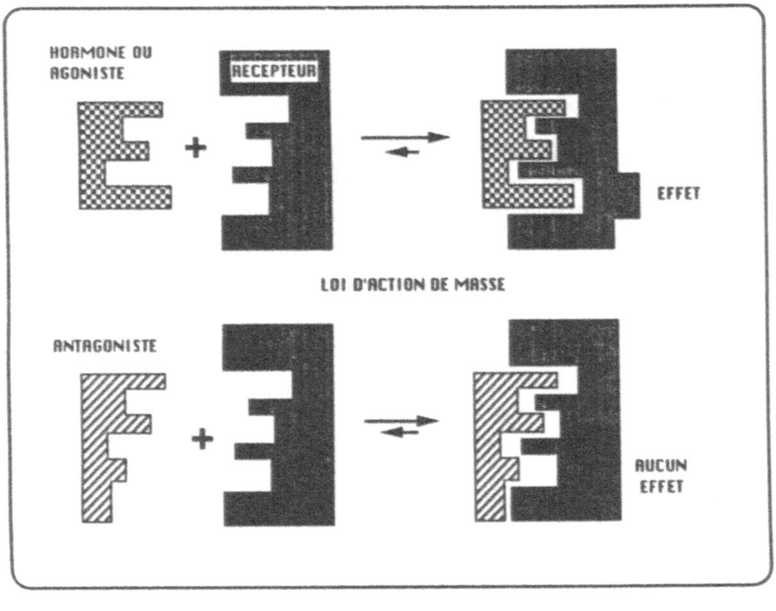

Figure 1.

L'approche rationnelle consiste à traiter ces problèmes purement physiques de structures moléculaires d'interactions moléculaires, de similitude moléculaire, de reconnaissance moléculaire, etc. Le calcul numérique y tient une place très importante, voici quelques thèmes :

La mécanique moléculaire : Calculer la structure 3D des molécules même simples devient très vite impraticable par la mécanique quantique. On lui a donc substitué la mécanique moléculaire qui est un procédé semi-empirique. L'idée est que les éléments géométriques, qui définissent la structure d'une molécule : longueurs des liaisons, angles de valence (angle entre deux liaisons partant d'un même atome), angles dièdres (angle de rotation autour d'une liaison), ont des valeurs vraies plus ou moins éloignées des valeurs standard. Ces écarts sont à l'origine d'énergies de tension qu'on exprime sous des formes très simples. Par exemple, l'énergie d'une liaison est de la forme $E = k(L - L_0)^2$ où k (constante de force) et L_0 (Longueur standard) dépendent du type de la liaison considérée. La Figure 2 montre quelles sont les variables utilisées en mécanique moléculaire et les expressions empiriques le plus souvent rencontrées pour les divers termes d'énergie.

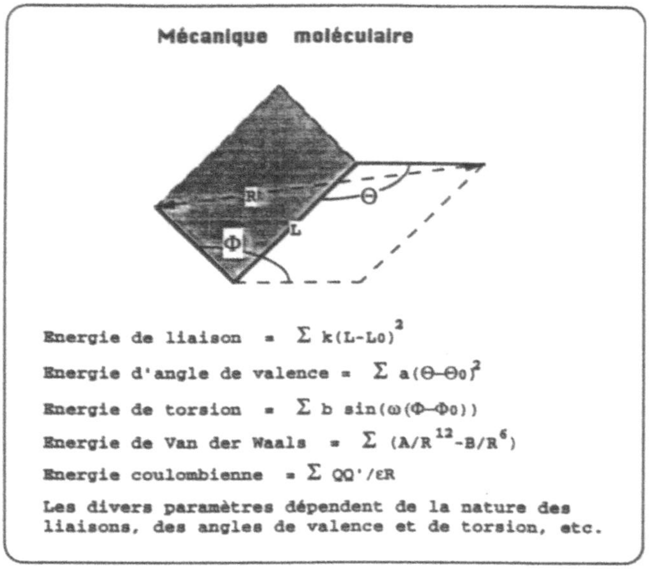

Figure 2.

L'analyse conformationnelle d'une molécule consiste donc à déterminer les principaux minima de la fonction énergie qui dépend de $3N-6$ variables si N est le nombre d'atomes. En réalité, ce sont surtout les angles dièdres qui déterminent la forme, il y en a à peu près N et c'est surtout la présence de cycles qui diminue encore la flexibilité de la molécule donc le nombre de minima à chercher. Finalement, avec l'aide d'un programme et d'un ordinateur, il n'y a pas trop de difficulté à déterminer les formes de moindre énergie pour la plupart des molécules dites “petites” (une centaine d'atomes). Il existe plusieurs programmes sur le marché.

Dynamique moléculaire et simulation de Monte Carlo. Plus les molécules deviennent grandes et plus la mécanique moléculaire devient lourde et impuissante à explorer tous les minima secondaires de la fonction énergie. C'est en particulier le cas des protéines, grosses molécules dont la structure primaire est un enchaînement d'acides aminés (voir figure) ; chaque acide aminé apporte en moyenne 3 à 4 rotations possibles. On cherche à décrire la flexibilité et les formes préférentielles de ces molécules. Il faut donc restreindre l'exploration du domaine conformationnel aux seules zones de basse énergie en ne perdant pas de temps sur les sommets. Il y a deux techniques pour cela :

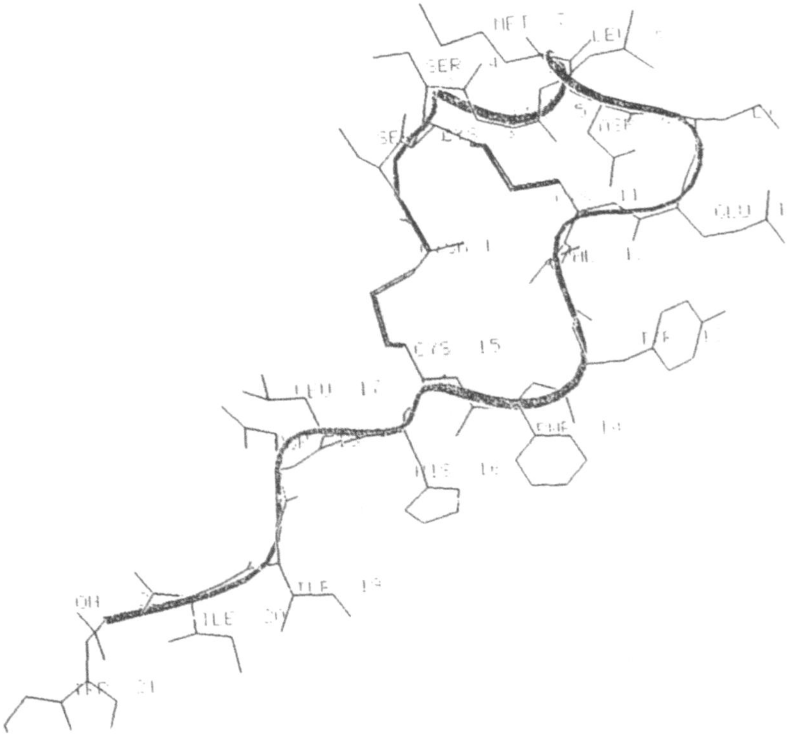

Figure 3. Structure 3D calculée de l'endothéline : cette petite protéine contient 21 acides aminés. L'extrémité N-terminale est une cystéine (CYSn1) qui est reliée à une autre cystéine (CYS15) par un pont disulfure (trait épais); un autre pont disulfure existe entre les positions 3 et 11. La séquence des 15 premiers aminoacides a donc une structure assez rigide. La trajectoire moyenne de la chaîne peptidique est représentée par une ligne courbe. La partie N-terminale (de HIS16 à TRP21) est très flexible, la forme représentée en est une parmi beaucoup d'autres.

Dans la dynamique moléculaire, on met la molécule dans une conformation initiale plausible, de basse énergie, et l'on donne aux atomes des vitesses initiales. L'évolution ultérieure est déterminée par la loi de la mécanique $F = M\Gamma$ et par la constance de l'énergie totale du système. Les forces s'exerçant sur les atomes sont calculées à partir des expressions des termes d'énergie conformationnelle, d'énergie électrostatique et d'énergie de van der Waals. On intègre $F = M\Gamma$ pour chaque atome dans des intervalles de temps très courts (fraction de picoseconde) pendant lesquels F est considérée comme constante.

Des programmes commerciaux existent pour ce genre de simulation. Les temps que l'on peut simuler dépendent de la taille de la molécule et varient de quelques picosecondes à quelques nanosecondes mais requièrent plusieurs heures d'ordinateur.

La simulation de Monte Carlo consiste à générer une séquence semi-aléatoire de conformations de telle sorte que l'on a, au bout d'un certain temps, un échantillonnage conformationnel qui obéit à la distribution de Boltzmann.

Dans ces deux types de simulation, il est courant d'utiliser des méthodes d'analyse de données pour trier la grande masse d'information produite.

Opération de docking. On désigne par ce terme l'opération qui consiste à placer deux structures 3D l'une contre l'autre de manière à obtenir la meilleure interaction, éventuellement en déformant l'une ou l'autre ou les deux molécules de manière à ce que le bilan final soit énergétiquement favorable. C'est un problème de complémentarité de forme et de propriétés qui conduit à des calculs lourds et compliqués à cause du grand nombre de degrés de liberté et de l'immensité du domaine à explorer.

Similitude de structure 3D. On ne connaît pas toujours la structure 3D de l'enzyme ou du récepteur pour lequel on cherche un inhibiteur ou un antagoniste. Mais on peut connaître la structure d'un substrat endogène, souvent peptidique lui-même. On peut penser qu'une molécule devant se fixer au même site doit ressembler à ce substrat endogène. Mais la ressemblance n'est pas une propriété physique : on sait dire si deux structures sont identiques, mais en-dehors de ce cas, sont-elles peu ou beaucoup différentes ? Les résultats peuvent dépendre énormément du procédé mathématique utilisé pour évaluer la ressemblance.

Perspectives futures

L'entrée du calcul numérique dans cette partie de la chimie biologique que sont les structures et les interactions moléculaires semble désormais irréversible, même si, pour le moment, on ne peut pas parler encore de grands et nombreux succès de cette approche rationnelle ; mais il semble bien qu'on soit à la veille ou au début d'une utilisation productive de cette méthode. Cette discipline restera, je crois, longtemps un mélange de rationnel et d'empirisme. Elle restera aussi longtemps tributaire des données expérimentales fournies par les cristallographes et les RMNistes et des connaissances acquises par les pharmacologistes et les biologistes.

Bibliographie

Autocorrélation ; structure de graphe des molécules :

[1] P. Broto, G. Moreau et C. Vandycke, *Eur. J. Med. Chem.* **19** (1984), 61–84.

[2] B. Mohar, D. Babic et N. Trinajstic, *J. Chem. Inf. Comput. Sci.* **33** (1993), 153–154.

Rational Drug Design :

[3] Malcolm D. Walkinshaw, "Rational Drug Design, A Three Dimensional Approach" *Drug News and Perspectives* **1** (1988), 329–337.

Similitude Moléculaire :

[4] W.G. Richards and E.E. Hodgkin, "Molecular Similarity", *Chemistry in Britain* (1988), 1141–1144.

[5] A.C. Good et W.G. Richards, *J. Chem. Inf. Comput. Sci.* **33** (1993), 112–116.

Gilles Moreau
Centre de Recherches Roussel UCLAF
102 à 111, Route de Noisy
F-93230 Romainville, France

Two Global Methods for Molecular Geometry Optimization

by Susana Gomez[1] and David Romero[2]
INRIA, Rocquencourt (France) [1], *IIMAS-UNAM, Obregon (Mexico)*[2]

1. Introduction

Mathematical models are often used to describe chemical processes. One of the most challenging problems is the determination of molecular conformations through energy minimization, the main difficulty being the enormous amount of local minima in the energy surface. In order to choose the appropriate method to find out minimum energy geometries, one has to consider the mathematical characteristics of the models used. We present here two global optimization methods, simulated annealing (SA) and the tunneling algorithm (TA), that have been successful in determining minimum energy conformations for small peptide molecules and atomic clusters respectively. In Section 2, after a brief presentation of the model for minimum energy conformation of peptides, we describe some results obtained with procedures that incorporate the SA concept. Section 3 is devoted to the TA and its application to the atomic cluster conformation problem. A summary of our experience is reported.

2. Peptide Conformation Through Simulated Annealing

In the realm of protein research, the variety of numerical optimization techniques proposed to locate molecular structures with minimum potential energy goes from Bremermann's method [1] to Monte-Carlo minimization [2], and includes variations of the powerful technique known as simulated annealing (SA) [3–6]. Since its inception [7], SA has proved to be an adequate tool to deal with many difficult combinatorial optimization problems [8–10].

A peptide molecule can be modelled as a mobile graph [4], where the nodes and the edges correspond, respectively, to the constituent atoms and to the atomic links between the atoms. Let Ω denote the set of integer values going from -180 to $+180$. Considering bond distances (atomic links) and bond angles (the angles between two adjacent atomic links) as fixed, the 3-dimensional shape of this graph is determined by $\Theta = (\theta_1, \ldots, \theta_n) \in \Omega^n$, where the θ_i are the values of the *dihedral* angles of both the main chain and the lateral chains. The potential energy $E(\Theta)$ of the corresponding molecule might be evaluated by means of ECEPP/2 [11], which is an empirical, pair interaction function, made up of Lennard–Jones, hydrogen bonding, electrostatics and torsional terms.

With this model on hand, the peptide conformational problem can be stated as that of finding $\Theta^* \in \Omega^n$ such that $E(\Theta^*) \leq E(\Theta)$, for any $\Theta \in \Omega^n$. In view of the huge, unknown number of local minima of $E(\Theta)$, the latter problem seems to be at least as difficult as the combinatorial optimization problems belonging to the *NP-complete* class, i.e., those problems for which all known algorithms that are able to solve them require a number of steps that grows exponentially with the number of variables, and for which there is little hope to ever find a polynomial solution algorithm.

A general SA procedure for this problem is now described. To every $\Theta \in \Omega^n$, associate a *neighborhood* $H(\Theta) \subset \Omega^n$ as the set of all conformations obtainable from Θ by arbitrarily modifying the value of one dihedral angle. SA starts with any conformation and with a nonnegative parameter T (called *temperature*) set to an appropriate high value. Given $\Theta \in \Omega^n$, a neighbor $\Theta' \in H(\Theta)$ is randomly generated with a uniform probability distribution. If $\Delta := E(\Theta') - E(\Theta)$ satisfies an *accepting rule*, then Θ' becomes the new current conformation, otherwise Θ' is rejected. The process is iterated until a condition of *steady state* is attained, in which case T is lowered according to a *cooling strategy* and the whole process is started anew. The algorithm halts when a *stop criterion* is satisfied: a local minimum with respect to $H(\Theta)$ has been reached. The most commonly used accepting rule is: Θ' is accepted if either $\Delta < 0$ or $x \leq \exp(-\Delta/T)$ (x is a random number uniformly distributed in [0, 1]). Instead, in a variation known as *threshold accepting* [12] the rule states that Θ' is accepted if $\Delta < T$.

In [4–5] experimentation with SA derived procedures to locate minimum energy conformations of peptides with up to 24 dihedral angles (N-acetyl-N′-methyl-glycineamide, N-acetyl-N′-methyl-alanineamide, N-acetyl-N′-methyl-aspartic acid, pentaglycine, Met^5-enkephalin, and Leu^5-enkephalin) was reported. To our knowledge, the obtained SA structures are the lowest available. Moreover, the threshold accepting version, as compared with the classical one, is simpler, requires less computing effort, and produces conformations with lower energy. With the largest peptides considered (24 dihedral angles), around 300,000 objective function evaluations were needed. This fact prevented trying bigger sizes, mainly because every evaluation was expensive in computing time. However, this drawback might be reduced through a more efficient implementation of the objective function evaluation.

SA procedures are easy to understand and require small effort in computer implementation. Complex objective functions, like ECEPP/2, do not prevent their use at all. Experiments can be performed in order to appropriately select the neighborhood, the cooling strategy, the steady state condition and the stop criterion. For instance, the best results are obtained when the neighborhood size is gradually reduced during the search process. Another useful device is to allow increases of T during the process whenever the remaining time for a run is large enough. In conclusion, SA seems to be an appropriate tool to conduct further research in molecular geometry optimizations with a high degree of confidence.

3. The Tunneling Method and Atomic Clusters

The name of the method describes its aim: to tunnel from one valley of the objective function to another, to find a sequence of local minima with decreasing values. There are two phases in the method: in phase 1, from a given initial point x^0, a local minimum (LM) x^* with $f^* = f(x^*)$ is found, using any local optimization method. From x^*, phase 2 obtains a point x^o in another valley with $f(x^o) \leq f^*$, which in turn will serve as initial point for phase 1. The problem in phase 2 can be reformulated as the problem of finding x^o such that

$$T_c(x^o) = \frac{f(x^o) - f^*}{\| x^o - x^* \|^{2\lambda^*}} \leq 0$$

or the more convenient problem of finding x^o such that [14]

$$T_e(x^o) = (f(x^o) - f^*)\exp(\frac{\lambda^*}{\| x^o - x^* \|^2}) \leq 0$$

The function T_c is called the *classical* [13] tunneling function (CTF) and T_e the *exponential* [14] tunneling function(ETF). These functions have a pole placed at x^* with strength λ^* (for the CTF λ^* is an integer that depends on the multiplicity of the minimum). To find x^o with $f(x^o) \leq f^*$, we can now use a local descent method. We have then found a point in another valley.

What we have just described is the main idea, but the algorithm has to deal with special features of different objective functions, and we will mention here the most important ones.

1) In practice T_c, or T_e, inherits the critical point behavior of the original function f, and in spite of the fact that $T(x)$ is positive (otherwise we have the solution x^0), a gradient type local descent search would be trapped at those critical points. To deal with this difficulty we can use the fact that increasing the strength parameter λ has a smoothing effect. Then by multiplying $T(x)$ by a term called a *mobile pole*: $\frac{1}{\|x^o - x^m\|^{2\lambda_m}}$ or $\exp(\frac{\lambda^m}{\|x^o - x^m\|^2}) \leq 0$ that the algorithm will automatically turn off ($\lambda^m = 0$) when $T(x)$ is smooth, or increase it ($\lambda^m = \lambda^m + \Delta\lambda$) to get a descent direction if it gets near a critical point, we locally smooth the problem.
2) For some objective functions f, the tunneling method might find many local minima at the same level f^*. This is expensive and in some applications these local minima are not of any particular interest. In this case, the algorithm can be easily modified to find x^0 satisfying $T(x^o) \leq -\epsilon$, to some desired tolerance ϵ.
3) In order to avoid going back to local minima already found at the same level, it suffices to multiply in T_c or T_e by terms that make the previous local minima poles.

4) An interesting comment here is that if x^* is the global solution, then x^0 does exist. Numerical experiments have shown that the global minimum is usually obtained with less than 20% of the total computing time, the other 80% being used to check that the candidate is indeed a global minimum. Given a candidate x^* the stopping test searches for x^0, starting from $2n$ different *neighbor points* generated using the cartesian directions (as in the SA method) i.e., $x_i = x^* + \epsilon r_i$, where r_i is the ith unitary vector (positive for the first n directions and negative for the remaining n), and $\epsilon << 1$. If after these $2n$ searches the method fails to find x^0, it stops. The best local minimum found so far is then said to be a good candidate to be the global minimum. Random directions can also be used if more than $2n$ searches are allowed, i.e., if the cost of the search permits it. In some applications, where getting minima with the value of the objective function less than a certain threshold is the aim, the expensive stopping rule can be avoided.

This method, in its present form, can only be used when gradient information is available and the function is C^2. It only handles continuous variables and it has also been implemented to deal with bounds [14]. The general constrained case has also been studied [15,16], and the method has been adapted to deal with least squares problems with singular jacobians (with rank $n-1$) [17,18]. The reader is referred to a general review [19]. The method is cheap, because it ignores all minima above the current local optimal level, and because it is not based on sampling, the number of function evaluations is generally small. Thus, it is especially suited for either expensive objective functions or functions with a large number of local minima (there are reported results with problems having 10^8 local minima [19]). The tunneling algorithm has been successfully applied to the problem of determining the structures of atomic clusters with minimum Lennard-Jones potential energy. Suppose the cluster conformation of n atoms is given by the cartesian coordinates $p_i = \{x_i, y_i, z_i\}$, $i = 1, \ldots, n$. The simplest assumption about interaction energy is that of Lennard-Jones two-body forces between atoms

$$V_n(p) = \sum_{i=1}^{n-1} \sum_{j>i}^{n} v(\| p_j - p_i \|_2),$$

where $v(r) = r^{-12} - 2r^{-6}$ is the potential energy of two atoms at a distance r from each other (the minimum of v for two atoms occurs when $r = 2^{1/6}$). Under this assumption, the atomic cluster conformation problem is then that of finding p^* such that $V_n(p^*)$ is minimal.

In order to choose an optimization method to solve this problem, the characteristics of the energy function V_n should be taken into account. We note first that V is continuous, differentiable and computationally very cheap to evaluate. The number of local minima is unknown and grows exponentially with the number n

of atoms in the cluster [20]. Moreover, and because of symmetries and relabelling, there is a multiplicity of local minima with the same energy level. The tunneling method can deal with this problem as explained before.

Ideally, the problem would be solved by a local minimizer that starts its iterations from any conformation lying in the valley of a global optimum. As, in general, there is no known method to determine such conformation, most reported results come from local minimizers coupled with physical intuition that provides the initial conformations. Another approach is that of using a global optimizer, able to pass from one valley to another, and, hopefully at the end, to a global optimum valley. Thus, the tunneling algorithm appears suitable to deal with the atomic cluster conformation problem.

Clusters of up to 72 atoms confined to a plane were studied [14]. From arbitrary initial conformations, the tunneling algorithm succeeded in finding solutions with atoms positioned at grid points corresponding to a hexagonal lattice packing, as predicted by other authors [21]. As expected, for $n = 7$ six atoms were positioned at the vertices of a regular hexagon of side size $r^* = 2^{1/6}$ plus one atom at its center. This shape can also be viewed as six equilateral triangles placed side by side. Clearly, 24 equal size equilateral triangles can be placed so as to form a regular hexagon of side size $2r^*$, and then 19 atoms can be positioned at the vertices of triangles in order to get a good candidate for optimality, compact conformation. The tunneling algorithm showed that this conformation is not optimal for $n = 19$, and the optimization process compressed it slightly distorting the inside triangles. This phenomenon was observed in every cluster of size $n \geq 19$ under study.

Several papers have been devoted to the 3-dimensional model [22–27]. The best reported results for small size clusters ($n \leq 147$) have been obtained with either SA techniques [25–26] or local optimization methods whose initial conformations follow an icosahedral growth scheme [27]. With one exception ($n = 38$), all these solutions were also obtained by the tunneling method [28], which starting from arbitrary initial points, passed from one valley of the energy function to another until it could not improve the energy value in the allowed time. To give an idea of the computational cost, for 8 atoms (24 optimization variables), starting at a random intial point, the method made 10570 function evaluations [28]. Comparing with the cost of the SA method reported in last section for 24 dihedral angles, TA is much cheaper, at the price of the uncertainty of being at the global minimum. In the case of the 38 atoms cluster, the tunneling algorithm found a conformation with lower Lennard-Jones potential energy than any other proposed elsewhere. This result is remarkable not only as a counterexample for the generally accepted icosahedral sequence proposed by Northby [27], but also because the obtained cluster shape corresponds to that of the *tetrakaidecahedron*, a beautiful semi-regular polyhedron studied by Archimedes in the third century B.C., and reported to be present in some crystals [29].

References

[1] H. Bremermann, *Math. Biosci.* **9** (1970), 1.
[2] Z. Li and H.A. Scheraga, *Proc. Nat. Acad. Sci. USA* **84** (1987), 6611.
[3] R. Garduño-Juárez, L.B. Morales, F. Pérez-Neri, and D. Romero, *J. Mol. Str.* (Theochem) **208** (1990), 279, and **222** (1992), 300.
[4] L.B. Morales, R. Garduño-Juárez, and D. Romero, *J. Biomol. Str. & Dyn.* **8** (1991), 721.
[5] L.B. Morales, R. Garduño-Juárez, and D. Romero, *J. Biomol. Str. & Dyn.* **9** (1992), 951.
[6] A. Nayeem, J. Vila, and H.A. Scheraga, *J. Comput. Chem.* **12** (1991), 594.
[7] S. Kirkpatrick, C.D. Gellat, and M.P. Vecchi, *Science* **220** (1983), 671.
[8] A. Drexl, *Comput.* **40** (1988), 1.
[9] D. Romero and A. Sánchez-Flores, *Comp. & Opns. Res.* **17** (1990), 465.
[10] E. Bonomi and J.-L. Lutton, *SIAM Rev.* **26** (1984), 551.
[11] G. Némethy, M.S. Pottle, and H.A. Scheraga, *J. Phys. Chem.* **87** (1983), 1883–1887.
[12] G. Dueck and T. Scheuer, *J. Comput. Phys.* **90** (1990), 161.
[13] A.V. Levy and A. Montalvo, *SIAM Slac.* (1985).
[14] S. Gómez and C. Barrón, *Rep. Inv. IIMAS* **1–3** (1991), submitted to *J. Global Opt.*
[15] S. Gómez and A.V. Levy, *Lecture Notes in Math.* Springer-Verlag **909** (1982), 34.
[16] S. Gómez and A.V. Levy, *Com. Tec. IIMAS* **231** (1980).
[17] S. Gómez, A.V. Levy, and A. Calderón, *Com. Tec. IIMAS* **370** (1987).
[18] S. Gómez, A.V. Levy, and A. Calderón, *Lect. Notes in Math.* Springer-Verlag **1230** (1987), 1.
[19] A.V. Levy and S. Gómez, *Numerical Opt. SIAM* (1985), 213.
[20] M.R. Hoare, *Adv. Chem. Phys.* **40** (1979), 49.
[21] G.L. Xue, R.S. Maier, and J.B. Rosen, *Proc. ACM Intl. Conf. Supercomp.* (1992).
[22] M.R. Hoare and P. Pal, *Adv. Phys.* **20** (1971), 161.
[23] J. Farges, M.F. De Feraudy, B. Raoult, and G. Torchet, *Surf. Sci.* **156** (1985), 370.
[24] J.D. Honeycutt and H.C. Andersen, *J. Phys. Chem.* **91** (1987), 4950.
[25] L.T. Wille, *Nature* **324** (1986), 46.
[26] L.T. Wille, *Chem. Phys. Lett.* **133** (1987), 405.
[27] J.A. Northby, *J. Chem. Phys.* **87** (1987), 6166.

[28] D. Romero, S. Gómez and M. Zúñiga, in preparation.

[29] A. Loeb, *Synergetics* by R.B. Fuller, Macmillan Publishing Company, New York (1982), 872.

Susana Gomez
INRIA
Domaine de Voluceau, B.P. 105
F-78153 Le Chesnay Cedex, France

David Romero
IIMAS-UNAM
Apto. Postal 20-726
Admon 20 del Alvaro
01000 Obregon, Mexico

Permanent Address
IIMAS-UNAM
Apto. Postal 20-726
Admon 20 del Alvaro
01000 Obregon, Mexico

Géométrie fractale et équations différentielles d'ordre non entier en chimie physique

par Alain Le Méhauté et Frédéric Héliodore
Alcatel Alsthom Recherche, Marcoussis (France)

La chimie physique est la science qui doit être mise en oeuvre dans les deux domaines clé que sont d'une part le génie chimique et d'autre part le génie des matériaux. Elle s'occupe des transferts d'extensité que sont la matière, la quantité de mouvement, la chaleur, le flux électromagnétique, etc. dans tous les dispositifs industriels [1,2]. Ces transferts ont une importance primordiale dans le monde économique qui est le nôtre car ils sont entachés de facteurs d'irréversibilité. Malgré leur complexité, on peut, toujours très heureusement représenter ces flux, quelle que soit l'extensité, par un champ de vecteurs et de trajectoires permettant de véhiculer l'extensité d'un endroit à un autre en un temps fini. Faute d'outils adaptés, la physico-chimie se limite le plus souvent à considérer exclusivement certains cas extrêmes de champs de transferts, par exemple des transferts avec ou sur des objets ponctuels (physique moléculaire), sur des variétés différentiables (relaxation) avec ou sans transport spatial (équation de type Laplacien) ou/et évolution temporelle (équation de type d'Alembertien). Toutes ces équations sont exprimées au moyen de liens entre formes différentielles de degré entier (0,1,2,3). Tel est le cas des équations de Maxwell, des équations de la mécanique des fluides et des équations de la thermodynamique phénoménologique. Le caractère différentiel étant primordial, il en résulte que la physico-chimie traditionnelle est le plus souvent désarmée pour traiter de phénomènes d'échanges en milieux non réguliers (turbulents et hétérogènes), des phénomènes de croissance et de cristallisation et des phénomènes chaotiques tels que le bruit chimique de réaction (singularités et brisures de symétrie).

L'introduction explicite de la géométrie en chimie-physique [3,4,5] permet de comprendre où se trouvent les difficultés et, au-delà avec en particulier l'introduction de la géométrie fractale, de les dépasser. On rappelle qu'une telle géométrie définit des interfaces le plus souvent auto-similaires ou auto-affines, de dimensions non entières et à caractéristiques non dérivables [6,7]. Si la géométrie peut se réduire à des surfaces équipotentielles (force : moment associé à un champ de vecteurs représentant le flux d'extensité) considérées comme des variétés différentiables, la statique et la dynamique du système physico-chimique peuvent être déduites du traitement d'équations différentielles dans le cadre d'une algèbre d'homologie faisant jouer un rôle prépondérant aux transformations naturelles (foncteurs) portant sur les relations flux/force [5]. Si, par contre, les "surfaces équipotentielles" sont des "surfaces" non rectifiables [5], les lois de comportement exigeront la prise en compte d'une régularisation de la "rugosité" par le processus physico-chimique conduisant à un feuilletage [3,5].

A titre d'exemple simple, afin de traiter la thermodynamique des surfaces des noirs de carbone et de silice, la géométrie fractale a été prise en compte pour la première fois par D. Avnir en 1983 [8]. Ces travaux montrèrent que la quantité de molécules test stockées sur les carbones n'est pas proportionnelle à leur taille caractéristique, mais à une puissance non entière de cette dernière. Il s'agit d'une propriété typiquement fractale dont les conséquences se manifestent dans le domaine de la catalyse. On doit ici se référer à la théorie de la mesure de frontières non rectifiables pour comprendre cette propriété [9]. Cette théorie offre en effet de pertinentes analogies avec les phénomènes d'absorption et de manière générale avec les phénomènes capacitifs [3]. Les implications de la théorie de la mesure en physico-chimie peuvent être poussées très avant. On est ainsi conduit à s'intéresser à la "température", non point des seules molécules sur le support, mais à celle du support lui même. Ce thème conduit, par exemple, aux analyses dues à M. Mendès France qui permettent d'associer la dimension non entière d'une courbe et la température au sens thermodynamique et statistique du terme [10,11,12,13,14].

Comme il est naturel, d'autres conséquences apparaissent dans le domaine de la dynamique. Tel est en particulier le cas du feuilletage. Lorsque par exemple le champ de vecteurs est la source d'un échange, le temps doit être introduit au niveau même de la prise en compte des singularités sur l'espace, car l'espace-temps n'est plus séparable. La nécessaire discrétisation du temps conduit à une algèbre de lacets permettant de distinguer le temps réversible du temps irréversible (le temps doit être défini sur C et non plus sur R). Cette algèbre permet, sous une autre forme, de déconvoluer le flux — représenté par les vecteurs de transfert — et la géométrie. Chacun des deux facteurs se trouve naturellement paramétré par une composante propre du temps. Les conditions de raccordement donnent accès aux équations de comportement [3,15,16]. Un tel traitement est particulièrement intéressant dans le cas d'une géométrie fractale car la paramétrisation conduit, pour l'utilisateur, au passage d'équations différentielles d'ordre entier n, à des équations différentielles d'ordre non entier α. Cet ordre est lui même lié à la dimension non entière de la géométrie. C'est ainsi que l'on est conduit à traiter des équations de type suivant [3,4,17] :

$$\frac{d^\alpha U}{dt^\alpha} \simeq f(U),$$

équations remarquablement adaptées à la maîtrise et à la compréhension des lois physico-chimiques en milieux hétérogènes. L'introduction de ces équations [6,7,9] et leur signification en terme de paramétrisation [4] et de géometrie de Riemann [17] constituent un vraie novation pour les sciences de l'ingénieur. Il est alors

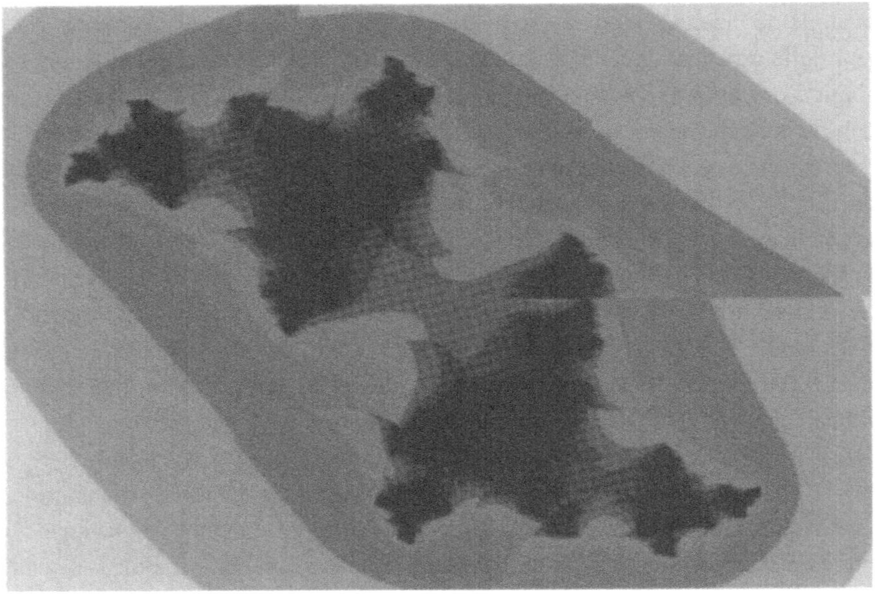

Figure 1. Ensemble de Julia fractionnaire donnant la stabilité de l'application $z \rightarrow g(z) = z^{\alpha} + k$ pour k fixé et $\alpha = 1.5$ dans C. Les dégradés de gris représente le temps d'itération nécessaire pour atteindre un seuil en module de $g(z)$, donné préalablement. Dans l'hypothèse où les valeurs de α sont entières, la figure présente des symétries. Les brisures de symétrie apparaissent à divers niveaux d'imbrication pour tout α n'appartenant pas à l'ensemble des entiers.

toujours possible [17] d'écrire la fonction de transfert de la dynamique en milieu fractal sous la forme d'une égalité entre une fonction harmonique et une fonction puissance agissant sur le corps des complexes :

$$\frac{Z + k_1}{Z + k_2} = W^{\alpha} \ avec \ W, Z, k_i \in C \ et \ \alpha = p/q$$

où α est l'approximation rationnelle à un ordre lié au niveau de régularisation de la géométrie fractale. Ce type d'équation, largement observé au plan expérimental [4] permet de remonter à des invariants fondamentaux dont l'importance pratique est considérable. Par ailleurs l'examen de cette égalité met en exergue deux faits mathématiquement intéressants : la dynamique est de type hyperbolique parce que la géométrie fractale est toujours associable à une arborescence et aux groupes libres qui en sont l'origine (les transformations de type Fuschien et/ou Kleinien sont les transformations de base). Cette arborescence fonde, pour partie seulement, le lien entre le phénomène et la géométrie. Le second terme relève de la paramétrisation de la géométrie. Son caractère en général irrationnel conduit à un comportement ergodique. La dynamique est représentable par une variété feuilletée, le nombre de feuillets étant associable à l'approximation rationnelle que nous affectons à α. La résolution pas à pas de ce type d'équation met en jeu des itérations g_n telles qu'en montre la figure donnée ci-desous (un ensemble de Julia fractionnarisé) :

$$z \to g(z) = z^\alpha + k \ .$$

Si α est entier ou rationnel, il apparaît des symétries ($g(z) = z$ à un ordre n fini). Dans le cas où α est irrationnel, il y a certes ergodicité mais l'approximation des solutions par le processus physico-chimique conduit à des brisures naturelles de symétries qui se manifestent au travers de *bilans énergétiques non bouclés* (absorption anormale d'énergie, potentiels additionnels, masse et charge résiduelle, ...), *l'apparition de singularités, l'émergence de nouveaux invariants* (associés aux groupes de dilatation) et *par des modifications des équations de jauge* [5,17,18], toutes modifications qui peuvent d'ores et déjà permettre d'imaginer un nouveau génie des matériaux (magnétisme artificiel, supraconducteur de synthèse, nouvelle électrodynamique, ...) et au-delà, de nouvelles méthodes de traitement de l'information et de l'énergie.

Bibliographie

[1] V.G. Levich, "Physicochemical Hydrodynamics", Prentice Hall, N.Y., 1962.

[2] R. Aris, and N.R. Amundson, "Mathematical methods in chemical engineering", Prentice Hall, N.Y. 1973.

[3] A. Le Méhauté et al, CR Acad. Sci. Paris 294-(II), (1982), 865

[4] A. Le Méhauté "Les géométries fractales" ed. Hermes, Paris 1990; "Fractal Geometries", Penton Press, Londres, 1991.

[5] A. Le Méhauté, F. Héliodore et D. Cottevieille, *Revue Tech. et Sci. de Défense*, Paris, avril/mai 1992.

[6] B. Mandelbrot, "Les fractals", Flammarion, Paris, 1975.

[7] B. Mandelbrot, "The fractal geometry of nature", Freeman, San Francisco, 1982.

[8] D. Avnir et al, *J. Chem. Phys.* **79** (1983), 3566; *Surf. Sci.* **126** (1983), 59; *Nature* **308** (1984), 261.

[9] C. Tricot, "La géométrie fractale", Springer-Verlag, Heidelberg,1993.

[10] M. Mendès-France et al., The Planck constant of curve in "Fractal geometry

and Analysis", J. Belair and S. Dubuc (eds.), Kluwer Acad. Publ., Amsterdam, 1991, p. 325.

[11] Y. Dupain, M. Mendès-France, C. Tricot, "Spirals and Phase Transitions", to be published.

[12] "Dimension et Entropie, les théories de la complexité autour de l'oeuvre de H. Atlan", Colloque de Cerizy, F. Fogelman Soulié (ed.), Seuil, Paris 1991, p. 78.

[13] M. Mendès-France, "From geometry to Euler Identities", *Theoretical Computer Sc.* **65** (1989), 213.

[14] M. Mendès-France, "Chaos implies confusion. In number theory and dynamical systems", Dodson Vickers (ed.), Cambridge University Press, *London Math. Soc. Lect. Notes* **134** (1989), 137.

[15] M. Mendés France, Courrier du CNRS, Image de la Physique, *Sup.* **51** (1983), 5.

[16] C. Tricot, *J. Chim. Phys.*, **85** (1988), 379.

[17] F. Héliodore, R. Nigmattulin, P. Riot et A. Le Méhauté, "Electrodynamique en milieu fractal", Hermès, Paris, 1993.

[18] A. Le Méhauté, "Les cascades d'hivers : Traité sur l'irréversibilité du temps", Hermès, Paris, 1993.

Alain Le Méhauté et Frédéric Héliodore
Alcatel Alsthom Recherche
F-91460 Marcoussis, France

Round Table P: July 9, 1992

MATHEMATICS IN MEDICINE AND BIOLOGY

by Robert Hiorns[1] and Bernard Prum[2]

University of Oxford, Oxford (UK) [1], *Université Paris V, Paris (France)* [2]

Introduction

Historically, the Life Sciences have provided stimulation for a number of important developments in different branches of mathematics. In particular, significant progress in Probability Theory is the result of problems that have been raised in the theory of population genetics (Mendel, Fisher, Kimura, and others). Furthermore, classical statistical theory arose largely because of the need to interpret data from biological and medical investigations. Today, there are newer topics such as morphogenesis, which is a source of problems that require new tools from Geometry and Complex Analysis. The biological use of fractal mathematics is also very common.

Mathematics and the Life Sciences have always interacted in both directions, and it is inevitable for the progress of science that they must do so. In the first place, mathematics represents an indispensable tool for the biologist or the physician. Important work is yet to be done in order to provide more elaborate and sophisticated mathematical methodologies capable of treating data and problems in the life sciences area. For example, the treatment of censored survival data to be correctly carried out needs an approach founded on martingale theory. The establishment of the human genome map requires the tools of complex combinatorics. In a similar manner, the analysis of more and more complex data demands ever-increasing computer support, including the methodology of expert systems to help with decision making, artificial intelligence, and medical image processing. Thus a whole new field opens itself to a collaboration between mathematicians and physicians.

The second and main point is that the Life Sciences are set to play the same role as did physics in the nineteenth century and early twentieth century for fundamental mathematics, that is to say, to raise to some questions. This is already happening, as for instance with the development of neural networks or again in the newer uses of mathematics in biochemistry. Mathematical models underlying many of the processes in the field of molecular biology are still to be devised. When they are, they will have to account for some extremely complex phenomena, such as the role of enzymes, phenomena occurring in membranes and retroactions. These models will certainly be linked to various mathematical studies such as the stability of the solutions to partial differential equations, bifurcations, models of chaos, etc.

For some 50 years we have experienced a transition from "smooth" mathematical tools (e.g., classical analysis, deterministic systems, differentiability, ...) to more "chaotic" tools (e.g., randomness, singularities, fractals, self-similarity, ...), which enables the mathematician to refocus a field of application from physics in a broad sense to biology.

The originality of the phenomena then considered introduces a very "decentralized" behaviour: the different components of a biological system, whether they be living creatures in a population or cells in an organ, determine their behaviour with a given independence. Their study requires the development of "*statistical biology*" similar to the established statistical mechanics; obviously this independence can be modelled by the same independence which underlies brownian motion, with bifurcations looking like phase transitions. Taking advantage of the development of modern computers, one can work to extend a whole speculative field, going from the more sophisticated theory of point processes to the stability of stochastic partial differential equations to explain some evolutions, or change point theory to express some imbalances.

Invited participants

Elja Arjas (Oulu)

Gilbert Chauvet (Angers)

Albert Goldbeter (Brussels)

Niels Keiding (Copenhagen)

Jonathan Swinton (London)

Petr Tautu (Heidelberg)

Medical Statistics: an Example from Diabetes Epidemiology

Niels Keiding
University of Copenhagen, Copenhagen (Denmark)

The purpose of this contribution is to introduce some concepts and methods of modern survival analysis such as those used in chronic disease epidemiology. This will be done through a concrete example, concerning the estimation of the incidence of diabetes and the death rate of diabetics based on a prevalent sample from Fyn county, Denmark.

Fyn county in Denmark contains about 450,000 inhabitants. On July 1, 1973, 1499 of them suffered from insulin-dependent diabetes mellitus ("diabetes" for short). This was ascertained by recording all insulin prescriptions in the National Health Service files for this county during a five month period covering the above date, and subsequently checking each patient's medical record at the general practitioner and, when relevant, the hospital. The data was collected by Green et al. (1981).

Two different problems were studied: estimating the age-specific mortality of diabetics (number of deaths) based on (*prospective*) follow-up of this prevalent population, and a *retrospective* estimation of diabetes incidence in the years before 1973, based on the information in the sample on (calendar time, age) at onset and additional historical information on diabetes mortality and population censuses. Our model is a non-homogeneous illness-death (Markov or semi-Markov) process for each person with intensities as illustrated in the diagram

$$
\begin{array}{ccc}
\text{healthy} & \xrightarrow{\alpha(a)} & \text{diabetes} \\
\mu(a) \searrow & & \swarrow \nu(a,d) \\
& \text{death} &
\end{array}
$$

where the incidence $\alpha(a)$ and the mortality $\mu(a)$ of the healthy (i.e., non-diabetic) depend on age a, whereas the mortality $\nu(a, d)$ of the diabetic (the death rate) may in addition depend on duration d, and all intensities may depend on further covariates, in particular sex and calendar time. A general discussion of the application of this framework for formalizing epidemiological concepts such as age-specific incidence and prevalence was given by Keiding (1991).

For the *prospective* study of the death-rate, the survival status for all patients was assessed by January 1,1982, using the Danish centralized person registry. The basic time variable was *age*. Additional time variables are duration of disease and calendar time. A special problem is that patients do not all enter the study at the same age.

Because of this problem of *delayed entry* (patients only being at risk for dying of diabetes after they have acquired the disease), we focus on the conditional survival distribution, given survival until a certain age, for which we use the standard non-parametric maximum likelihood estimator, the Kaplan–Meier estimator with delayed entry (left truncation). Additional covariates may be included in a regression model for survival data; this is illustrated by incorporating disease duration in a Cox regression model (with time- (i.e., age-) dependent covariate). A survey of problems in studying delayed entry was given by Keiding (1992).

The death-rate in a seven-year follow up period was evaluated by Green and Hougaard (1984). An extended discussion of the analysis of this data was given by Andersen et al. (1992).

The *second problem* regarded the possibility of exploiting the retrospective information in the prevalent sample of July 1, 1973 on (calendar time, age) at onset for each patient to assess diabetes incidence $\alpha(t, a)$, $1933 \leq t \leq 1973$, $0 \leq a \leq 31$ years.

Obviously, the prevalent sample of July 1, 1973 represents a selected sample of patients, containing only the survivors. Heuristically, each surviving patient represents not only his/herself but also the "ghosts" (Turnbull,1976) who died before sampling. The expected number of patients necessary to generate one survivor is p^{-1} where $p =$ this patient's survival probability from onset to July 1, 1973. To estimate p (for $0 \dagger a \dagger 30$ years) we may apply the analysis by Ramlau-Hansen et al. (1987) who fitted a Cox regression model for the relative mortality (Andersen et al., 1985) to historical hospital data on diabetes death-rate. Data on population censuses is available in official Danish vital statistics.

A smooth estimate of diabetes incidence $\alpha(t, a)$ in the *Lexis diagram* (the calendar time x age plane) may now be obtained using bivariate kernel smoothing (Keiding et al. 1989, Keiding 1990, McKeague and Utikal 1991). The basic trends show increase with calendar time for a given age (slightly more for males than for females) and increase with age for given calendar time, with an indication of a peak at about the time of puberty. Viewed on a cohort basis, consistent increase of diabetic incidence with age are observed in the later cohorts, but this pattern is not very obvious in the earlier cohorts.

References

[1] Andersen, P.K., Borch-Johnsen, K., Deckert, T., Green, A., Hougaard, P., Keiding, N., and Kreiner, S., "A Cox regression model for the relative mortality and its application to diabetes mellitus survival data", *Biometrics* **41** (1985), 921–932.

[2] Andersen, P.K., Borgan, O., Gill, R.D., and Keiding, N., *Statistical Models Based on Counting Processes*, Springer, New York, 1992 (in press).

[3] Green, A., Hauge, M., Holm, N.V., and Rasch, L.L., "Epidemiological studies of diabetes mellitus in Denmark. II. A prevalence study based on insulin prescriptions". *Diabetologia* **20** (1981), 468–470.

[4] Green, A. and Hougaard, P., "Epidemiological studies of diabetes mellitus in Denmark: 5. Mortality and causes of death among insulin-treated diabetic patients", *Diabetologia* **26** (1984), 190–194.

[5] Keiding, N., "Statistical inference in the Lexis diagram", *Phil. Trans. Roy. Soc. London A* **332** (1990), 487–509.

[6] Keiding, N., 'lq Age-specific incidence and prevalence: a statistical perspective (with discussion)". *J. Roy. Statist. Soc. A* **154** (1991), 371–412.

[7] Keiding, N., "Independent delayed entry (with discussion)". In: *Survival Analysis: State of the Art*, J.P. Klein & P.K. Goel (eds.), Kluwer, 1992, 309–326.

[8] Keiding, N., Holst, C., and Green, A., "Retrospective estimation of diabetes incidence from information in a current prevalent population and historical mortality", *Amer. J. Epidemiol.* **130** (1989), 588–600.

[9] McKeague, I.W. and Utikal, K.J., "Inference for a nonlinear counting process regression model", *Ann. Statist.* **18** (1990), 1172–1187.

[10]] Ramlau-Hansen, H., Jespersen, N.C.B., Andersen, P.K., Borch-Johnsen, K., and Deckert, T., "Life insurance for insulin-dependent diabetics", *Scand. Actuar. J.* (1987), 19–36.

[11] Turnbull, B.W., "The empirical distribution function with arbitrarily grouped, censored and truncated data", *J. Roy. Statist. Soc. B.* **38** (1976), 290–295.

Niels Keiding
Statistical Research Unit
Faculty of Medicine
University of Copenhagen
Copenhagen, Denmark

Point Process Approach of Life Histories: Causal Chains

by Elja Arjas
University of Oulu, Oulu (Finland)

1. Introduction

The aim of this short contribution is to illustrate how certain notions from the modern theory of point processes can be used in describing complicated evolutions of individuals or populations, over time. As an illustration, we make an attempt to formulate a fundamental question in substantive medical and biological research, that of *cause* and *effect*, in this probabilistic framework. The presentation is based on Arjas and Eerola (1992).

Two important reservations must be made at once. First, even most sophisticated mathematical formulations cannot "solve" questions regarding causality in substantive research, even less eliminate the various philosophical issues relating to the nature, and even existence, of such a relationship. Second, any meaningful description based on probabilities should also address the statistical question: "How can probabilities be obtained from empirical evidence?" There is no ultimate solution to this question either, and we will not even make an attempt to dwell on the statistical inference methods from longitudinal data. We will only make a general claim that the general theory of filtered point processes provides a very natural framework for a unified statistical theory (e.g., Andersen et al. (1992)).

Hypotheses in areas of substantive research are often phrased in the simple form "C causes E". Much empirical work, based on observational or experimental data, goes into finding evidence in favor or against such claims. Unlike statistical dependence or association, causal dependence is directed, i.e., "C causing E" or the occurrence of E depending on the occurrence of I. In some cases an event can have both the role of a cause and that of an effect: being first an effect, E may later act as a cause of another effect. This is the typical structure of a *causal chain*.

Two connotations of causality are of particular interest. First, causation corresponds to the idea that a cause *explains* the occurrence of the effect, and second, a cause has *predictive power* concerning the effect. These two aspects of causality are interrelated, but they correspond to different ways of conditioning on the observed history. In the former, one is looking for an explanation to a particular event which is known to have occurred ("What was (were) the cause(s) of E?").

In the latter, one considers generic events and statements of the type "C causes E" (e.g., smoking causes lung cancer), meaning that "normally" or "frequently" C is followed by E. Here we restrict ourselves to this predictive aspect of causality.

Most research hypotheses in empirical work are probabilistic rather than deterministic by nature, that is, of the above form "C is frequently followed by E", and then the assumption of *constant conjunction* ("C is always followed by E") in classical deterministic causality (Hume, 1748) is unrealistic. In theories of *probabilistic causality* this requirement is replaced by the assumption that a cause raises the probability of the effect. Suppes' (1970) causal theory is qualitative and uses the difference $P(E|C) - P(E)$. For C to be a *prima facie* cause for E, it should precede E, and $P(E|C) - P(E)$ should be positive. Obviously, $P(E|C) - P(E) > 0$ implies that $P(E|C) > P(E) > P(E|\neg C)$, $\neg C$ being the complement of C, so that qualitatively it makes no difference whether one considers $P(E|C) - P(E|\neg C)$ or $P(E|C) - P(E)$. Often it is reasonable to condition explicitly on some "background" event B, leading to conditional probabilities $P(E|BC)$, $P(E|B)$, etc.

Finally, we mention the *counterfactual* tradition in causal analysis. The second part of Hume's definition implies that if the cause is not present then there is no effect (i.e., $\neg C$ to $\neg E$). In probabilistic counterfactual analysis this leads to considering the difference $P(E|BC) - P(E|B\neg C)$.

In these classical formulations of probabilistic causality, time is present only implicitly. However, natural questions like "How does the time of occurrence of a cause influence the occurrence of the effect?"; or, "Is the causal response immediate or is there a time lag between the cause and the effect?", can only be answered in a longitudinal setting in which the times of occurrence are taken into account. It also seems unrealistic to expect that in more complicated causal chains dependence on the history of the process could have the form of a simple Markov structure (cf. the discussion in Suppes, 1990). Here we consider briefly a general framework in which both the time aspect and the conditioning history are accounted for explicitly.

2. Marked Point Processes and Their Prediction Processes

In order to use a concrete causal problem as a reference, we consider two simple causal chain models which are encountered frequently in epidemiology.

Scheme 1. Chain: healthy $\rightarrow$ C (exposed) $\rightarrow$ E (diseased).

The following questions arise then naturally: to what extent can exposure be viewed as a cause to the disease? Does early exposure have a different effect (if any) than late exposure?

Scheme 2. By adding a final "dead" state to the above model, we are led to the chain

Chain: healthy $\rightarrow$ C (exposed) $\rightarrow$ E or C (diseased) $\rightarrow$ E (dead).

This gives rise to at least the following additional questions: to what extent can exposure and/or disease be viewed as cause(s) of death? What is the role of onset time in this respect? Do different exposure time/onset time combinations lead to different conclusions? These simple causal chains can be described by the following graphs:

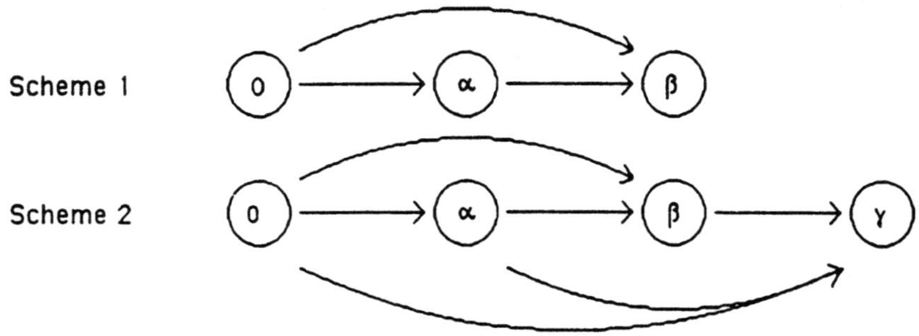

In particular, we could then choose 0 = healthy, α = exposed, β = diseased, γ = dead. For a complete description of the development from state 0 to state γ we should also consider "transition times":

$$T_\alpha \text{ (resp. } T_\beta, T_\gamma) = \text{time at which } \alpha \text{ (resp. } \beta, \gamma) \text{ is reached.}$$

(Here we need certain conventions, such as: if $T_\beta < T_\alpha$ then $T_\alpha = \infty$, etc.)

Somewhat similar models have been used in modeling the evolution of cancer by the transition of normal cells into malignant cells via intermediate stages. For example, in a two-stage model $NC \to IC \to MC$ of Moolgavkar and Knudson (1981) a cell can either die, divide into two new cells, or move to the next stage during each stage with a certain transition rate. The incidence of cancer at age t is a function of these transition rates. Exogenous exposure can then be modelled as a time-dependent intervening cause of transition.

In general, the progression of events in a longitudinal study is conveniently modelled as a stochastic process. Here we use *marked point processes* (MPP's).

Let $(T, X) = \{(T_n, X_n); n \geq 1\}$ be such that $0 \leq T_1 \leq T_2 \leq \ldots$ is an ordered sequence of times of occurrence (with strict inequalities between finite times), whereas X_n is a description of what happens at T_n. Formally, the X_n's are random variables taking values in a (countable, non-empty) set Δ. In Schemes 1 and 2, which describe the evolution of a single individual, a convenient choice would be $\Delta = \{\alpha, \beta, \gamma\}$.

Let $H_t = \{(Tn, Xn)\ ; Tn \leq t\}$ be the pre-t history of (T, X). Then $(H_t)_{t\geq 0}$ forms a stochastic process with values in $(H, \mathcal{H})$, where H is the set of finite collections of Δ-marked points and $\mathcal{H}$ is the corresponding family of Borel sets. In Scheme 1 we could choose

$$H = \{\Delta,\ \{(v, \alpha)\}\ ,\quad \{(w, \beta)\}\ ,\quad \{(v, \alpha), (w, \beta)\}\ ;\ 0 < v < w\}\ ,$$

and in Scheme 2 we should add to H histories of the form

$$\{(u, \gamma)\}\ ,\ \{(v, \alpha), (u, \gamma)\}\ ,\ \{(w, \beta), (u, \gamma)\}$$

and

$$\{(v, \alpha), (w, \beta), (u, \gamma)\}\ ,\ (0 < v < w < u)\ .$$

The observation (= knowledge) of the process (H_t) in time leads one to consider progressive conditioning on the σ-fields

$$\mathcal{F}_t = \sigma\{H_t\}\ , \quad t \geq 0\ .$$

Therefore a probability (on the path space of (T, X)) is often most conveniently defined by specifying the $(\mathcal{F}_t)$-conditional hazards of the MPP. In the absolutely continuous case this amounts to specifying the mark- specific stochastic intensities. More generally, one is led to consider the $(\mathcal{F}_t)$-compensators $(\Lambda_t(x))$, $x \in \Delta$, of the counting processes $(N_t(x))$, $x \in \Delta$ where

$$N_t(x) = \sum_{n \geq 1} 1\{T_{n \geq t}, X_n = x\}$$

(see e.g., Arjas (1989) and the references given there). For our purposes it suffices to know that the compensators have the probabilistic "hazard" interpretation $d\Lambda_t(x) = P(dN_t(x) = 1|\mathcal{F}_t-)$. In particular, this concerns possible jumps: $\Delta\Lambda_t(x) = P(\Delta N_t(x) = 1|\mathcal{F}_{t-})$, where $\Delta\Lambda_t(x) = \Lambda_t(x) - \Lambda_{t-}(x)$ and $\Delta N_t(x) = Nt(x) - N_{t-}(x)$. The exact mathematical definition of $(\mathcal{F}_t)$-compensators is based on martingale theory.

In Scheme 1, and assuming absolute continuity, this would mean specifying the following three sets of functions (which could also depend on individual covariates):

for $t > 0 : \lambda_\alpha(t) =$ hazard rate of α for an individual in state 0;
for $t > 0 : \lambda_\beta(t) =$ hazard rate of β for an individual in state 0;
for $t > v > 0 : \lambda_{\beta|\alpha}(t|v) =$ hazard rate of β for an individual with $T_\alpha = v$.

In Scheme 2 we need additionally the functions

for $t > 0 : \lambda_\gamma(t)$
for $t > v > 0 : \lambda_{\gamma|\alpha}(t)$
for $t > w > 0 : \lambda_{\gamma|\beta}(t)$
for $t > w > v > 0 : \lambda_{\gamma|\alpha,\beta}(t|v, w)$

with the obvious interpretation. For example, the last function could be interpreted as the (death) hazard rate at t of an individual with exposure time $T_\alpha = v$ and disease onset time $T_\beta = w$. It is important to note that no simple Markov or semi-Markov property related to the states α, β and γ is assumed here.

We now explain how this mathematical framework can be connected to causal problems. Consider, therefore, some event or random variable of interest (*a response*). Suppose that the observation of (H_t) covers both the relevant *background information* and the appearance of a potential *cause*, and ultimately, also the *response*. Under such circumstances our aim is to find out how at time t, when the response is not yet known, the corresponding prediction depends on the observed pre-t history H_t.

A more formal expression of this idea is as follows: Let Y be an (F_∞-measurable) random variable of interest, taking values in $(Y, \mathcal{Y})$, say. Consider the prediction process $(\mu_t)_{t\geq 0}$, (see Norros, 1985) defined by

$$\mu_t(A) = P(Y \in A|\mathcal{F}_t)\ ,\ \ A \in Y\ .$$

It can be shown that there exist regular versions of transition probabilities

$$\mu_t^*(\bullet;\bullet) : H \to \mathcal{Y}\ ,\quad t \geq 0\ ,$$

such that we can use the representation

$$\mu_t(A) = \mu_t^*(H_t; A)$$

(i.e., $P(Y \in A|\mathcal{F}_t)(\omega) = \mu_t^*(Ht(\omega); A))$. We view Y here as the response, and therefore $\mu_t^*(H_t; \bullet)$ is the conditional distribution of the response given the pre-t history H_t. It is convenient to extend the definition from actual pre-t histories to all of $(H, \mathcal{H})$ in the following way: if two histories H and H' agree on $[0, t-]$, then $\mu_t^*(H; \bullet) = \mu_t^*(H'; \bullet)$.

In this framework a prediction, in terms of the process (μ_t^*), has always the following three ingredients:

(i) given (t, H), the distribution $A \to \mu_t^*(H; A)$ forms "in a given situation, a prediction concerning the value of Y";

(ii) given (H, A), the function $t \to \mu_t^*(H; A)$ expresses "how the probability of $\{Y \in A\}$ is updated in time, based on the progressive observation of (H_t)" (learning effect);

(iii) given (t, A), the function $H \to \mu_t^*(H; A)$ expresses "how the probability of $\{Y \in A\}$ depends on what has happened before t".

We remark that in (i), for certain combinations of (t, H), the distribution of Y can be the Dirac measure (point mass), i.e., the value of Y is actually known because H_t has been observed.

The contrasts $P(E|BC) - P(E|B\neg C)$ and $P(E|BC) - P(E|B)$ considered in the introduction have their equivalents in the present dynamic framework. To see this, we first let the event $\{Y \in A\}$ have the role of the effect E. In conditioning the probabilities, we view $H_{t-} = (T_n, X_n) : T_n < t$, the strict pre-$t$ history of (T, X), as the background event B. The remaining element, the cause C, is expressed in terms of a "new" marked point (t, x) appearing at time t. By adjoining such a point to a strict pre-t history H_{t-}, a pre-t history $H_{t-} \cup \{(t, x)\}$ is formed. Since the mark x can take any value in the mark space Δ, we can consider more than a single possible cause. Finally, we express the situation corresponding to "non-cause" $\neg C$ by the requirement that there is no point at time t, i.e., the pre-t history at t is actually the same as H_{t-}.

By combining these elements we are led to consider $\mu_t^*(H_{t-} \cup \{(t, x)\}; A)$ *in place of the probability* $P(E|BC)$, *and* $\mu_t^*(H_{t-}; A)$ *in place of* $P(E|B\neg C)$. We denote for simplicity their difference by

$$C_t(x) = \mu_t^*(H_{t-} \cup \{(t, x)\}; A) - \mu_t^*(H_{t-}; A)$$

which therefore has the same role as the difference $P(E|BC) - P(E|B\neg C)$ in the classical formulation.

In order to find a similar dynamic reformulation to the difference $P(E|BC) - P(E|B)$ note first that the probability $P(E|B)$ can be written as $P(E|B) = P(C|B)P(E|BC) + P(\neg C|B)P(E|B\neg C)$, a weighted mean of $P(E|BC)$ and $P(E|B\neg C)$.

Intuitively, the left limit $\mu_{t-}(A) = \mu_{t-}^*(H_{t-}; A)$ is seen to have the same role as $P(E|B)$ in the classical analysis, and it has a weighted mean expression similar to that found for $P(E|B)$ above. For this purpose we consider the behaviour of the process $\mu_t(A) = E(1_A(Y)|\mathcal{F}_t)$, $t \geq 0$, in particular its jumps $\Delta\mu_t(A) = \mu_t A) - \mu_{t-}(A)$. Since $\mu_t(A) = \mu_t^*(H_t; A)$, it is clear that $\mu_t(A)$ equals $\mu_t^*(H_{t-} \cup \{(t, x)\}; A)$ if there is at t a marked point with mark x, and it equals $\mu_t^*(H_{t-}; A)$ if there is no point at t. (These are the two terms in the expression of $C_t(x)$ above.) A concise way of writing this is

$$\mu_t(A) = \sum_x \Delta N_t(x)\mu_t^*(H_{t-} \cup \{(t, x)\}; A) + \left(1 - \sum_x \Delta\Lambda_t(x)\right)\mu_t^*(H_{t-}; A)\,.$$

Similarly, considering the left limit $\mu_{t-}(A)$, one can show that

$$\mu_{t-}(A) = \sum_x \Delta\Lambda_t(x)\mu_t^*(H_{t-} \cup \{(t, x)\}; A) + (1 - \sum_x \Delta\Lambda_t(x))\mu_t^*(H_{t-}; A)\,.$$

(Recall that $(\Lambda_t(x))$, $x \in \Delta$, are the $(\mathcal{F}_t)$-compensators of the counting processes $(N_t(x))$.) In other words, $\mu_{t-}(A)$ can be expressed as a conditional weighted mean, or expected value, of the possible values $\mu_t(A)$ can take when H_{t-} is given. Observe, in particular, that if all compensators are continuous (in such a case the marked points are called completely $(\mathcal{F}_t)$-unpredictable), we have $\mu_t - (A) = \mu_t^*(H_{t-}, A)$. The other extreme case is when $\Delta\Lambda_t(x) = 1$ for some $x \in \Delta$, i.e., $\Delta N_t(x) = 1$ is completely predictable. Then $\mu_{t-}(A) = \mu_t^*(H_{t-} \cup \{(t, x)\}; A)$.

The above reasoning shows the exact correspondence between the difference $P(E|BC) - P(E|B)$ considered by Suppes (1970) and others, and the expression $\Delta\mu_t(A) = \mu_t(A) - \mu_{t-}(A)$ arising from our MPP formulation, provided that $\Delta N_t(x) = 1$ and the connection between point (t, x) and cause C is made. Also note that the process $(\mu_t A))$ is an $(\mathcal{F}_t)$-martingale, and therefore the above analysis concerns the jump structure of a point process martingale. It can be shown that in this case $\mu_t(A)$ admits an integral representation of the form

$$\mu_t(A) = P(Y \in A) + \sum_x \int_0^t C_s(x) dN_s(x) - d\Lambda_s(x)) \, ,$$

where $C_t(x)$, considered already above, is called the *innovation gain* from observing the marked point (t, x). This shows, in explicit terms, how the recursive updating in (ii) above takes place. In particular, we see again that if all marked points at t are completely unpredictable from $(\mathcal{F}_t)$, and $\Delta N_t(x) = 1$ occurs (a "complete surprise"), the jump of the prediction probability equals the innovation gain: $\Delta\mu_t(A) = C_t(x)$. If, on the other hand, the event $\Delta N_t(x) = 1$ can be predicted with certainty in the sense that $\Delta\Lambda_t(x) = 1$, we have (almost surely) $\Delta\mu_t(A) = 0$. In other words, if a cause is certain to occur, its actual occurrence does not change the prediction concerning the effect.

This completes our dynamic reformulation of the contrasts $P(E|BC) - P(E|B\neg C)$ and $P(E|BC) - P(E|B)$. We close by some complementary remarks.

First recall the counterfactual tradition referred to in the introduction. The implication $\neg C$ to $\neg E$ requires the specification of the truth value of C, which then leads to the ill-defined concept of the similarity of the "actual" and "possible" worlds. In the present formulation there is no need for such a specification. The history H_{t-} is a free parameter in the prediction probabilities, and therefore we can consider predictions at time t conditioned on different histories (i.e., H_{t-} and $H_{t-} \cup \{(t, x)\}$) without any conceptual difficulties.

Second, the considered MPP and the corresponding history process (H_t) are usually thought to arise directly from what is observed. However, the level of observation need not be uniquely defined in any given context, as we may think of different levels of accuracy in which the development of interest is registered by an observer. By the same token, the above predictive probabilities depend on the available information. The contribution of a marked point, measured in terms of an innovation gain, can even assume a different sign if the level of observation is changed. This corresponds to the well-known Simpson's paradoxin the classical theory.

References

[1] Andersen, P.K., Borgan, O., Gill, R.D. and Keiding, N., *Statistical models based on counting processes*, (1992), Monograph, Springer, New York, (in press).

[2] Arjas, E., "Survival models and martingale dynamics", *Scand. J. Statist.* **16** (1989), 177–225.

[3] Arjas, E. and Eerola, M., "On predictive causality in longitudinal studies", (1992), *Journal of Statistical Planning and Inference* (to appear).

[4] Dempster, A.P., "Causality and Statistics", *Journal of Statistical Planning and Inference* **25** (1990), 261–278.

[5] Hume, D., *The Treatise of Human Nature*, 1748.

[6] Moolgavkar, S.H. and Knudson,A.G., "Mutation and cancer: a model for human carconogenesis", *JNCI* **66** (1981), 1037–1052.

[7] Norros, I., "Systems weakened by failures", *Stochast. Process. Appl.* **20** (1985), 181–196.

[8] Salmon, W., "Probabilistic causality", *Pacific Philos. Quarterly* **61** (1980), 50–74.

[9] Suppes, P., *A Probabilistic Theory of Causality*, North-Holland, Amsterdam, 1970.

[10] Suppes, P., "Probabilistic Causality in Quantum Mechanics", *Journal of Statistical Planning and Inference* **25** (1990), 293–302.

[11] Wiener, N., "The theory of prediction". In" *Modern Mathematics for the Engineer*, E. Beckenbach (ed.), McGraw-Hill, New York, 1956, pp 165–190.

Elja Arjas
Department of Applied Mathematics and Statistics
University of Oulu
Linnanmaa
90570 Oulu, Finland

Two Signal Cell Activation Explored with Simple Mathematics

by Jonathan Swinton
Imperial College, London (United Kingdom)

Abstract

This paper explores the effect of a two-signal activation requirement on immune system cells. By using simple mathematical models I compare different hypotheses about the signalling process in the context of available data. In particular I demonstrate that it is possible to explain the phenomenon of high zone tolerance in this way.

Introduction

There is a great and increasing variety of cells and chemicals known to be involved in the generation of an immune response against infection [1]. The class of cells known as T helper lymphocytes is believed to play a central role in this complex set of interactions, and an understanding of the way such cells respond to stimulation is vital, not only when attempting to build mathematical models of the T helper cell response, but also when trying to understand the immune response as a whole. A characteristic feature of T helper cell stimulation is high zone tolerance: the suppression of proliferation at high doses of specific antigen [2,3,4] (Figure 1).

A number of systems have been characterised where anergy, or unresponsiveness to further stimulation, is induced in T cells following delivery of a stimulus normally expected to induce proliferation. A second, co-stimulatory, signal is required to activate the cell and avoid anergy [5,6]. If the cell does become anergic, it not only fails to proliferate, but becomes refractory to further stimulation, no matter how well presented.

It is clear that this two-signal mechanism imposes a limit to proliferation at high doses, since when the antigen is in excess, the second, antigen-independent signal, becomes the rate-limiting step. My primary aim in this paper is to determine whether the same costimulatory requirement is also sufficient to reduce the proliferation observed at high doses. Further, I wanted to explore, and compare, additional hypotheses about the provision of signals to T cells in the context of available data. In fact, at least one of my apparently reasonable, and relatively similar, models does not conform with the data. Thus it becomes possible to draw conclusions about the induction mechanism from the modelling process. For more details about the background to this work the reader is referred to a forthcoming expansion of the material covered here [7].

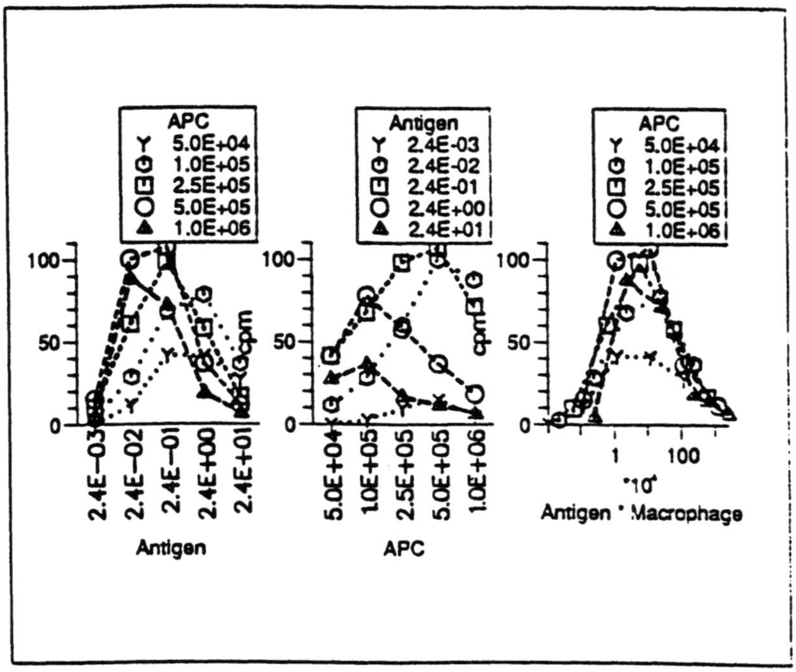

Figure 1. Proliferative responses of T cells to varying numbers of antigen presenting cell and dose of antigen redrawn from Matis et al. [14]. The left hand graph shows the data as a function of antigen dose, the second as a function of antigen presenting cell density, and the third as a function of the product of these two numbers.

Mathematical Formulation

My goal is to create simple mathematical models with discriminatory power among different assumptions. In order to do this I have deliberately ignored many possible complicating factors, including questions of density dependence, toxicity and finite proliferative capacity. The broad framework of the models is shown in Figure 3. In all the models, the first step is for a naive T cell to get a first signal from an antigen, presented by an antigen presenting (APC) or macrophage. Once it has received this first signal it moves into a second class in which it either receives the correct second signal and moves into a proliferating class, or fails to receive it and becomes nonfunctional. The second signal is always in contact with an antigen presenting cell.

The differences among the models emerge when we consider how delivery of the second signal is asserted to fail.

In the first model there is a fixed average time in which cells can receive the second signal, while in the second model their responsiveness is terminated by a repeat signal of the presented antigen. The third model follows the first in

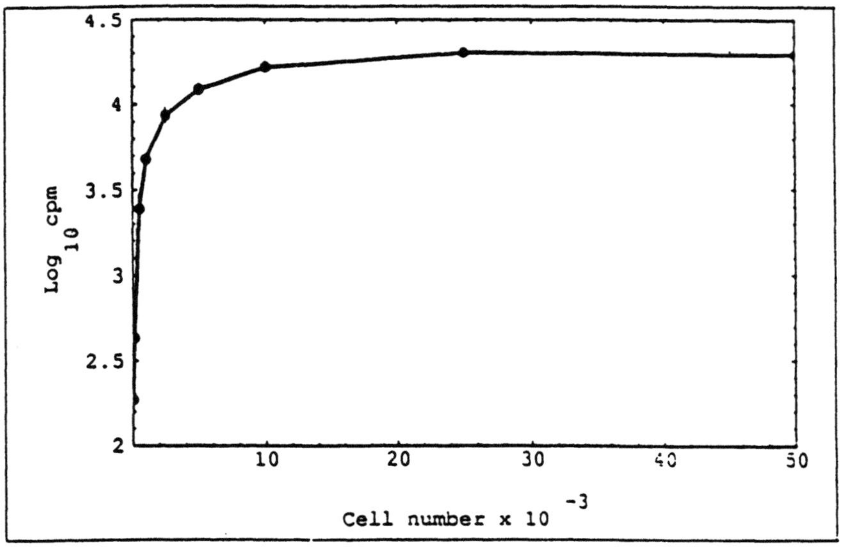

Figure 2. The proliferative response of a T cell clone as a function of macrophage density in the presence of a constant antigen dose. Redrawn from Lamb et al. [2].

assuming that cells require the two signals to move into the proliferating class, but it is when they receive the signal in the proliferating class that they switch to nonresponsiveness.

I can now give the mathematical formulation corresponding to these assumptions.

Model 1: Naive unstimulated cells T_0 are stimulated by antigen (A) and macrophages (M) at a rate σ to move temporarily into the T_r class. Once there they are either stimulated by an APC at rate μ into the proliferating class T, or decay at a rate κ into a nonfunctional class.

$$\begin{aligned}\dot{T}_0 &= -\sigma MAT_0 \\ \dot{T}_r &= \sigma MAT_0 - \mu MT_r - \kappa T_r \\ \dot{T} &= \mu MT_r + \lambda T\end{aligned}$$

I start with an initial number τ of naive cells, so that at $t = 0$, $T_0 = \tau$, $T_r = T = 0$.

Model 2: The second model is identical to the first save that the rate at which cells move out of the T_r class is now dependent on the density of presented antigen and equal to κMA.

$$\begin{aligned}\dot{T}_0 &= -\sigma MAT_0 \\ \dot{T}_r &= \sigma MAT_0 - \mu MT_r - \kappa MAT_r \\ \dot{T} &= \mu MT_r + \lambda T\end{aligned}$$

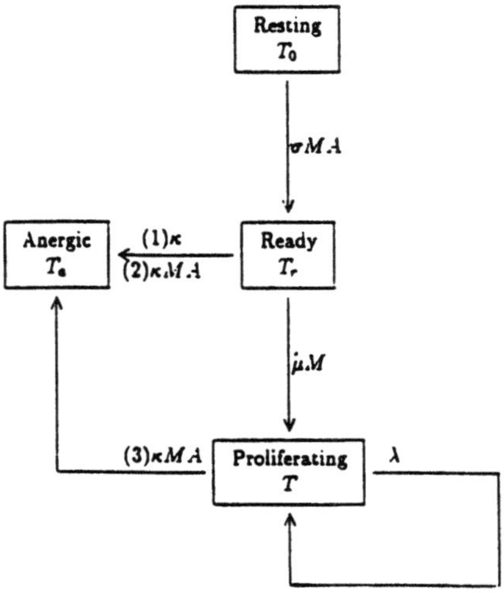

Figure 3. Diagrammatic representation of the models.

Model 3: The previous models are modified so that recontact with antigen moves the cells out of the proliferating class.

$$\dot{T}_0 = -\sigma MAT_0$$
$$\dot{T}_r = \sigma MAT_0 - \mu MT_r$$
$$\dot{T} = \mu MT_r + (\lambda - \kappa MA)T$$

Results

These models are all linear and have explicit if inelegant solutions. This means that it is simple to discover the qualitative behaviour of the models independently of any particular choice of parameter values. Nevertheless it is easiest to illustrate the results with numerical solutions at specific parameter values. All of the qualitative features discussed in this section can be shown to hold, not merely for the particular parameter values for which they are plotted, but also in general, by appropriate mathematical manipulations of the known analytic solutions [7].

Model 1 (Figure 4), perhaps the simplest, fails to reproduce the key feature of high zone tolerance. However, for Model 2 (Figure 5), the dose response curve does come down for large values of A. Moreover, the macrophage dependence is predicted to saturate. Model 3 (Figure 6) also creates the right kind of dose response curve. The macrophage curve is also suppressive at high levels.

I can now compare these qualitative results with those found in experimental data. Matis et al. observed a set of dose-response curves clearly showing high-zone

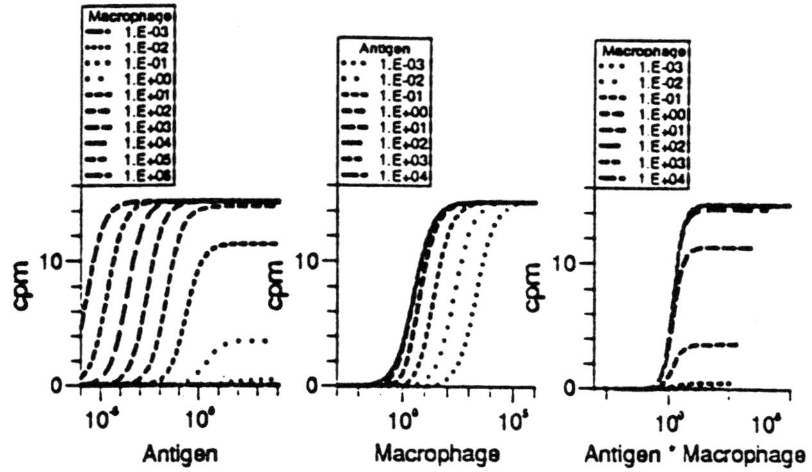

Figure 4. The dose-response curves generated by model 1. This model fails to generate high-zone tolerance. The response was taken to be the proliferation rate λ T—t—. By suitable rescalings of M, A and time we can set the rates $\mu = \kappa = \sigma = 1$. Since the system is linear, the initial condition τ is irrelevant and we can also set it to 1. This leaves us with the three free parameters λ, M and A, together with a choice for the time t at which the assay is to be performed. Of these, A and M are varied over wide logarithmic ranges in the plots, leaving us with the growth rate λ and the time of assay t. The effect of changing λ t is to change the absolute magnitude of the dose response curves, but not to the relative shapes, provided that we do not take λt too small. I choose λ =2 and t=1.

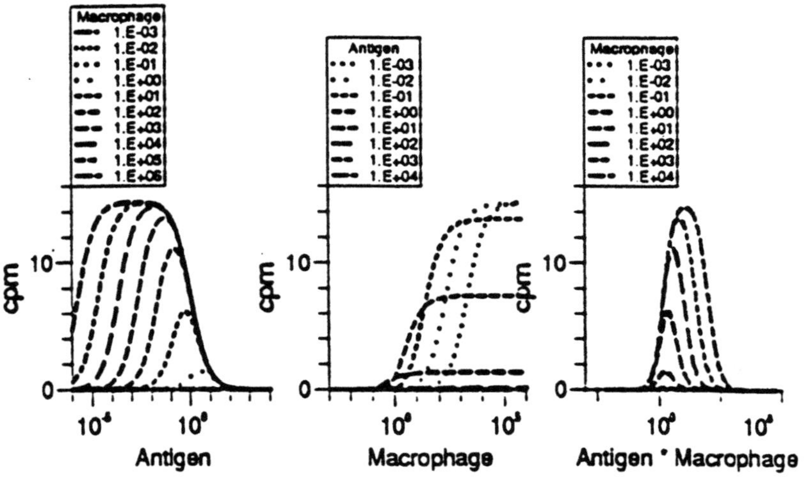

Figure 5. Dose response curves generated by Model 2. This model does generate high zone tolerance, and high levels of macrophage also suppress proliferation. Parameter values as Figure 4.

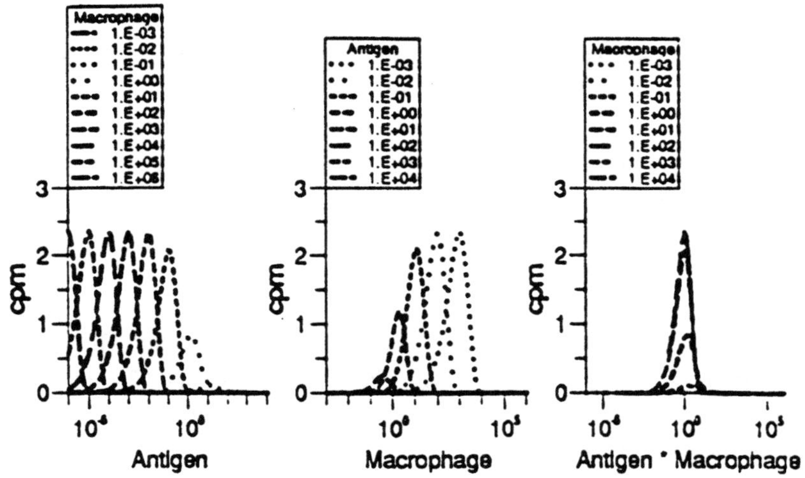

Figure 6. Dose response curves generated by Model 3. This model also generates high zone tolerance, but high levels of macrophage do not suppress proliferation. Parameter values as Figure 4.

tolerance [4] (Figure 1). These can be found in many other papers, including one by Lamb et al. [2] which also, interestingly from the current perspective, has a macrophage number-T cell response curve (Figure 2) which appears to saturate at high doses in contrast to the result of Matis et al [4]. Thus model 1 is inconsistent with the observed data, while models 2 and 3 both have data to support them. Given that the observations of the qualitative shape of the macrophage response curve can distinguish between my different models it may prove fruitful to pursue further research in this area.

Discussion

I have shown, perhaps unsurprisingly, that two-signal models can indeed provide a consistent explanation of high-zone tolerance. More interestingly, if this is indeed a valid explanation we can draw conclusions about the precise stage at which non-proliferation is induced.

My conclusion is that, under the conditions considered here, it is only when a cell has been successfully stimulated and is proliferating that it is vulnerable to proliferative inhibition by restimulation. This accords with earlier speculation [3] and with recent work on possible biochemical pathways for the inhibition of proliferation [8]. Models in which anergy is only induced at an early stage by the lack of a second signal are incapable of producing the kind of dose-response curves so commonly seen.

While I have explicitly concentrated on events early on in immune responses, it will be possible to use this framework to explore the more general anergic effects found upon later restimulation of the cells. It would of course also be possible to include factors such as density-dependent growth into the models, but the added complications and the loss of explicit solutions seem unlikely to be repaid with increased insight. A further refinement would be to the very structure of the model. It may be that my division of the cell population into unactivated and activated cells is artificial, and a more appropriate model would be to represent the mean activation level of the cells in a population (Charlotte Hetzel, Imperial College, unpublished data).

Acknowledgements. I thank the Wellcome Trust for a fellowship. This work has been informed by long discussions with Nicky Schweitzer and Charlotte Hetzel, and by the enthusiasm of Roy Anderson. Figure 3 was drawn with the macro package *diagram* of Paul Taylor.

References

[1] A.K. Abbas, A.H. Lichtman, and J.S. Pober, *Cellular and Molecular Immunology*. Saunders, Philadelphia, 1991.

[2] J. Lamb, B. Skidmore, N. Green, J. Chiller, and M. Feldmann, "Induction of tolerance in influenza virus-immune T lymphocyte clones with synthetic peptides of influenza hemagglutinin", *Journal of Experimental Medicine* **157** (1983),1434–1447 .

[3] G. Suzuki, Y. Kawase, I. Yahara, Y. Kobayashi, and R.H. Schwartz, "Antigen-induced suppression of the proliferative response of T cell", *The Journal of Immunology* **140** (1988),1359–1365.

[4] L.A. Matis, L.H. Glimcher, W.E. Paul, and R.H. Schwarz, "Magnitude of response of histocompatibility-restricted T-cell clones is a function of the product of the concentrations of antigen and Ia molecules", *Proceedings of the National Academy of Sciences U.S.A* **80** (1983), 6019–6023.

[5] M.K. Jenkins, "The role of cell division in the induction of clonal anergy", *Immunology Today* **13** (1992), 69–73.

[6] D.L. Mueller, M.K. Jenkins, and R.H. Schwartz, "Clonal expansion versus functional clonal inactivation: a costimulatory signalling pathway determines the outcome of T cell antigen receptor occupancy", *Annual Reviews of Immunology* **7** (1989), 445–480.

[7] J. Swinton and A.N. Schweitzer, "Two signal cell activation as an explanation of high zone tolerance: a mathematical exploration of the nature of the second signal", preprint, Imperial College, 1992.
[8] H.L. Cooper, R.Fuldner, E.McDuffie, and R.Braverman, "T cell receptor activation induces rapid phosphorylation of prosolin, which mediates down-regulation of DNA synthesis in proliferating peripheral lymphocytes", *The Journal of Immunology* **146** (1991), 3689–3696.

Jonathan Swinton
Wellcome Research Centre for Parasitic Infections
Department of Biology
Imperial College
London SW7 2BB, United Kingdom

A Theory of the Functional Organization of Biological Systems Applied to Neural Networks

by Gilbert Chauvet
Université d'Angers (France) and University of Pittsburgh (USA)

Abstract

A theory of the functional organization of biological systems, based on the concept of *functional interaction*, is proposed with its two aspects: the *topology* of the system represented by a *hierarchical* graph, and the dynamics of the process, including the *geometry* of the system. The variational study of the topology leads to an *optimum principle* for a class of biological systems. A *non-symmetric* and *non-local n-level field theory* allows us to study the dynamics of the processes. Results are given for the nervous system, and a comparison between biological systems and physical systems is offered. Specific mathematical problems are suggested from this theory.

Introduction

Our aim in this short paper is to discuss the nature of a possible specific formalized biological theory that can provide some criterion of evolution in time of a biological system. Specifically, the *functional organization* of the biological system is considered to be its major expression, supported of course by its *structure*, i.e., the corresponding *physical* system. It will be shown that such a description leads to various mathematical problems.

General biological organization has been the subject of much research from various points of view, based on well-established mathematical or physical theories (Thom, 1972; Nicolis and Prigogine, 1977; Prigogine, 1972). Some other formalisms, such as transformation systems (Delattre, 1971), compartmental analysis (Conrad, 1972; Walter, 1980), general and hierarchical systems (Arbib, 1972; Pattee, 1970), automaton theory (Kaufman, 1985), graph theory (Rashevsky, 1961; Rosen, 1958; Levins, 1970), neural networks (Von Foerster, 1967), information theory (Atlan, 1972), and statistical mechanics (Kerner, 1972; Demetrius, 1982).

However, an integrated description of physiological phenomena from the cell level to the organ level must include a unique formalized definition of the biological subsystem within a mathematical framework. To reach for this aim, several major difficulties arise: (i) the passage from one level of organization to another in a system that includes many levels; (ii) the use of different formalisms for the elements and the population of these elements (as in the physical description of the kinetics of gases); (iii) the formalization of hierarchical systems whose functional organization does not necessarily coincide with the corresponding structural organization. Therefore, specific biological concepts have to be introduced, which lead to new mathematical objects.

The approach proposed is very similar to the methods used in theoretical physics, and could be an introduction to an integrative theoretical physiology. Here we shall make a concise presentation of the concepts developed elsewhere (Chauvet, 1990; 1993, I, II, III), and discuss some related mathematical questions.

The Problem of the Representation of Biological Systems

1. Three examples: operon, nervous tissue, respiratory system

The integrated regulation of a living system such as a mammal, involves four physiological functions (energetic or digestive, renal, cardio- vascular, and respiratory) and two regulation systems (hormonal and nervous). The result of this integration can be illustrated by two large equilibria: the acid-base and water-ions balances. Because of the intricacy of the molecular level where biosynthesis of proteins is realized, with the highest levels, e.g., the mechanics of lungs, a good interpretation of the regulation properties, basic for the functioning of the living system, is extremely difficult and leads to major theoretical problems. Let us consider three examples, from the "lowest" to the "highest" structural level.

The first example concerns the dynamics of the macromolecules. The *operon*, thoroughly described in the case of bacteria (Jacob and Monod, 1961), includes three kind of genes, regulator, operator and structural genes, the specific role of which is the regulation of the biosynthesis of proteins (Fig.1a). The aporepressor aRe is emitted by the mRNA (messenger) that is transcripted from the regulator gene. Following the concentration of the product P, aRe is activated, or not, into a repressor Re. In the active state, Re acts on an operator gene GO which then acts on the structural genes GS. After transcription, these GS provide the mRNA which, by combination with ribosomes R, give the polysomes Y. Then, the Y combine with tRNA (transfer RNA) already linked with aminoacids (T-AA) according to a given sequence of codons to provide enzymes E_i. The enzymes enter into a sequence of chemical reactions that transform a substrate S_{i-1} into S_i. The operon could be viewed as a single structural unit whose function is the controlled biosynthesis of the protein P, all the actions described above being functional interactions. Together they constitute a hierarchical system (Fig.1b) with Re, X, Y, E as functional interactions, and G (operon), R, T-AA, and S, as structural units.

The second example is the nervous system. We will consider specifically a monosynaptic pathway, i.e., axons of neurons that are "connected" with other neurons by means of synapses. Synapses are anatomical formations between the first neuron at the level of dendrites and the second neuron at the level of the dendrites and the soma. A potential travelling along the axon after an injection

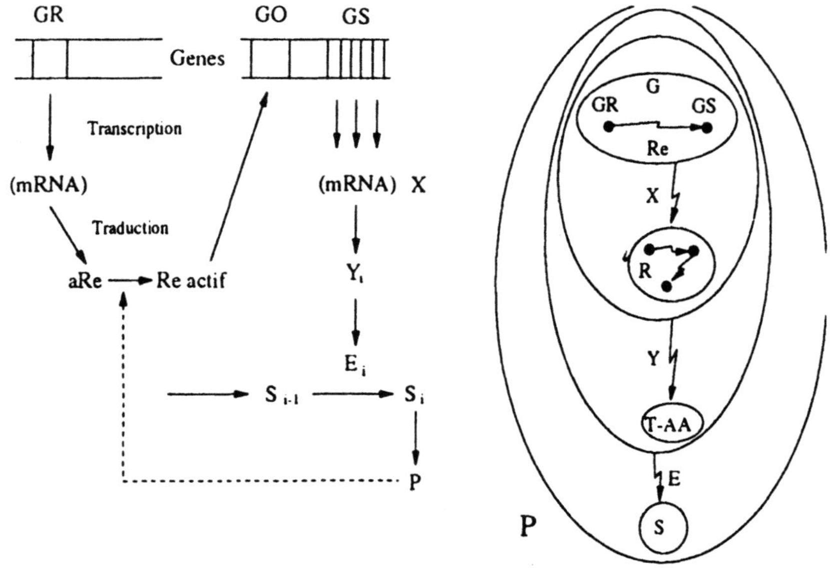

Figure 1. *Left:* Regulation of the biosynthesis of protein P by the operon constituted by genes GR, GO, GS. *Right:* The same system represented as a hierarchical system. See text for an explanation of the symbols.

of current (on the cell body or on the axon) terminates at a synapse, causes the release of a transmitter that binds to a receptor located on the second neuron, and modifies the local potential. If the sum of all these postsynaptic potentials is larger than a threshold value, then the cell fires, i.e., emits an action potential that propagates along the axon, and the same process starts again. Because of the enormous number of synapses per neuron, and the very large number of neurons, the neurons of the nervous tissue become active, i.e., fire, if some conditions are fulfilled, particularly the ability of synapses to transmit the "message". A common experiment in neurobiology consists in recording the electric potential in the extracellular space of the postsynaptic neurons (called the "extracellular field potential") after stimulating presynaptic neurons at several increasing values of intensity (Fig.2a). The waveform obtained (Fig.2b) is the result of two phenomena: (i) the firing of the neuron implying the negativity of the curve, (ii) the activation of the large number of synapses. A quantitative interpretation of such a curve has not yet been given. With the theory proposed here, we have been able to give a quantitative interpretation of the monosynaptically evoked extracellular field potential (Chauvet and Berger, 1990).

The third example concerns the respiratory system, one of the most complicated systems in the organism. The task of the respiratory system is to maintain

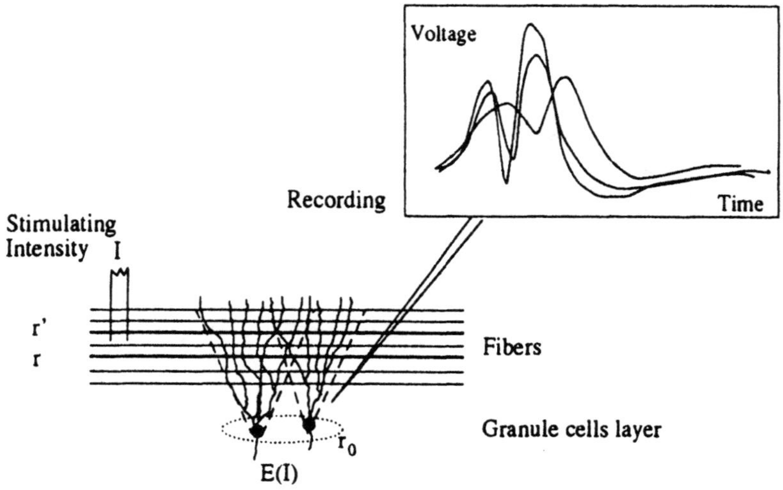

Figure 2. *Bottom*: Fibers are connected with neurons of the granule cell layer by means of synapses. For a stimulating intensity I corresponding to a cylinder of fibers (limited by bold lines), a domain E(I) is activated. *Top*: The electric potential recorded in the extracellular space is represented in the frame.

the constant level of the concentration of H_+ and CO_2. To reach this objective, it integrates almost all the other systems: the nervous, metabolic, cardiovascular, and renal systems (Fig.3). Three kinds of factor are involved in the regulation: (i) the chemical control of extracellular space, (ii) the neurochemical control, and (iii) the neuromechanical control from mechano-receptors located in the lungs, the bronchi and the muscles.

2. Functional interactions, formal biological systems and physiological functions

From the simplest to the most complicated of these examples, we can try to extract the common, abstract principles of construction. If a physical structure can be viewed as a combination of *forces*, a natural question arises from the study of all the subsystems of an organism (Chauvet, 1987), one which concerns the nature of the physiological system: what is the biological equivalent of the concept of force in physics, the combinatoric of which leads to the dynamics of the system? What we have called "*functional interaction*" is the ability of one structural unit to act on another at a distance, with a finite velocity. A functional interaction is defined as the action of a product emitted by one structural unit on another: The emitted product P_2 synthesized by the source u_1 is transferred towards a sink u_2. This process is described by the relation: $P_2 = \Psi_{12}(P_1; r)$. For example, a molecule emitted by a specific cell acts on another cell within which it induces a series of

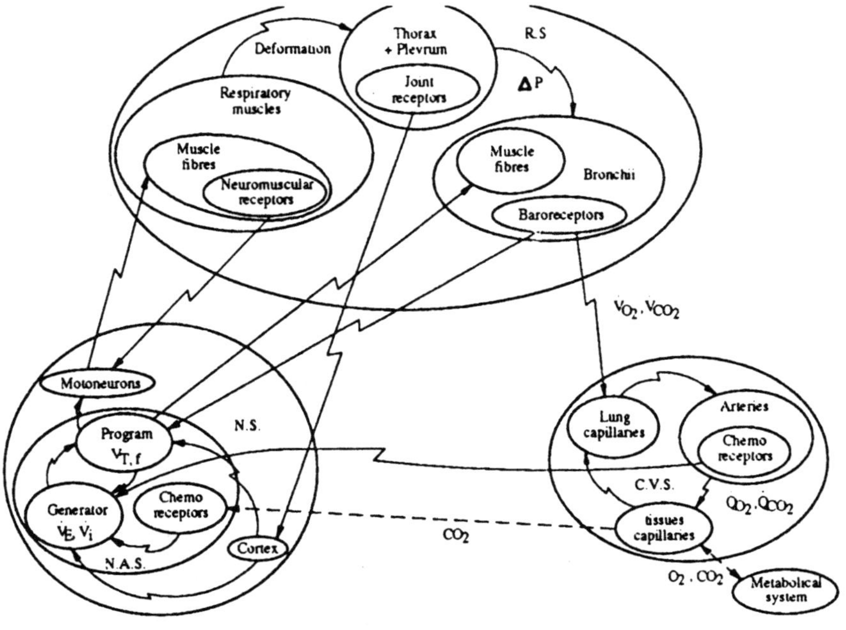

Figure 3. Representation of the respiratory system as a hierarchical system. Each "balloon" is a structural unit. Arrows represent the functional interactions, i.e., actions of one structural unit on another.

transformations.

The functional interaction has two important properties, the *non-symmetry* and the *non-locality*, which result in the remarkable and specific properties of a biological system, as shown below. The combinatoric of the non-symmetric and non-local functional interactions results in specific laws of evolution of the physiological systems, different from those deduced from the combination of symmetric and (generally) local forces in physics, which are the bases of thermodynamics, and classical and quantum mechanics. A comparison between biological and physical systems is therefore possible.

The concept of the Formal Biological System (FBS) was introduced to provide a mathematically defined biological system with specific properties as similar as possible to those of a real biological system. An FBS is characterized by two such properties: (i) it is an "equipotent system", i.e. all the potentialities of the genome are identical at each level of organization; and (ii) it is a "mutational system" in which the structural units undergo independent, random modifications.

Definition 1: *A structural unit is a structural equivalence class constituted by units that are identical with regards to their structure, and independent with regard to their function (for certain criteria). An elementary physiological function is* *the collective behavior (cooperation in some tasks) of at least one functional interaction. A physiological function (a biological system) is the collective product of a set of structural units which can be hierarchically classified according to their elementary interactions.*

The construction of the set of functional interactions, constituting the *topology* of the biological system, is based on a very simple — though non-trivial — hypothesis, called the *self-association hypothesis*, i.e., an interaction between two structural units becomes necessary when the elementary function of one of them is destroyed, i.e., when a source becomes a sink. The topological system (O-FBS) composed of such functional interactions represents the functional organization of the FBS. The (D-FBS) is the dynamical system constructed according to what we have called the principle of vital coherence. After a given perturbation applied to a node of the graph, i.e., a structural unit, the (O-FBS) indicates whether a perturbation exists at another node, and the (D-FBS) gives the intensity of this perturbation.

Study of the (O-FBS) System: Hierarchical Graph

The functional organization of the biological system at level l is defined by the distribution $\left(n_\alpha^{(l)}\right)_{\alpha=1,\mu^{(l)}}$ of functional links between structural units at this level. This distribution is called a state of organization. The (O-FBS) consists in functional interactions, i.e., structural units and elementary functions, the topology of which is the functional organization. $n_\alpha^{(l)}$ is the number of sinks for the function P_α of the system.

Let (S) be a set of ν structural units all having the same potentialities. According to the self-association hypothesis, each unit is either a source or a sink. If a source could not become a sink it would die following an alteration in the function. Suppose that n of the ν units do not yield the product necessary for their survival. These n units will therefore have to be coupled to the other $(\nu - n)$ units. A number of possibilities exists for the associations that create the topology of an (O-FBS). We propose a function $\prod(n)$, called the "*potential of functional organization*", representing the reservoir of possibilities during the lifetime of the system. This mathematical function is determined if the following properties are verified: (i) $\prod$ gives a measure of the number of potential functional interactions; (ii) $\prod$ leads, under some conditions, to a hierarchy of the system; (iii) the hierarchy found is such that the value of $\prod^l$, calculated for a level l, decreases from this level l to the next higher $(l + 1$; (iv) the value of $\prod$, calculated for the total hierarchical system, increases, whereas a reorganization of the system, as defined below, implies a decrease in this value.

Definition 2: Potential of functional organization. *The potential $\prod^l$ at level l is the logarithm of the number o^l of all functional organizations that are available* *(and possible) for the observed state of organization, at its ith-level. It is assumed that all structural units have identical potentialities.*

A simplified form would be

$$\prod - \sum_{\text{levels}} \sum_{\text{products}} [\text{sinks}]\, ln\, [\text{sources}]. \tag{1}$$

We give now the most important properties, without proof, deduced from the definition of $\prod(x)$.

Property 1: *The state of maximal organization (called the organizational state) for the maximal potential $y_M = \prod(x_m)$ is an attractor for the dynamics of the organization which tends to x_M, either by decreasing or by increasing values of x when the time t tends towards t_M.*

This basic property allows us to give a mathematical definition of a FBS, which can be experimentally verified, and the following extremum hypothesis defines a class of formal biological systems:

Extremum hypothesis: *A biological system evolves from initial conditions such that the number of sinks either decreases or increases, i.e., corresponds to a monotonous time function, and therefore reaches a state where the potential of organization is maximum.*

The validity of the proposed theory, which is finally based on this extremum hypothesis can be experimentally tested. The definition of functional interactions makes it possible to identify sources and sinks, and to verify the decreasing monotonous property of the number of sinks, *even if the explicit dynamics of the system is unknown.* Another possibility would be to check the validity of the consequences of the theory developed below. The evolution of the system is governed by a function called "*orgatropy*", deduced from the function $\prod$:

Property 2: The concept of "orgatropy". *If the degree of organization ν^l changes with a variation in the number of structural units in the level l without re-organization, then the system with maximum potential satisfies the criterion of evolution for the function F called orgatropy:*

$$ddF^l(\nu^l) \geq 0 \tag{2}$$

and defined by

$$F : \nu \rightarrow x_M \rightarrow y_M = F(\nu) = \prod_{\max} o\, h(v)\ .$$

By defining "*specialization*" as follows:

Definition 3: Specialization. *Given, at initial time $t^{(0)}$, $n_1^{(0)}$ sinks for the product P_1, and $n_2^{(0)}$ sinks for the product P_2, such that $\nu^{(0)} > n_1^{(0)}+n_2^{(0)}$, the transformation at this l-level, called specialization of the l-level at final time $t^{(f)}$, corresponds to a partition into two subsets of $n_1^{(f)}$ and $n_2^{(f)}$ units such that*

$$\nu^{(f)} = n_1^{(f)} + n_2^{(f)} \tag{3}$$

It can be shown that a function $\Omega(t)$, called the "*functional order*" is:

Definition 4: Functional order. *The functional order of a biological hierarchical system is defined by the state function:*

$$\Omega(t) = -\sum_{l=1}^{L} F^l(\nu^l)$$

describes the variation in time of the hierarchical system.

Theorem: On the time evolution of an (O-FBS). *Let a biological system be defined by the extremum hypothesis:*

$$\begin{aligned} \prod (n^l_M)_{l=l,L}\ maximal \Leftrightarrow \forall\alpha = 1, \mu^l \quad & (dn^l_\alpha(t) < 0\ \ if\ \ n^l_\alpha > n^l_M \\ or \quad & (dn^l_\alpha(t) = 0\ \ if\ \ n^l_\alpha = n_M) \\ or \quad & (dn^l_\alpha(t) > 0\ \ if\ \ n^l_\alpha < n_M) \end{aligned}$$

for all $l = 1, L$. If its degree of organization, ν^l, is modified such that the criterion of specialization is satisfied for products P_α and P_β:

$$dN_\alpha^\beta(t) = 0$$

then the self-organization of the system is such that:

(1) At most two structural equivalence classes are created, each being specialized in the production of a given product, according to a bi-unitary process (by the association of two units):

$$\delta\nu^l < 0\,.$$

(2) Its functional order increases and corresponds to the emergence of one level of organization:

$$\delta\Omega \geq 0$$

and simultaneously its orgatropy decreases:

$$\delta F \leq 0\,.$$

(3) The potential of organization $\prod$ decreases:

$$ {}^{(0)}\prod^{l} \rightarrow \prod^{l+1} = {}^{(f)}\prod^{l} . $$

(4) At the limit, in the state of maximum specialization, $\prod$ is minimum and its value is:

$$ \prod^{l+p} = 2\mu^{l} $$

with $p = \mu/2$ if μ^{l} is even, and:

$$ \prod^{l+p} = 2(\mu^{l} - 1) + \prod_{(t)}^{l} $$

with $p = (\mu^{l} - 1)/2$ if μ^{l} is odd, where μ^{l} is the number of products in the l-level before the complete specialization of the units, and p is the step of the last product synthesized.

Study of the (D-FBS) System: n-level Field Theory

The physiological processes expressed by functional interactions related to the geometry of the structure can be conceived as *the transport of a field variable submitted to the action of a field operator*. Let $\Psi(r, t)$ be the field variable defined in the r-space, and let H be the field operator which depends on Ψ and on successive derivatives $\Psi^{(n)}$ with respect to time and space coordinates. The field equation can be written in a general form as follows:

$$ H(\Psi, \Psi^{(n)}, n = 1, 2, \ldots)\Psi(r, t) = \Gamma(r, t) \tag{4} $$

where Γ is the source term. In this equation, H describes the propagation of the field variable Ψ from r to r_0 , the local transformation in r_0 is represented by $\Gamma(r_0, t_0)$. Because the operator acts from one point of space to another, it must take into account the distance between these two points, and therefore must include an *interaction operator*. More generally, the influence of the location of points, i.e., the role of geometry, on the dynamical processes, can be studied with a field theory. The dynamical processes that express the behavior of the related functional interactions continuously occur in space and time with a finite velocity. Therefore, what is observed at point (r_0, t_0) results from what was emitted at point (r, t) where $t_0 = t + (\|r - r_0\|)/\nu_\Psi$ and ν_Ψ is the velocity of the interaction.

The finiteness of the velocity of functional interactions with a different value at each level determines the non-locality, because of the delay $d^{r}(r)/\nu_\Psi$ for the transport of functional interaction Ψ. This fact leads to dynamics with different characteristics at each level:

Property 3: Existence of non-locality. *The following invariance of the functional interaction Ψ in a hierarchical system,*

$$\Psi : V \nearrow\searrow V_0 \Leftrightarrow \Psi : r \nearrow\searrow r_0 \tag{5}$$

expressed in the physical space $\mathbb{R}^3$ and in the space of units D_r, implies non-locality.

All the specific concepts introduced above, including the property of non-symmetry of functional interactions, can be summarized as follows:

Property 4: Structurality of a biological system. *A biological system is characterized by the finiteness of the velocity of interactions and therefore by the dynamics that occur on a specific time scale, and which constitute the collective behavior of sources and sinks. The distinct values of time scales allow the construction of the hierarchy of the system. In such a hierarchical system, the interaction operator is non-local and non-symmetric. There is no relation between the ranks of levels of organization and time scales.*

The large number of elements in one level relatively to the number of elements in the higher one leads to the use of a different formalism, described by the following property:

Property 5: Statistical equation of the states of the fields. *The action of the fields on the population of structural units is described by a statistical distribution function of the states $f^1(\Psi^0\Psi^1, \ldots, \Psi^l)$ which is a solution of the equation:*

$$\frac{\partial f^1}{\partial t} = Q(t) f^1 + \sum_{i=0}^{l} \left(\frac{\partial f^1}{\partial t}\right)_{\Delta\Psi^i} \tag{6}$$

where $Q(t)$ is a rate constant, positive for an "attractive" population, negative for a "repulsive" population. Specific parameters are included in the terms $\Delta f^l_{\Psi^i} = (\partial f^l/\partial t)_{\Delta\Psi^i}\Delta t$ that describe the state transitions.

Application to Biological Neural Networks

1. Potential of organization for a monosynaptic pathway: Functional organization of a neural network.

Given n as the number of neuron-targets, $N = \nu - n$ the number of neuron-sources, because a sink cannot be a source for the same product, and $g(k)$ the number of synapses for the k-th target-neuron, $k = 1, n$, Eq. (1) can be generalized:

$$\prod = \sum_{k=1}^{n} \ln(C_N^1 + C_N^2 + \ldots + C_N^{g_k}) \,. \tag{7}$$

It is possible to prove that a monosynaptic pathway in the nervous tissue belongs to the class of the biological systems that satisfy the extremum hypothesis. In other words, according to Property 1 these systems lead to monotonic dynamics of the number of sinks. On the other hand, it is known (see, for example, Hopkins and Brown, 1984) that two physiological processes exist during neural development: (i) a proliferation of neurons per cell division, and (ii) a reorganization of synaptic connections such that the number of synapses decreases. Because these two facts are interpreted by the present results, we can conclude that the potential of functional organization remains maximum during development.

2. Dynamics of the biological neural network: Neural field equations.

Two field variables define the dynamics of the neural network, which can be: (i) At the synaptic level in the time scale $\{t\}$, the *synaptic efficacy* μ, which takes into account local and non-local effects, and depends on quantities for the presynaptic neuron in r' and for the postsynaptic neuron in r_0; (ii) at the level of the network, in the time scale $\{T\}$, the instantaneous local somatic activity $X(r_0, T_0)$ or the *local membrane potential* $\Psi(r_0, T_0)$; at the synaptic level, another field variable could be: $\Phi_0(s, t)$, the postsynaptic local membrane potential. Therefore, the passage from one level of organization (synapses) to the other (neurons) with Ψ and μ as local somatic depolarization and synaptic efficacy respectively, implies the measurement of the Ψ- and μ-fields in two different spaces whose points are denoted by (r, T) and (s, t), respectively. The determination of the non-local operator leads to the following μ-field equation and Ψ-field equation:

$$\begin{aligned}\frac{\partial \mu(s,t)}{\partial t} &= \nabla_s(D^s \nabla_s \mu(s,t)) \\ &+ \int_{D_R(r_0)} \rho(r^l) \int_{D_s(r',r_0)} \mu_0(s',t')\pi(s',r';r_0)A(s,s')ds'dr' \\ &+ \Gamma_\mu(s,t)\end{aligned} \qquad (8-1)$$

$$\begin{aligned}\frac{\partial \Psi(r_0,T_0)}{\partial T} &= \nabla_r(D^r \nabla_r \Psi(r_0,T_0)) \\ &+ \int_{D_R(r_0)} \rho(r)\Psi(r,T) \int_{D_s(r,r_0)} \mu(s,t)\pi(s,r;r_0)B(r_0,T_0,r,T)dsdr \\ &+ \Gamma_\Psi(r_0,T_0) \\ s \equiv s(r_0)\ ,&\quad D_R(r_0) = \bigcup_{r'} D_s(r',r_0)\end{aligned}$$

$$(8-2)$$

$D_R(r_0)$ is the reunion of subspaces $D_s(r, r_0)$, i.e., the space of synapses in neurons that are localized in r_0 and correspond to neurons in r. A and B are attenuation functions of potentials along the membrane between two points. Equations (8) constitute the neural 2-level field equations for the activity of neural tissue. The neural tissue is characterized by two geometrical functions: the density of neurons ρ and the density-connectivity of synapses π.

Finally the *extracellular potential* for all elements dr in the space of neurons is created when a stimulating intensity I is applied. Let $E(I)$ be the space of stimulated neurons that includes a certain number of activated neurons. The extracellular field potential is the sum of two contributions due to the firing and the synaptic activation:

$$V_E(t; I) = V_E^{fir}(t; I) + V_E^{syn}(t; I) \tag{9}$$

where

$$\begin{aligned} V_E^{fir}(t; I) &= \int_{E(I)} Q_{i\to e}(r) I \; field \; (r, t)\rho_0(r)dr \\ V_E^{syn}(t; I) &= \int_{E(I)} Q_{syn\to e}(r) V^{syn}(r, t)\rho_0(r)dr \end{aligned} \tag{10}$$

where ρ_0 is the density of postsynaptic neurons. Coefficients $Q_{syn\to e}$ and $Q_{i\to e}(r)$ describe the interactions between the neuron and the extracellular space (Costalat et al., 1991). With this method, it has been possible to give a physiological interpretation of the waveform recorded in a monosynaptic pathway, specifically, the dentate gyrus in hippocampus (Chauvet and Berger, 1990, 1992).

Conclusion: Biological and Physical Systems

If we accept that the general functioning of a biological system may be conceived as the dynamics of a set of hierarchically organized functional interactions, and described by activator-inhibitor fields at each level of organization with their own time scale, then the present theory suggests that the transformations, for instance from amoeba to mammals, correspond to an "over-organization", i.e., an increase of functional order. Within this approach, this *principle of functional order from hierarchy* describes the natural trend towards the time-decoupling of the physiological functions, and, in that sense, towards a simplification of the dynamics, i.e., an order of the system, expressed in terms of the hierarchical functional organization.

In summary, a living system is characterized by two complementary sets of mathematical laws: the first set governs the physical structure with an optimum principle concerning thermodynamical entropy (Prigogine's criterion of evolution: Prigogine, 1947; Glansdorff and Prigogine, 1971), including living organisms (Prigogine et coll., 1972); whereas *the second set governs the organization of physiological functions with an optimum principle concerning the potential of organization. A law concerning functional order results from this second optimum principle.*

The proposed theory, merely outlined here, is based on new concepts such as: functional interaction, potential of functional organization, orgatropy, functional order and multiple fields. These require the study of hierarchical graphs and non-local fields from a mathematical point of view, and the mathematical development of these objects independently of the biological context. For example, the demonstration of the theorem regarding the evolution in time of the (O-FBS) has been proved for real numbers, this property being not satisfied for the very low values of integers. A mathematical theory of hierarchical systems with the concept of emergence of a level of organization is needed.

References

[1] Atlan, H., *L'organisation biologique et la théorie de l'information*, Hermann, Paris, 1972.

[2] Arbib, M.A., Organizational principles for theoretical neurophysiology. In: *Towards a Theoretical Biology*, vol. 4: essays, C.H. Waddington (ed.), Edinburgh, University Press, 1972.

[3] Chauvet G.A., Hierarchical functional organization of a formal biological system: A dynamical approach. I. The increase of complexity by self-association increases the domain of stability of a biological system, *Phil. Trans. R. Soc. Lond. B.* **339** (1993), 425–444.

[4] Chauvet G.A., Hierarchical functional organization of a formal biological system: A dynamical approach. II . The concept of non-symmetry leads to a criterium of evolution deduced from an optimum principle of the (O-FBS) sub-system, *Phil. Trans. R. Soc. Lond. B* **339** (1993), 445–461.

[5] Chauvet G.A., Hierarchical functional organization of a formal biological system: A dynamical approach. III. The concept of non-locality leads to a field theory describing the dynamics at each level of organization of the (D-FBS) sub-system, *Phil. Trans. R. Soc. Lond. B* **339** (1993), 463–481.

[6] Chauvet, G.A., *Traité de Physiologie Théorique, tome III: Physiologie intégrative, Champ et Organisation fonctionnelle.* Masson, Paris, 1990; to appear in English translation by Pergamon Press.

[7] Chauvet, G.A., Berger, T.W., Two-level field theory interpretation of hippocampal extracellular field potentials, *Soc. Neurosci. Abstr.* **16**:1 (1990), 739.

[8] Chauvet, G.A., Non-locality in biology results from hierarchy. Application to nervous system, *J. Mathematical Biol.*, 1992, to appear.

[9] Conrad, M., Statistical and hierarchical aspects of biological organization. In: *Towards a theoretical Biology, 4 essays.* University Press, Edinburgh, 1972

[10] Costalat R., Genet S., Thiels E., Berger T.W., Chauvet G., Application of a model of the interaction between two neurons to extracellular field potentials of populations of hippocampal pyramidal cells, *Soc. Neurosci. Abstr.*, 1991.

[11] Delattre, P., *Système, structure, fonction, évolution.* Coll. Rech. Interdisc,

Maloine, Paris, 1971.

[12] Demetrius, L., Self-organization in macromolecular systems: The notion of adaptive value. *Proc. Natl. Acad. Sci. USA*, **81** (1984), 6068–6072.

[13] Von Foerster, H., Computation in neural nets. *Curr. Mod. Biol.*, **1** (1967), 47–93.

[14] Glansdorff, P. & Prigogine, I., *Thermodynamic Theory of Structure, Stability and Fluctuations*, Wiley, New York, 1971.

[15] Hopkins, W.G. & Brown, M.C., *Development of Nerve Cells and their Connections*, Cambridge University Press, Cambridge, 1984.

[16] Jacob, F., Monod, J., Genetic regulation mechanisms in the synthesis of proteins, *J. Mol. Biol.* **3** (1961), 318–356.

[17] Kaufman, S.A., Boolean systems, adaptive automata, evolution. In: *Disordered systems and biological organization*, E. Bienenstock, F. Fogelman-Soulié, G. Weisbuch (eds.), F20 NATO Series, 1985.

[18] Kerner, E.H., *Gibbs Ensemble: Biological Ensemble,* Gordon and Breach, New York, 1972.

[19] Levins, R., Complex systems. In *Towards a theoretical Biology*, 3. drafts, University Press, Edinburgh, 1970.

[20] Nicolis, G. and Prigogine, I., *Self-organization in non-equilibrium systems*, J. Wiley and Sons, New York, 1977.

[21] Pattee, H.H., The problem of biological hierarchy. In: *Towards a Theoretical Biology*, 3. Drafts, University Press, Edinburgh, 1970.

[22] Prigogine, I., *Etude thermodynamique des phénomènes irréversibles*, Desoer, Liège, 1947.

[23] Prigogine, I., Nicolis, G. and Babloyantz, A., Thermodynamics of evolution, *Physics Today* **25**: 11 (1972), 23–28.

[24] Prigogine, I., Thermodynamics of Evolution. *Physics today*, Nov. & Dec1972.

[25] Rashevsky, N., *Mathematical principles in biology and their applications*, C.C.Thomas, Springfield, 1961.

[26] Rosen, R., A relational theory of biological system. *Bull. Math. Biophys.* **20** (1958), 245.

[27] Thom, R., *Stabilité structurelle et morphogenèse*, Edisciences, Paris, 1972.

[28] Walter, G.G., Stability and structure of compartmental models of ecosystems, *Math. Biosci.* **51** (1980), 1–10.

Gilbert Chauvet
Institut de Biologie Théorique
Université d'Angers
10, rue A. Boquel
49100 Angers, France

Dept. of Behavioral Neuroscience
University of Pittsburgh
Pittsburgh, PA
USA

Stochastic Vistas to the Analysis of Physical Mapping of the Genome

by Petre Tautu
German Cancer Research Center, Heidelberg (Germany)

Abstract

The physical mapping of the genome is considered as a one-dimensional Boolean model with stationary semi-Markovian random closed sets in the sense of G. Matheron.

1. Introduction

It is now well-known that the genetic information contained in the cells of all living beings actually appears in the chromosomes as long sequences of desoxyribonucleotidic acid (DNA), consisting of 4 nucleotides (A for adenin, C for cytosin, G for guanin and T for thymin). Part of them called the genes, code for the proteins, which fulfill all the functions necessary for life. Molecular genetics deals with two problems: finding the sequence of the DNA (some 10'000 "letters" for a virus, 5 million for a bacterium, 3 billion for man) and establishing a relation between the genes and the function of the related proteins, in particular in terms of diseases occurrences.

Hence, the main aim of contemporary molecular genetics is the global analysis of complex genomes, with the remote purpose of isolating disease genes and understanding the genome function. The first step is the realization of a genomic map, that is, the molecular representation of the relationships among landmarks, organized according to a defined coordinate system [33]. Two basic types of maps are then considered: (i) a genetic map, based primarily on linkage data, and (ii) a physical map (cytogenetic or molecular), based on the study of cloned segments of DNA. The physical molecular map directly characterizes large pieces of DNA by establishing molecular landmarks (e.g., restriction endonuclease sites, sequence tagged sites) with the aid of techniques that size and order larger fragments of DNA.

Several strategies have been described, most of them utilizing a large number of randomly cloned DNA segments; these segments are in the form of shorter phage vectors, cosmids, or yeast artificial chromosomes, sufficiently to create a continuous array of overlapping assemblages of clones, referred to as contigs. The ordering of clones may be carried out by (i) restriction mapping, (ii) fingerprinting, and (iii) anchoring. Ultimately one obtains "contig maps" which should represent the physical structure of contiguous regions of the genome by specifying the overlap relationships among a set of clones. The generation and most uses of these maps depend on detailed analysis of individual clones [26]. The natural language of physical mapping is sequence information.

2. Probabilistic Postulates

The stochastic approach to the physical mapping of the genome relies upon the classical problem of coverage, e.g., the coverage of a line with intervals of random length. Indeed, the technical objective is to cover the entire genome by mapping a large number of cloned DNA fragments, each of which constitute a small fraction of the genome. In readily understandable terms, the problem can be formulated as follows [2]: place paper strips randomly along a line and then place staples randomly along the line to join the strips into larger units. Obviously, the line is the genome, the strips are the DNA clones and the staples are the anchors, i.e., short single-copy DNA sequences (restriction fragment length polymorphisms, arbitrarily primed polymerase chain reaction products or sequence-tagged sites). The ultimate large units are large assemblies of DNA clones called "islands" [21].

The reader is referred to [19] for a special chapter on coverage problems, but also for simple problems of nonoverlapping intervals on $R = (-\infty, \infty)$: the occurrence of such intervals could be regarded as the realization of a random process.

In principle, a stochastic coverage process might be thought of as any random mechanism governing "the positioning and configuration of random sets on the line, or in the plane, or in space, or indeed in any metric space" [13]. Formally, the random covering problem is expressed as follows: Let X be a given set and C a collection of subsets which are chosen according to some random procedure. P is a given probability measure on C. Let $\Omega = C \times C \times \ldots \times C \times \ldots$ be the sample space corresponding to the process of choosing independently subsets according to the probability law P. The system of subsets $C_1, C_2, \ldots$ is said to cover the set X, if each element of X belongs to at least one of the sets C_i. If we consider a fixed number n of intervals, X fixed and $\{C_i\}$ a sequence of subsets in the same space as X, the coverage probability is defined as $P_n = P\{X \subset \cup_i^n C_i\}$. Define a "stopping point" x as the first point x such that $X \subset \cup_i^x C_i$. Then $P_n = P\{x \leq n\}$.

In the present paper, the coverage process is a stochastic process indexed by R. The paper will deal with "true" coverage, because the "apparent" coverage [3] is realized by a particular experiment strategy where overlap detection rules differ and the mapping progress is limited by some error factors [27].

The reader is referred for some mathematical developments to P. Hall [13] and D. Aldoud [1]; specific results for physical mapping are given in [21], [2], [7], [4], and [35].

The present paper is confined to the framework of stochastic geometry, in the sense that the original problem is transformed in one concerning a random set on an appropriate manifold (the straight line in our case) [18].

3. Assumptions

In the following assumptions the word *segment* (i.e., genomic segment) denotes the (DNA) islands and the word *gap* (or spacing) is the substitute for ocean.

(A1) The genetic material is linearly disposed.

(A2) The segments occur on the real line, generated by a (homogeneous) Poisson process with incidence rate θ. (An additional assumption is included if the anchoring strategy is taken into account: the anchors occur according to a Poisson process of rate θ'.)

(A3) The segments are random sets in R, where a Lebesgue measure λ governs their (random) locations and an arbitrary measure m governs their (random) lengths.

(A4) The segments lengths $\{L_i, i \geq 1\}$ are i.i.d. random variables with common probability distribution G.

Assumption (A1) is based on Benzer's paper [5] where the theory of interval graphs has been applied (see, e.g., the theory of linear maps in [25]). It is worthwhile noting that interval graphs are employed in seriation [17], a technique proposed for constructing preliminary (multilocus) genetic maps [6].

There is a difference in the procedures by which the random segments are placed on the line (i.e., how the space is randomly filled). The best way to understand the differences would be a comparison of similar problems in graph theory, that is, edge covering and, respectively, edge packing. If one considers a finite and connected graph $G = (V, E)$ and an integer k, the edge covering (packing) problem is if G does have a subset of at most (at least) k edges such that every vertex $v \in V$ is incident with at least (at most) one of them. The answer is affirmative if there exists a matching of cardinality $|V| - k$ (resp., k). The representation of restriction maps by "filling" has been suggested in [20].

Assumption (A2) is generally admitted. Let $(x_i), i \in Z$, be the points of a homogeneous Poisson process $\{\zeta\}$ on R, ordered such that $x_i < x_{i+1}$ for all $i \in Z$ and $x_{-1} < 0 < x_0$. For each point x_i an open interval $C_i = (x_i, x_i + h)$ of random length, L_i is chosen. Then the coverage process is the sequence of covering intervals

$$C_i = \{x_i + L_i\} . \tag{1}$$

The random interval $(x_i + h,\ x_i + h + x_{i+1})$ is a gap or spacing. Intuitively, the "instants" $x_i,\ i \leq 1$, can be interpreted as the arrival times of customers in a stationary service system $M/M/\infty$ (see [12], [32]) or $M/D/\infty$ (see [11]): the interval lengths L are the corresponding service times and the gaps represent the idle periods (vacancy) of the queueing system. Then the busy time in $[0, t]$ corresponds to the coverage.

The stochastic coverage process can completely be described by an indispensable measure-theoretical approach due to L.A. Shepp [29]. Let ζ be a homogeneous Poisson process on $R \times R^+ = \{(x, y),\ x, y \in R,\ y > 0\}$ which is generated by the product measure $\lambda x \mu$: λ is the Lebesgue measure on R and μ is a given Borel measure on R^+. The Poisson process models the occurrence of the left-hand endpoints of a segment. To each $y > 0$ one associates a measurable subset $A_y \subset R$ (with Lebesgue measure $\lambda(A_y) = y$), so that to each point (x, y) of ζ corresponds an interval $(x + A_y)$. Let $\mathcal{U}$ denote the (a.s. countable) union of the random intervals associated with the random points of the driving Poisson process, i.e., $\mathcal{U} = \cup_{(x,y)\in\zeta}(x + Ay)$. Then $\mathcal{U} = R$ w.p.1 if

$$\int_0^1 dx \exp\left\{\int_x^\infty (y - x)\mu(dy)\right\} = \infty$$

holds (see also [37]). This is a necessary and sufficient condition on μ for the union of intervals associated with each point of ζ to cover all of R w.p.1 (Eq.1 in [29]).

If a sequence $\{L_n\}$, arranged in decreasing order, $1 = L_1 \geq L_2 \geq \ldots L_n \to 0$ is taken into consideration, with the assumption that μ is a measure assigning unit mass to each L_n, then $\mathcal{U} = R$ a.s. iff the series

$$\sum_{n=1}^{\infty} \exp\{L_1 + \ldots + L_n\}$$

diverges (see Theorem 3.2. in [13]). This condition is the same as those found for the union of independent and uniformly distributed arcs of length L_n to cover a unit circumference (see also [15]). The measure μ gives a low frequency coverage iff (2) holds with measure μ replaced by

$$\mu_1(\{dy\}) := \mu([1,\ ]\ \{dy\})\ .$$

P. Hall [13] pointed out that if series (3) converges for all $\theta > 0$, then the number of points in R that are not covered is uncountably infinite.

In [15] it is noticed that the covering motion of a 1-dimensional Brownian motion is rendered evident by the fine analysis of its paths. The heuristic can be found in [1; see D2]. For coverage results with immediate applications the reader is referred to (H12 in [1] and (Theorem 2.6 in [13]).

Assumption (A4) allows the use of renewal theory arguments: formula (2.22) in [13] for the complete coverage probability can be derived by applying the renewal theorem (see also appropriate renewal arguments in [2], [7] and [21]).

A result in [8; i.e., Problem VI.13.12 and Example XIV.2(a)] is noteworthy: if in a Poisson process n renewal epochs (arrivals) occur in $[0, t]$, their conditional distribution is *uniform*. (This confirms once more the remark that two properties are predominant in geometrical probability, namely uniformity and independence [18].) The probability that no gap of length 1 appears follows from a covering theorem given by W. Feller (Theorem 3,I.9 in [8]).

4. The Boolean Model

The view of the physical molecular mapping as a Boolean model is a consequence of assumption (A3): by definition, a simple linear Boolean model is a sequence of possibly overlapping segments on the line, whose lengths are i.i.d. random variables and whose left-hand endpoints derive from a homogeneous stationary Poisson process ζ. If we take into consideration the segments C_i in (1) as random compact sets (in the sense of G. Matheron [23], [24], the union

$$Y\zeta = \bigcup_{i=1}^{\infty}(x_i + C_i)$$

is a Boolean model (see also [34]). It formalizes the idea of "throwing sets down at random" [1]. (Recall Robbins' 1944 theorem on random sets discussed in §5.13 in [19].)

The following notations and definitions will be introduced: C is the space of all closed sets in E (a locally compact, Hausdorff, and separable space), C' is the space of non-empty closed sets in E, $\mathcal{K}$ is the set of compact subsets of E, and $\mathcal{K}'$ is the space of non-empty compact subsets of E. The Fell topology on C makes it compact, separable and Hausdorff [22].

(D1) A random compact set is a Borel measurable mapping A from an abstract probability space (Ω, F, P) into the measure space (C, Σ), where Σ is the σ-algebra generated by C^K, $K \in \mathcal{K}$ [24].

(D2) Stationarity and isotropy: A random closed set A is said to be stationary if A and the translated sets $A + x$ all have the same distribution; A is isotropic if it is randomly and uniformly oriented.

(D3) Convexity: A random closed set A in the Euclidean space is convex if x, y A implies $\lambda x + (1 - \lambda)y$ A for $0 \leq l \leq 1$.

(D4) Separation: Two compact sets K and K' are separated by another compact set A if for any $x \in K$ and $x' \in K'$, the segment $[x, x'] = \{\lambda x + (1 - \lambda)x'$, $0 \leq l \leq 1\}$ hits the compact set A (Definition 4.2.1 in [24]). If K and K' are separated by A, a convex set cannot hit K and K' without hitting A.

(D5) A functional Q on $\mathcal{K}$ is defined by $Q(K) = P(C^K) = P(\{A \cap K = \emptyset\})$, $K \in \mathcal{K}$. This functional plays the same role for a random closed set as the distribution function for a random variable. In fact, a random closed set on R is a random variable.

Thus, a stationary random closed set C on R is said to be semi-Markovian if

$$Q(K \cup K' \cup A)Q(A) = Q(K \cup A)Q(K' \cup A) ,$$

for any K, K' and $A \in \mathcal{K}$ such that K and K' are separated by A (definition 5.1.1 in [24] and relation 2.1 in [23]).

Then, a variant of assumption A3) can be suggested as follows:

A3′ The physical molecular mapping process is a stationary semi-Markovian random closed set Y_ζ on R.

The interpretation is that Y as a stochastic process starts from scratch each time that the point x leaves the set Y (see XI.8 in [8]). In other words, Y is a stationary two-state renewal process on R, or a stationary partition of R by consecutive segments [24].

These consequences of space dimension does not considerably simplify the problem because the fine structure of the genomic map might present differences between the covered and the uncovered sets. It is already proved [9] that the uncovered sets V (vacancy) can be interpreted as the closure of the finite image of a *subordinator*. Or,

(a) A subordinator is an increasing right-continuous R-valued process with stationary independent increments, and these increments have compound Poisson distributions (see VI.4 in [8].) Actually, a compound Poisson approximation is deduced for the total length of uncovered intervals (H12 in [1]).

(b) V is identified with a *regenerative* set, and every regenerative set is equivalent to the image of a subordinator. Let the measure μ determine the probabilistic law under which open subintervals of $(0, \infty)$ are chosen to be covering intervals. The random set V of uncovered points in $[0, \infty)$ is a regenerative set with potential measure Φ given by $\Phi[0, x]$ 1, if

$$\int_0^1 dx \exp \{\mu(y, \infty] dy\} = \infty .$$

(Theorem 1 in [9]; see also Theorem 2 for the extension on the whole line).

(c) A regenerative set is essentially the closure of a strong Markov set, and also the range of a subordinator is a Markov set.

(d) The set of uncovered points in $[0, \infty)$ is a.s. the union of closed intervals of positive length iff $\mu(0, \infty] < \infty$ (Corollary 2 in [9]).

It is now possible to join together these concepts in the framework of the theory of regenerative systems on the real line [16]: the theory unifies the notions of strong Markov processes indexed by R, renewal processes on R, regenerative processes, semi-Markov processes, etc. It is the aim of this paper to make a point of the theoretical investigation of an important practical problem and to pinpoint the potency of the model approach in stochastic geometry.

The interpretation of the coverage process if in assumption (A2) the driving process would be non-Poissonian brings large difficulties about. It must be mentioned that the Poisson process is the only process which possesses the order statistic property and belongs to the exponential family of stochastic processes having a likelihood function of a general form [31].

References

[1] Aldous, D., *Probability Approximations via the Poisson Clumping Heuristics*, Springer, New York, 1989.

[2] Arratia, R., Lander, E.S., Tavaré, S., Waterman, M.S., *Genomic mapping by anchoring random clones: A mathematical analysis.* Genomics **11** (1991), 806–827.

[3] Balding, D.J., Torney, D.C., "Statistical analysis of DNA fingerprint data for ordered clone physical mapping of human chromosomes", *Bull. Math. Biol.* **53** (1991), 853–879.

[4] Barillot, E., Dausset, J., Cohen, D., "Theoretical analysis of a physical mapping strategy using random single-copy landmarks", *Proc. Natl. Acad. Sci. USA* **88** (1991), 3917–3921.

[5] Benzer, S., "On the topology of the genetic fine structure", *Proc. Natl. Acad. Sci. USA* **45** (1959), 1607–1620.

[6] Buetow, K.H., Chakravarti, A., "Multipoint gene mapping using seriation. I. General methods", *Amer. J. Human Genet.* **41** (1987), 180–188.

[7] Ewens, W.J., Bell, C.J., Donnelly, P.J., Dunn, P., Matallana, E., Ecker, J.R., "Genome mapping with anchored clones: Theoretical aspects", *Genomics* **11** (1991), 799–805.

[8] Feller, W., *An Introduction to Probability Theory and Its Applications. Vol. II*, Wiley, New York, 1966.

[9] Fitzsimmons, P.J. Fristedt, B., Shepp, L.A., "The set of real numbers left uncovered by random covering intervals", *Z. Wahrscheinlichkeitstheorie verw. Geb.* **70** (1985), 175–189.

[10] Glaz, J., Naus, J., "Multiple coverage of the line", *Ann. Probab.* **7** (1979), 900–906.

[11] Greenberg, I., "The moments of coverage of a linear set", *J. Appl. Probab.* **17** (1980), 865–868.

[12] Hall, P., "Macroscopic properties of a linear mosaic", *Adv. Appl. Probab.* **17** (1985), 330–346.

[13] Hall, P., *Introduction to the Theory of Coverage Processes*, Wiley, New York, 1988.

[14] Higgins, M.J., Turmel, C., Noolandi, J., Neumann, P.E., Lalande, M., "Construction of the physical map for three loci in chromosome band 13q14: Comparison to the genetic map", *Proc. Natl. Acad. Sci. USA* **87** (1990), 3415–3419.

[15] Kahane, J.-P., "Recouvrements aléatoires", *Gazette des Mathématiciens* **53** (1992), 115–129.

[16] Kaspi, H., Maisonneuve, B., "Regenerative systems on the real line", *Ann. Probab.* **16** (1988), 1306–1332.

[17] Kendall, D.G., "Incidence matrices, interval graphs, and seriation in archaeology", *Pacific J. Math.* **28** (1969), 565–570.

[18] Kendall, D.G., "An introduction to stochastic geometry", In: *Stochastic Geometry*, E.F. Harding, D.G. Kendall (eds.), Wiley, London, 1974, pp 3–9.

[19] Kendall, M.G., Moran, P.A.P., *Geometrical Probability*, Charles Griffin, London, 1963.

[20] Krawczak, M., "Algorithms for the restriction-site mapping of DNA molecules", *Proc. Natl. Acad. Sci. USA* **85** (1988), 7298–7301.

[21] Lander, E.S., Waterman, M.S., "Genomic mapping by fingerprinting random clones: A mathematical analysis", *Genomics* **2** (1988), 231–239.

[22] Mase, S., "Random compact convex sets which are infinitely divisible with respect to Minkowski addition", *Adv. Appl. Probab.* **11** (1979), 834–850.

[23] Matheron, G., "Ensembles fermés aléatoires, ensembles semi-Markoviens et polyédres Poissoniens", *Adv. Appl. Probab.* **4** (1972), 508–541.

[24] Matheron, G., *Random sets and integral geometry*, Wiley, New York, 1975.

[25] Mirkin, B.G., Rodin, S.N., *Graphs and Genes*, Springer, Berlin, 1984.

[26] Olson, M., Hood, L., Cantor, C., Botstein, D., "A common language for physical mapping of the human genome", *Science* **245** (1989), Springer-Verlag, Berlin, 1434–1435.

[27] Palazzolo, M.J., Sawyer, S.A., Martin, C.H., Smoller, D.A., Hartl, D.L., "Optimized strategies for sequence-tagged-site selection on genome mapping", *Proc. Natl. Acad. Sci. USA* **88** (1991), 8034–8038.

[28] Serra, J., *Image Analysis and Mathematical Morphology*, Academic Press, London, 1982.

[29] Shepp, L.A., "Covering the line with random intervals", *Z. Wahrscheinlichkeitstheorie verw. Geb.* **23** (1972), 163–170.

[30] Siegel, A.F., "Random space filling and moments of coverage in geometrical probability", *J. Appl. Probab.* **15** (1978), 340–355.

[31] Sorensen, M., "On sequential maximum likelihood estimation for exponential families of stochastic processes", *Intern. Statist. Rev.* **54** (1986), 191–210.

[32] Stadje, W., "Coverage problems for random intervals", *SIAM J. Appl. Math.*

49 (1989), 1538–1551.

[33] Stephens, J.C., Cavanaugh, M.L., Gradie, M.I., Mador, M.L., Kidd, K.K., "Mapping the human genome: Current status", *Science* **250** (1990), 237–244.

[34] Stoyan, D., Kendall, W.S., Mecke, J., *Stochastic Geometry and its Applications*, Wiley, Chichester, 1987.

[35] Torney, D.C., "Mapping using unique sequences", *J. Mol. Biol.* **217** (1991), 259–264.

[36] Weil, W., Wieacker, J.A., "Densities for stationary random sets and point processes", *Adv. Appl. Probab.* **16** (1984), 324–346.

[37] Wschebor, M., "Sur un théoréme de Léonard Shepp", *Z. Wahrscheinlichkeitstheorie verw. Geb.* **27** (1973), 179–184.

Petre Tautu
Dept. of Mathematical Models
German Cancer Research Center
Heidelberg, Germany

Rhythms and Chaos in Biological Systems

by Albert Goldbeter
Université Libre de Bruxelles, Brussels (Belgium)

Rhythmic phenomena are ubiquitous in biology and occur with periods ranging from a few milliseconds (nerve cells) to seconds (heart), minutes (oscillatory enzyme reactions), hours (periodic hormone secretion), days (circadian rhythms), and years (annual rhythms, and predator-prey oscillations in ecology). This list is by no means exhaustive; moreover, new rhythms are being uncovered at the cellular or supracellular levels. The reason why oscillations are so common in biological systems is primarily related to the fact that these systems are governed by nonlinear evolution equations, owing to the fact that they are regulated by a multiplicity of positive and negative feedback processes. The analysis of mathematical models for regulated biological systems confirms that oscillations are a conspicuous property of living systems, regardless of the level of biological organization at which these rhythms occur [1].

Mathematical models for biological rhythms are of two sorts: one class aims at providing a rough caricature whose very simplicity is well-suited to the analytical approach; such an approach allows us to obtain detailed information on the dynamics of the system. Models of the second kind are closely related to experimental data. The underlying equations for such models are often more complex, so that analytical tools have only limited use and one has to resort to numerical simulations.

Most oscillations in biological systems are of the limit cycle type. In recent years, however, the emphasis has shifted to more complex modes of oscillatory behavior [2]–[4]. Among these, particular interest has been devoted to complex periodic oscillations (*bursting*) and to aperiodic oscillations (*chaos*). Such complex phenomena have been studied in different biological contexts, particularly in neurobiology; bursting indeed characterizes the activity of many nerve cells [3]. Another field where simple and complex oscillatory phenomena have been studied by means of theoretical models is biochemistry [5]. The occurrence of rhythms and chaos in biological systems will be illustrated here by some examples taken from this field.

Limit Cycle Oscillations

Several experimental examples of sustained oscillations in biochemical systems are known, the prototype being that of glycolytic oscillations in yeast cells. These oscillations provided one of the first examples of a biological rhythm demonstrated in vitro; they occur when a sugar, injected constantly at an appropriate rate into a suspension of yeast cells or into yeast extracts, is transformed periodically into ethanol and CO_2 (see [5] for further details and references). The oscillations, which have a period of some 5 minutes, originate from an enzymatic reaction transforming a substrate (S) into a product (P); the reaction is catalyzed by phosphofructokinase (PFK). As in many instances in other fields of chemistry or biology, periodic behavior originates from positive feedback, i.e., autocatalysis: the product of PFK indeed activates the enzyme and thereby enhances its own production.

A two-variable model for the autocatalytic enzyme reaction has been proposed [5]; it is governed by the system of ordinary differential equations (1):

$$\begin{aligned} \frac{d\alpha}{dt} &= v - \sigma\phi \\ \frac{dg}{dt} &= q\,\sigma - k_s\,\gamma \end{aligned} \tag{1}$$

with

$$\phi = \frac{\alpha(1+\alpha)(1+\gamma)^2}{L + (1+\alpha)^2(1+\gamma)^2} \tag{2}$$

Function ϕ denotes the nonlinear expression of the rate of the product-activated enzyme reaction; α and γ denote, respectively, the dimensionless concentrations of substrate and product; v, s, q, k_s and L are parameters (see [5] for further details).

The main result of the analysis of equation (1) is that these equations admit a single steady-state solution which becomes unstable in a range of substrate injection rate (v) bounded by two critical values of parameter v associated with Hopf bifurcations. In that range, the system evolves toward sustained oscillations corresponding to a limit cycle in the (α, γ) phase plane. Phase plane analysis proves particularly useful in establishing this result which accounts for the main experimental observation on glycolytic oscillations in yeast.

Birhythmicity

The question arises as to whether the stable limit cycle is always unique for a given set of parameter values — as is the case for equation (1) — or whether multiple stable limit cycles may sometimes coexist. A conjecture based on phase plane analysis leads to the following system of two differential equations, representing a direct extension of equation (1):

$$\begin{aligned} \frac{d\alpha}{dt} &= v + \frac{\sigma_i\gamma^n}{K^n + \gamma^n} - \sigma\phi \\ \frac{d\gamma}{dt} &= q\sigma\phi - k_s\gamma - \frac{q\sigma_i\gamma^n}{K^n + \gamma^n} \end{aligned} \tag{3}$$

where ϕ remains given by equation (2). In equation (3), the new term refers to a reaction of nonlinear recycling of product into substrate, characterized by the threshold constant K; the maximum rate of this process is denoted by s_i. The analysis of the model indicates that in a certain range of s_i values, the system admits two stable limit cycles separated by an unstable cycle [5]; this phenomenon has been referred to as *birhythmicity* [6]. Birhythmicity has been demonstrated experimentally in several chemical reactions but has not yet been observed in biological systems.

Bursting and Chaos

More complex modes of oscillatory behavior require the interaction of at least three variables. To allow for the occurrence of more complex oscillatory phenomena, the model described by equation (1) was again extended to consider a sequence of two autocatalytic enzyme reactions coupled in series [5]–[7]: the first enzyme, E_1, transforms substate S into product P_1; the latter is transformed into product P_2 by the second enzyme, E_2. Both reactions are subjected to positive feedback as E_1 and E_2 are activated by P_1 and P_2, respectively. This system is governed by the three differential equations (4) (see [5]–[7] for further details):

$$\begin{aligned} \frac{d\alpha}{dt} &= v - \sigma_1\phi(\alpha, \beta) \\ \frac{d\beta}{dt} &= q_1\sigma_1\phi(\alpha, \beta) - \sigma_2\eta(\beta, \gamma) \\ \frac{d\gamma}{dt} &= q_2\sigma_2\eta(\beta, \gamma) - k_s\gamma \end{aligned} \tag{4}$$

where the rate functions f and h of the enzymes E_1 and E_2 are given, respectively, by expressions (5) and (6):

$$\phi(\alpha, \beta) = \frac{\alpha(1+\alpha)(1+\beta)^2}{L_1 + (1+\alpha)^2(1+\beta)^2} \tag{5}$$

$$\eta(\beta, \gamma) = \frac{\beta(1+\gamma)^2}{L_2 + (1+\gamma)^2} \tag{6}$$

In these equations, α, β and γ represent the normalized concentrations of S, P_1 and P_2, respectively. In addition to the modes of behavior exhibited by equations (1) and (3), depending on parameter values, the system governed by equations (4–6) displays the following dynamic properties:

(i) coexistence between a stable limit cycle and a stable steady state;

(ii) coexistence between two or even three stable limit cycles (*bi-* and *trirhythmicity*);

(iii) bursting oscillations, in which a constant number of spikes of b and g occur over a period;
(iv) aperiodic oscillations (*chaos*) corresponding to the evolution toward a strange attractor;
(v) coexistence between a stable limit cycle and a stable strange attractor.

The richness in behavioral modes originates in system (4) from the coupling of two instability-generating mechanisms. This result is analogous to obtaining bursting and chaos in oscillatory systems driven by some periodic input [2]–[4]; in contrast, chaos in system (4) is autonomous as it occurs in the absence of periodic forcing.

From periodic behavior to chaos in a cellular system: cAMP oscillations in *Dictyostelium* amoebae

Another well-known example of biochemical oscillation occurs during the life cycle of *Dictyostelium* amoebae. After starvation, these cells aggregate in a wavelike manner by a chemotactic response to pulses of cyclic AMP (cAMP) emitted periodically by aggregation centers; the period of cAMP pulses and of the associated waves is of the order of 5 minutes (see [5] for further references). A model for the signaling system has been proposed [8]; it is described by a system of four nonlinear differential equations. While reductions to three or even two variables account for the periodic synthesis of cAMP, the four-variable and three-variable versions also give rise to chaotic oscillations [5]. As in the system described by equations (4)–(6), chaos occurs as a result of the coupling of two endogenous oscillatory mechanisms. The domain of aperiodic oscillations in parameter space remains, however, much reduced as compared to the domain of simple or complex periodic oscillations.

Although originating from mathematical models based on biochemical kinetics, these results are of general significance and pertain to the occurrence of rhythms and chaos in other biological systems. The transitions between the various modes of dynamic behavior indeed remain largely independent of the detailed form of the underlying, nonlinear evolution equations.

References

[1] Winfree A.T., *The Geometry of Biological Time*, Springer, New York, 1980.
[2] Olsen L.F. and Degn H., *Quart. Rev. Biophys.* **18** (1985), 165–225.
[3] Holden A. (ed.)., *Chaos*, Manchester Univ. Press, 1986.
[4] Glass L. and Mackey M.C., *From Clocks to Chaos: The Rhythms of Life*, Princeton Univ. Press, Princeton, 1988.
[5] Goldbeter A., *Rythmes et chaos dans les systèmes biochimiques et cellulaires*, Masson, Paris, 1990.

[6] Decroly O. and Goldbeter A., *Proc. Natl. Acad. Sci. USA* **79** (1982), 6917–6921.
[7] Decroly O. and Goldbeter A., *J. Theor. Biol.* **124** (1987), 219–250.
[8] Martiel J.L. and Goldbeter A., *Biophys. J.* **52** (1987), 807–828.

Albert Goldbeter
Faculté des Scineces
Université de Bruxelles
Campus Plaine, C.P. 231
B-1050 Brussels, Belgium

Discussion and Conclusions

1. Introduction to the General Discussion

The presented papers show how mathematicians can work to solve problems raised by biologists and clinicians. Furthermore, they all succeed in portraying to some degree the special flavour of mathematical applications in these fields. All round tables have been asked to consider how to develop their fields of interest, and this discussion tries to establish the issues involved in this exciting and developing area of applicable mathematics and it concludes with an outline of the type of support which should be provided for future activities. The discussion includes contributions from the audience as well as the invited speakers, and although it has not been possible to identify all contributors by name, the suggestions and experiences which are summarized below were found to be most valuable and gratefully acknowledged.

We shall encounter major issues concerning mathematicians' general lack of knowledge of other fields outside the physical sciences and the consequent difficulties of communication with non-physical scientists. Let us begin, however, by considering what makes mathematics and its applications successful in the fields of medicine and biology. First, we need to know what precisely is the biological or medical question to be answered and then we must consider how it may be answered by a statistical analysis of observational data or by a mathematical analysis of a model of the real system or some combination of these two approaches. Together with the specification of the problem, acceptable limits for a solution must be determined. This requires care in these fields of application where variability causes special problems, which are different from those found in, e.g., the physical sciences. Furthermore, the level of generality of the scientific results, and presumably the impact that they will make, may be more limited. The lack of good data or restrictions arising from the limitations of the available techniques for analysis can adversely affect the value of the work.

In order to consider the development of a good mathematical application the level of contextual knowledge on the part of the mathematician will be relevant. Equally, for a successful and fruitful collaboration to be achieved, there must be some appreciation and understanding of the properties of mathematical solutions by the biologist or clinician. Above all, good communication has to take place between collaborators with different backgrounds. Sometimes mathematicians underestimate this and feel that communication is a one-off and one-way process. The unfortunate fact is that many mathematicians are unsuitable as collaborators because they are only comfortable when communicating with other mathematicians! This is less of a problem for most medical and biological scientists who are trained to exchange ideas because a reliance on the technical skills of others is necesary for their survival.

In the life sciences, rarely do problems arise in a suitable form for mathematical analysis and work is needed before techniques can be applied to give solutions. Sometimes the problem may be uninteresting to the mathematician because of the triviality of the analysis required, but the biologist may see a new and deep penetration of understanding of biology from what is commonplace mathematics. The classic example of this is the Hardy–Weinberg law which in generality states that almost any amount of disruption of the gene pool by the systematic removal of parents will, after random mating, produce a simple binomial distribution for the genotypes of the offspring. The message here is that the mathematician is not usually the best judge of the importance of his contribution. If Hardy was not able to judge, should we expect any one else to do so?!

At the other extreme, the mathematician becomes preoccupied by the mathematical problem, which may have become completely detached from reality. Solutions in abstract form are incomprehensible and therefore useless to the biologist. There is no hope unless a genuine two-way dialogue can take place to address real issues in the applied field. The biologist must be able to interpret mathematical results in order to be able to write them into his or her own sphere of knowledge. The contribution to the frontiers of science is nil if the impact of the mathematics cannot be conveyed or communicated. The biologist must be left with the feeling that the mathematician has not been indulging in an exclusive game but that they have both been involved as a team in a game against nature. To win this latter game brings satisfaction and benefits to all members of the team rather than just to the mathematician!

The aim of good collaborative work is that through respectable mathematics, with rigour and style, the biologist or clinician makes a new contribution to learning which is substantial and not merely an embellishment of that which is already known. Moreover, the biologist should be prompted to make new observations because the past observations are not complete and because the mathematics shows that there are other solutions that need to be explored and confirmed by future observation. Where outcomes from the model form predictions that may surprise the biologist or clinician we are on the way to real achievement. At the very least, future observations may reveal when the model diverges from reality. Some of these future experiments may only differ slightly from those which gave rise to the initial problem. Sometimes science advances in very small steps and there may need to be much hard and painstaking measurement before a particular mathematical model may be confirmed or rejected. All of this requires great patience from mathematicians who easily become bored!

As the ultimate simple criterion for success, we suggest that the mathematician has to make the biologist or clinician stop and think in a new way about their world!

We have to accept that a mathematician cannot know all about the field of application. However, sufficient knowledge usually resides in the collaborative partnership. It is not realistic to think that a mathematician could go off to some desert island and in isolation acquire the knowledge required to return to contribute meaningfully to biology at the frontiers of research. The amount of diverse knowledge that's now needed to make any worthwhile contribution to science is so considerable that only *close and continuous collaboration* throughout the duration of a project will produce successful research. Nowadays people work in teams and mathematicians clinicians and biologists from different European countries are already beginning to work together. This is the way forward and needs further encouragement.

Before moving to the general discussion, a few points made at a pre-meeting of the invited speakers with the organisers are briefly mentioned at this point. In particular, the mood was optimistic, and it was felt that there is a refreshing degree of biological reality in current published research and this must be preserved (RWH is pleased to note this and to confirm that this is deliberate editorial policy, at least for the *IMA Journal*!) There was plenty of evidence that mathematicians were not just having fun on a superficial level of application but getting to the roots of biology in a serious and dedicated way. The point was made that applications were so richly varied that it is wrong to look for a cohesive approach to all situations in biology and medicine. More unifying was the point that there is often a deeper subtlety underlying substantive questions in these applied fields and to note that there was no correspondence in general between intuitive biology and intuitive mathematics. The need for sophistication of mathematical approaches was needed often to deal with the complexity of causation mechanisms and problems of prediction.

Finally, it should be remembered that the talks given are only a taste of the wide range of topics and fields in which important work is currently undertaken. A few of these important topics which are not represented here but which stimulate many mathematicians are: the development of pattern and form, ecological models including harvesting of natural populations, genetic-disease association and pedigree analysis and models for the spread of human infectious disease including HIV.

2. General Discussion

Albert Goldbeter proceeded to elaborate upon the interface between the mathematician and the biologist. He declared himself to be a chemist, not a mathematician, doing theoretical biology which is typically a field which requires a multidisciplinary approach. Contacts with experimental biologists are essential in developing models closely related to experimental data. He felt that it will be a long time before the theoretical approach will make a great impact on the experimental field. However, the position is much improved in the last twenty years, to judge for example by the development of new journals devoted to mathematical biology and theoretical biology. Furthermore, papers on mathematical biology are now being published in experimental journals in biology. That used to be very rare but now it's more and more the case.

On the subject of education, he thought it essential to give courses in mathematical biology to biologists and clinicians but also that biology should be taught at least as an optional course in mathematical departments. In his university at Brussels, mathematicians only get a course in biology if they do so on their own initiative. Furthermore, there are few courses in mathematics for biologists, and those that exist are only introductory ones. There are countries in Europe where mathematical biology is being taught, e.g., in Holland where there are chairs in theoretical biology in biological departments in Amsterdam, Utrecht and Leiden. He believed that others should follow their example.

Continuing the theme of the appropriate mathematical education necessary to do theoretical biology, Gilbert Chauvet remarked that this was a difficult matter. In France, he was obliged to obtain a complete education in mathematics and theoretical physics before he could turn towards medicine. So he thought it necessary in future to have special courses which could mix mathematics and physics together with medicine or biology. The real problem was the difficulty, after a course of limited length, in satisfying several specialized examiners each with a rigorous background in their own discipline. He thought that it should be possible to overcome this problem.

The difficulty of communicating between the substantive scientists on the one hand and mathematicians and statisticians on the other was raised by Elja Arljas. When people working in applied fields hear about some new mathematical or statistical model they become excited and everyone starts using that model, often uncritically. He felt that models should be considered in the right order as a hierarchy with continuous filtering of ideas both ways. It is important to start with a qualitative description of the problem. In his formal presentation, the first step was to specify some basic random variables of interest together with a very rough description of the problem. In other situations, observational data and models are used together with some specific statistical techniques based upon some statistical principles. The hierarchical process has different levels and information filters to the computational level and then back as a qualitative understanding of the problem emerges. We can then start again and develop our ideas in this process. He noted that communication is a major challenge for all mathematicians. Asking qualitative questions, and trying to get the applied scientist to raise questions in a way that would be a good beginning is the most difficult part of the process.

His experience as a statistician having worked with medical people for many years had proved to Niels Keiding that the idea of one person encompassing many specialities is out-dated. It is not possible to be competent in all areas and good mathematical biology, such as is described in the formal presentations to this round table, will usually be the result of collaboration between different people with different backgrounds. Collaboration between people from different areas is an enterprise mathematicians do not usually know too much about. They are used to working in isolation and this is in principle the ideal working condition for most mathematicians. Either they work in complete isolation or talk to people with exactly the same background as themselves. But this is not the way the results we have been hearing about at this round table have been produced. Here people with different backgrounds have to interact and to respect one another. The mathematician has to respect the biologist and vice versa and this is no trivial matter in many cases. There was an additional enriching human aspect, he indicated, amounting to more than just the scientific value, in the very rewarding experience for all those involved.

Peter Tautu commented that in medicine and biology there are different fields with different evolution and different interests and the same is true for mathematics. Epidemiology needs more statistics and many of the concepts used are well-developed and established ones. By contrast, in molecular genetics a new age has arrived with a new field of research in which young biologists are cloning and cutting DNA but are dependent upon computers. He noted further the change from the use of mathematics and computers in medical physics to their use in biology, and particularly in genetics. These uses of computers embraced much more mathematics than is necessary for biologists to know. He stressed the importance of including physicists, whom he found to be particularly creative in his field and who had helped him reproduce cancer growth using interacting particles. Others can help too, and we must look to all specialists in this complex world, and have what Pasteur called 'un esprit préparé'. He warned that this is not easy, as he found in Heidelberg, where the experimental sciences dominate and theoretical collaborative partners can be hard to find. However, he is full of hope because of the newer biologists and the new questions for which there were no immediate observational answers, e.g., concerning AIDS.

By comparison, Jonathan Swinton declared himself to be new to these issues and that he was still learning lessons which others had learnt. He felt that we need to recognise that mathematics does not have an entirely positive image outside of mathematics. Everyone is talking about how important communication is, but we cannot really expect biologists and others to come to us. It is not really good enough to say that they have to learn mathematics. He reminded us of those mathematical seminars, where the speaker takes no pains to explain the problem or any background or even why they are talking about it. As he had moved from a mathematical environment to a biological one this rarely happened. Biologists are much more aware of the importance of communication with their peers as they are doing science rather than mathematics. If we want to break into that area we have to be very aware of this fact. Its not really good enough to give a talk and say we need some data on this and then ask 'Are there any biologists in the audience?' We need to distinguish between mathematics and mathematical modelling. Mathematical modelling is a very different discipline to mathematics. When he became a mathematical biologist the thing that he had to learn most was not biology but mathematical modelling as a science or an art. We need to be more aware of the needs of the modellers than those of the mathematicians.

In the ensuing discussion, Dominic Lombert reminded us that in mathematics and physics there exists a very fruitful relationship because every physical concept can be translated into mathematical language. It may be that sometime in the future we shall have the same situation with a complete mathematical formalization of biological concepts.

However, it was argued by another contributor, that we have not only to apply mathematical methods but to encourage new mathematical constructs. We can imagine now that a new biological problem forces theoretical biologists to create new concepts or a new mathematical structure. Physics was once like this in its relation to mathematics, In order to develop the same situation in biology, theoretical biologists have to forge new structures in order to attract the attention of the mathematicians.

A collaboration which had proved to be a painful experience was then described by a mathematician, Miles Dobson, who had written a paper with three biologists on the use of fractals to explain the distribution of body sizes of insects in vegetation. He had felt that his mathematics was not incorrect but that there was a better way of doing it. Although the mathematical ideas were simple, it was difficult to make them fit into the biological framework. He didn't feel he had to know that much biology but came away with the feeling biology had now advanced from a merely stamp collecting stage and had become nearer to engineering and physics. That impression was confirmed by the talks at this round table. Hence, biologists should do mathematics as a part of the language of science. A lot of the mathematics, e.g., differential equations, could be taught to biology students, freed of analytical detail and presented very much as a tool. There has been really productive interrelationship or interaction between engineering and mathematics and it can be the same with biology. In science there are other pressures and there are always going to be certain conflicts between the interest and motives of a mathematician and those of a biologist or an engineer. Most engineers want to have something 99% right, perhaps a biologist would be happy with 75% right and a mathematician wants it 100% right. Maybe that is a difference one has to accept and live with, but it must not be forgotten.

A physical chemist, Edgar Soulié, then explained by analogy that as an end-user of manufactured products you don't expect your dealer to be an expert in mechanics but you do expect your TV set to reach a standard of performance. Translating this to the present context, it may not be realistic for the mathematician to become a biologist nor for the biologist to become a mathematician. Last year, at an international conference, one communication used a range of mathematical techniques in a study of breast cancer. The relevance of the application and the quality of the mathematics were impressive but no medical doctor could master enough of the mathematics (simultaneously understanding piece-wise linear programming and neural nets) and also be competent in his speciality, i.e., the diagnosis of patients and cancer therapy.

3. Conclusions and Suggestions

Following the lively and helpful discussion at the Round Table, we can perhaps move forwards to prepare for the future. The contributors to the discussion have helped by giving their views based upon their collectively vast experience. There is a surprising degree of consensus and agreement on major issues such as the type of mathematical education and the realization of the need for collaborative effort. There is much to learn from the parallels with physics and engineering applications of mathematics, but also there is a distinctiveness about the life sciences which is precious and raises new questions on how to portray variabililty and deal with it properly to advance our understanding.

Below is a framework for discussion. Not all of the parts of the outline will be equally useful but all need to be considered. Clearly, much good work is being done already and it is important to integrate new effort and activties with the work of existing societies and organizations that cater for mathematical applications in medicine and biology. The way forward is towards a series of discussions which might focus on these suggestions.

a. Training events or workshops

1. For biologists and clinicians, to indicate to them what mathematics can do, covering *a broad appreciation of the potential of modelling rather than a training in techniques.*
2. For mathematicians, teaching modelling by selecting techniques which are known to be useful in biological or medical applications. Teaching to include the appreciation of the special nature of the life sciences as an application area and the particular problems of communication outside of the physical sciences.

b. Joint conferences with mathematicians and biologists represented in equal proportions with deliberately extended discussions at the end formal presentations.

c. Visits

1. Short-term visits to introduce workers from different disciplines and discuss a range of applied problems from which collaborative projects might emerge; and

2. Long-term visits to undertake substantial collaborative work of high quality on one or more projects leading to jointly authored papers intended to form landmark contributions to the field?

d. Publications:

1. A newsletter describing the activities under the above headings and giving mailing addresses of all collaborating workers, including e-mail details.
2. Journals
3. Lecture notes or monographs

e. Education

1. The types of courses that are provided by institutions throughout Europe, particularly reviewing mathematics degrees with biological options and the mathematical content of biological degrees.
2. The organization and logistics of departments in universities to encourage medical and biological uses of mathematics. Chairs in appropriate subjects and research fellowships are possibilities which have been mentioned.

Taken together, these may be seen as a platform upon which future developments may grow and a means of strengthening the support for those already involved in projects. Some of the above functions have been undertaken, e.g., by the European Society for Theoretical and Mathematical Biology and by the Mathematics in Medicine and Biology Forum of the Institute of Mathematics and its Applications. The IMA holds a series of conferences at Oxford on the Dynamics of Biological Populations and the Forum publishes a regular newsletter. There have been NATO workshops on specific topics such as the development of spatial pattern. The existing Journals are probably sufficient at this time, including *Statistics in Medicine, Biometrics, Bulletin of Mathematical Biology, IMA Journal of Mathematics in Medicine and Biology, Mathematical Biosciences, Journal of Theoretical Biology, Theoretical Population Biology*, and others. Only if a specifically European approach in such applications develops should a new journal be considered, but this is not seen as an immediate prospect.

Index of Authors of Volume III

Al-Amrani A., 381
Allavena Marcel, 489
Arjas Elja, 520
Auton Tim R., 257
Barrat Pierre, 397, 412
Bayer-Fluckiger Eva, 37
Bérard Pierre, 347
Bernard Claude, 461
Berg Christian, 277
Bessenrodt Christine, 39
Bos Henk, 165
Bricogne Gérard, 473
Bungartz Paul, 101
Byrne C. 407, 423, 424

Campos L.M.B.C., 263
Camus Jacques, 75, 89
Cercignani Carlo, 260, 345
Chatelet Gilles, 217
Chauvet Gilbert, 534
Chiappori Pierre-André, 229, 232
Chemla Karine, 168
Chevallard Yves, 129
Chouchan Michèle, 1
Cifoletti Giovana, 161
Cornet Bernard, 431
Correa Rafael, 383

Dewilde W, 80
Dubucs Jacques, 210
Eells J, 369
Engl H.W., 264

Gabay Daniel, 321
Gasteiger Joachim J., 466
Gilg A., 259
Gispert Hélène, 172
Godino Juan D., 150
Goldbeter Albert, 559
Goldstein Catherine, 157
Goldstein L., 401
Golfier Michel, 477
Gomez Susana, 503
Granger Gilles-Gaston, 185
Gray Mary, 40
Grugnetti Lucia, 145
Guesnerie Roger, 229
de Guzman Miguel, 33

Hanusse Patrick, 455
Héliodore Frédéric, 510
Hiorns Robert, 515
Holgate Philip, 28
Hoyles Celia, 134
Hunt Julian C.R., 257

Ibarrola Pilar, 411
Kahane Jean-Pierre, 1
Kantor Jean-Michel, 23
Keiding Niels, 517
Keitel Christine, 138
Kersten Ina, 39
Kowalski Oldrič̌h, 26
Kubilius Jonas, 30
Kuksin S, 400
Kuku Aderemi O., 356

Labarca Rafael, 359
Laksov Dan, 35
Lance Christopher, 345
Lemaire Jean-Michel, 370
Le Méhauté Alain, 510
Lindstrom Tom, 32
Longo Giuseppe, 202

Martins J. S., 368

Menger Pierre-Michel, 238
Meyer Paul-André, 386
Miatello Roberto J., 388
Moreau Gilles, 497
Munkholm Hans J., 277, 345

Narashimhan M.S., 353
Netuka Ivan, 277
Neunzert Helmut, 257
Norbury John, 267

Ortiz Eduardo, 176

Pelczar A., 367
Petitot Jean, 192
Pham Ngoc Anh, 375
Puka Llukan, 24
Pareignis Bodo, 27
Porter Timothy, 115
Prum Bernard, 515

Ritter James, 157
Romero David, 503
Roszkowska Barbara, 40
Rouchier André, 129, 131
Roy Marie-Françoise, 40

Salinger David, 277
Schiffels G., 374
Sharp R.Y., 378
Simis Aron, 390
Sinaceur Hourya, 183
Souček Vladimir, 277
Soulié Edgar, 451
Stoufflet, 258
Strasser Rudolf, 101
Sureau Geneviève, 397
Swinton Jonathan, 528
Szendrei A., 403

Tautu Petre, 550
Tedeschini Lalli Laura, 39

Valentin Patrick, 345
Verjovsky Alberto, 365
Villani Vinicio, 110

Zweig Liliane, 397

Progress in Mathematics

Edited by:

J. Oesterlé
Départment de Mathématiques
Université de Paris VI
4, Place Jussieu
75230 Paris Cedex 05, France

A. Weinstein
Department of Mathematics
University of California
Berkeley, CA 94720
U.S.A.

Progress in Mathematics is a series of books intended for professional mathematicians and scientists, encompassing all areas of pure mathematics. This distinguished series, which began in 1979, includes authored monographs, and edited collections of papers on important research developments as well as expositions of particular subject areas.

We encourage preparation of manuscripts in such form of TeX for delivery in camera-ready copy which leads to rapid publication, or in electronic form for interfacing with laser printers or typesetters.

Proposals should be sent directly to the editors or to: Birkhäuser Boston, 675 Massachusetts Avenue, Cambridge, MA 02139, U.S.A.

1 GROSS. Quadratic Forms in Infinite-Dimensional Vector Spaces
2 PHAM. Singularités des Systèmes Differentiels de Gauss-Manin
4 AUPETIT. Complex Approximation
5 HELGASON. The Radon Transform
6 LION/VERGNE. The Weil representation Maslov index and Theta series
7 HIRSCHOWITZ. Vector bundles and differential equations
10 KATOK. Ergodic Theory and Dynamical Systems I
11 BALSLEY. 18th Scandinavian Congress of Mathematicians
12 BERTIN. Séminaire de Théorie de Nombres, Paris 79-80
13 HELGASON. Topics in Harmonic Analysis on Homogeneous Spaces
14 HANO. Manifolds and Lie Groups
15 VOGAN JR. Representations of Real Reductive Lie Groups
16 GRIFFITHS/MORGAN. Rational Homotopy Theory and Differential Forms
17 VOVSI. Triangular Products of Group Representations and Their Applications
18 FRESNEL/VAN DER PUT. Géometrie Analytique Rigide et Applications
19 ODA. Periods of Hilbert Modular Surfaces
20 STEVENS. Arithmetic on Modular Curves
21 KATOK. Ergodic Theory and Dynamical Systems II
22 BERTIN. Séminaire de Théorie de Nombres, Paris 80-81
23 WEIL. Adeles and Algebraic Groups
24 LE BARZ. Enumerative Geometry and Classical Algebraic Geometry
25 GRIFFITHS. Exterior Differential Systems and the Calculus of Variations
27 BROCKETT. Differential Geometric Control Theory
28 MUMFORD. Tata Lectures on Theta I
29 FRIEDMANN. The Birational Geometry of Degenerations
30 YANO/KON. Submanifolds of Kaehlerian and Sasakian Manifolds
31 BERTRAND. Approximations Diophantiennes et Nombres Transcendant
32 BROOKS. Differential Geometry

33 Zuily. Uniqueness and Non-Uniqueness in the Cauchy Problem
34 Kashiwara. Systems of Microdifferential Equations
35/36 Artin/Tate. Vol. 1 Arithmetic. Vol. 2 Geometry
37 Boutet. Mathématique et Physique
38 Bertin. Séminaire de Théorie de Nombres, Paris 81-82
39 Ueno. Classification of Algebraic and Analytic Manifolds
40 Trombi. Representation Theory of Reductive Groups
41 Stanley. Combinatorics and Commutative Algebra
42 Jouanolou. Théorèmes de Bertini et Applications
43 Mumford. Tata Lectures on Theta II
45 Bismut. Large Deviations and the Malliavin Calculus
47 Tate. Les Conjectures de Stark sur les Fonctions L d'Artin en s=0
48 Fröhlich. Classgroups and Hermitian Modules
49 Schlichtkrull. Hyperfunctions and Harmonic Analysis on Symetric Spaces
50 Borel et al. Intersection Cohomology
51 Séminaire de Théorie de Nombres, Paris 82-83
52 Gasqui/Goldschmidt. Déformations Infinitésimales desStructures Conformes Plates
53 Laurent. Théorie de la 2ième Microlocalisation dans le Domaine Complexe
54 Verdier. Module des Fibres Stables sur les Courbes Algébriques
55 Eichler/Zagier. The Theory of Jacobi Forms
56 Shiffman/Sommese. Vanishing Theorems on Complex Manifolds
57 Riesel. Prime Numbers and Computer Methods for Factorization
58 Helffer/Nourrigat. Hypoellipticité Maximale pour des Opera-teurs Polynomes de Champs de Vecteurs
59 Goldstein. Séminaire de Théorie de Nombres, Paris 83-84
60 Arbarello. Geometry Today
62 Guillou. A la Recherche de la Topologie Perdue
63 Goldstein. Séminaire de Théorie des Nombres, Paris 84-85
64 Myung. Malcev-Admissible Algebras
65 Grubb. Functional Calculus of Pseudo-Differential Boundary Problems
66 Cassou-Nogues/Taylor. Elliptic Functions and Rings of Integers
67 Howe. Discrete Groups in Geometry and Analysis
68 Robert. Autour de l'Approximation Semi-Classique
69 Faraut/Harzallah. Analyse Harmonique: Fonctions Speciales et Distributions Invariantes
70 Yager. Analytic Number Theory and Diophantine Problems
71 Goldstein. Séminaire de Théorie de Nombres, Paris 85-86
72 Vaisman. Symplectic Geometry and Secondary Characteristic Classes
73 Molino. Riemannian Foliations
74 Henkin/Leiterer. Andreotti-Grauert Theory by Integral Formulas
75 Goldstein. Séminaire de Théorie de Nombres, Paris 86-87
76 Cossec/Dolgachev. Enriques Surfaces I
77 Reyssat. Quelques Aspects des Surfaces de Riemann
78 Borho/Brylinski/McPherson. Nilpotent Orbits, Primitive Ideals, and Characteristic Classes
79 McKenzie/Valeriote. The Structure of Decidable Locally Finite Varieties
80 Kraft/ Schwarz/Petrie (eds.) Topological Methods in Algebraic Transformation Groups
81 Goldstein. Séminaire de Théorie des Nombres, Paris 87–88
82 Duflo/Pedersen/Vergne (eds.) The Orbit Method in Representation Theory
83 Ghys/de la Harpe (eds.) Sur les Groupes Hyperboliques d'après M. Gromov

84 Araki/Kadison (eds.) Mappings of Operator Algebras
85 Berndt/Diamond/Halberstam/Hildebrand (eds.) Analytic Number Theory
89 Van der Geer/Oort/Steenbrink (eds.) Arithmetic Algebraic Geometry
90 Srinivas. Algebraic K-Theory
91 Goldstein. Séminaire de Théorie des Nombres, Paris 1988-89
92 Connes/Duflo/Joseph/Rentschler. Operator Algebras, Unitary Representations, Enveloping Algebras, and Invariant Theory. A Collection of Articles in Honor of the 65th Birthday of Jacques Dixmier
93 Audin. The Topology of Torus Actions on Symplectic Manifolds
94 Mora/Traverso (eds.) Effective Methods in Algebraic Geometry
95 Michler/Ringel (eds.) Representation Theory of Finite Groups and Finite–Dimensional Algebras
96 Malgrange. Equations Différentielles à Coefficients Polynomiaux
97 Mumford/Norman/Nori. Tata Lectures on Theta III
98 Godbillon. Feuilletages, Etudes géométriques
99 Donato/Duval/Elhadad/ Tuynman. Symplectic Geometry and Mathematical Physics. A Collection of Articles in Honor of J.-M. Souriau
100 Taylor. Pseudodifferential Operators and Nonlinear PDE
101 Barker/Sally. Harmonic Analysis on Reductive Groups
102 David. Séminaire de Théorie des Nombres, Paris 1989-90
103 Anger/Portenier. Radon Integrals
104 Adams/Barbasch/Vogan. The Langlands Classification and Irreducible Characters for Real Reductive Groups
105 Tirao/Wallach. New Developments in Lie Theory and Their Applications
106 Buser. Geometry and Spectra of Compact Riemann Surfaces
107 Brylinski. Loop Spaces,Characteristic Classes and Geometric Quantization
108 David. Séminaire de Théorie des Nombres, Paris 1990-91
109 Eyssette/Galligo. Computational Algebraic Geometry
110 Lusztig. Introduction to Quantum Groups
111 Schwarz. Morse Homology
112 Dong/Lepowsky. Generalized Vertex-Algebras and Relative Vertex Operators
113 Moeglin/Waldspurger. Décomposition Spectrale et Series d'Eisenstein
114 Berenstein/Gay/Vidras/Yger. Residue Currents and Bezout Identities
115 Babelon/Cartier/Kosmann-Schwarzbach (eds.) Integrable Systems. The Verdier Memorial Conference
116 David (ed.) Séminaire de Théorie des Nombres, Paris, 1991-1992
117 Audin/Lafontaine (eds.) Holomorphic Curves in Symplectic Geometry
118 Vaisman. Lectures on the Geometry of Poisson Manifolds
121 Joseph/Mignot/Murat/Prum/Rentschler (eds.) First European Congress of Mathematics (Paris, July 6-10, 1992). Volume III: Round Tables
122 Guillemin. Moment Maps and Combinatorial Invariants of Hamiltonian T^n-spaces
123 Brylinski/Brylinski/Guillemin/Kac (eds.) Lie Theory and Geometry: In Honor of Bertram Kostant
124 Aebischer/Borer/Kälin/Leuenberger/Reimann. Symplectic Geometry. An Introduction based on the Seminar in Bern, 1992
125 Lubotzky. Discrete Groups, Expanding Graphs and Invariant Measures
126 Riesel. Prime Numbers and Computer Methods for Factorization

GPSR Compliance

The European Union's (EU) General Product Safety Regulation (GPSR) is a set of rules that requires consumer products to be safe and our obligations to ensure this.

If you have any concerns about our products, you can contact us on ProductSafety@springernature.com

In case Publisher is established outside the EU, the EU authorized representative is:

Springer Nature Customer Service Center GmbH
Europaplatz 3
69115 Heidelberg, Germany

Batch number: 10725820

Printed by Printforce, the Netherlands